Fluorine in Pharmaceutical and Medicinal Chemistry

From Biophysical Aspects to Clinical Applications

Molecular Medicine and Medicinal Chemistry

Published:

MicroRNAs in Development and Cancer
edited by Frank J. Slack (Yale University, USA)

Merkel Cell Carcinoma: A Multidisciplinary Approach
edited by Vernon K. Sondak, Jane L. Messina, Jonathan S. Zager, and Ronald C. DeConti (H Lee Moffitt Cancer Center & Research Institute, USA)

DNA Deamination and the Immune System: AID in Health and Disease
edited by Sebastian Fugmann (National Institutes of Health, USA), Marilyn Diaz (National Institutes of Health, USA) and Nina Papavasiliou (Rockefeller University, USA)

Antibody Drug Discovery
edited by Clive R. Wood (Bayer Schering Pharma, Germany)

Molecular Exploitation of Apoptosis Pathways in Prostate Cancer
by Natasha Kyprianou (University of Kentucky, USA)

Fluorine in Pharmaceutical and Medicinal Chemistry: From Biophysical Aspects to Clinical Applications
edited by Véronique Gouverneur (University of Oxford, UK) and Klaus Müller (F Hoffmann-La Roche AG, Switzerland)

Volume 6

Molecular Medicine and
Medicinal Chemistry

Fluorine in Pharmaceutical and Medicinal Chemistry

From Biophysical Aspects to Clinical Applications

Véronique Gouverneur
University of Oxford, UK

Klaus Müller
F Hoffmann-La Roche AG, Switzerland

Editors

Imperial College Press

Published by

Imperial College Press
57 Shelton Street
Covent Garden
London WC2H 9HE

Distributed by

World Scientific Publishing Co. Pte. Ltd.
5 Toh Tuck Link, Singapore 596224
USA office: 27 Warren Street, Suite 401-402, Hackensack, NJ 07601
UK office: 57 Shelton Street, Covent Garden, London WC2H 9HE

British Library Cataloguing-in-Publication Data
A catalogue record for this book is available from the British Library.

Molecular Medicine and Medicinal Chemistry — Vol. 6
FLUORINE IN PHARMACEUTICAL AND MEDICINAL CHEMISTRY
From Biophysical Aspects to Clinical Applications

ISBN-13 978-1-84816-634-9
ISBN-10 1-84816-634-6

Typeset by Stallion Press
Email: enquiries@stallionpress.com

Printed by FuIsland Offset Printing (S) Pte Ltd Singapore

Foreword

François Diederich

For almost a century after the first preparation of elemental F_2 by Moissan in 1886, synthetic fluorine chemistry was pursued and developed by a small community of experts capable of handling the aggressive gas using special laboratory equipment. Important technological developments resulted from this work, such as the bulk-scale preparation of fluorinated hydrocarbons for refrigerators and other cooling devices, which, however, later became banned due to their atmospheric greenhouse effects and the depletion of the ozone layer. Nonetheless, lasting successful applications resulted, for example, from the development of fluorinated polymers such as Teflon®, of volatile gases for anesthesia, and of the separation of uranium isotopes using UF_6 centrifuges, for the production of nuclear fuel for use in powerplants.

The development of fluorine-containing drugs started in 1957 and was in the following years strongly aided by the increasing availability of commercial fluorinating agents allowing the safe and selective introduction of organofluorine, i.e. C–F bonds, using common laboratory equipment. This has resulted in the introduction of over 150 fluorinated drugs to the market, and currently nearly 20% of all pharmaceuticals and 40% of all agrochemicals in development contain organofluorine.

The reasons for this explosive growth in interest in the introduction of organofluorine are multiple. While beneficial effects on ADME (absorption, distribution, metabolism, and excretion) and safety were recognized earlier on, the interest in organofluorine has lately focused on more atom-based properties, such as distinct conformational and stereoelectronic properties, modulation of the pK_a-value of neighboring

Brønstedt acid/base centers, polarity, and the influence on lipophilicity as expressed by the distribution coefficient log*D* and the partition coefficient log*P* (both for the octanol/water system). Additionally, attention has shifted on intermolecular interactions of organofluorine, such as H-bonding and dipolar interactions, and it has been shown that selective organofluorine interactions with protein residues can be used to substantially enhance protein–ligand binding affinity and selectivity. New fluorinated building blocks are emerging at an increasing speed and are introduced into innovative drugs and agrochemicals. Substituting C–H by C–F bonds clearly benefits from the fact that the size of organofluorine is only slightly larger than the size and volume of the hydrogen substituent and that consequently no particular steric hindrance is encountered in most H/F replacements. All of this is extensively documented in the various chapters of this timely monograph, which prepares the chemists in modern drug discovery research and in crop protection sciences in great depth for using organofluorine in an appropriate way to tune and improve the properties of their actives and leads.

Another contemporary area in pharmaceutical and biomedical research involving organofluorine is the development of new fluorinated probes for use in solid-state ^{19}F NMR investigations and in non-invasive clinical and molecular imaging. Furthermore, the introduction of ^{18}F-radiolabels is increasingly competing with ^{11}C-radiolabels for the preparation of probes for positron emission tomography (PET) imaging. These biomedical applications are also covered in great depth in this monograph.

The twelve chapters in the monograph are written by leaders in the field and are grouped into three sections. The first section describes the synthesis of fluorinated biomolecules and how the introduction of organofluorine alters and enhances physicochemical and molecular recognition properties. The first chapter *Synthesis and Properties of Fluorinated Nucleobases in DNA and RNA* by H. Gohlke, J. Bozilovic, and J. W. Engels starts with an overview on organofluorine in molecular recognition, which focuses on C–F···H–N interactions in the context of fluorinated nucleobase analogs. The incorporation of the corresponding nucleotides into oligonucleotides and their interactions with complementary native nucleobases in RNA, ribozymes, and siRNA are subsequently reviewed. Analysis of these interactions is based on a multi-dimensional approach combining X-ray data, results from thermodynamic studies, and computer simulations. Additionally, the synthesis of selected fluorinated nucleoside analogs is covered. The nature and polarity of molecular environments is critical for organofluorine interactions in proteins, as described in the second chapter *Molecular Interactions of Fluorinated Amino Acids within the Hydrophobic Core of a Coiled Coil Peptide* by T. Vagt, M. Salwiczek, and B. Koksch. α-Helical coiled coils are investigated as model systems to decipher the interactions of organofluorine within a native protein environment. This is achieved by introducing amino acids with hydrophobic fluorinated side chains of different volume and polarity into the folding peptides. The studies reveal that the effect of fluorinated amino acids strongly depends on the immediate microenvironment: the helical peptide model systems are selectively stabilized by interactions of organofluorine, in particular of CF_3 groups, with the lipophilic amino acids Leu, Ile, and Val. The CF_3 group is at the center of the chapter *Probing the Binding Affinity and Proteolytic Stability of Trifluoromethyl Peptide Mimics as Protease Inhibitors* by M. Zanda, A. Volonterio, M. Sani, and S. Dall'Angelo. α,α-Difluoro- and α,α,α-trifluorocarbonyl residues are fully hydrated in aqueous solution and these hydrates, as part of peptidomimetic ligands, are good entities to bind to the catalytic Asp dyad in aspartic proteases. In the meanwhile, α,α,α-trifluoroacetyl groups have been recognized as general binding elements for biological targets with polar active sites and their introduction into ligands for proteolytic enzymes, such as endopeptidases and matrix metalloproteases, as well as their specific intermolecular interactions with the proteins are described. The last chapter in Section 1, entitled *Trifluoromethyl-Substituted α-Amino Acids as Solid-State ^{19}F-NMR Labels for Structural Studies of Membrane-Bound Peptides*, written by V. S. Kubyshkin, I. V. Komarov, S. Afonin, P. K. Mykhailiuk, S. L. Grage, and A. S. Ulrich presents the synthesis of trifluoromethyl-substituted natural and unnatural α-amino acids as ^{19}F-NMR labels to study membrane-associated polypeptides in the solid state. The chapter starts by outlining biostructural applications of solid-state ^{19}F NMR methods and subsequently focuses on the synthesis of the probes and their incorporation into peptides. Challenges in the preparation of hitherto missing probes, such as F_3C–substituted proline, are identified.

Section 2 deals with the introduction of organofluorine into biomedical leads and drugs and their use against various biological targets. The chapter by S. Swallow on *Fluorine-Containing Pharmaceuticals* starts with a general survey of organofluorine in drug discovery and development. It subsequently presents several interesting case studies that highlight the effects of H/F substitutions on the development of commercial drugs. Beneficial organofluorine contributions are established and confirmed in revealing structure–activity relationships (SARs). The range of these benefits is indeed quite impressive and extends from improved potency to more favorable ADME, pharmacokinetic, and safety properties. A more focused chapter by J. T. Welch describes *Applications of Pentafluorosulfanyl Substitution in Life Sciences Research.* While popular for quite some time in agrochemicals, this "super-trifluoromethyl" group, with a size slightly smaller than a *t*-butyl group, has in recent years also found increasing application in pharmaceuticals development. As SF_5–substituted building blocks become rapidly commercially available, there is usually no need for direct fluorination. The chapter *Strategic Incorporation of Fluorine into Taxoid Anticancer Agents* by A. Pepe, L. Sun, and I. Ojima illustrates how the metabolic stability of taxoid anticancer drugs is improved and their general cytotoxicity reduced by introduction of organofluorine. It also describes the use of solid-state ^{19}F NMR spectroscopy to elucidate the bioactive conformations of taxoids. A comprehensive and useful coverage of *Synthesis and Antiviral, Antitumour Activities of Fluorinated Sugar Nucleosides* is provided by F. Zheng, X.-L. Qiu, and F.-L. Qing. They present the preparation of a large variety of nucleoside building blocks with fluorinated ribose moieties and discuss the conformational effects resulting from organofluorine introduction. M. Winkler and D. O'Hagan in their chapter on *Synthesis of Fluorinated Neurotransmitter Analogues* report on the development of non-peptidic fluorinated small molecules that find application in biomedical ^{19}F NMR and ^{18}F PET studies. Fluorinated adrenaline and dopamine analogs are covered as well as a diversity of other compounds binding to central neuroreceptors such as the glutamine, histamine, acetylcholine, and serotonin receptors.

The third and final section deals with the use of ^{19}F probes in NMR and of ^{18}F-radiolabeled probes in PET imaging applications. An authoritative survey of *^{18}F-Radionuclide Chemistry* is provided by R. Bejot and V. Gouverneur. The introduction of the radiolabels into probes for PET studies requires special protocols for synthesis and purification due to the limited half-life of the radionucleus, and these protocols are covered in an informative way. *^{18}F-Labelled Tracers for PET Oncology and Neurology Applications* by S. K. Luthra and E. G. Robins describes the protocols for the preparation of specific PET probes for *in vivo* imaging to elucidate disease-based mechanisms in oncology and neurology. The authors cover the synthesis of ^{18}F-labeled nucleosides, RGD (Arg–Gly–Asp) sequences, peptides that bind to specific biological targets (and the application of these probes to *in vivo* imaging of tumor angiogenesis), apoptosis, and amyloid plaque formation. The final chapter by V. D. Kodibagkar, R. R. Hallac, D. Zhao, J.-X. Yu, and R. P. Mason on *^{19}F NMR: Clinical and Molecular Imaging Applications* discusses the use of fluorinated probes in non-invasive clinical and molecular imaging to investigate enzyme activities and cell tracking in various diseases. It connects well to the earlier chapters reporting the synthesis of such probes.

All chapters are carefully selected and contribute to a unique, well-rounded monograph. Learning is fully ensured, as I can certify from the preparation of this foreword. I am not aware of any other monograph covering organofluorine applications in such depth and diversity. It will be of great practical use to scientists in industry — both pharmaceutical and agrochemical — and in academia. Both experts and novice practitioners will benefit from the reading. The monograph should also find use as a basis for advanced courses on organofluorine applications in biomedical research in masters and doctoral degree programs. The chosen format of individual chapters, namely comprehensive coverage of both modern synthetic methodology and *in vitro* and *in vivo* biological applications of the resulting building blocks and ligands, is highly attractive. It becomes quite clear that there is lots of room for further developments of innovative fluorinated building blocks and investigations of their physicochemical and biological properties. There is no doubt that this monograph will stimulate much future research on organofluorine in pharmaceutical and biomedical chemistry.

François Diederich
Laboratorium für Organische Chemie
ETH Zurich, Hönggerberg, HCI
CH-8093 Zurich, Switzerland
Zurich, November 1, 2011

Preface

Véronique Gouverneur and Klaus Müller

Fluorine has a distinctive place in the periodic table and has absorbed the attention of numerous scientists over many decades. Fluorine chemistry today is a well-established branch of modern sciences, which has tremendously benefited various research areas from material to medical sciences. The synthesis of fluorinated compounds has been extensively explored. Today, this field of research still stretches to the limit the creativity of chemists eager to develop new concepts for both selective fluorination and clever design and manipulation of fluorinated building blocks. The increased availability of fluorinated compounds has led to insightful studies aimed at deciphering the effects of fluorine substitution on physicochemical properties. Aspects of fluorine chemistry have been competently discussed in numerous books and reviews. This monograph is intended for a broad readership of professionals and researchers particularly interested in life sciences and medicine. An effort has been made to integrate chemistry, biology, drug discovery and medicine in a way that gives the reader an appreciation of how fluorine has enriched the life sciences in many respects. Molecules substituted with fluorine have improved our understanding of the molecular mechanisms of disease states and are continuously contributing to the advancement of drug discovery and diagnostic imaging. These aspects are covered in this book, which is organised around three sections. The first part provides answers on the fundamental question of how the introduction of fluorine modulates the physicochemical and molecular recognition properties of biologically relevant molecules. This is followed by an in-depth coverage of the impact that fluorine has made on drug discovery and development. The last section gives the reader informative accounts on the use of ^{19}F-spinlabelled and ^{18}F-radiolabelled probes for imaging by nuclear magnetic resonance and positron emission tomography, respectively.

In this multi-authored monograph, industrial and academic experts in the field bring the reader up to date with twelve chapters discussing all aspects of their respective research areas from essential background information to the most recent developments. In the process of editing this book, we have come to appreciate the enormous amount of talent of 'fluorine scientists' that has enabled spectacular advances in molecular medicine. We wish to express our most sincere gratitude and thanks to the authors of this monograph (Holger Gohlke, Jelena Bozilovic, Joachim W. Engels, Toni Vagt, Mario Salwiczek, Beate Koksch, Matteo Zanda, Alessandro Volonterio, Monica Sani, Sergio Dall'Angelo, Vladimir S. Kubyshkin, Igor V. Komarov, Sergii Afonin, Pavel K. Mykhailiuk, Stephan L. Grage, Anne S. Ulrich, Steve Swallow, John T. Welch, Antonella Pepe, Liang Sun, Iwao Ojima, Feng Zheng, Xiao-Long Qiu, Feng-Ling Qing, Margit Winkler, David O'Hagan, Romain Bejot, Véronique Gouverneur, Sajinder K. Luthra, Edward G. Robins, Vikram D. Kodibagkar, Rami R. Hallac, Dawen Zhao, Jian-Xin Yu and Ralph P. Mason), to François Diederich who has kindly agreed to comment on this monograph and to the countless chemists, biologists, physicists, physicians and clinicians around the world who have contributed to advancing life sciences and medicine over the years using fluorine as an enabling element. Heartfelt thanks to the members of the Gouverneur research group for helping with the proofreading (Matthew Tredwell, Matthew Hopkinson, Jamie Wolstenhulme, Charlotte Hollingworth, George Blessley, Ida Sofia Stenhagen and Guy Giuffredi).

We very much hope that this monograph will inspire many dedicated scientists and stimulate further developments relying on fluorine, with even more key discoveries in and for the future.

Véronique Gouverneur and Klaus Müller
Oxford, 9 November 2011

Contents

Section 1

1

Synthesis and Properties of Fluorinated Nucleobases in DNA and RNA

Holger Gohlke, Jelena Bozilovic*[†] *and Joachim W. Engels*[†]

1.1 Introduction

The stability of nucleic acid structures is predominantly governed by hydrogen bonding, base stacking, and solvation. To probe these interactions, a common approach is to replace the native bases adenine (**A**), uracil (**U**)/thymidine (**T**), guanosine (**G**) and cytosine (**C**) with analogues in which functional groups are added, deleted, blocked or rearranged. The size and shape of the analogues should be preserved as closely as possible to native bases. Such 'non-polar nucleoside isosteres' (NNIs) then allow detection of the predominant forces within nucleic acid structures without introducing steric effects. For DNA, this concept was introduced by Kool and coworkers in 1994 (Schweitzer and Kool, 1994). Initially, these molecules were intended to be used as probes of the importance of hydrogen bonding and base stacking in the formation of stable DNA duplex structures (Schweitzer and Kool, 1994; Kool and Sintim, 2006). In the context of DNA replication, it was later

* Institute of Pharmaceutical and Medicinal Chemistry, Heinrich-Heine-Universität, Universitätsstr. 1, 40225 Düsseldorf, Germany.
[†] Institute of Organic Chemistry and Chemical Biology Goethe-University, Max-von-Laue-Str. 7, 60438 Frankfurt, Germany. E-mail: joachim.engles@chemie.uni-frankfurt.de

concluded by these authors that steric effects rather than hydrogen bonding was the chief explanation for replication fidelity (Kool, 1998; Kool and Sintim, 2006). In 1999, the concept of NNI was introduced into the RNA world by Engels and coworkers, initially using fluorinated benzenes and benzimidazoles as pyrimidine and purine base analogues, respectively (Parsch and Engels, 1999, 2000). Syntheses and crystallographic studies of these and other NNIs as well as thermodynamic analyses and computer simulations of model RNAs incorporating them are reviewed here. These combined studies have proved invaluable for probing the physical forces that govern the stability of RNA and shedding light on the role of fluorine in molecular recognition. As a biological application, the incorporation of fluorinated NNI into ribozymes and siRNA is finally described.

1.2 Fluorine in Molecular Recognition

Substituting hydrogen by fluorine in organic compounds influences a variety of the molecule's properties. In medicinal chemistry, fluorine substitution has long been known as a means of enhancing metabolic stability, modifying chemical reactivity and conformational equilibria, and improving transportation and absorption characteristics of pharmaceuticals (Müller *et al.*, 2007; Bégué and Bonnet-Delpon, 2008; Hagmann, 2008). In contrast, the role of 'organic fluorine' in influencing molecular recognition properties, i.e. specific bonding between two or more molecules through non-covalent interactions, is much less understood. Here we focus on influences that are particularly important in the context of fluorinated NNIs.

The properties of the C–F bond provide a starting point for appreciating some of these influences. Replacing F for H is considered the most conservative substitution for hydrogen on steric grounds, although a fluorine atom is closer in size (and bond length) to oxygen than hydrogen (O'Hagan, 2008). Hydrogen and fluorine are also quite different regarding their electronic influences. The high electronegativity of fluorine imparts a less covalent and more electrostatic character to a highly polarized C–F bond, allowing interactions between the C–F bond dipole and other dipoles in close proximity (O'Hagan, 2008). In turn, fluorine's three lone pairs are held tightly, as manifested by the atom's high ionization potential and low polarizability, and so are reluctant to get involved in resonance or act as hydrogen bonding acceptors.

As an immediate consequence, fluorine substitution leads to the seemingly orthogonal effects of increasing local polarity *and* molecular hydrophobicity (Guerra and Bickelhaupt, 2003; Biffinger *et al.*, 2004; DiMagno and Sun, 2006): as electrostatic and, in particular, time-dependent interactions of C–F bonds are of minimal importance in polar heteroatom solvents, C–F dipoles and lone pairs are poorly solvated in water. Removing this group from water is thus energetically favourable. Accordingly, for a set of 293 pairs of compounds where a single H/F exchange had been performed, an increase of the log*D* value by on average 0.25 log units upon fluorine substitution was observed (Bohm *et al.*, 2004). Notably, for a series of singly and multiply fluoro-substituted benzenes, no impact of the molecular dipole moment on the partition coefficient was found, suggesting that surface area arguments suffice to explain hydrophobicity trends (DiMagno and Sun, 2006).

Due to the electron-withdrawing effect of fluorine, fluorination has an important indirect impact on interactions of an aromatic ring to which fluorine is attached. On going from benzene to hexafluorobenzene, the quadrupole moment of the molecules inverts (Hernandeztrujillo and Vela, 1996), favouring eclipsed face-to-face π-stacks in crystallized 1:1 mixtures (Patrick and Prosser, 1960). Similarly, using the dangling-end method for determining the energetics of aromatic π-stacking of fluorinated benzene analogues of DNA bases, a large stabilizing interaction is observed with a 2,3,4,5-tetrafluorophenyl dangling nucleotide. Yet, the overall correlation between the number of fluorine groups and the stabilization is only weak (Lai *et al.*, 2003). The σ-inductive effect of fluorine also affects neighbouring functionalities. With respect to fluorinated NNIs, the influence on the acidity of neighbouring C–H groups is of prime importance: the C–H group is known to be a hydrogen-bond donor (Desiraju, 1997) and a C–H···X interaction is strengthened by a more positively polarized C–H group (Thalladi *et al.*, 1998). A particularly strong polarization is expected for a hydrogen in bis-*ortho* position to two fluorine atoms, e.g. in 1,3-difluorobenzene or 1,3,5-trifluorobenzene (Razgulin and Mecozzi, 2006). Not surprisingly, in the crystal structure of 1,3,5-trifluorobenzene, each H and F atom is involved in the formation of two C–H···F interactions

(Thalladi *et al.*, 1998), leading to a close similarity to the classical structure of 1,3,5-triazine (Coppens, 1967). Obviously, the C–H···F interaction is favoured by multipolar interactions between the C–H / F–C bond dipoles, as also demonstrated by numerous O=C···F–C contacts found in the Cambridge Structural Database (Olsen *et al.*, 2003).

In the presence of competing heteroatom acceptors, covalently bound fluorine hardly ever acts as an acceptor for available Brønsted acidic sites, owing to its low proton affinity and weak polarizability (Murray-Rust *et al.*, 1981; Dunitz and Taylor, 1997; Dunitz, 2004; Kool and Sintim, 2006). Attractive C–F···H–X dipolar interactions have been described, however, for well-structured molecular environments in which heteroatom acceptors are excluded, such as the thrombin active site (Olsen *et al.*, 2003; Bohm *et al.*, 2004) or engineered crystals (Desiraju, 2002; Reichenbacher *et al.*, 2005).

1.3 Synthesis of Fluoro-Substituted Benzenes, Benzimidazoles and Indoles, and their Incorporation into Model RNA

1.3.1 *Chemical syntheses of fluoro-substituted benzenes, benzimidazoles and indoles*

When we started to synthesize fluorobenzene nucleosides **1–4** (Fig. 1.1) some ten years ago we were able to directly introduce the lithiated fluorobenzenes to the protected ribonolactone.

The method is based on a similar procedure introduced by Krohn *et al.* (1992), where a bromo-fluorobenzene is added to a benzylated ribonolactone. The resulting lactol is reduced and yields directly the pure β-ribonucleoside in high yield.

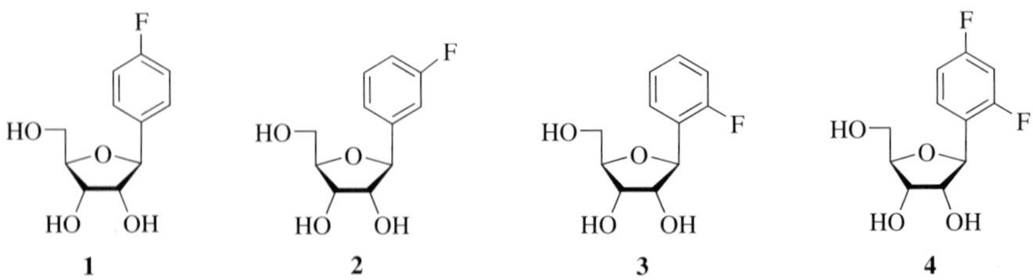

Figure 1.1. Initial set of fluorobenzene ribosides.

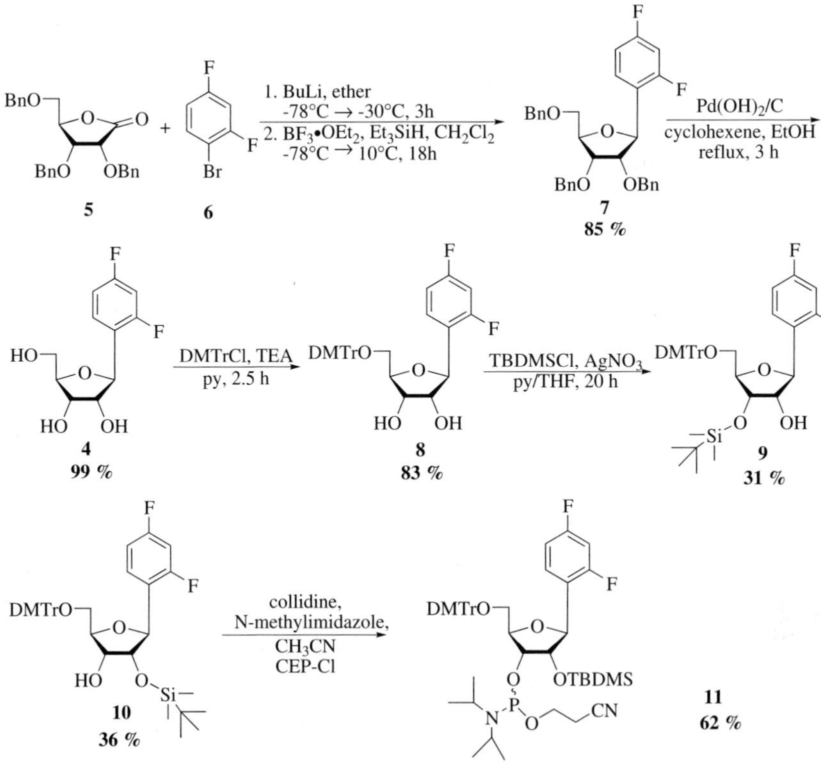

Figure 1.2. Synthesis of a fluorobenzene nucleoside phosphoramidite.

In Fig. 1.2, the synthesis of 2,4-difluorobenzene riboside as phosphoramidite building block is shown as an example (Parsch and Engels, 2002). The unprotected riboside **4** is tritylated with dimethoxytritylchloride in pyridine with high regioselectivity for the 5′-position, followed by silylation with *tert*-butyldimethylsilyl chloride for protecting the 2′- or 3′-positions. After a chromatographic separation, the 2′-silylated nucleoside **10** is coupled with chloro-cyanoethyl-diisopropylphosphoramidite (CEP-CL). The overall yields for the four nucleobases prepared in an analogous way range from 10 to 20%.

In Fig. 1.3, as an example, the synthesis of 1′-deoxy-1′-(4-fluoro-benzimidazol-1-yl)-β-D-ribofuranose **21** is shown, following the glycosylation procedure of Vorbruggen (Vorbruggen and Hofle, 1981).

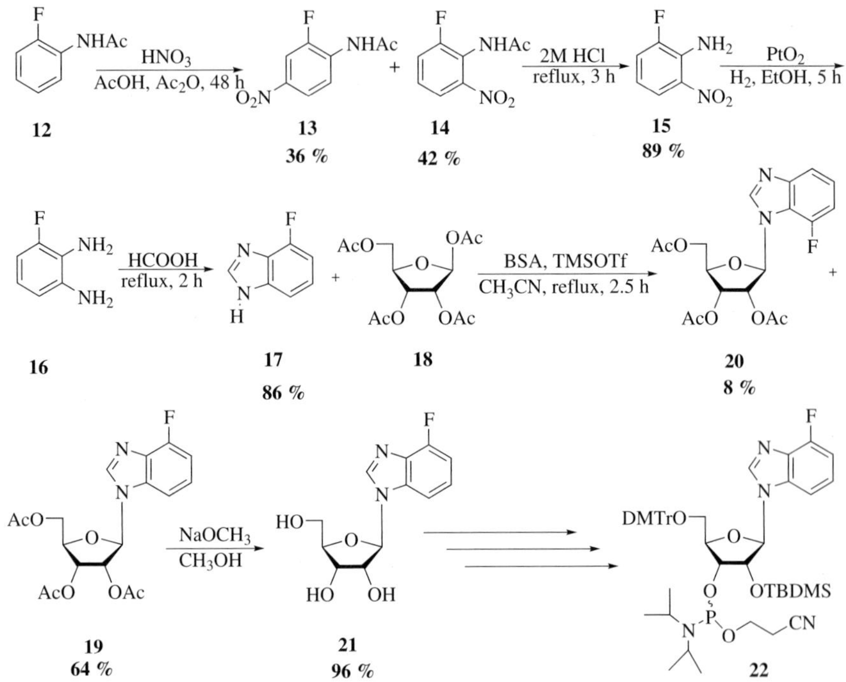

Figure 1.3. Synthesis of a fluorobenzimidazole nucleoside phosphoramidite.

Refluxing two equivalents of 4-fluorobenzimidazole **17** with *N*,*O*-bis(trimethylsilyl)acetamide and subsequent reaction of the silylated base with one equivalent of 1,2,3,5-tetra-*O*-acetyl-β-D-ribofuranose **18** in the presence of the Lewis acid trimethylsilyl-trifluoromethanesulfonate afforded the desired 2′,3′,5′-tri-*O*-acetyl-1′-deoxy-1′-(4-fluorobenzimidazol-1-yl)-β-D-ribofuranose **19** in 56% to 64% yield. Using microwave reaction conditions we could improve the yields (Nikolaus *et al.*, 2007). Deprotection of the acetylated nucleoside **19** furnished 1′-deoxy-1′-(4-fluorobenzimidazol-1-yl)-β-D-ribofuranose **21** in 90% to 96% yield (Bats *et al.*, 2000; Parsch and Engels, 2000, 2001). The phosphoramidites were obtained in an analogous way as described above and the yields ranged from 10 to 15% (Parsch and Engels, 2000, 2002).

Syntheses of the fluorobenzimidazole ribosides **21** and **39–41** started with the appropriately substituted fluoroacetanilides, followed by nitration,

Table 1.1. Partition coefficients (log$P$ values) of nucleosides containing fluoro-substituted nucleobase analogues (for abbreviations, see Fig. 1.4.)

Nucleoside	Partition coefficient	Nucleoside	Partition coefficient
B	1.050	4,6DFBI	4.240
4FB	1.497	4TFM	6.762
2,4DFB	1.695	5TFM	6.757
2,4,6TFB	1.398	6TFM	6.688
2,4,5TFB	1.825	I	0.72
PFB	2.156	4FI	2.781
BI	0.149	5FI	2.252
4FBI	1.789	6FI	4.420
5FBI	1.802	7FI	2.383
6FBI	1.799	4,6DFI	5.540

reduction, and ring closure with formic acid (Parsch and Engels, 2000, 2002) (see Fig. 1.4 for the structures of all fluorobase analogues). The trifluoromethyl-benzimidazoles **42–45** were analogously prepared starting from the trifluoromethyl acetanilides (Moore *et al.*, 2004). The log$P$ values measured in octanol/water for the mono-substituted benzimidazoles are identical and the orientation of the fluoro-substituent does not seem to have a major influence here. The difluorobenzimidazole riboside 4,6DFBI **41** is significantly more lipophilic (log$P$ = 4.2) (Table 1.1).

Since the indole structure is present in a variety of natural compounds, such as amino acids and alkaloids, we decided to synthesize and evaluate a similar pattern of substitution on fluoroindoles. Furthermore, the charge distribution and dipole moments between fluoroindole and fluorobenzimidazole compounds differ significantly, which we also expected to give rise to differences in stability of RNA oligonucleotides containing these building blocks.

To the best of our knowledge, ribofluoro-indole compounds had not been synthesized and evaluated before. We reported the successful syntheses of fluoroindole building blocks and the X-ray crystal structures of all synthesized fluoroindole ribonucleosides **23–27** (Fig. 1.5) (Bozilovic and Engels, 2007).

The synthesis of fluoroindoles, which are not commercially available, was achieved very efficiently in a four-step procedure as shown in Fig. 1.5

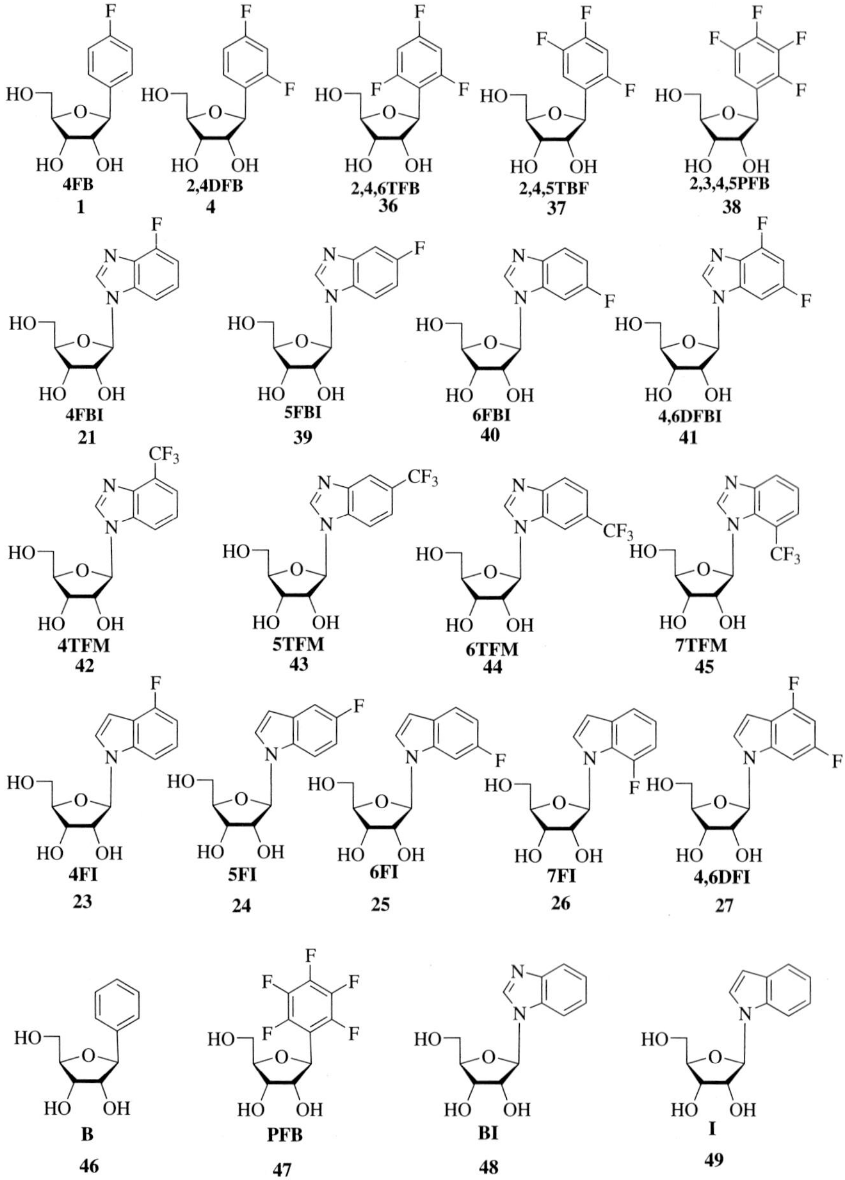

Figure 1.4. Summary of fluorobase analogues comprising fluorobenzenes, fluorobenzimidazoles, fluoroindoles, and trifluoromethyl benzimidazoles.

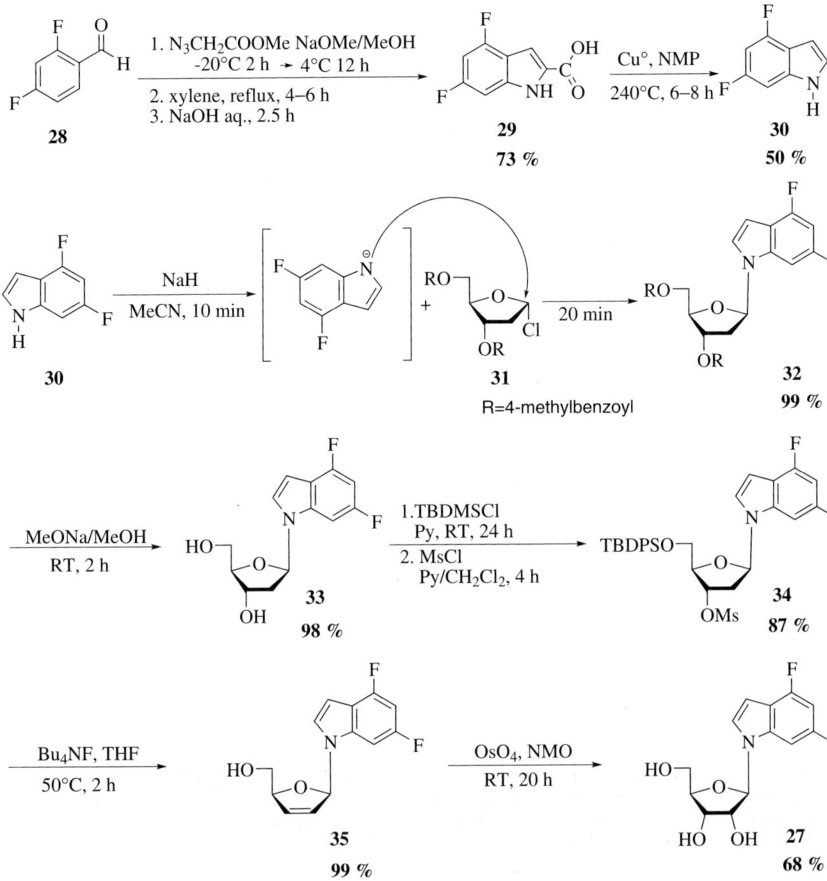

Figure 1.5. Synthesis of fluoroindole riboside 27.

for the 4,6-difluoroindol-β-D-ribofuranoside **27**. For the 4,6-difluoroindole **30**, methyl azidoacetate, an intermediate in indole synthesis, was synthesized by a literature procedure (Da Rosa *et al.*, 2003). Prolonged reaction time and lower reaction temperature improved this procedure (Bozilovic and Engels, 2007). The hydrolysis of the methyl ester is followed by a decarboxylation, where the main problem lies in the very high temperature at which the reaction occurs. Prolonging the reaction time does not lead to higher yields (Bozilovic and Engels, 2007). Since direct glycosylation of the indole moiety with the ribose failed, synthesis of the deoxyriboside and subsequent transformation to the ribose was chosen as an alternative. Although this is a longer procedure, the individual steps could be optimized. Direct S_N2 substitution of the indole anion with the chlorosugar **31** yielded **32** quantitatively. After deprotection, the 5′-OH is protected again with tert-butyl dimethylsilyl chloride (TBDMSCl) and 3′-OH with mesylchloride (MsCl). Deprotection of 5′-OH and elimination of mesylate gave **35** in one step, which on dihydroxylation via osmium tetroxide resulted in the formation of indole riboside **27**. For the RNA building block, the phosphoramidite was prepared as outlined above (Fig. 1.2). The synthesis of the indole ribosides **23–26** followed the same procedure. The total set of fluorobase analogues derived from fluoro-modified benzene, benzimidazoles and indoles is outlined in Fig. 1.4.

1.3.2 *Synthesis of 12-mer RNA duplexes that incorporate fluoronucleosides*

The phosphoramidites **11** and **22** (Figs. 1.2 and 1.3) as well as the phosphoramidites of all other fluorobenzenes, fluorobenzimidazoles and fluoroindoles were incorporated into a defined 12-mer RNA duplex to investigate their influence on RNA stability. We chose the purine rich sequences 5′-CUUUUCXUUCUU-3′ paired with 3′-GAAAAGYAAGAA-5′, where the central bases X or Y contained modified fluoronucleosides. This yielded either mono-modified duplex RNA, where a fluorinated NNI is paired against a natural base, or double-modified duplex RNA containing fluorinated NNIs paired against each other. Solid-phase RNA synthesis followed the Caruthers DNA/RNA cycle (Fig. 1.6) (Caruthers, 1985).

In short, a starting riboside, attached to controlled-pore glass (CPG) support on the 3′-end, is stepwise extended via tetrazole activation with an appropriately protected amidite of **A**, **C**, **G**, or **U**. In the case of X or Y, the fluorinated NNI is added and the sequence further extended until the 5′-end. After successful additions, which are monitored by the trityl colour, the full size oligoribonucleotide is liberated from its protecting groups and the crude product is purified. In most cases high performance liquid chromatography on ion exchange columns or sometimes reversed-phase material gives sufficiently pure oligoribonucleotides, which are characterized by mass spectrometry. Notably, the CD spectra of the RNA

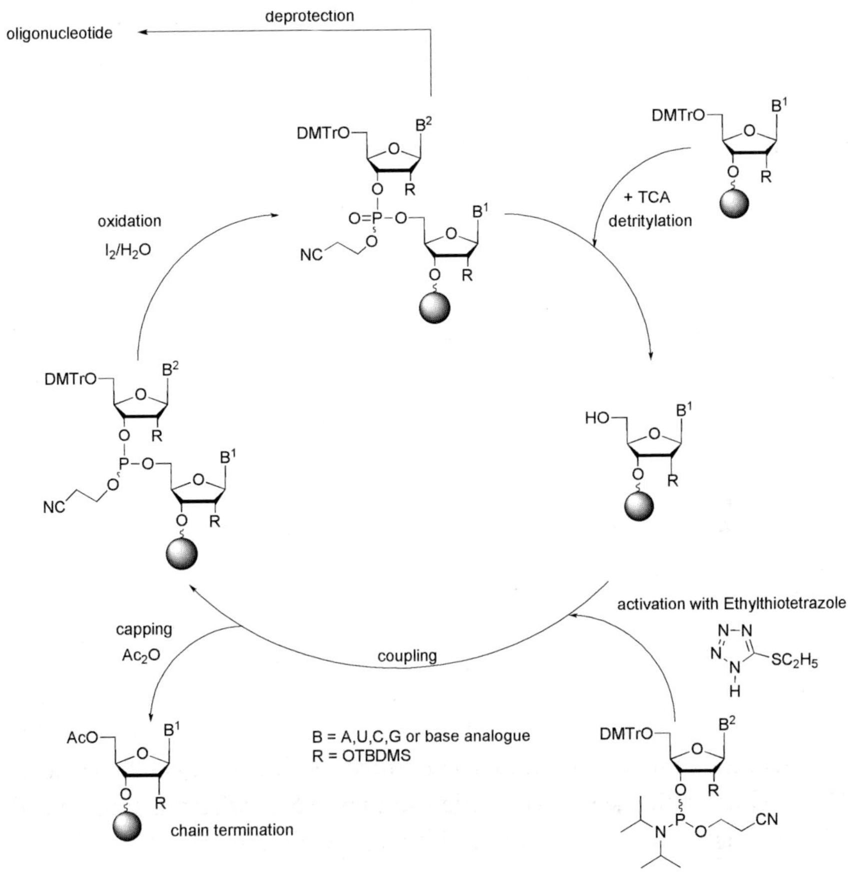

Figure 1.6. RNA synthesis: Caruthers cycle with which the base analogues B = **1–4**, **21**, **23–27**, **39–41** are incorporated. (TCA = trichloroacetic acid.)

duplexes with fluorinated NNI showed the typical curves for an α-type helix (Fig. 1.7), indicating that the structure of the duplex RNA is not disturbed by incorporation of the modified nucleosides.

1.3.3 *RNA melting studies and thermodynamic data*

Thermodynamic stabilities of the modified RNA duplexes were determined by thermal denaturation as monitored by UV absorbance in a phosphate buffer (10 mM Na_2HPO_4, 10 mM NaH_2PO_4, pH 7) containing NaCl

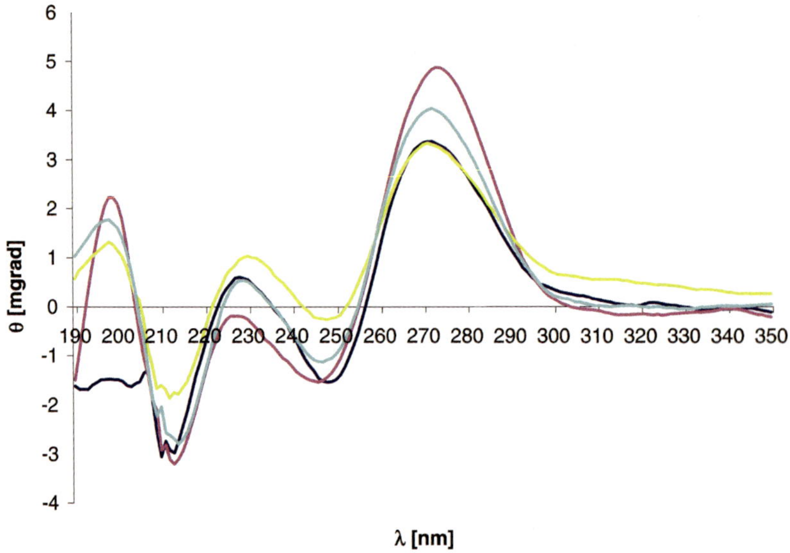

Figure 1.7. CD spectra of the model 12mer RNA containing fluoroindole nucleosides at position X: **23** (blue), **25** (pink), **26** (yellow), **27** (cyan).

(140 mM). Thermodynamic data were extracted from the melting curves by means of a two-state model for the transition from duplex to single strands.

Melting temperatures (T_m) and free energies of duplex stability (ΔG) for all fluorobenzenes, -benzimidazoles and -indoles paired against all four natural bases are summarized in Table 1.2. Thermodynamic analysis of the T_m furthermore yielded enthalpic (ΔH) and entropic ($T\Delta S$) contributions to the duplex stability (data not shown) (Parsch and Engels, 2002).

Two results stand out from these studies. First, compared to the natural **U·A** base pair (T_m = 37.8°C) or the wobble base pair **U·G** (T_m = 38.6°C), pairing of fluoronucleosides against natural bases in general decreases RNA duplex stability, as demonstrated by T_m values that are lower by 2–15°C. The lowest T_m were measured for **36** **2,4,6**TFB and **47**PFB due to steric effects resulting from a bis-*ortho* fluoro substitution. Conversely, pairing **40** **6**FBI or **25** **6**FI against **A** almost restored the duplex stability to that of the natural **U·A** base pair, with T_m values of 34.0°C and 37.1°C, respectively. Second, and more strikingly, several of these fluoro-benzenes, -benzimidazoles and -indoles showed a lack of discrimination against the

Table 1.2. Thermodynamic stabilities of model 12mer RNA containing fluorobenzenes (FB), fluorobenzimidazoles (FBI), trifluoromethylbenzimidazoles (TFM) and fluoroindoles (FI) paired with natural bases **A**, **C**, **G**, and **U**.

	Y = A		Y = C		Y = G		Y = U	
X	T_m [a]	ΔG [b]	T_m [a]	ΔG [b]	T_m [a]	ΔG [b]	T_m [a]	ΔG [b]
4FB	23.8	7.9	24.1	8.0	24.2	8.0	25.6	8.4
2,4DFB	27.4	9.0	27.3	8.9	27.6	9.0	27.9	9.1
2,4,6TFB	23.3	7.8	20.6	7.1	22.8	7.7	22.9	7.9
2,4,5TFB	25.6	8.4	26.7	8.8	28.7	9.3	27.5	8.9
PFB	22.9	7.7	21.7	7.2	23.0	7.7	23.1	7.5
4FBI	28.0	9.1	27.5	8.9	28.7	9.3	28.5	9.2
5FBI	28.9	9.1	31.0	9.2	31.7	9.3	28.2	8.9
6FBI	34.0	7.5	31.3	10.1	32.4	10.3	28.6	9.3
4,6DFBI	28.4	9.2	28.7	9.2	29.4	9.5	29.3	9.5
4TFM	30.1	11.0	27.4	9.9	29.2	9.7	25.1	9.3
5TFM	30.2	9.1	28.1	8.8	29.3	8.7	24.1	8.1
6TFM	30.1	10.9	28.0	8.9	27.1	9.6	24.8	8.2
4FI	25.1	8.3	24.1	8.0	21.4	7.6	21.0	7.2
5FI	32.8	9.7	26.8	8.6	29.3	9.3	25.0	8.2
6FI	37.1	10.6	34.6	15.4	35.5	11.3	31.7	9.7
7FI	34.4	9.9	24.8	8.2	24.6	8.3	29.8	9.5
4,6DFI	33.8	8.4	32.0	9.2	30.2	9.4	29.0	9.7

[a] In °C. [b] In kcal/mol.

natural bases, thus acting as universal bases (Loakes, 2001). In the case of **4** 2,4DFB and **41** 4,6DFBI, the T_m values differ by less than 1°C, yielding these fluoronucleosides as the best universal bases.

In a second series, the stability of RNA duplexes containing fluorobenzene self-pairs at positions X and Y were determined (Table 1.3). Not unexpectedly (Parsch and Engels, 2002; Lai and Kool, 2004), the measurements demonstrated that the pairing preference of fluorinated NNI is higher in self-pairs than in pairs with natural bases. Furthermore, the duplex stability increases incrementally with the number of fluorine substituents in the NNI. Surprisingly, this leads to RNA duplex stabilities with self-paired bases **4** 2,4DFB, **37** 2,4,5TBF and **38** 2,3,4,5PFB(T_m = 35.2, 35.8, 38.0°C) that are similar to or exceed that of the natural **U·A** base pair.

Table 1.3. Thermodynamic stabilities of model 12mer RNA containing self-pairs of benzene- and fluoro-substituted benzenes (FB).

X	Y	T_m [a]	ΔG [b]
Benzene	Benzene	26.4	8.5
4FB	4FB	32.5	10.2
2,4DFB	2,4DFB	35.2	11.6
2,4,5TFB	2,4,5TFB	35.8	11.8
2,3,4,5TetFB	2,3,4,5TetFB	38.0	12.2

[a] In °C. [b] In kcal/mol.

1.4 Origin of the Molecular Recognition Properties of Fluorinated Nucleobases

1.4.1 *Stacking and desolvation: Insights from thermodynamic analyses*

RNA stability is predominantly governed by base stacking, solvation forces and hydrogen bonding. The above data provide a rich source for analysing individual contributions of these forces, in particular, in view of the role of fluorine in molecular recognition.

The influence of base stacking was determined by comparing stabilities of duplex RNA with a respective base at position X, whereas the Y position is empty (abasic site) (Parsch and Engels, 2002). Compared to X = U, incorporation of the NNI **46** increases the duplex stability by 1.3°C (0.2 kcal/mol), demonstrating, as already found for 'dangling end' residues in the context of DNA (Guckian *et al.*, 2000), that NNIs stack more strongly than their natural counterparts. Fluorine substitution then leads to another gain in stability through base stacking in that X = **1** increases the duplex stability by 2.7°C (0.5 kcal/mol) and X = **4** by 4.4°C (1.1 kcal/mol). Both an increase in the molecular dipole moment (**46**: 0.3 D, **1**: 2.4 D, **4**: 2.2 D) (Lai *et al.*, 2003), resulting in increased van der Waals dispersive forces with the neighbouring bases (Lai *et al.*, 2003), and a higher lipophilicity (**46**: log$P$ = 1.05, **1**: 1.50, **4**: 1.70) account for this. Compared to X = **G** as a reference, X = **41** even leads to a stabilization by 5.4°C (1.2 kcal/mol), demonstrating that stacking interactions become more favourable with increasing size of the NNI, similar to the finding that a purine base stacks on a duplex more strongly than a smaller pyrimidine base (Petersheim and Turner, 1983). In the case of the fluoroindoles this could be determined by the dangling end method; for **25** a T_m increase of 11°C (2.7 kcal/mol) and for **27** of 4°C (0.9 kcal/mol) was measured (Bozilovic, 2008).

An X = **U**, Y = **U** mismatch (T_m = 30.1°C) was used as a reference for investigating the influence of desolvation of the hydrogen bond donors and acceptors of the natural bases during formation of the base pair with an NNI (Parsch and Engels, 2002). This desolvation effect was found to destabilize the RNA duplex in the case of X = **46**, Y = **U** by −8.3°C (−2.2 kcal/mol). Notably, fluorine substitution reduced this effect to about −6.6°C (−1.7 kcal/mol) for both the X = **4** and **41** cases. A similar trend was found when base pairing between natural nucleobases and universal fluorinated NNIs was investigated by potential of mean force calculations (Koller *et al.*, 2010), indicating interaction differences between paired bases, with more attractive interactions in the case of **4** and **41** than in the case of **46** (see below).

1.4.2 *C–H···F–C interactions: Crystallographic analysis of fluoro-substituted NNIs*

When paired against natural bases, fluorinated NNIs destabilize DNA and RNA helices, and can exhibit universal base properties (Table 1.2) (Parsch and Engels, 2002; Lai and Kool, 2004). These observations make Watson–Crick base pairing involving hydrogen bonds to fluorine unlikely (Parsch and Engels, 2002; Somoza *et al.*, 2006). When paired opposite one another, however, a considerable degree of stability is regained (Table 1.3) and a selective pairing of fluorinated NNIs in the context of nucleic acids is observed (Parsch and Engels, 2002; Lai and Kool, 2004; Kopitz *et al.*, 2008). Weak C–F···H–C dipolar interactions have been implicated to act as stabilizing forces in this case (Lai and Kool, 2004; Kopitz *et al.*, 2008). These findings relate to the present discussion as to whether 'organic fluorine' can act as a hydrogen bond acceptor and, if so, under which conditions (Dunitz and Taylor, 1997; Evans and Seddon, 1997; Guckian *et al.*, 2000; Dunitz and Schweizer, 2006).

Following the idea that crystal packing information allows information to be deduced about non-bonded interactions (Velec *et al.*, 2005), initially crystal structures of fluorobenzene nucleosides **1–4** were determined. To our surprise the well-known herringbone pattern of benzene nucleosides was only present in nucleoside **2** (Matulicadamic *et al.*, 1996). In contrast, for nucleosides **1**, **3** and **4**, when crystallized from water, the C–F···H–C distance between neighbouring aromatic rings of **1** is 238 pm, which is significantly shorter than the sum of the van der Waals radii of 255 pm (Rowland and Taylor, 1996). Similarly, when **1** was crystallized from methanol, a distance of 230 pm was found between the fluorine and a hydrogen atom in *ortho* position to F in a neighbouring aromatic ring (Parsch and Engels, 1999; Bats *et al.*, 2000). The C–F···F–C arrangement shows a nearly linear configuration with a C–F···F angle of 158° (Fig. 1.8). We interpreted these findings in terms of weak attractive C–F···H–C dipolar interactions. The crystal structure of **3** revealed that the shortest C–F···H–C distance amounts to exactly the sum of the van der Waals radii of 255 pm of fluorine and hydrogen. This distance is found between the fluorine and H5′ of the sugar. Compared to the aromatic C–H bond adjacent to a fluorine-bound carbon in **1**, the less polar aliphatic C–H5′ bond apparently provides less attraction for the aromatic fluorine.

Subsequently, we were able to determine the crystal structures of several fluorinated NNIs, including difluorobenzimidazoles and mono- and difluoroindoles (Parsch and Engels, 2002; Zivkovic and Engels, 2002;

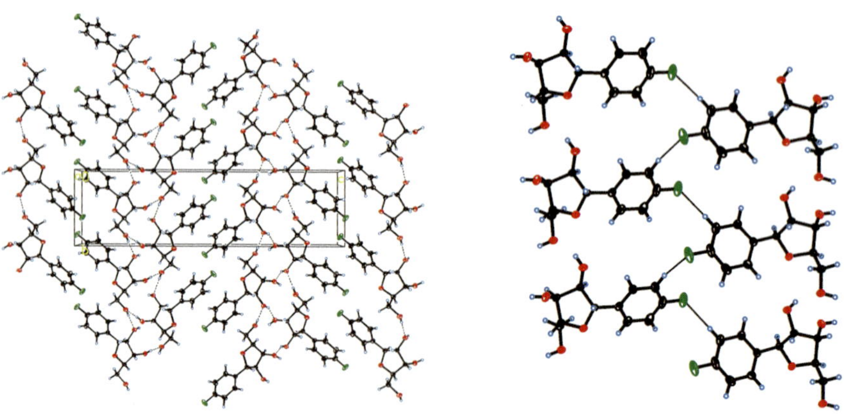

Figure 1.8. Crystal packing of 1′-deoxy-1′-(4-fluorophenyl)-β-D-ribofuranose **1**.

Bozilovic *et al.*, 2007b). With respect to the latter, in **23** the distance between fluorine and a hydrogen in *ortho* position to F in a neighbouring molecule was larger than the sum of the van der Waals radii (269 pm). In turn, the shortest distance between those two atoms was found for **26** at 230 pm with a C–H···F angle of 124°. The crystal structures of **24**, **25**, and **27** showed intermediate distances of 246 pm, 240 pm and 239 pm, respectively (Bozilovic *et al.*, 2007b).

Plotting the H···F distances and C–H···F angles according to Desiraju and coworkers (Thalladi *et al.*, 1998) finally revealed that some of the C–H···F–C interactions found between fluorinated nucleosides have the shortest H···F distances reported so far for sp^2-centred H···F interactions (Fig. 1.9).

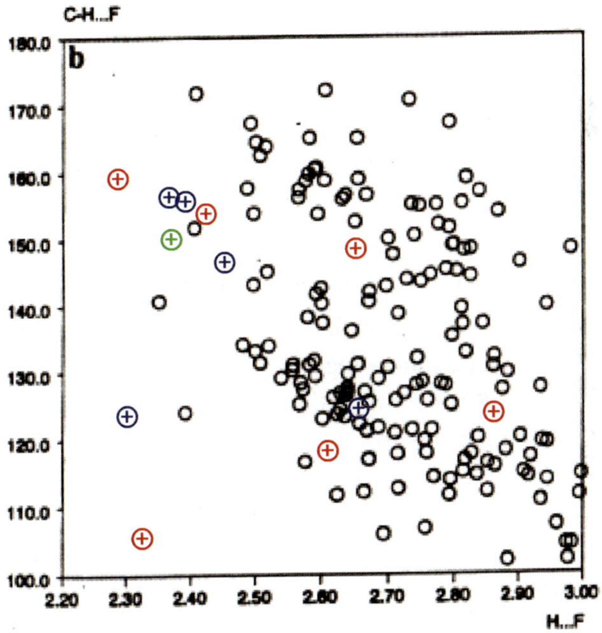

Figure 1.9. F···H distances and C–F···F angles of C–F···H–C contacts found in crystal packings according to Thalladi *et al.* (1998) (black circles) as well as in crystal packings of fluorobenzene nucleosides (red crosses) (Parsch and Engels, 1999; Zivkovic and Engels, 2002), difluorobenzimidazole nucleosides (green cross) (Zivkovic and Engels, 2002), and fluoroindole nucleosides (blue crosses) (Bozilovic *et al.*, 2007a).

1.4.3 *Molecular dynamics simulations and free energy calculations*

Additional insights at an atomic level into the stability determinants of RNA incorporating fluorinated nucleosides are provided by computer simulations. State-of-the-art molecular dynamics (MD) simulations and binding free energy calculations together with a structural component analysis were performed for RNA duplexes containing fluorobenzene self-pairs at positions X and Y, resulting in deviations between experimental and computed relative binding free energies of less than 0.4 kcal/mol (Kopitz *et al.*, 2008). Notably, these calculations revealed different origins for the incremental increase in duplex stability with increasing number of fluorine substituents in the NNI (Table 1.3): for the transitions **46**→**1** and **37**→**38** the binding free energy changes are dominated by favourable solvent contributions, resulting in an *indirect* effect of fluorine substitution. In contrast, for the transitions **1**→**4** and **4**→**37**, changes of interactions within the RNA contribute favourably to the observed stability gain, showing a *direct* effect of fluorine substitution.

How can one explain these findings? Interestingly, global molecular properties such as the lipophilicity of the nucleosides or the molecular dipole moment of the NNI were inappropriate to explain the differences. Rather, the observed trend parallels differences in surface area regions contributed by fluorine atoms that are buried upon duplex formation (Fig. 1.10). This points to a local influence of fluorine substitution and can be explained by the poor aqueous solvation of C–F dipoles, yielding a hydrophobic character of these regions (Guerra and Bickelhaupt, 2003).

As for the direct RNA contributions to duplex stability, weak attractive C–F···H–C interactions between the self-pairs were identified as stabilizing forces, with more short-range interactions present in **4** and **37** than in **1** (Fig. 1.11). This interpretation is corroborated by the analysis of the occupancy of C–F···H–C interactions: C–F···H–C interactions prevail for a larger fraction of time in the **4** (0.76) and **37** (0.70) cases than in the case of **1** (0.46).

Apparently, the fluorobenzene self-pairs in the context of duplex RNA constitute a well-structured supramolecular system, which leads to favourable C–F···H–C interactions between self-pairs of **4** and **37**, as was found for other well-structured molecular environments such as enzyme active sites (Olsen *et al.*, 2003; Olsen *et al.*, 2004) or crystals (Desiraju, 2002; Reichenbacher *et al.*, 2005).

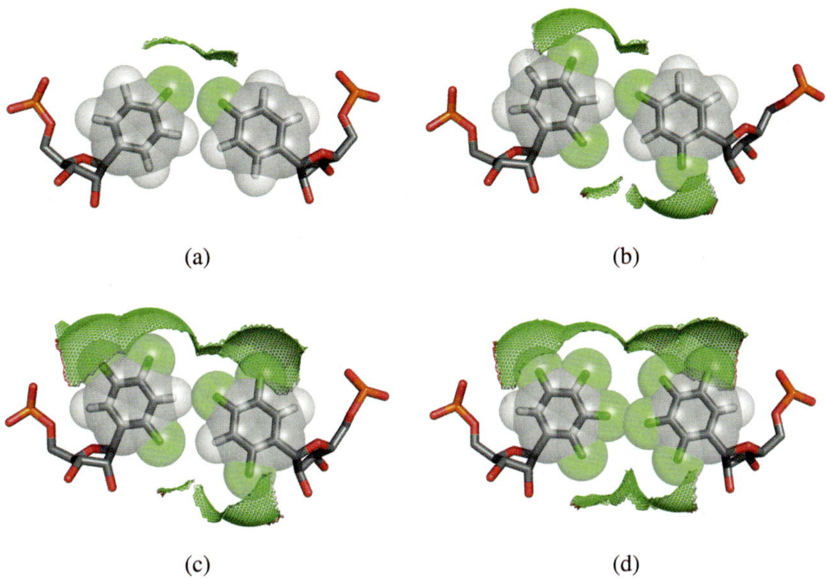

Figure 1.10. Averaged structures obtained from MD trajectories of model 12mer RNA containing the self-paired fluorobenzene nucleotides **1** (a), **4** (b), **37** (c), and **38** (d). The solvent-accessible surfaces of the fluorine atoms are depicted as green meshes.

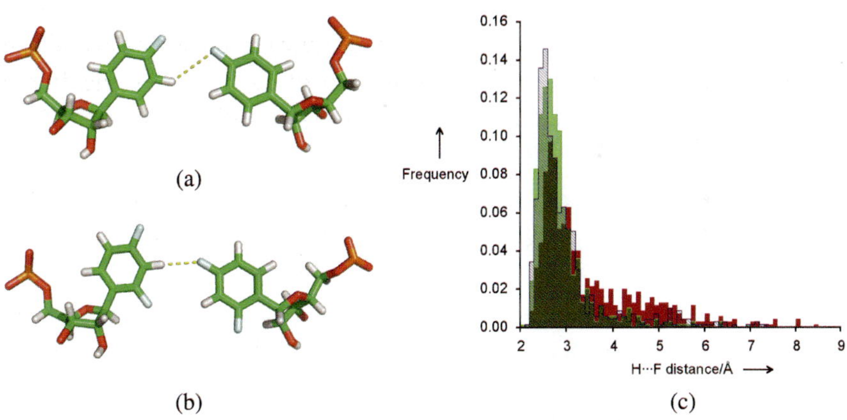

Figure 1.11. Averaged structures obtained from MD trajectories of model 12mer RNA containing the self-paired fluorobenzene nucleotides **1** (a) and **4** (b). (c) Frequency distributions of the distances between the H3 atom of the bases at position X and the F4 atoms of the bases at position Y for **1** (red), **2** (green), and **3** (blue). The distances computed for **1** and **4** are marked by a yellow dashed line in (a) and (b).

In another study, the 'universal base' character of fluorinated NNI was investigated by potential of mean force calculations of base pairing between natural bases and fluorobases (Koller *et al.*, 2010). In agreement with previous studies (Stofer *et al.*, 1999), Watson–Crick base pairing was computed to be favourable by about 1.5–2 kcal/mol per hydrogen bond formed. In contrast, pairing between **4** or **41** and the natural bases **A** or **C** was found to be unfavourable by 0.55–1.01 kcal/mol, in agreement with experiment (Parsch and Engels, 2002) and also in another MD study (Zacharias and Engels, 2004). Yet, for a given fluorinated NNI, the differences in the base pairing free energies with either one of the native bases are between 0.14 and 0.38 kcal/mol, supporting the universal base pairing properties. Finally, pairing between natural bases and **46** is more unfavourable by 0.6–1.0 kcal/mol than if a fluorinated base is used instead. Apparently, more attractive pairing interactions prevail in the case of **4** or **41** compared to **46**. They arise from dipole–dipole interactions involving the C–F bond of the fluorobases and the exocyclic amine group of **A** or **C** and weak hydrogen bonds between N1 (N3) of **A** (**C**) and H–C3(5) of **4** (**41**).

1.5 Incorporation of Fluoro-Substituted NNI into the Hammerhead Ribozyme and siRNA Constructs and their Acceptance by Polymerases

1.5.1 *Hammerhead ribozyme*

RNA, as a central molecule in the chemistry of life, is involved in the cellular process of gene expression and protein biosynthesis (Gesteland *et al.*, 1999). RNA exhibits a great structural diversity and its secondary as well as tertiary structure is mainly stabilized by hydrogen bonding and base stacking. Many efforts have already been undertaken to elucidate structural changes via incorporation of artificial nucleosides.

Here, we intended to incorporate fluoro-substituted NNI into a hammerhead ribozyme (Fig. 1.12) that is directed against the integrase region of the human immunodeficiency virus (HIV), a system we had introduced earlier (Klebba *et al.*, 2000; Müller-Kuller *et al.*, 2009).

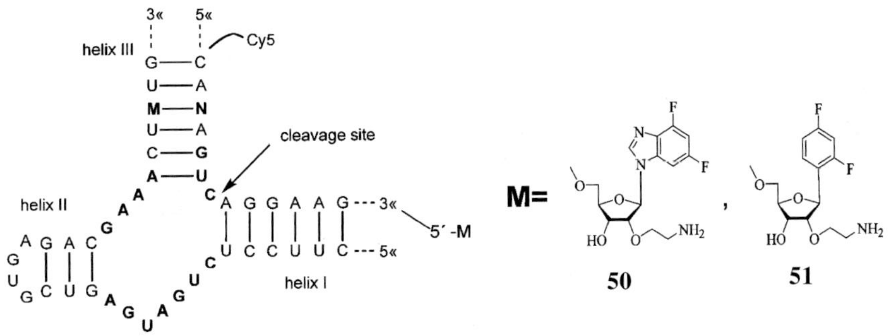

Figure 1.12. Secondary structure of a hammerhead ribozyme where the fluoronucleosides M = **50** and **51** are incorporated. N = A, C, G, U.

Hammerhead ribozymes are catalytically active ribonucleic acids and interfere with gene expression through hydrolysis of the complementary mRNA. This makes them potential therapeutic agents for gene therapy (Lewin and Hauswirth, 2001; Scott, 2007). Recognition of viral mRNA and catalytic activity is dependant on Watson–Crick base pairing and decreases dramatically in the presence of point mutations ('hot spots') in the target region, since ribozymes are inactivated by mutating bases in the catalytic region. We reasoned that incorporation of universal bases in these positions should allow toleration of escape mutants in HIV (Klöpffer and Engels, 2003, 2004, 2005).

The choice of the universal base was based on our investigations of fluorinated NNI in a model 12mer RNA, where the disubstituted fluoro-benzene and -benzimidazole derivatives **4** and **41** appeared to be ideal universal bases (Loakes, 2001). These NNIs did not differentiate thermodynamically between the four natural nucleosides **A**, **C**, **G** or **U**. Furthermore, an enhanced base stacking ability upon additional fluorination was observed. This suggested choosing analogues **1** and **41** for the biological study. However, a destabilization of the modified RNA 12mer duplexes was noticed, which is due to the lack of hydrogen bonding interactions between the modified and natural bases. In an effort to compensate this reduced duplex stability, we synthesized the 2′-β-aminoethyl-substituted fluorinated NNI **50** and **51** and investigated their ability to stabilize RNA duplexes. As the primary amino group is protonated under physiological conditions, the intermolecular electrostatic interaction with the negatively charged RNA backbone of the second RNA strand increased the stability of the RNA duplexes, as expected, probably due to charge–charge interactions (Klöpffer and Engels, 2004).

The hammerhead ribozyme was then modified with the universal bases **50** and **51** (Fig. 1.12). In order to investigate the ribozyme's ability to tolerate point mutations in the target sequence without losing its catalytic activity, we analysed the kinetics of cleavage reactions. As expected, no discrimination with respect to the individual base pairing partner was found: although the overall efficiency was reduced by about one order of magnitude, resulting in a cleavage rate of 1 per 5 minutes, the rates towards **A**, **U**, **C** and **G** in the cognate mRNA were identical within 30%. The difference between the fluorobenzene and fluorobenzimidazole ribosides was in favour of the benzimidazole, probably due to better stacking. Later on it was found that the hammerhead design can be improved using additional loop stabilization, but this was not pursued any further due to the identification of siRNA as potential gene regulators (Canny *et al.*, 2004).

1.5.2 *Fluorobenzene and benzimidazoles in RNA interference and siRNA*

Since the recognition of RNA interference (RNAi) in 1998, the process by which specific mRNAs are targeted and degraded by complementary short-interfering RNAs (siRNA) became a powerful tool to control gene function (Fire, 2007; Mello, 2007). The generally accepted mechanism of RNAi can be divided into two main steps. In the first step, double-stranded RNA (dsRNA) is cleaved into short 21–24nt siRNAs (Elbashir *et al.*, 2001). This process is catalysed by Dicer, an endonuclease of the RNase III family. The resultant siRNA duplexes have 3′-overhangs of 2nt with 3′-hydroxyl termini and a 5′-phosphate at both ends. In the second step, siRNAs are incorporated into the RNA-induced silencing complex (RISC). A helicase in RISC unwinds the duplex siRNA, which then pairs to messenger RNAs (mRNAs) that bear a high degree of sequence complementarity to the siRNA. In humans the degradation of the target mRNA is mediated by the Argonaute 2 protein associated with RISC. The target mRNA is cleaved in the complementary region at the phosphodiester bond that lies across from nucleotides **10** and **11** of the 5′-end of the siRNA (Fig. 1.13). For

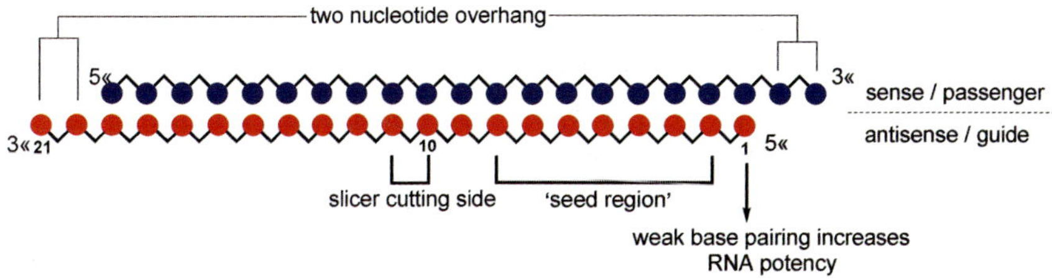

Figure 1.13. siRNA construct.

RNAi-mediated mRNA cleavage and degradation to be successful, a 5′-phosphate must be present on the antisense strand and the double helical antisense-target mRNA duplex must be in the A-form. The X-ray structure of Ago2 and either a single or double stranded oligonucleotide highlights the structural situation (Wang *et al.*, 2008).

Chemically modified nucleosides have been shown to be of great importance for antisense strategies and are now being applied for RNA interference-mediated gene silencing (Bramsen *et al.*, 2009). Since the incorporation of 2′-amino modified NNIs had been shown to yield active ribozymes, the analogous constructs were designed for siRNAs. During our studies several publications appeared showing the use of fluorobenzene ribosides in this context. First, Kool and coworkers used compound **4** in a full seed walk (Fig. 1.13) of siRNA against *Renilla* luciferase (Somoza *et al.*, 2006). From these data it is obvious that the 3′- and 5′-ends tolerate the fluorobenzene well. The central position 10 is highly discriminative. Surprisingly, position 7 gave a high activity too, indicating a possible interaction with the RISC complex (Fig. 1.13). A similar study by the Alnylam group was in good agreement, even though they incorporated the ribo-difluorotoluyl nucleotide (Xia *et al.*, 2006). In our hands, the central positions from 9 to 11 are also less tolerant for fluorobenzene **4** and other tested fluorobenzimidazole nucleosides. When incorporating fluorobenzimidazole **41** at position 21 of siRNA the activity was still preserved, which indicates a successful phosphorylation at the 5′-end. This proved to be particularly interesting because these fluorobenzimidazole ribosides can be easily derivatized by a sequence of Michael addition with acrylonitrile followed by Raney-nickel reduction yielding the aminopropyl derivative (Haas and Engels, 2007).

1.5.3 *Polymerase acceptance of fluorobenzimidazoles*

In collaboration with R. Kuchta at Boulder, Colorado, we tested our fluorinated benzimidazole nucleoside analogue of **41** (deoxynucleoside) against polymerases. These were DNA polymerase α (pol α) and Klenow fragment (exo-) of DNA polymerase I (*Escherichia coli*). Both pol α and Klenow fragment exhibit a remarkable inability to discriminate against these analogues as compared to their ability to discriminate against incorrect natural deoxynucleotide triphosphates (dNTPs). Neither polymerase shows any distinct electronic or steric preferences for analogue incorporation (Kincaid *et al.*, 2005).

Another set of analogues was designed to examine human DNA primase, which synthesizes short RNA primers that DNA polymerase α further elongates. Primase readily misincorporates the natural nucleotide triphosphates (NTPs), which generates a wide variety of mismatches. In contrast, primase exhibited a remarkable resistance to polymerizing NTPs containing NNI. This is different from other polymerases where the shape concept, put forward by Kool, is more applicable (Kool and Sintim, 2006). We tested bases whose shape was almost identical to the natural bases (4-aminobenzimidazole and 4,6-difluorobenzimidazole) (Klöpffer and Engels, 2005), bases with very different shapes compared to natural bases (5- and 6-trifluoromethylbenzimidazoles **43** and **44**), bases much more hydrophobic than natural bases (4- and 7-trifluoromethylbenzimidazole **42** and **45**), bases with hydrophobicities similar to natural bases but with the Watson–Crick hydrogen bonding groups in unusual positions (7-β-D-guanine) and bases capable of forming only one Watson–Crick hydrogen bond with the template base (purine and 4-aminobenzimidazole). Primase was found only to polymerize NTP analogues capable of forming Watson–Crick hydrogen bonds, which explains the failure for the incorporation of **42**–**45**.

1.6 Conclusion

We have described the syntheses of fluorinated benzene, benzimidazole, and indole nucleobase analogues and their incorporation into model RNA, ribozymes and siRNA. The analogues act as NNIs and allow probing of the physical forces that govern the stability of RNA. Notably, several of these NNIs showed a lack of discrimination against natural bases and thus behave as universal bases. Furthermore, the stability of model RNA incorporating self-pairs of fluorinated NNIs increased with the number of fluorinated substituents and reached that of a natural base pair. Combined crystallographic studies, thermodynamic analyses and computer simulations furthermore shed light on the role of organic fluorine in molecular recognition. These studies demonstrated an intricate influence of the molecular environment in this case. As a consequence, it may generally not be sufficient to discuss the molecular recognition properties of organic fluorine in terms of global molecular descriptors. Rather, analyses at an atomic level are required.

Acknowledgements

We are grateful for the tremendous efforts of our coworkers on this project. This work was funded by the Deutsche Forschungsgemeinschaft within the Sonderforschungsbereich 579 ('RNA–ligand interactions') and by a grant to J.W.E. within the EU-FP6 RIGHT project (LSHB-CT-2004-005276). H.G. gratefully acknowledges support by the Frankfurt Centre for Scientific Computing.

References

Bats, J.W., Parsch, J. & Engels, J.W. (2000) 1-deoxy-1-(4-fluorophenyl)-beta-D-ribofuranose, its hemihydrate, and 1-deoxy-1-(2,4-difluorophenyl)-beta-D-ribofuranose: structural evidence for intermolecular C–H···F–C interactions. *Acta Crystallogr C* 56 (Pt 2), 201–205.

Bégué, J.-P. & Bonnet-Delpon, D. (2008) *Bioorganic and Medicinal Chemistry of Fluorine,* John Wiley & Sons, Hoboken, NJ.

Biffinger, J.C., Kim, H.W. & DiMagno, S.G. (2004) The polar hydrophobicity of fluorinated compounds. *Chembiochem* 5, 622–627.

Bohm, H.J., Banner, D., Bendels, S. *et al.* (2004) Fluorine in medicinal chemistry. *Chembiochem* 5, 637–643.

Bozilovic, J. & Engels, J.W. (2007a) Synthesis of fluorinated indoles as RNA analogues. *Nucleos Nucleot Nucl* 26, 869–871.

Bozilovic, J., Bats, J.W. & Engels, J.W. (2007b) Synthesis and crystal structures of fluorinated indoles as RNA analogues. *Coll Symp Series* 7, 385–386.

Bozilovic, J., Bats, J.W. & Engels, J.W. (2007c) Synthesis and structure of fluoroindole nucleosides. *Can J Chem* 85, 283–292.

Bozilovic, J. (2008) Synthese und Einfluss von fluorierten Indolen-, 7-*N*-Purin und 9-*N*-Deazapurin-Nukleosiden auf die RNA Stabilität, *Dissertation*, Goethe-University, Frankfurt.

Bramsen, J.B., Laursen, M.B., Nielsen, A.F. *et al.* (2009) A large-scale chemical modification screen identifies design rules to generate siRNAs with high activity, high stability and low toxicity. *Nucleic Acids Res* 37, 2867–2881.

Canny, M.D., Jucker, F.M., Kellogg, E. *et al.* (2004) Fast cleavage kinetics of a natural hammerhead ribozyme. *J Am Chem Soc* 126, 10848–10849.

Caruthers, M.H. (1985) Gene synthesis machines: DNA chemistry and its uses. *Science* 230, 281–285.

Coppens, P. (1967) Comparative X-ray and neutron diffraction study of bonding effects in *S*-triazine. *Science* 158, 1577–1579.

Da Rosa, F.A.F., Rebelo, R.A. & Nascimento, M.G. (2003) Synthesis of new indolecarboxylic acids related to the plant hormone indoleacetic acid IAA. *J Braz Chem Soc* 14, 11–15.

Desiraju, G.R. (1997) Crystal gazing: structure prediction and polymorphism. *Science* 278, 404–405.

Desiraju, G.R. (2002) Hydrogen bridges in crystal engineering: interactions without borders. *Acc Chem Res* 35, 565–573.

DiMagno, S.G. & Sun, H.R. (2006) The strength of weak interactions: aromatic fluorine in drug design. *Curr Top Med Chem* 6, 1473–1482.

Dunitz, J.D. & Taylor, R. (1997) Organic fluorine hardly ever accepts hydrogen bonds. *Chem-Eur J* 3, 89–98.

Dunitz, J.D. (2004) Organic fluorine: Odd man out. *ChemBioChem* 5, 614–621.

Dunitz, J.D. & Schweizer, W.B. (2006) Molecular pair analysis: C–H···F interactions in the crystal structure of fluorohenzene? And related matters. *Chemistry* 12, 6804–6815.

Elbashir, S.M., Harborth, J., Lendeckel, W. *et al.* (2001) Duplexes of 21-nucleotide RNAs mediate RNA interference in cultured mammalian cells. *Nature* 411, 494–498.

Evans, T.A. & Seddon, K.R. (1997) Hydrogen bonding in DNA — a return to the status quo. *Chem Commun*, 2023–2024.

Fire, A.Z. (2007) Gene silencing by double-stranded RNA (Nobel Lecture). *Angew Chem Int Edit* 46, 6966–6984.

Gesteland, R.F., Cech, T.R. & Atkins, J.F. (1999) *The RNA World*, Cold Spring Harbor Laboratory Press, New York.

Guckian, K.M., Krugh, T.R. & Kool, E.T. (2000) Solution structure of a nonpolar, non-hydrogen-bonded base pair surrogate in DNA. *J Am Chem Soc* 122, 6841–6847.

Guerra, C.F. & Bickelhaupt, F.M. (2003) Orbital interactions and charge redistribution in weak hydrogen bonds: the Watson–Crick AT mimic adenine-2,4-difluorotoluene. *J Chem Phys* 119, 4262–4273.

Haas, J. & Engels, J.W. (2007) A novel entry to 2′-*O*-aminopropyl modified nucleosides amenable for further modifications. *Tetrahedron Lett* 48, 8891–8894.

Hagmann, W.K. (2008) The many roles for fluorine in medicinal chemistry. *J Med Chem* 51, 4359–4369.

Hernandez-Trujillo, J. & Vela, A. (1996) Molecular quadrupole moments for the series of fluoro- and chlorobenzenes. *J Phys Chem* 100, 6524–6530.

Kincaid, K., Beckman, J., Zivkovic, A. *et al.* (2005) Exploration of factors driving incorporation of unnatural dNTPS into DNA by Klenow fragment (DNA polymerase I) and DNA polymerase alpha. *Nucleic Acids Res* 33, 2620–2628.

Klebba, C., Ottmann, O.G., Scherr, M. *et al.* (2000) Retrovirally expressed anti-HIV ribozymes confer a selective survival advantage on CD4+ T cells *in vitro*. *Gene Ther* 7, 408–416.

Klöpffer, A.E. & Engels, J.W. (2003) The effect of universal fluorinated nucleobases on the catalytic activity of ribozymes. *Nucleos Nucleot Nucl* 22, 1347–1350.

Klöpffer, A.E. & Engels, J.W. (2004) Synthesis of 2′-aminoalkyl-substituted fluorinated nucleobases and their influence on the kinetic properties of hammerhead ribozymes. *ChemBioChem* 5, 707–716.

Klöpffer, A.E. & Engels, J.W. (2005) Synthesis and purification of fluorinated benzimidazole and benzene nucleoside-5′-triphosphates. *Nucleos Nucleot Nucl* 24, 651–654.

Koller, A.N., Bozilovic, J.W. Engels, J.W. *et al.* (2010) Aromatic N vs. aromatic F: bioisosterism discovered in RNA base pairing interactions leads to a novel class of universal base analogues. *Nucleic Acids Res* 38, 3133–3146.

Kool, E.T. (1998) Replication of non-hydrogen bonded bases by DNA polymerases: a mechanism for steric matching. *Biopolymers* 48, 3–17.

Kool, E.T. & Sintim, H.O. (2006) The difluorotoluene debate — a decade later. *Chem Comm* 35, 3665–3675.

Kopitz, H., Zivkovic, A., Engels, J.W. *et al.* (2008) Determinants of the unexpected stability of RNA fluorobenzene self pairs. *ChemBioChem* 9, 2619–2622.

Krohn, K., Heins, H. & Wielckens, K. (1992) Synthesis and cytotoxic activity of C-glycosidic nicotinamide riboside analogs. *J Med Chem* 35, 511–517.

Lai, J.S., Qu, J. & Kool, E.T. (2003) Fluorinated DNA bases as probes of electrostatic effects in DNA base stacking. *Angew Chem Int Edit* 42, 5973–5977.

Lai, J.S. & Kool, E.T. (2004) Selective pairing of polyfluorinated DNA bases. *J Am Chem Soc* 126, 3040–3041.

Lewin, A.S. & Hauswirth, W.W. (2001) Ribozyme gene therapy: applications for molecular medicine. *Trends Mol Med* 7, 221–228.

Loakes, D. (2001) The applications of universal DNA base analogues. *Nucleic Acids Res* 29, 2437–2447.

Matulicadamic, J., Beigelman, L., Portmann, S. *et al.* (1996) Synthesis and structure of 1-deoxy-1-phenyl-β-D-ribofuranose and its incorporation into oligonucleotides. *J Org Chem* 61, 3909–3911.

Mello, C.C. (2007) Return to the RNAi world: rethinking gene expression and evolution (Nobel lecture). *Angew Chem Int Edit* 46, 6985–6994.

Moore, C.L., Zivkovic, A., Engels, J.W. *et al.* (2004) Human DNA primase uses Watson–Crick hydrogen bonds to distinguish between correct and incorrect nucleoside triphosphates. *Biochemistry* 43, 12367–12374.

Müller, K., Faeh, C. & Diederich, F. (2007) Fluorine in pharmaceuticals: looking beyond intuition. *Science* 317, 1881–1886.

Müller-Kuller, T., Capalbo, G., Klebba, C. *et al.* (2009) Identification and characterisation of a highly efficient anti-HIV Pol hammerhead ribozyme. *Oligonucleotides*, 19, 265–271.

Murray-Rust, P., Stallings, W.C., Monti, C.T. *et al.* (1983) Intermolecular interactions of the C–F bond — the crystallographic environment of fluorinated carboxylic-acids and related structures. *J Am Chem Soc* 105, 3206–3214.

Nikolaus, N.V., Bozilovic, J. & Engels, J.W. (2007) Microwave-assisted ribosylation of modified heterocyclic bases by Vorbruggen method. *Nucleos Nucleot Nucl* 26, 889–892.

O'Hagan, D. (2008) Understanding organofluorine chemistry. An introduction to the C–F bond. *Chem Soc Rev* 37, 308–319.

Olsen, J.A., Banner, D.W., Seiler, P. *et al.* (2003) A fluorine scan of thrombin inhibitors to map the fluorophilicity/fluorophobicity of an enzyme active site: evidence for C–F···C=O interactions. *Angew Chem Int Edit* 42, 2507–2511.

Olsen, J.A., Banner, D.W., Seiler, P. *et al.* (2004) Fluorine interactions at the thrombin active site: protein backbone fragments H–C-alpha-C=O comprise a favorable C–F environment and interactions of C–F with electrophiles. *ChemBioChem* 5, 666–675.

Parsch, J. & Engels, J.W. (1999) C–F···H–C hydrogen bonds in crystals of fluorobenzene ribonucleosides. *Coll Symp Series* 2, 11–14.

Parsch, J. & Engels, J.W. (2000) Synthesis of fluorobenzene and benzimidazole nucleic-acid analogues and their influence on stability of RNA duplexes. *Helv Chim Acta* 83, 1791–1808.

Parsch, J. & Engels, J.W. (2001) Stacking and stability of RNA duplexes containing fluorobenzene and fluorobenzimidazole nucleosides. *Nucleos Nucleot Nucl* 20, 815–818.

Parsch, J. & Engels, J.W. (2002) C–F···H–C hydrogen bonds in ribonucleic acids. *J Am Chem Soc* 124, 5664–5672.

Patrick, C.R. & Prosser, G.S. (1960) Molecular complex of benzene and hexafluorobenzene. *Nature* 187, 1021–1021.

Petersheim, M. & Turner, D.H. (1983) Base-stacking and base-pairing contributions to helix stability — thermodynamics of double-helix formation with CCGG, CCGGP, CCGGAP, ACCGGP, CCGGUP, and ACCGGUP. *Biochemistry* 22, 256–263.

Razgulin, A.V. & Mecozzi, S. (2006) Binding properties of aromatic carbon-bound fluorine. *J Med Chem* 49, 7902–7906.

Reichenbacher, K., Suss, H.I. & Hulliger, J. (2005) Fluorine in crystal engineering — 'the little atom that could'. *Chem Soc Rev* 34, 22–30.

Rowland, R.S. & Taylor, R. (1996) Intermolecular nonbonded contact distances in organic crystal structures: comparison with distances expected from van der Waals radii. *J Phys Chem* 100, 7384–7391.

Schweitzer, B.A. & Kool, E.T. (1994) Aromatic nonpolar nucleosides as hydrophobic isosteres of pyrimidine and purine nucleosides. *J Org Chem* 59, 7238–7242.

Scott, W.G. (2007) Ribozymes. *Curr Opin Struct Biol* 17, 280–286.

Somoza, A., Chelliserrykattil, J. & Kool, E.T. (2006) The roles of hydrogen bonding and sterics in RNA interference. *Angew Chem Int Edit* 45, 4994–4997.

Stofer, E., Chipot, C. & Lavery, R. (1999) Free energy calculations of Watson–Crick base pairing in aqueous solution. *J Am Chem Soc* 121, 9503–9508.

Thalladi, V.R., Weiss, H.C., Blaser, D. *et al.* (1998) C–H···F interactions in the crystal structures of some fluorobenzenes. *J Am Chem Soc* 120, 8702–8710.

Velec, H.F., Gohlke, H. & Klebe, G. (2005) DrugScore(CSD)-knowledge-based scoring function derived from small molecule crystal data with superior recognition rate of near-native ligand poses and better affinity prediction. *J Med Chem* 48, 6296–6303.

Vorbrüggen, H. & Hofle, G. (1981) Nucleoside syntheses. 23. On the mechanism of nucleoside synthesis. *Chem Ber* 114, 1256–1268.

Wang, Y.L., Sheng, G., Juranek, S. *et al.* (2008) Structure of the guide-strand-containing Argonaute silencing complex. *Nature* 456, 209–234.

Xia, J., Noronha, A., Toudjarska, I. *et al.* (2006) Gene silencing activity of siRNAs with a ribo-difluorotoluyl nucleotide. *ACS Chem Biol* 1, 176–183.

Zacharias, M. & Engels, J.W. (2004) Influence of a fluorobenzene nucleobase analogue on the conformational flexibility of RNA studied by molecular dynamics simulations. *Nucleic Acids Res* 32, 6304–6311.

Zivkovic, A. & Engels, J.W. (2002) Crystal structures of fluoro- and chloro-modified benzene nucleosides. *Coll Symp Series* 5, 396–398.

2

Molecular Interactions of Fluorinated Amino Acids within the Hydrophobic Core of a Coiled Coil Peptide

Toni Vagt, Mario Salwiczek* and Beate Koksch**

2.1 Introduction

The specificity of peptide-mediated interactions, their structural diversity and their important role in the regulation of essential life functions increasingly brings peptides into the focus of pharmaceutical research (Sato *et al.*, 2006; Hüther and Dietrich, 2007). Besides their classical use as vaccines and hormones for the treatment of metabolic dysfunctions, peptides are also promising pharmaceutical leads suitable for the treatment of serious diseases such as AIDS, cancer or neurodegenerative diseases (Permanne *et al.*, 2002; Matthews *et al.*, 2004). In particular, their small size and high specificity, combined with the fact that peptides are less toxic and rarely cause an immune response, make them powerful tools for drug design. On the other hand, the main disadvantages of peptide-based drugs are their low bioavailability, low membrane permeability and rapid, enzyme-catalysed cleavage. Therefore, the establishment of efficient methods to

* Department of Biology, Chemistry and Pharmacy, Institute of Chemistry and Biochemistry, Takustr. 3, 14195 Berlin, Germany. E-mail: koksch@chemie.fu-berlin.de.

modify and stabilise peptides is one of the main challenges in the development of therapeutically applicable drugs.

The structural mimicry of peptides using non-natural amino acids has proven to be a useful method to enhance the *in vivo* stability of peptides. To this end, β-amino acids and D-amino acids have been used for the construction of proteolytically stable structures (Welch *et al.*, 2007; David *et al.*, 2008). Another strategy is the application of amino acids with modified side chains that carry additional side chain functionalities. These amino acids can be used to artificially expand the repertoire of naturally occurring amino acids, thereby providing greater chemical diversity (Dougherty, 2000). In this context, the incorporation of fluorine into peptides, in the form of fluorinated amino acids, is of interest. The unique properties of the fluorine atom often bring about a significant increase in bioavailability and create a wide variety of new interaction patterns. However, despite numerous attempts to elucidate the influence of fluorination on the interactions of amino acid side chains within a native protein environment, the basic physical and chemical properties of fluorinated amino acids within such an environment are not fully understood. In particular, the formation of C–F···H–X type hydrogen bonds, the steric effects of fluoroalkyl groups, as well as the so-called 'fluorous effect' are still topics of debate. A detailed understanding of these properties, however, is essential for the successful application of fluorinated amino acids to peptide and protein design. Especially comprehensive information about specific interactions between these building blocks and the canonical amino acids is needed for the construction of peptides and proteins that engage in specific types of molecular recognition. In this respect, model systems that allow sensitive detection of the influence of fluorinated amino acids on peptide conformation and stability are required.

One peptide folding motif that provides such a model system is the α-helical coiled coil (Hakelberg and Koksch, 2007). Over the past decade, different groups have incorporated various fluorinated amino acids into different coiled coil systems and investigated their effect on stability and interaction specificity. Some of the most interesting results of these studies, from our perspective, are highlighted in this chapter.

2.2 The α-Helical Coiled Coil as a Model System to Investigate Fluorinated Amino Acids within a Native Protein Environment

Coiled coils, also called 'leucine zippers', are small helical proteins that are composed of two to seven α-helices forming left-handed super-coiled oligomers (Woolfson, 2005). They have a broad range of biological functions (Mason and Arndt, 2004), making them attractive candidates for the development of peptide-based drugs. As structural components of many DNA-binding proteins, e.g. JUN, FOS and GCN4, they play an important role in gene transcription. Long coiled coils may serve as molecular springs and as scaffolds that make up the cytoskeleton (Rose and Meier, 2004). Moreover, coiled coil structures are also involved in the membrane fusion of viruses such as HIV-1 with host cells (Esté and Telenti, 2007). Figure 2.1 shows the X-ray crystal structure of the GCN4 leucine zipper as an example of a biologically relevant coiled coil (O'Shea *et al.*, 1991).

The predictable quarternary structure of coiled coils derives from a rather sophisticated primary structure that consists of a repetitive alignment

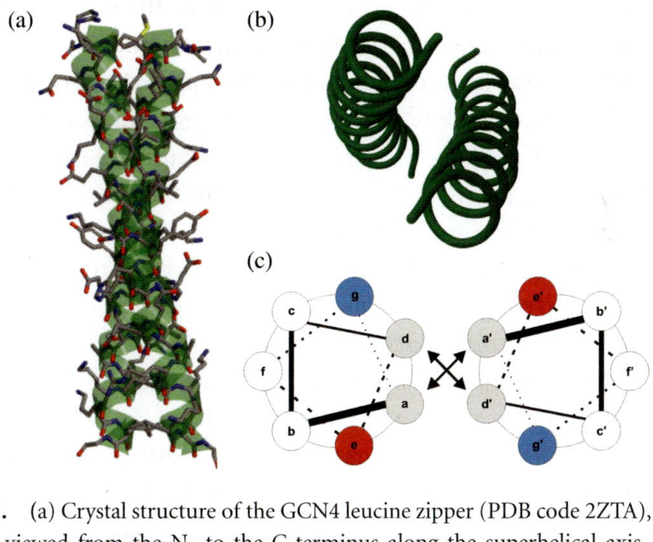

Figure 2.1. (a) Crystal structure of the GCN4 leucine zipper (PDB code 2ZTA), (b) GCN4 backbone viewed from the N- to the C-terminus along the superhelical axis, (c) helical wheel representation of (b) depicting the heptad repeat pattern (reproduced from M. Salwiczek 2010 with permission, Copyright © 2011 Mario Salwiczek).

of seven residues termed $(abcdefg)_n$, where n indicates the number of these heptad repeats. In dimeric coiled coils the **a**- and **d**-positions are mainly occupied by bulky hydrophobic residues, such as leucine, isoleucine and valine that provide the driving force for folding and oligomerisation. The remaining residues are usually hydrophilic, with the **e**- and **g**-positions mostly containing charged amino acids. These form interhelical salt-bridges that further contribute to stability and, in addition, mediate folding specificity (parallel *vs.* antiparallel and homo- *vs* heterooligomerisation). In contrast to higher-order oligomers, the **b**-, **c**- and **f**-positions in dimeric coiled coils are not involved in interhelical interactions. In the unfolded peptides, the hydrophobic **a**- and **d**-residues are evenly distributed over the peptide sequence. However, in the α-helical conformation, which effectively comprises 3.5 residues per turn in the case of coiled coils, the **a**- and **d**-positions are localised to one face of the helix. The other mainly hydrophilic residues make up the other faces of the helix. This spatial separation of hydrophobic and hydrophilic residues renders the molecules highly amphiphilic. As the **a**- and **d**-residues interact unfavourably with water, oligomerisation is the consequence of the formation of a hydrophobic core.

The ability to predict coiled coil structures based on their primary sequence, their facile synthetic accessibility, sufficient sensitivity to substitutions and our knowledge of high-resolution structures (see Moutevelis and Woolfson (2009) for a periodic table of coiled coil structures) make them excellent models with which to study the interactions of non-natural amino acids, including those carrying fluorinated side chains, within a natural protein environment (Jäckel *et al.*, 2004, 2006; Salwiczek *et al.* 2009). Accordingly, an important part of our knowledge about the impact of fluorinated amino acids on the folding and stability of peptides and proteins originates from investigations using coiled coils and helix bundles.

In 2001, the laboratories of Kumar and Tirrell incorporated several trifluoromethyl-containing amino acids into different leucine zipper peptides. Substitution of leucine residues in the **d**-positions of the peptide GCN4-p1 by 5,5,5-trifluoroleucine (5-TFL) increased the thermal stability of the coiled coil structure. The leucine zipper also showed greater resistance to chaotropic denaturants, while retaining its DNA-binding ability (Tang *et al.*, 2001b). Additionally, substitution of 4,4,4-trifluorovaline (4-TFV) for valine at three **a**-positions yielded a coiled coil peptide with a complete fluorous core and increased the thermal stability of this structure by 15°C (Bilgiçer *et al.*, 2001a). A similar effect was observed after substitution of six **d**-positions within a 74-residue coiled coil peptide by 5-TFL (Tang *et al.*, 2001a). The increase in thermal stability caused by this modification was enhanced by an additional 9°C when 5,5,5,5′,5′,5′-hexafluoroleucine (HFL) was placed in position **d** instead of 5-TFL (Tang *et al.*, 2001c). As the exclusion of hydrophobic leucine side chains from the aqueous environment is the main driving force for coiled coil formation (Grigoryan *et al.*, 2008), the replacement of these residues by highly hydrophobic, fluorinated analogues significantly increases the coiled coil stability, which is consistent with the more stabilising effect of HFL compared to TFL. Similar effects were observed for hydrophobic core fluorination of four-helix bundles. Marsh and coworkers have incrementally incorporated HFL into the hydrophobic core of an antiparallel four-helix bundle, resulting in a gradual increase in thermal stability that directly correlates with the content of HFL-residues within the hydrophobic core (Lee *et al.*, 2004, 2006).

However, denaturation studies on GCN4 variants carrying 5,5,5-trifluoroisoleucine (5-TFI) or 4-TFV residues at four of the **a**-positions led to remarkable results (Son *et al.*, 2006). While the melting temperature increases by 27°C when isoleucine is replaced by 5-TFI, the difference in melting temperature is only 4°C between peptides containing 4-TFV and those containing valine. The same trend in stability was observed by chemical denaturation. On the other hand, the affinity and specificity of DNA binding were essentially similar to the unfluorinated peptides. The authors attributed the distinct difference observed for 5-TFI and 4-TFV to entropic as well as packing effects of the amino acid side chains. Because of its increased steric bulk, the trifluoromethyl group of 4-TFV provokes steric clashes between the helix backbone and valine's γ-methyl group and probably compensates the gain in stability which is caused by its enhanced hydrophobicity. Furthermore, the special side chain packing of isoleucine in **a** and **a′** positions of parallel coiled coil dimers possibly allows more complete burial of the trifluoromethyl group of 5-TFI.

In addition to the structural stabilisation imparted by fluorination of homooligomeric coiled coils, the successful construction of self-sorting peptides could also demonstrate a unique property of fluorinated amino acids, often called the 'fluorous effect'. Related to the phase behaviour of perfluorocarbons, their hydrophobicity and lipophobicity drive fluoroalkylated side chains to separate in an aqueous as well as a lipid environment. Such behaviour was shown for HFL, which was incorporated into a *de novo* designed 30-AA coiled coil peptide (Bilgiçer *et al.*, 2001, 2002). In the peptide denoted **H**, seven of eight **a**- and **d**-positions are occupied by leucine. These residues were changed to HFL, forming peptide **F**. Both peptides were equipped with a Cys-Gly-Gly linker to enable disulfide bond formation between both helices and disulfide bridged **HF** heteromers were formed. Incubation under appropriate redox conditions decreased the concentration of **HF**-heteromers while an increase in the formation of **HH** and **FF** homomers was observed. The formation of homomers was also preferred when reduced **H** and **F** monomers were mixed in redox buffer (Fig. 2.2(a)). Besides the preferred formation of homomers, a significant increase in the melting point of **FF** dimers (82°C) compared to the **HH** homomers (34°C) and **HF** heteromers (36°C) was observed.

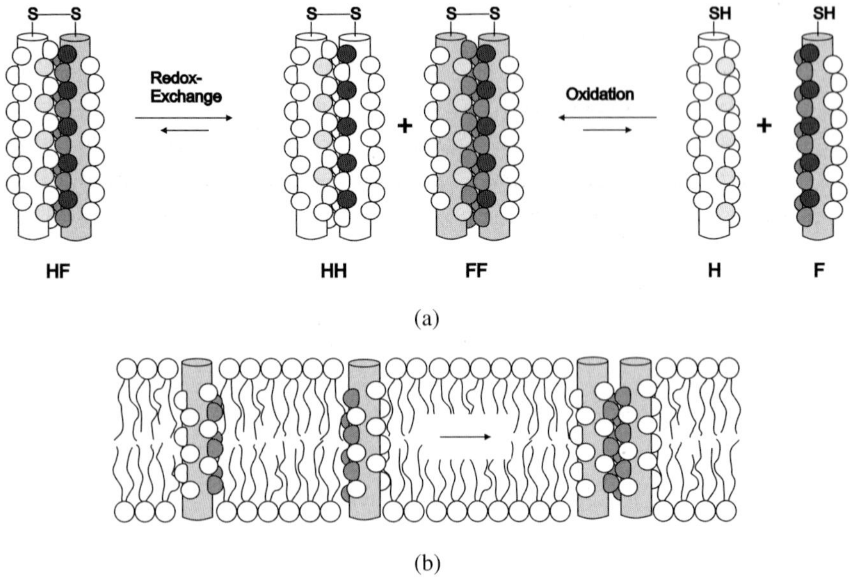

Figure 2.2. (a) Self-sorting of fluorinated coiled coil peptides. HFL in position **a** and **d** is highlighted in grey; lysine is highlighted in dark grey, glutamic acid in light grey (according to Bilgiçer *et al.*, 2001b). (b) Self-sorting of fluorinated transmembrane helices. HFL is highlighted in grey (according to Bilgiçer *et al.*, 2004).

The directed self-assembly of highly fluorinated helical peptides was also demonstrated in a membrane environment by the same group (Bilgiçer *et al.*, 2004). Incorporation of HFL into the **a**- and **d**-positions of the unpolar transmembrane region of a 29-AA peptide resulted in dimerisation of the previously monomeric helices within a membrane-like environment (Fig. 2.2(b)).

Both experiments not only impressively show the possibility to induce self-aggregation of peptides by the incorporation of fluorinated amino acids, but also demonstrate that an increase in hydrophobicity is not the sole driving force for stabilisation of coiled coil structures. Although other studies did not provide any evidence for specific interactions between fluorinated amino acids (Lee *et al.*, 2004, 2006), there is proof that the fluorous effect plays an important role in structurally stabilising fluorinated coiled coils.

However, the approaches described so far were focused on the effects that a spacious fluorination has on coiled coil structures. Because specific interactions between fluoroalkyl groups and their tendency to self-assemble seem to significantly affect the properties of highly fluorinated peptides, such studies yield limited insight into the effects fluorinated amino acids show when embedded in a native protein environment. While the extended fluorination of coiled coil structures is generally associated with a remarkable increase in stability, the incorporation of a single fluoroalkylated amino acid often shows different effects (Jäckel *et al.*, 2004). Investigations of non-natural amino acids within a defined protein structure provide useful information about their biophysical properties, such as size or polarity. Such an investigation, in which single fluoroamino acid substitutions within the hydrophobic core of a coiled coil structure were studied, is described in detail in the following section.

2.3 Single Fluoroamino Acid Substitutions within a Heterodimeric Coiled Coil

2.3.1 *The α-helical coiled coil as a model for a natural protein environment*

The coiled coil shown in Fig. 2.3 serves as a model system for the study of the effects of single fluoroamino acid substitutions at two different positions within the hydrophobic core. It has been designed to form parallel heterodimers where peptide VPE is an exclusively native interaction

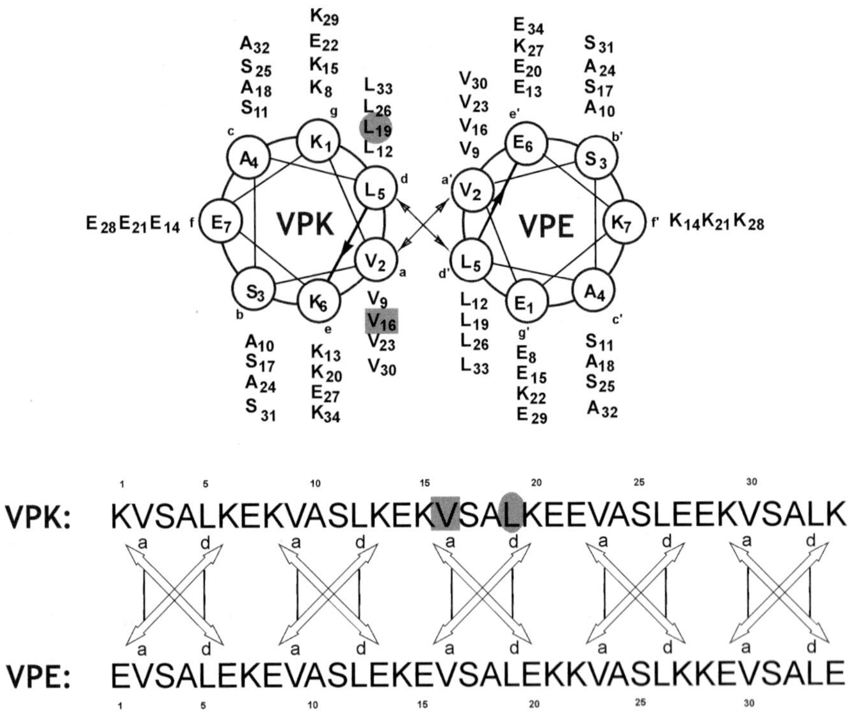

Figure 2.3. Helical wheel and sequence representation of the parental VPE–VPK dimer. The substitution positions $\mathbf{a}_{16}$ and $\mathbf{d}_{19}$ in VPK are highlighted with a grey square and a grey circle, respectively.

partner for different analogues of VPK, which carries fluorinated substitutions either at position $\mathbf{a}_{16}$ or $\mathbf{d}_{19}$ within the hydrophobic core. VPK peptides with leucine at the substitution position serve as references.

Although the calculated van der Waals (VdW) volumes of the isopropyl and the trifluoromethyl group (56.2 Å^3 *vs* 39.8 Å^3, according to Zhao *et al.* (2003)) clearly disprove isosterism, their steric effects were shown to be nearly identical (Bott *et al.*, 1980; de Riggi *et al.*, 1995). Thus, leucine at the respective substitution position was replaced by (*S*)-2-aminobutanoic acid (Abu) and its fluorinated analogues MfeGly, DfeGly and TfeGly, with the latter carrying a trifluoromethyl group instead of an isopropyl group at the β-carbon (Fig. 2.4). A further increase in spatial demand was achieved by substituting DfpGly for Leu.

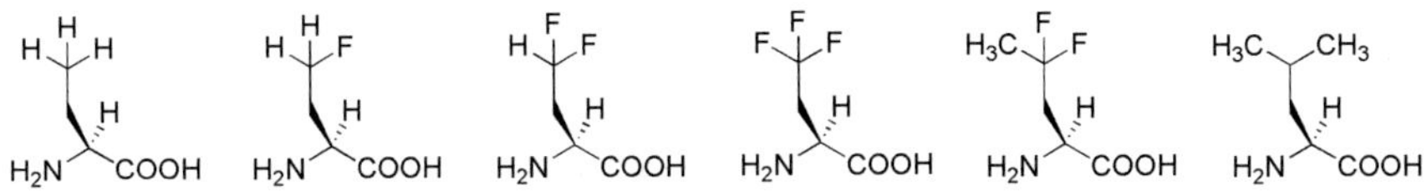

Figure 2.4. Chemical structures of (*S*)-2-aminobutanoic acid (Abu), its fluorinated analogues (*S*)-2-amino-4-fluorobutanoic acid (MfeGly), (*S*)-2-amino-4,4-difluorobutanoic acid (DfeGly) and (*S*)-2-amino-4,4,4-trifluorobutanoic acid (TfeGly) as well as (*S*)-2-amino-4,4-difluoropentanoic acid (DfpGly) and leucine (Leu). The abbreviations of the fluorinated amino acids derive from regarding them as ethyl and propyl substituted glycines, respectively.

2.4 Biophysical Characterisation of the Interactions

2.4.1 *Hydrophobicity of the fluorinated amino acids*

The impact of fluorination of organic molecules on their hydrophobicity generally depends on the molecular structures under investigation (Smart, 2001). Although aromatic and olefinic fluorination as well as the perfluorination of alkyl side chains increases hydrophobicity, a contrary effect is often observed when alkyl side chains are partially fluorinated and still carry hydrogen atoms in close proximity to the fluorine substitutions.

An RP-HPLC assay combined with theoretical calculations reveals that, in comparison to their hydrocarbon analogues, the partitioning into aqueous environment of the fluorinated amino acids used here is more favoured (Samsonov *et al.*, 2009). Table 2.1 summarises the retention times of Fmoc-protected analogues of the canonical aliphatic amino acids Gly, Ala, Val, Ile, Leu and Abu as well as of all the fluorinated amino acids. As shown in Fig. 2.5, the retention times for amino acids carrying hydrocarbon side chains correlate very well with the VdW volume of their side chains. The data for the fluorinated analogues of Abu (MfeGly, DfeGly and TfeGly) as well as DfpGly do not fit into the plot shown in Fig. 2.5. Fluorination clearly increases the volume of the side chain. However, due to the strong polarisation of the adjacent hydrogen atoms and the strong dipole moment of the C–F bond, the fluorinated amino acids exhibit unique electronic properties. If side chain volume was the sole factor in determining hydrophobicity, one would expect MfeGly, DfeGly and DfpGly to be less polar than they are; TfeGly could be considered slightly hyper-hydrophobic. Nevertheless, a

Table 2.1. Retention times of the Fmoc-protected native aliphatic amino acids, Abu and its fluorinated analogues.

Amino acid	VdW volume[a] (Å^3)	rt[b] (min)
Hydrocarbon side chains		
Gly	7.2	8.6
Ala	24.5	9.8
Abu	41.8	11.4
Ile	76.4	15.3
Leu	76.4	15.6
Fluorinated side chains		
MfeGly	47.6	10.7
DfeGly	53.9	12.1
TfeGly	60.1	13.6
DfpGly	71.3	12.8

[a] The VdW volume corresponds to the (fluoro)alkyl groups attached to the β-carbon of the amino acid and is based on calculations according to literature procedures (Zhao *et al.*, 2003).
[b] Retention time on a Capcell-PAK C_{18} column (5 μm) applying a linear gradient from 40% to 70% acetonitrile in water containing 0.1% TFA with a flow rate of 1 mL/min at room temperature.

closer look at the Abu analogues reveals that the retention time of fluorinated amino acids increases more by stepwise fluorination than it does by simply increasing the volume via elongation or branching of the side chains.

Thus, two effects must be considered: (1) fluorination polarises the side chain and renders the amino acids more polar and (2) this effect becomes progressively less pronounced with increasing fluorine content. These findings are also reflected by the theoretically calculated differences in hydration energies ($\Delta\Delta E^{hydr}$) between the fluorinated amino acids and structurally equivalent non-fluorinated counterparts (Table 2.2). Negative values for $\Delta\Delta E^{hydr}$ indicate a more favourable interaction of the fluorinated amino acids with water. In agreement with the retention times, this difference decreases as the fluorine content increases. This observation can be explained by fluorine being a weak hydrogen bond acceptor (Dunitz *et al.*, 2006) and by the impact of fluorination on the overall polarisability of the molecules. Because fluorine is itself a very weakly polarisable atom and very electronegative, it also lowers

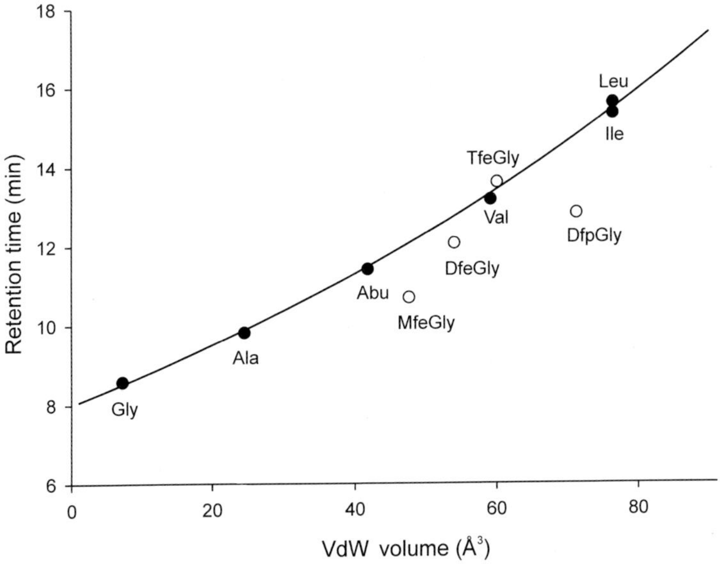

Figure 2.5. Retention times of the Fmoc-protected amino acids plotted against the VdW volume of the side chains. Non-fluorinated amino acids are represented by closed circles and the correlation between side chain volume and retention time is shown as a solid black curve that can be described as an exponential equation ($rt = 8 \cdot e^{0.009 \cdot \text{VdW volume}}$). The fluorinated amino acids are represented by open circles.

Table 2.2. Hydration energy differences between fluorinated and non-fluorinated amino acids.

Fluorinated residue	Non-fluorinated residue	$\Delta\Delta E^{hydr}$ (kcal mol^{-1})[a]
MfeGly	Abu	−2.5
DfeGly	Abu	−0.9
TfeGly	Abu	−0.5
DfpGly	Propyl glycine	−0.9

[a] Difference between hydration energies of the fluorinated and the non-fluorinated residues in TIP3 water (Samsonov *et al.*, 2009).

the polarisability of adjacent C–C and C–H bonds (Biffinger *et al.*, 2004). As a consequence, hydrogen bond, dipole–dipole as well as London dispersion interactions of the side chains with water become progressively weaker as the number of fluorine atoms increases.

The hydrophobic effect and, accordingly, the hydrophobic residues within a polypeptide chain, strongly affect its folding and stability. Having summarised the general properties of the fluorinated amino acids in this respect, the next two sections will provide detailed insight into the consequences of the stereoelectronic effects of fluorination on the interactions of fluorinated amino acids with native residues in the hydrophobic environment of the above described heterodimeric coiled coil.

2.4.2 *The impact of fluorine substitutions on coiled coil structure*

The effects of single amino acid substitutions on the structure of a protein can be easily probed by applying CD spectroscopy. Helical proteins show characteristic CD spectra with two distinct minima, at 208 nm and 222 nm (Venyaminov and Yang, 1996), the latter of which is used to calculate the helical fraction (f_H) given as a percentage according to Eq. 2.1 (Chen *et al.*, 1974).

$$f_H = \frac{[\theta]\cdot 100}{-39{,}500\cdot\left(1-\frac{2.57}{n}\right)} \qquad (2.1)$$

Here, $[\theta]$ is the experimentally determined mean residue ellipticity at 222 nm and n the number of residues. Figure 2.6 shows the CD spectra of all the VPE–VPK dimers (Fig. 2.3) substituted at either position $\mathbf{a}_{16}$ or position $\mathbf{d}_{19}$ (Salwiczek *et al.*, 2009).

The CD spectra are similar regarding their characteristic shape, but the differences in intensity indicate that the substitutions do have an impact on the structure of the VPE–VPK dimer that is quantified by the helical fractions given in Table 2.3. The helical content of the different dimers ranges from 49% to 83% for substitutions at position $\mathbf{a}_{16}$ and 59% to 72% for those at position $\mathbf{d}_{19}$. With the exception of MfeGly, which generally yields the most helical dimer, the fluorinated analogues of Abu (DfeGly and TfeGly) reduce helicity. The helical content of the DfpGly substituted peptides is nearly equal to those carrying leucine in the respective positions. Nevertheless, there are no clear trends pointing to a correlation of side chain volume, hydrophobicity or fluorine content and helicity of the coiled coil. It has been shown that some highly fluorinated amino acids, such as hexafluoroleucine and TfeGly exhibit lower helix propensities than their

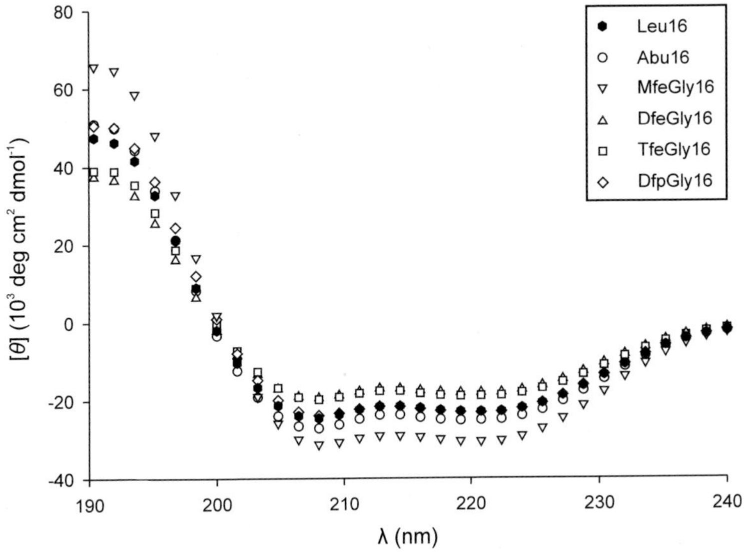

(a)

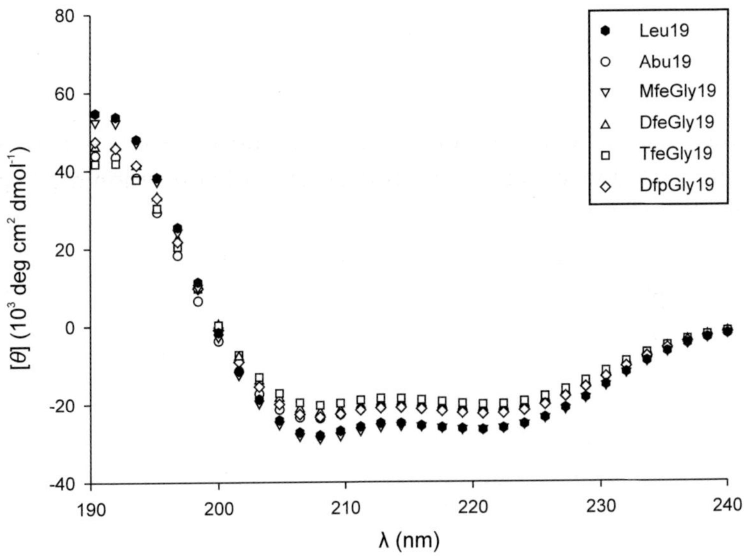

(b)

Figure 2.6. CD spectra of 20 μM solutions of the different VPE–VPK dimers at 20°C, pH 7.4 (100 mM phosphate buffer): (a) substitutions at position $\mathbf{a}_{16}$ and (b) substitutions at position $\mathbf{d}_{19}$.

Table 2.3. Helical content of the VPE–VPK dimers.

Amino acid	$[\theta]_{222nm}$ (deg cm^2 dmol^{-1})	f_H (%)
Position a_{16}		
Val	−26 274	72
Leu	−22 937	63
Abu	−25 004	68
MfeGly	−30 462	83
DfeGly	−18 119	49
TfeGly	−18 900	52
DfpGly	−23 100	63
Position d_{19}		
Leu	−26 274	72
Abu	−21 929	60
MfeGly	−26 288	72
DfeGly	−21 582	59
TfeGly	−20 122	55
DfpGly	−22 449	61

hydrocarbon analogues, which means that their preference to adopt a conformation that is best suited for α-helices is low (Chiu *et al.*, 2006, 2007).

However, the general conclusion that fluorinated amino acids disfavour a helical conformation and consequently have a negative impact on the stability of the coiled coil discussed here is clearly disputable for two reasons. First, coiled coil folding is mainly driven by the hydrophobic effect that forces the **a**- and **d**-positions of the adjacent monomers to form the hydrophobic core. Coiled coil helix formation is a consequence of these interactions rather than an intrinsic property of the monomeric peptides. Second, if helix propensity as measured in single helices was the sole property in determining whether an amino acid prefers a helical conformation in the context of a coiled coil, then one would expect the same trend in helicity at both substitution positions. However, a comparison of the helical content especially for the fluorinated residues reveals different trends.

While DfeGly yields the least helical dimer at position $\mathbf{a}_{16}$, the trend is reversed for position $\mathbf{d}_{19}$ where TfeGly yields the least helical dimer. Also, according to the literature, Abu exhibits a roughly 0.1 kcal mol^{-1} higher helix propensity than leucine (Chiu *et al.*, 2006). Within the coiled coil, Abu as a substitute for Leu indeed increases helical content at position $\mathbf{a}_{16}$ but it decreases it at position $\mathbf{d}_{19}$. These findings suggest that the effects of substitutions on the coiled coil structure cannot primarily be explained by helix propensity. Rather, they depend on the immediate environment of the substitution and the molecular interactions within the dimer.

As discussed in the next section, helicity does not correlate with the stability of these coiled coil structures.

2.4.3 *The impact of fluorine substitutions on the thermodynamic stability of the dimer*

The CD signal at 222 nm is a quantitative measure of helicity and thus structural integrity. Recording the temperature dependence of this signal allows monitoring of the unfolding of the coiled coil structure. Assuming a two-state equilibrium unfolding, the resulting melting curves can be mathematically fitted to yield the thermodynamic parameters T_m, ΔH_m, ΔG that describe the unfolding process. Here, T_m is the midpoint of the thermal unfolding transition (melting temperature) defined as the point where 50% of the coiled coil is unfolded, ΔH_m is the enthalpy change at the melting temperature and ΔG° is the standard free energy of unfolding extrapolated from ΔG and is the direct measure of stability. The equilibrium for the dimer to monomer transition can be described as follows:

$$\Delta G^\circ = \Delta G - RT \ln K_m \tag{2.2}$$

where K_m is the equilibrium constant at T_m. Applying the Gibbs–Helmholtz equation to describe ΔG and specifying K_m for a dimer to monomer transition yields Eq. 2.3:

$$\Delta G^\circ = \Delta H_m \cdot \left(1 - \frac{T}{T_m}\right) + \Delta C_p \cdot \left\{T - T_m - T \cdot \ln\left(\frac{T}{T_m}\right)\right\} - RT \ln 2[D_0] \tag{2.3}$$

The fitted melting curves of all the VPE–VPK variants are shown in Fig. 2.7. To enable qualitative and quantitative comparison of both substitution

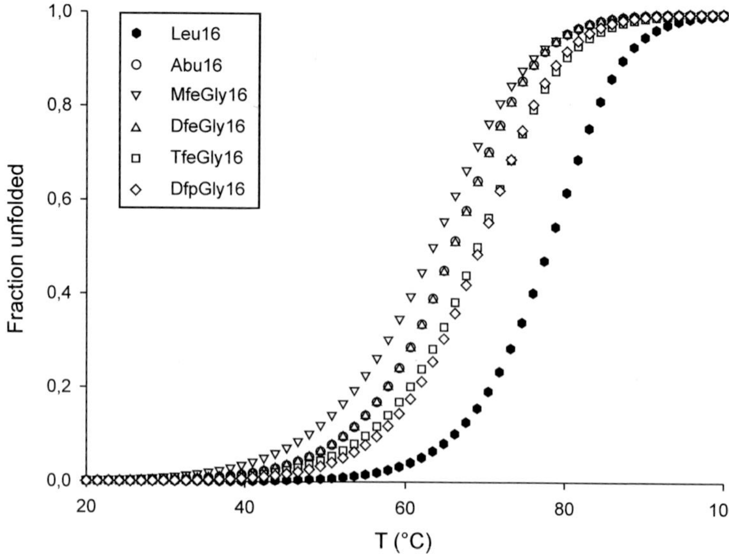

(a)

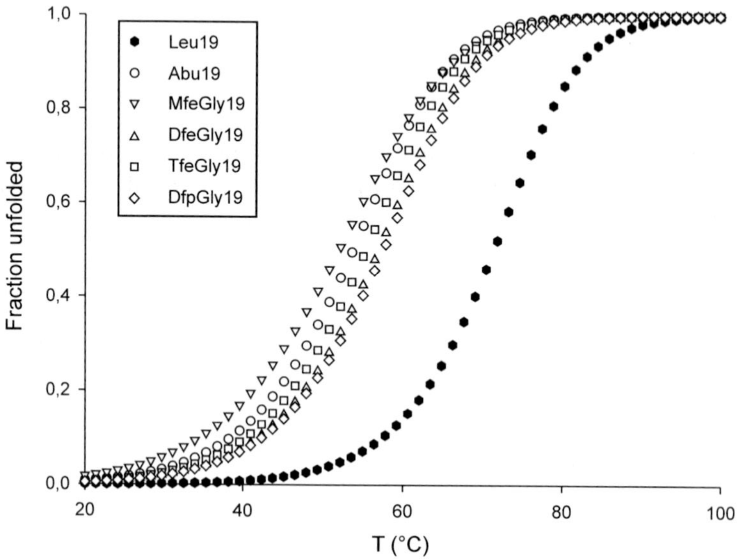

(b)

Figure 2.7. Melting curves of the different VPE–VPK dimers fitted to a monomer–dimer equilibrium at 20 μM overall peptide concentrations (pH 7.4, 100 mM phosphate buffer): (a) substitutions at position $\mathbf{a}_{16}$ and (b) substitutions at position $\mathbf{d}_{19}$.

Table 2.4. Melting temperature and thermodynamic stability of the VPE–VPK dimers.

Amino acid	T_m (°C)	$\Delta G°$(kcal mol^{-1})[a]
Position a_{16}		
Leu	77.9	13.8
Abu	65.9	11.5
MfeGly	63.5	10.3
DfeGly	66.9	11.5
TfeGly	69.0	11.5
DfpGly	69.3	12.3
Position d_{19}		
Leu	71.3	11.7
Abu	53.7	9.6
MfeGly	52.0	8.9
DfeGly	56.9	10.0
TfeGly	55.3	9.9
DfpGly	57.9	10.0

[a] Standard state: 1 M, 101325 Pa, 298 K at pH 7.4 (100 mM phosphate buffer).

positions, we use the peptides that carry Leu at the respective substitution position as references. At both position $\mathbf{a}_{16}$ and position $\mathbf{d}_{19}$, the substitution of Leu by Abu, its mono-, di- and trifluorinated analogues as well as DfpGly decreases the thermodynamic stability of the coiled coil (Table 2.4).

At position $\mathbf{a}_{16}$, the notable reduction of hydrophobic side chain volume going from Leu to Abu, although increasing helicity, decreases stability by more than 2 kcal mol^{-1}. Monofluorination of the Abu-side chain results in a pronounced increase in helicity. However, because MfeGly represents the most polar substitution in this series, hydrophobic interactions within the core are strongly disturbed, which is shown by a further reduction in stability of approximately 1 kcal mol^{-1}. This example shows that the interactions within the coiled coil are more important for stability than helix propensity *per se*. The increasing side chain volumes and hydrophobicities of DfeGly and TfeGly increase stability, to the level of the Abu peptide, although they are the least helical variants in this series. The DfpGly substituted peptide, closest to Leu in helicity and side chain volume, though more stable, is still less stable than the Leu peptide by more than 1 kcal mol^{-1}. In this case, the increased volume of the side chain and the higher helicity of the oligomer seem to outweigh the disadvantageous property of DfpGly being markedly more polar than Leu and TfeGly.

At position $\mathbf{d}_{19}$, the effects of the substitutions are generally less pronounced than at position $\mathbf{a}_{16}$. The replacement of Leu by Abu again, albeit to a lesser extent, destabilises the coiled coil structure. Comparable to position $\mathbf{a}_{16}$, monofluorination of the Abu side chain at position $\mathbf{d}_{19}$ increases helicity but reduces stability. Some stability is recovered by introducing the decreasingly polar and bulkier DfeGly and TfeGly residues. In contrast to position $\mathbf{a}_{16}$, the incorporation of DfpGly, in spite of having similar helical content to that of the Leu peptide, does not lead to stabilisation of the coiled coil.

The general decrease in stability that is shown for all of the fluorinated peptides may be explained by various arguments. Helix propensity is a factor that surely affects the stability of a helical fold. However, if we only take the example of the MfeGly substituted peptides, the most helical but least stable, we can conclude that helix propensity is a less important factor here. The main driving force for oligomerisation and stability of coiled coils is the hydrophobicity and spatial demand of the residues at the **a**- and **d**-positions (Wagschal *et al.*, 1999; Tripet *et al.*, 2000). The fluorinated amino acids used here are smaller and less hydrophobic than leucine, which serves as the reference. This explains the generally observed decrease in stability. The differences between position $\mathbf{a}_{16}$ and position $\mathbf{d}_{19}$, with regard to stability trends, can only be explained by their significantly different packing characteristics (Monera *et al.*, 1993). Figure 2.8 demonstrates an example of the packing of TfeGly at position $\mathbf{a}_{16}$ and position $\mathbf{d}_{19}$ that was generated from molecular dynamics simulations (Salwiczek *et al.*, 2009). In parallel coiled coils, the **a**-positions pack against the **a′**-positions of the opposite strand in a parallel manner, causing the C_α–C_β vectors to point away from each other and out of the hydrophobic core. In contrast, the parallel packing of the **d**-positions against **d′**-positions of the opposite strand results in these vectors pointing towards each other and into the hydrophobic core. These dissimilarities in side chain packing result in unequal stability trends for both substitution positions. There is a slight

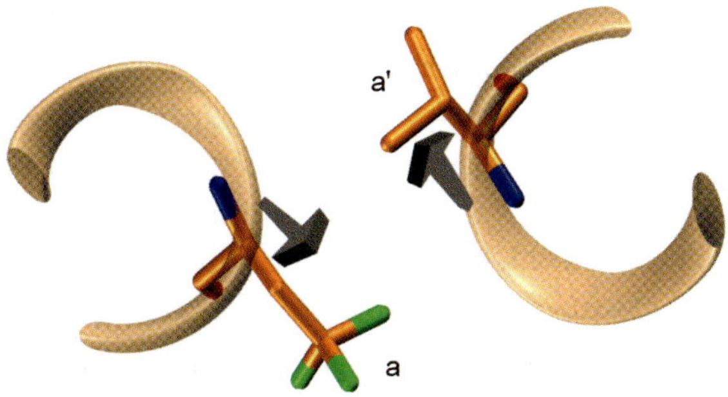

(a)

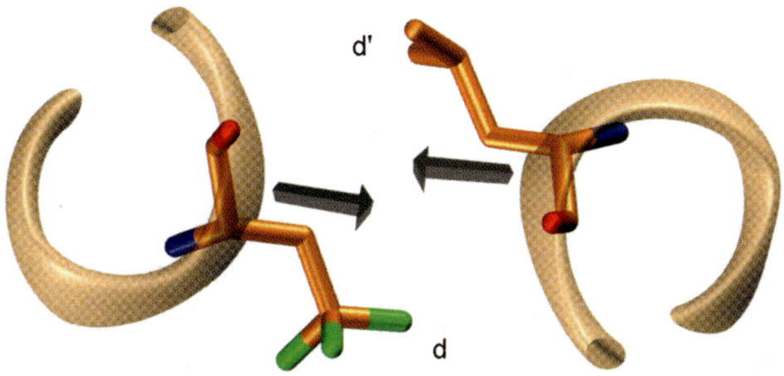

(b)

Figure 2.8. Representation of the different packing of the residues at (a) position $\mathbf{a}_{16}$ and (b) position $\mathbf{d}_{19}$ using the example of TfeGly. The grey arrows indicate the C_α–C_β vectors that have a significantly different orientation towards each other at both substitution positions. (Reproduced from Salwiczek *et al.*, 2009.)

correlation of stability with spatial demand of the fluorinated residues at position $\mathbf{a}_{16}$ because DfpGly, with a side chain rather similar to Leu, yields the most stable fluorinated coiled coil. At position $\mathbf{d}_{19}$, however, the increase in side chain volume has no effect on stability and all the fluorinated peptides, except for those carrying the highly polar MfeGly, are equal. An explanation can be found if we compare the fluorinated amino acids regarding the position of the fluorine atoms. All of them carry fluorine substitutions at the γ-carbon atom, which strongly polarises the respective β-methylene groups. As explained above, these polarised β-methylene groups point away from the β-methylene groups of the interaction partner at position **a** while they point towards them at position **d**.

Also, the C_β–C_β' distance is roughly 1 Å shorter at the **d**-positions. As a consequence, the interaction partners of the amino acids at position $\mathbf{d}_{19}$ may experience a stronger impact of these polar methylene groups than the interaction partners of position $\mathbf{a}_{16}$ that may in turn outweigh the advantage of increasing the spatial demand of the side chain.

In summary, investigations using this heteromeric coiled coil dimer as a model for a natural protein environment suggest that the impact of fluorine-induced polarity in aliphatic amino acids highly depends on the immediate microenvironment of the substitutions. In quaternary structures such as the coiled coil, the predefined backbone conformation determines to a great extent the orientation of the side chain and therefore the dipoles that are associated with fluorination. However, the final conformation and thus the impact of fluorination remain hard to predict and to generalise. An approach that may help in finding the ideal interaction partners for fluorinated amino acids is to screen for fluorophilic residues within the pool of genetically encoded amino acids by applying phage display technology. A phage display based screening system that relies on the same coiled coil model has also been established and will be described in the next section.

2.5 Screening for Native Interaction Partners

Phage display has proven to be a powerful tool for the investigation of molecular recognition (Kehoe and Kay, 2005). Besides the screening of distinct interactions, such as protein–DNA, protein–protein and protein–peptide interactions, this approach has also been successfully applied to determine specifically interacting coiled coil structures (Lai *et al.*, 2004; Hagemann *et al.*, 2008). In general, a peptide (or protein) is displayed on the surface of a bacteriophage (a virus that infects bacteria) particle, while the DNA which encodes for this peptide is contained within the particle. Saturation mutagenesis enables randomisation of defined sections of this peptide and the construction of comprehensive phage-displayed peptide libraries. By virtue of specific interactions between a ligand of choice and the displayed peptide, an individual phage can be enriched and selected from the library. Subsequently, the selected phage can be amplified by infection of bacteria and the primary structure of the displayed peptide can be determined by sequencing the DNA.

Within the scope of investigating the interaction pattern of fluorinated amino acids, this technique was applied in our laboratory only recently. Based on the VPK–VPE model, a screening system for the evaluation of specific interactions within the coiled coil hydrophobic core was established (Vagt *et al.*, 2009). While VPK carries the non-natural amino acid, VPE was presented on the surface of bacteriophage M13 and used for library construction. Non-natural amino acids were investigated at position $\mathbf{a}_{16}$ within VPK (Vagt *et al.*, 2010). According to this, a VPE library was constructed in which the interaction partners of the fluorinated amino acids, which are located in VPE, constitute the randomisation positions for the library (Fig. 2.9). For construction of the library, the amino acid positions $\mathbf{d'}_{12}$, $\mathbf{a'}_{16}$ and $\mathbf{d'}_{19}$ in VPE, which represent the direct interaction partners within the hydrophobic core packing, as well as position $\mathbf{g'}_{15}$, which can also interact with the fluoroalkyl substituted amino acid in $\mathbf{a}_{16}$ were randomised.

The resulting VPE library was fused to the minor core protein pIII on the surface of bacteriophage M13. The VPK variants, which are substituted

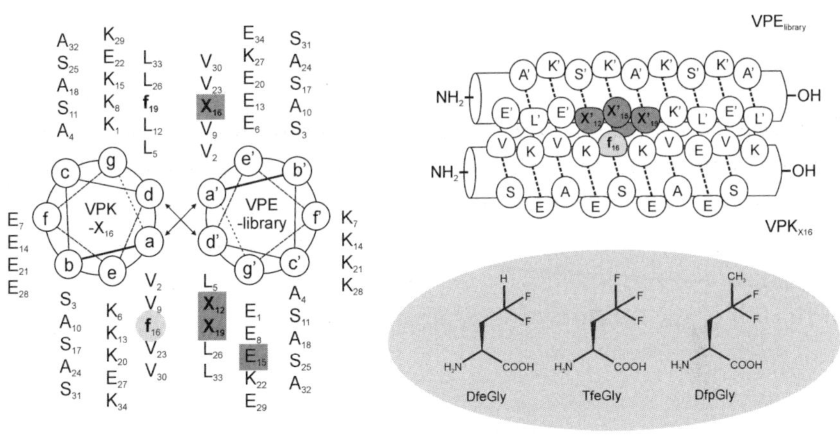

(a) (b)

Figure 2.9. Schematic representation of the VPE library used for the selection of preferred binding partners of fluorinated amino acids in position $\mathbf{a}_{16}$ of VPK. (a) Helical wheel presentation. Substitution position $\mathbf{a}_{16}$ in VPK is marked by a grey circle while the randomised positions in the corresponding VPE library are highlighted by grey squares. (b) Schematic side view (according to Vagt *et al.*, 2010).

in position $\mathbf{a}_{16}$, were elongated by biotin-Gly-Ser-Gly to allow immobilisation on streptavidin coated surfaces and served as targets in library screenings. Coiled coil pairing selectivity was used to select for individual binding partners for the different substituted VPK peptides of the corresponding VPE library. Irrespective of the nature of the fluorinated amino acid within the hydrophobic core of VPK, very similar interaction partners were selected. In each case, predominantly hydrophobic amino acids were found in all of the randomised positions.

The peptide sequences selected by phage display as binding partners for the $\mathbf{a}_{16}$ modified VPK peptides follow the pattern Leu(12)Leu(15) Ile(16)Tyr(19) or Leu(12)Tyr(15)Ile(16)Leu(19) (Table 2.5), which match the VPE variants selected as binding partners for the VPE wild type. The selection of leucine, already present in the **d**-positions of wild type VPE, is in agreement with the general preference of this amino acid in **d**-positions of the parallel coiled coil dimer. The selection of isoleucine in position $\mathbf{a}_{16}$ can be easily explained by the fact that hydrophobic, β-branched amino acids are the most stabilising amino acids in **a**-positions of parallel coiled coil peptides. However, the finding of tyrosine in $\mathbf{g}_{15}$ and $\mathbf{d}_{19}$ was somewhat unexpected. We consider the possibility of cation–π interactions between tyrosine and lysine in the opposite **g**-positions of VPK, which favour selection of this amino acid (Gallivan and Dougherty, 1999).

The three most frequently selected VPE peptides were chemically synthesised and all combinations of coiled coil pairs were investigated by CD spectroscopy. As expected, thermal denaturation revealed a distinct increase in thermal stability for the selected VPE variants in combination with each of the substituted VPK peptides. However, while the frequency

Table 2.5. Amino acids selected in the randomised positions in $VPE_{library\text{-}a16}$ as preferred binding partners for DfeGly, DfpGly and TfeGly in position $\mathbf{a}_{16}$ of VPK.

	Position $\mathbf{d}'_{12}$	Position $\mathbf{g}'_{15}$	Position $\mathbf{a}'_{16}$	Position $\mathbf{d}'_{19}$
Val	Leu	Tyr or Leu	Ile	Tyr or Leu
DfeGly	Leu	Hydrophobic amino acids	Ile	Tyr or Leu
DfpGly	Leu	Leu	Ile	Leu or Tyr
TfeGly	Leu	Leu or Tyr	Ile	Tyr or Leu

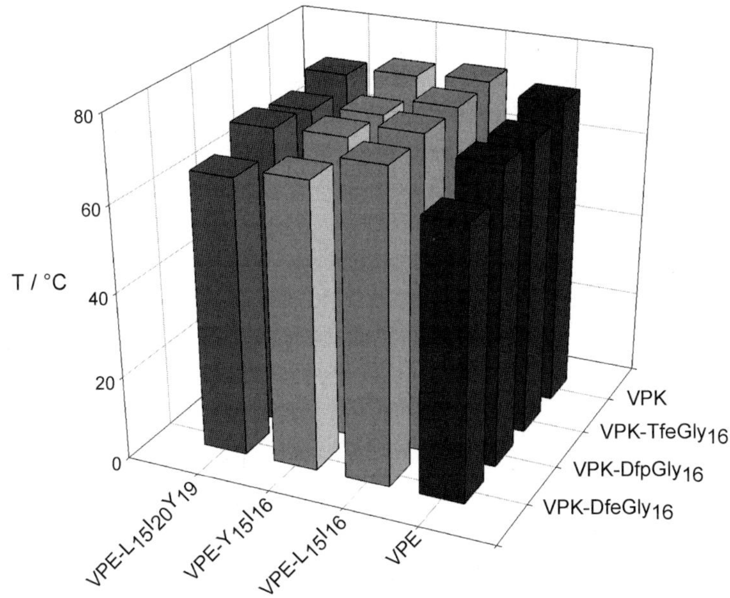

Figure 2.10. Comparative illustration of the thermal stability of the different VPK peptides substituted in position $\mathbf{a}_{16}$ by DfeGly, DfpGly and TfeGly in combination with the VPE variants selected by phage display. For melting points see Table 2.6 (according to Vagt *et al.*, 2010).

with which the different peptides were selected as interacting partners for DfeGly, DfpGly and TfeGly slightly differed, each VPK variant, including the wild type, formed the most stable coiled coil in combination with VPE-$L_{15}I_{16}$ (Fig. 2.10). This result demonstrates that, despite the differences in size and hydrophobicity of these building blocks, DfeGly, DfpGly, TfeGly and Val prefer the same amino acids as interaction partners within the hydrophobic core of a parallel coiled coil.

Figure 2.10 illustrates the thermal stability of the VPK variants substituted in position $\mathbf{a}_{16}$ in combination with the selected VPE peptides. Besides the loss in stability resulting by incorporation of a single fluorinated amino acid within the hydrophobic core, the compensation of this effect by rearrangement of the corresponding amino acid positions in VPE is conspicuous. The substituted VPK peptides clearly prefer the same VPE peptide as the wild type.

Table 2.6. Thermal stability (melting points in °C) of the different VPK peptides substituted in position $\mathbf{a}_{16}$ by DfeGly, DfpGly and TfeGly in combination with the VPE variants selected by phage display.

	VPE-$L_{15}I_{16}$	**VPE-$Y_{15}I_{16}$**	**VPE-$L_{15}I_{16}Y_{19}$**	**VPE**
Val	74.9	74.0	72.1	72.4
TfeGly	74.4	70.3	69.0	69.4
DfpGly	74.2	70.9	70.4	70.1
DfeGly	73.3	67.5	65.3	63.9
Peptide				
VPE-$L_{15}I_{16}$	H-EVSALEKEVASLEKLISALEKKVASLKKEVSALE-OH			
VPE-$Y_{15}I_{16}$	H-EVSALEKEVASLEKYISALEKKVASLKKEVSALE-OH			
VPE-$L_{15}I_{16}Y_{19}$	H-EVSALEKEVASLEKLISAYEKKVASLKKEVSALE-OH			

As for the three hydrophobic core positions, mostly hydrophobic amino acids were selected in position $\mathbf{g}'_{15}$. The resulting extension of the hydrophobic interface possibly minimises the effect caused by the investigated fluorinated amino acids by an additional stabilisation of the coiled coil structure. It could also alter the oligomerisation state of the coiled coil structure which complicates analysis and comparison of the stability of the different coiled coil couples. According to this, a second VPE library was constructed in which exclusively the three amino acid positions within the hydrophobic core of VPE ($\mathbf{d}'_{12}$, $\mathbf{a}'_{16}$, $\mathbf{d}'_{19}$) were randomised. Screening of this library yielded the same VPE variant as best binding partner for the substituted VPK peptides and the VPK wild type respectively. The amino acids in the variable amino acid positions of this VPE variant follow the pattern Leu(12)Ile(16)Leu(19), which, aside from position $\mathbf{g}'_{15}$, in principle match the peptide sequence selected from the first library. However, the phenotypes selected from the decreased library for the DfeGly, DfpGly and TfeGly substituted VPK peptides practically show no variance. Exclusively leucine occurred in position $\mathbf{d}'_{12}$ and $\mathbf{d}'_{19}$, while for the extended VPE libraries also tyrosine was selected in these positions (Vagt *et al.*, 2010). This observation confirms the assumption that the selection of aromatic amino acids in position $\mathbf{d}'_{19}$, as observed during the panning using the extended library, only correlates with the extension of the hydrophobic interaction surface and is not characteristic for the different fluorinated amino acids incorporated into VPK.

In summary, substitution of DfeGly, DfpGly or TfeGly in a central **a** position of a parallel coiled coil peptide results in the same arrangement of the corresponding peptide strand as for the naturally occurring amino acids valine and leucine. Despite differences in steric demand and hydrophobicity of these building blocks, their preference for interaction partners seems to be the same as that of canonical aliphatic amino acid side chains.

2.6 Conclusions and Outlook

In this chapter, the systematic study of the influence of fluorinated amino acids on peptide secondary structure using an α-helical coiled coil peptide model system is reviewed. As the hydrophobic recognition domain of this folding motif reports on even minor alterations with sufficient sensitivity, it serves as a versatile model for the systematic study of non-natural amino acids within a native protein environment. Extensive incorporation of trifluoromethyl containing amino acids generally causes structural stabilisation of the α-helical coiled coil. The higher hydrophobicity of these amino acids as well as the specific interactions of fluorinated side chains with each other is responsible for this effect. Thus, fluorinated analogues of naturally occurring amino acids can be used to effectively direct the nature and stability of peptide folding.

On the other hand, single substitutions of fluorinated amino acids reveal different properties and often exert a destabilising effect. In comparison to canonical aliphatic amino acids, the side chains of their fluorinated analogues are often characterised by both hydrophobicity and polarity. Because the impact of each effect can hardly be evaluated separately, it is not possible to predict how the incorporation of fluorinated amino acids will affect peptide and protein stability.

Moreover, the influence of single non-natural amino acids strongly depends on the peptide and protein environment itself. Mutagenesis studies using a parallel coiled coil have shown that partially fluorinated amino acids (not more than three fluorine atoms) prefer the same interaction partners as canonical aliphatic amino acids. However, as a well defined quarternary structure, the α-helical coiled coil motif requires several primary structure features for its formation, which may limit the range of potential interaction partners. Besides significant differences in size, hydrophobicity and polarity, the fluorinated amino acids investigated so far obviously behave similarly to the canonical aliphatic amino acids like leucine or valine within the hydrophobic environment of the chosen model system.

Consequently, the rational design of peptides containing fluorinated amino acids for medicinal use is still challenging. However, the information gained from the model studies described in this chapter is useful and, in combination with the evaluation of further properties such as protease stability, may enable targeted application of these new building blocks in peptide and protein design.

Acknowledgements

Our work was financially supported by Deutche Forschungsgemeinschaft (KO 1976/2-2 and Graduate School #788). Furthermore we would like to thank Prof. Dr Günter Haufe for providing 4-fluoroamino butyric acid and Dr Allison Berger for proofreading the manuscript.

References

Biffinger, J.C., Kim, H.W. & DiMagno, S.G. (2004) The polar hydrophobicity of fluorinated compounds. *ChemBioChem* 5, 622–627.

Bilgiçer, B., Fichera, A. & Kumar, K. (2001a) A coiled coil with a fluorous core. *J Am Chem Soc* 123, 4393–4399.

Bilgiçer, B., Xing, X. & Kumar, K. (2001b) Programmed self-sorting of coiled coils with leucine and hexafluoroleucine cores. *J Am Chem Soc* 123, 11815–11816.

Bilgiçer, B. & Kumar, K. (2002) Synthesis and thermodynamic characterization of self-sorting coiled coils. *Tetrahedron* 58, 4105–4112.

Bilgiçer, B. & Kumar, K. (2004) *De novo* design of defined helical bundles in membrane environments. *Proc Natl Acad Sci USA* 101, 15324–15329.

Bott, G., Field, L.D. & Sternhell, S. (1980) Steric effects. A study of a rationally designed system. *J Am Chem Soc* 102, 5618–5626.

Chen, Y.-H., Yang, J.T. & Chau, K.H. (1974) Determination of the helix and β form of proteins in aqueous solution by circular dichroism. *Biochemistry* 13, 3350–3359.

Chiu, H.-P., Suzuki, Y., Gullickson, D. *et al.* (2006) Helix propensity of highly fluorinated amino acids. *J Am Chem Soc* 128, 15556–15557.

Chiu, H.-P. & Cheng, R.P. (2007) Chemoenzymatic synthesis of (*S*)-hexafluoroleucine and (*S*)-tetrafluoroleucine. *Org Lett* 9, 5517–5520.

David, R., Günther, R., Baumann, L. *et al.* (2008) Artificial chemokines: combining chemistry and molecular biology for the elucidation of interleukin-8 functionality. *J Am Chem Soc* 130, 15311–15317.

di Riggi, I., Virgili A., de Moragas, M. *et al.* (1995) Restricted rotation and NOE transfer: a conformational study of some substituted (9-anthryl)carbinol derivatives. *J Org Chem* 60, 27–31.

Dougherty, D.A. (2000) Unnatural amino acids as probes of protein structure and function. *Curr Opin Chem Biol* 4, 645–652.

Dunitz, J.D. & Taylor, R. (2006) Organic fluorine hardly ever accepts hydrogen bonds. *Chem Eur J* 3, 89–98.

Esté, J.A. & Telenti, A. (2007) HIV entry inhibitors. *Lancet* 370, 81–88.

Gallivan, J.P. & Dougherty, D.A. (1999) Cation-π interactions in structural biology. *Proc Natl Acad Sci USA* 96, 9459–9464.

Hagemann, U.B., Mason, J.M., Müller, K.M. *et al.* (2008) Selectional and mutational scope of peptides sequestering the Jun–Fos coiled-coil domain. *J Mol Biol* 381, 73–88.

Hakelberg, M. & Koksch, B. (2007) Coiled coil model systems as tools to evaluate the influence of fluorinated amino acids on structure and stability of peptides. *Chim Oggi* 25, 48–53.

Hüther, A. & Dietrich, U. (2007) The emergence of peptides as therapeutic drugs for the inhibition of HIV-1. *AIDS Rev* 9, 208–217.

Jäckel, C., Seufert, W., Thust, S. *et al.* (2004) Evaluation of the molecular interactions of fluorinated amino acids with native polypeptides. *ChemBioChem* 5, 717–720.

Jäckel, C., Salwiczek, M. & Koksch, B. (2006) Fluorine in a native protein environment — how the spatial demand and polarity of fluoroalkyl groups affect protein folding. *Angew Chem Int Edit* 45, 4198–4203.

Kehoe, J.W. & Kay, B.K. (2005) Filamentous phage display in the new millennium. *Chem Rev* 105, 4056–4072.

Lai, J.R., Fisk, J.D., Weisblum, B. *et al.* (2004) Hydrophobic core repacking in a coiled-coil dimer via phage display: insights into plasticity and specificity at a protein–protein interface. *J Am Chem Soc* 126, 10514–10515.

Lee, K.-H., Lee, H.-Y., Slutsky, M.M. *et al.* (2004) Fluorous effect in proteins: *de novo* design and characterization of a four-α-helix bundle protein containing hexafluoroleucine. *Biochemistry* 43, 16277–16284.

Lee, H.-Y., Lee, K.-H., Al-Hashimi, H.M. *et al.* (2006) Modulation protein structure with fluorous amino acids: increased stability and native-like structure conferred on a 4-helix bundle protein by hexafluoroleucine. *J Am Chem Soc* 128, 337–343.

Mason, J.M. & Arndt, K.M. (2004) Coiled coil domains: stability, specificity and biological implications. *ChemBioChem* 5, 170–176.

Matthews, T., Salgo, M., Greenberg, M. *et al.* (2004) Enfuvirtide: the first therapy to inhibit the entry of HIV-1 into host CD4 lymphocytes. *Nat Rev Drug Discov* 3, 215–225.

Monera, O.D., Zhou, N.E., Kay, C.M. *et al.* (1993) Comparison of antiparallel and parallel two-stranded α-helical coiled coils: design, synthesis, and characterization. *J Biol Chem* 268, 19218–19227.

Moutevelis, E. & Woolfson, D.N. (2009) A periodic table of coiled coil protein structures. *J Mol Biol* 385, 726–732.

O'Shea, E.K., Klemm, J.D., Kim, P.S. *et al.* (1991) X-ray structure of the GCN4 leucine zipper, a two-stranded, parallel coiled coil. *Science* 254, 539–544.

Permanne, B., Adessi, C., Saborio, G.P. *et al.* (2002) Reduction of amyloid load and cerebral damage in a transgenic mouse model of Alzheimer's disease by treatment with a β-sheet breaker peptide. *FASEB J* 16, 860–862.

Rose, A. & Meier, I. (2004) Scaffolds, levers, rods and springs: diverse cellular functions of long coiled coil proteins. *Cell Mol Life Sci* 61, 1996–2009.

Salwiczek, M., Samsonov, S., Vagt, T. *et al.* (2009) Position-dependent effects of fluorinated amino acids on the hydrophobic core formation of a heterodimeric coiled coil. *Chem Eur J* 15, 7628–7636.

Samsonov, S., Salwiczek, M., Anders, G. *et al.* (2009) Fluorine in protein environments: a QM and MD Study. *J Phys Chem B* 113, 16400–16408.

Sato, A.K., Viswanathan, M., Kent, R.B. *et al.* (2006) Therapeutic peptides: technological advances driving peptides into development. *Curr Opin Biotechnol* 17, 638–642.

Smart, B.E. (2001) Fluorine substituent effects (on bioactivity). *J Fluorine Chem* 109, 3–11.

Son, S., Tanrikulu, I.C. & Tirrell, D.A. (2006) Stabilization of *bzip* peptides through incorporation of fluorinated aliphatic residues. *ChemBioChem* 7, 1251–1257.

Tang, Y., Ghirlanda, G., Petka, W.A. *et al.* (2001a) Fluorinated coiled-coil proteins prepared *in vivo* display enhanced thermal and chemical stability. *Angew Chem Int Edit* 40, 1494–1496.

Tang, Y., Ghirlanda, G., Vaidehi, N. *et al.* (2001b) Stabilization of coiled coil peptide domains by introduction of trifluoroleucine. *Biochemistry* 40, 2790–2796.

Tang, Y. & Tirrell, D.A. (2001c) Biosynthesis of a highly stable coiled coil protein containing hexafluoroleucine in an engineered bacterial host. *J Am Chem Soc* 123, 11089–11090.

Tripet, B., Wagschal, K. Lavigne, P. *et al.* (2000) Effects of side chain characteristics on stability and oligomerization state of a *de novo* designed model coiled coil: 20 amino acid substitutions in position 'd'. *J Mol Biol* 300, 377–402.

Vagt, T., Jäckel, C., Samsonov, S. *et al.* (2009) Selection of a buried salt bridge by phage display. *Bioorg Med Chem Lett* 19, 3924–3927.

Vagt, T., Nyakatura, E., Salwiczek, M. *et al.* (2010) Towards identifying preferred interaction partners of fluorinated amino acids within the hydrophobic environment of a dimeric coiled-coil peptide. *Org Biomol Chem* 40, 1382–1386.

Venyaminov, S.Y. & Yang, J.T. (1996) 'Determination of protein secondary structure', in Fasman, G.D. (Ed.), *Circular dichroism and the conformational analysis of biomolecules*, Plenum Press, New York, pp. 69–107.

Wagschal, K. Tripet, B., Lavigne, P. *et al.* (1999) The role of position a in determining the stability and oligomerization state of α-helical coiled coils: 20 amino acid stability coefficients in the hydrophobic core of proteins. *Protein Sci* 8, 2312–2329.

Welch, B.D., VanDemark, A.P., Heroux, A. *et al.* (2007) Potent D-peptide inhibitors of HIV-1 entry. *Proc Natl Acad Sci USA* 104, 16828–16833.

Woolfson, D.N. (2005) The design of coiled coil structures and assemblies. *Adv Protein Chem* 70, 79–112.

Zhao, Y.H., Abraham, M.H. & Zissimos, A.M. (2003) Fast calculation of van der Waals volume as a sum of atomic and bond contributions and its application to drug compounds. *J Org Chem* 68, 7368–7373.

3

Probing the Binding Affinity and Proteolytic Stability of Trifluoromethyl Peptide Mimics as Protease Inhibitors

Matteo Zanda,[*,†] *Alessandro Volonterio,*[†] *Monica Sani*[†,‡] *and Sergio Dall'Angelo*[‡]

3.1 Introduction

Modification of peptides by the introduction of fluorine-containing functions on the backbone is becoming a very popular and reliable strategy to improve their binding affinity and metabolic stability (Jäckel and Koksch, 2005). The resulting fluorinated peptide mimics often have peculiar biological, structural and binding properties, largely due to the special characteristics of fluorine (Kirk, 2006). In this arena, the trifluoromethyl group has undoubtedly the capacity to bring about the most intriguing and potentially useful properties, and has therefore been used for several

* Institute of Medical Sciences, University of Aberdeen, Foresterhill, Aberdeen AB25 2ZD, UK. E-mail: m.zanda@abdn.ac.uk.
† C.N.R.-I.C.R.M. and Dipartimento di Chimica, Materiali ed Ingegneria Chimica 'G. Natta' del Politecnico di Milano, via Mancinelli 7, I-20131 Milano, Italy.
‡ KemoTech s.r.l., Parco Scientifico e Tecnologico della Sardegna, Edificio 3, Loc.Piscinamanna, 09010 Pula (CA), Italy.

conceptually different applications in the field of peptidomimetics (Zanda, 2004). The high chemical stability and the xenobiotic nature of the trifluoromethyl group are responsible for its outstanding metabolic stability. In addition, its strongly electron-withdrawing character is responsible for the increased electrophilicity of neighbouring functions, such as carbonyl groups, as well as for the decreased basicity of proximal amino groups. More subtle effects, such as the local hydrophobicity, poor hydrogen-bond acceptor ability, and capacity to undertake multipolar interactions with positively charged species are responsible for the interesting conformational behaviour of CF_3-peptides and the binding to receptor pockets (Smart, 1994; Mikami *et al.*, 2004; Ma and Cahard, 2008 and references cited therein). Finally, a rather controversial aspect is represented by the bioisosterism of the trifluoromethyl group, namely its capacity to act as a replacement of groups having similar sizes or shapes without substantially altering key biological properties such as binding affinity. While the success of this situation seems to vary from case to case, recent evidence (Müller *et al.* 2007) supports the substantial bioisosterism between the CF_3 and the ethyl group, while the previously thought isosteric isopropyl group is suggested to be larger (Leroux, 2004; Jagodzinska *et al.*, 2009). This chapter will discuss several classes of CF_3-containing peptide mimics as protease inhibitors, with an emphasis on the actual effect and role of the trifluoromethyl group.

3.2 Peptidyl Trifluoro-Ketones

The peptidyl trifluoro-ketone moiety was introduced by Abeles and coworkers as a key tool for the inhibition of a number of esterases and proteases (Gelb *et al.*, 1985). The field has been extensively reviewed and only the basic concepts and most recent uses of trifluoromethyl ketones as protease inhibitors will be discussed herein (Sani *et al.*, 2006). In water trifluoromethyl ketones are known to exist predominantly in their hydrated form. They are also known to have a strong electrophilic character. Indeed, they reversibly form rather stable hemiketals with nucleophiles (Fig. 3.1), including those in the active sites of serine and cysteine proteases. Thus, trifluoromethyl ketones can behave as tetrahedral transition-state analogues by reacting with a hydroxy or thiol

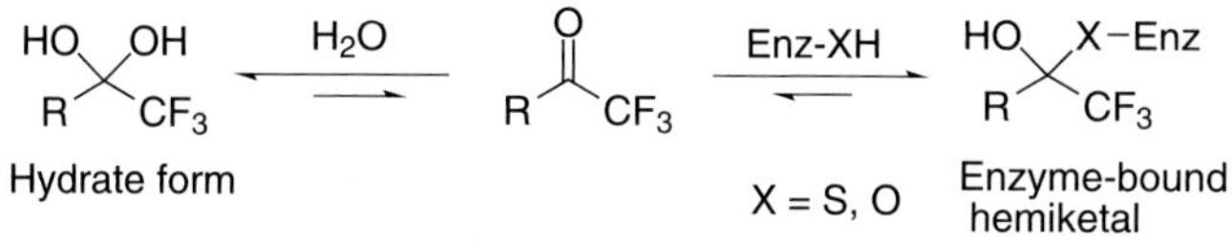

Figure 3.1. Trifluoromethyl ketones as protease inhibitors.

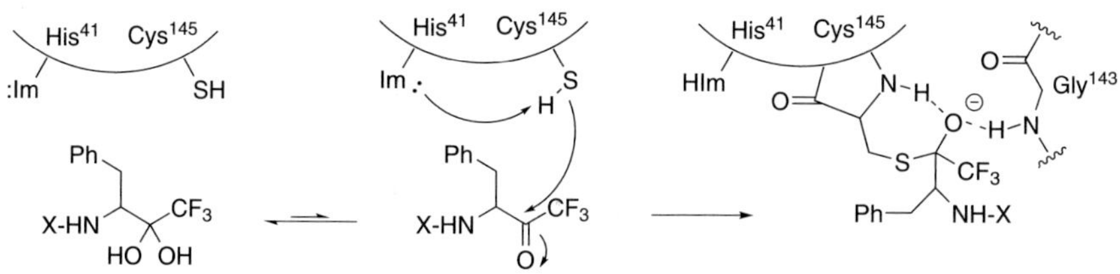

X = Cbz-Ala-Val-Leu

Figure 3.2. Proposed mechanism of inhibition of SARS-CoV 3CL protease trifluoroketone inhibitors.

function of serine and cysteine proteases, respectively, to form in a reversible manner a tetrahedral hemiacetal intermediate covalently bound to the enzyme active site.

Among the recent applications of trifluoroketones as protease inhibitors, the design of severe acute respiratory syndrome-associated coronavirus (SARS-CoV) 3CL protease inhibitors is of particular interest (Shao *et al.*, 2008). Several peptidyl trifluoromethyl ketones displayed time-dependent inhibition with K_i values as low as 0.3 μM after 4 hours of incubation. The proposed mechanism of action of the SARS-CoV protease (Fig. 3.2), which has a Cys–His catalytic dyad (Cys-145 and His-41), is a covalent attack mechanism peculiar to this class of inhibitors.

Other structurally related peptidyl trifluoromethyl ketones were also reported as SARS-CoV 3CL protease inhibitors (Sydnes *et al.*, 2006; Regnier *et al.*, 2009).

A peptidyl trifluoromethyl ketone was reported to be a submicromolar inhibitor of the Dengue virus NS3 protease (Fig. 3.3) (Yin *et al.*, 2006), a serine protease, which represents a potential therapeutic target against

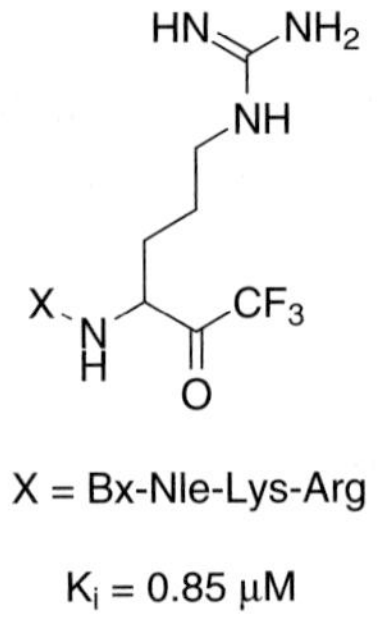

Figure 3.3. Dengue virus NS3 serine protease inhibitor.

the Dengue fever, is a viral disease endemic to most tropical and subtropical regions throughout the world.

3.3 Peptidomimetics Containing the Trifluoroethylamine Function as Peptide Bond Replacement

Peptide drugs clear faster from the body than non-peptidic small molecule drugs due to their low bioavailability, which in turn is due to rapid hydrolysis by endogenous proteases, and usually need to be injected rather than administered orally. Moreover, peptidic drugs are generally more expensive to synthesize and less stable than any small molecule counterparts. However, they can be much more potent than small molecules, show higher specificity, and have fewer toxicology problems. They also do not accumulate in organs or face drug–drug interaction challenges that affect small molecules (Loffet, 2002).

For the reasons above, the design and synthesis of metabolically stable peptide analogues that can either mimic or block the bioactivity of natural peptides or enzymes is an important area of medicinal chemistry and drug discovery. Replacement of a scissile backbone peptide bond with a hydrolytically stable replacement constitutes a viable and effective approach to the rational design of peptide mimics and surrogates. Indeed, a number of therapeutically useful peptidomimetics incorporating any of the peptide isosteres that are currently available have been described (Fletcher and Campbell, 1998).

Although, many amide replacements are known which retain the geometry of the amide bond or maintain the hydrogen bond-accepting properties of the amide, only a few of them are capable of preserving the hydrogen bond-donating properties of the amide: sulfonamides, anilines, secondary alcohols, hydrazines and certain heterocycles. The main issue for identifying a truly effective NH amide replacement can be recast as how to minimize the basicity of an NH donor so that a NH_2^+ moiety is not formed at physiological pH. In fact, such charged groups are poorly tolerated deep in the active site of a protein where binding interactions cannot compensate for the energetic cost of desolvation (Black and Percival, 2006).

Some years ago, our group proposed the stereogenic trifluoroethylamine function as a mimic of the peptide bond (Sani *et al.*, 2007). Indeed, a trifluoroethyl group can replace the carbonyl of an amide and generate a metabolically stable, essentially non-basic amine that maintains the typical hydrogen bond donor ability of an amide. The main properties featured by the trifluoroethylamino group are: (1) low NH basicity, (2) a $CH(CF_3)NHCH$ backbone angle close to 120°, (3) a $C–CF_3$ bond substantially isopolar with the C=O, (4) structural analogy with the tetrahedral proteolytic transition state. Furthermore, the sp^3 hybridization of all the atoms forming the stereogenic trifluoroethylamine moiety should allow better adaptation of the atoms to receptors' active sites, thus optimizing the energetically favourable interactions (hydrogen bonds, van der Waals, hydrophobic, etc.).

Our initial work in the field focused on a group of peptidomimetic structures incorporating a trifluoroethylamine unit replacing the retro-peptide bond (Volonterio *et al.*, 2000, 2001, 2002, 2003; Sani *et al.*, 2002). We then focused on the stereocontrolled synthesis of a new generation of peptidomimetics, very close to natural peptides, having a fluoroalkyl backbone modification: $\Psi[CH(CF_3)NH]$Gly-peptides (Fig. 3.4, R=H) (Molteni *et al.*, 2003, 2009). In this case, the trifluoroethylamine function replaces a native peptidic amide-bond. These peptidomimetics were obtained by means of a stereocontrolled aza-Michael reaction involving α-amino acid esters as nucleophiles and activated fluorinated olefins as Michael acceptors.

More recently this work has been expanded to include novel peptidomimetics featuring $–CH(R_F)NH-$ units with different degrees of fluorination

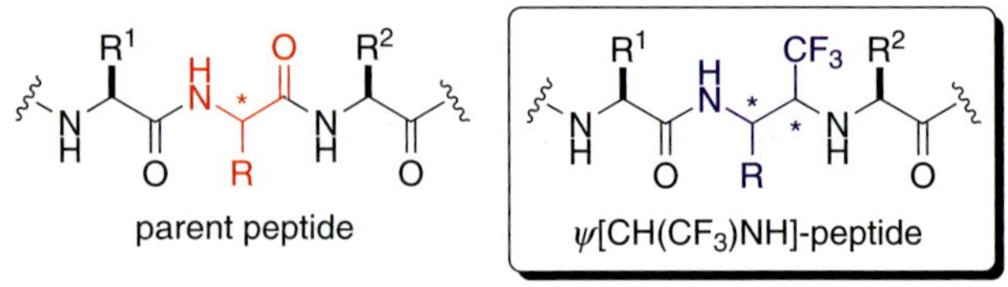

Figure 3.4. Peptidomimetics incorporating a trifluoroethylamine surrogate of the native peptide bond.

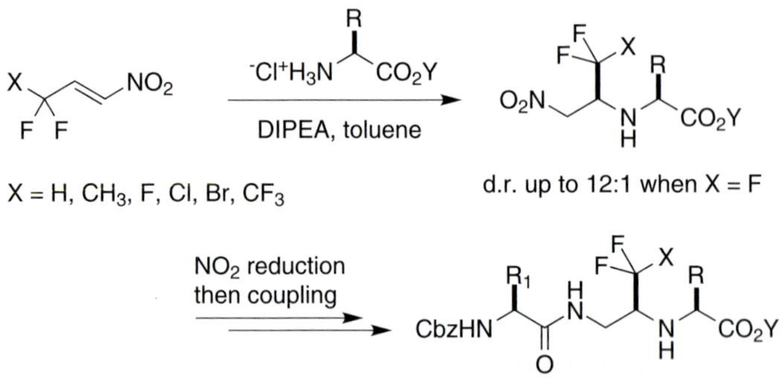

Figure 3.5. Synthesis and elaboration of fluoroethylamine aza-Michael adducts into the target peptidomimetics.

(Bigotti *et al.*, 2008a,b). The key step in the synthesis was a stereoselective aza-Michael addition of chiral α-amino acid esters to β-fluoroalkyl-α-nitroethenes (Fig. 3.5). The diastereoselection of the process was influenced by the electronegativity, rather than by the steric bulk of the fluorinated residue R_F in the β-position of the nitroalkene acceptors. Replacement of a single F atom of R_F by a hydrogen or methyl group resulted in a dramatic drop of stereocontrol, whereas Br, Cl and CF_3, albeit bulkier than F, provided results only slightly worse in terms of stereocontrol.

Very recently, we found that a trifluoroethylamine surrogate with the appropriate stereochemistry can be successfully used to replace the native peptide-bond in opioid peptides (Sinisi *et al.*, 2009). Enkephalins are endogeneous opioid pentapeptides having the sequence H-Tyr-Gly-Gly-Phe-Xaa-OH where the C-terminal amino acid Xaa can be Leu (Leu-enkephalin) or Met (Met-enkephalin) (Hughes *et al.*, 1975). These

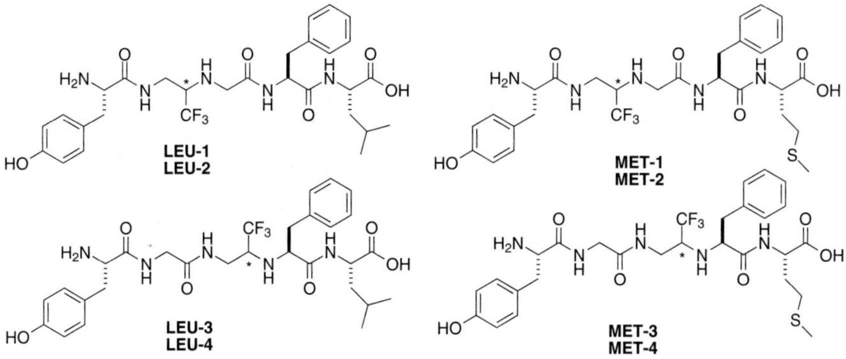

Figure 3.6. Trifluoroethylamine analogues of Leu- and Met-enkephalins.

compounds are regulators of pain and nociception (the perception of noxious stimuli), in the body. The receptors for enkephalins are the opioid receptors, which are G-protein-coupled receptors (Dhawan *et al.*, 1996). Enkephalins preferentially bind to the δ-receptor, with a significant affinity also for the μ-receptor. Notwithstanding their remarkable pain-alleviating activity, enkephalins are unsuitable as analgesic drugs because of their low metabolic stability, rapid *in vivo* degradation and substantial inability to penetrate the blood–brain barrier (Janecka *et al.*, 2008). With the aim of identifying more stable enkephaline mimics with good opioid receptor affinity and improved metabolic stability, we embarked on a study of eight different enkephalin analogues (Fig. 3.6). All compounds feature a stereogenic trifluoroethylamine function replacing the Gly^2–Gly^3 peptide bond (**LEU-1**, **LEU-2** epimers for Leu-enkephalin, and **MET-1**, **MET-2** epimers for Met-enkephalin), as well as the Gly^3–Phe^4 peptide bond (**LEU-3,4** and **MET-3,4**).

The synthesis of **LEU-3** is shown in Fig. 3.7. The key reaction is the aza-Michael addition of H-Phe-Leu-O*tert*Bu dipeptide to 3,3,3-trifluoro-1-nitropropene, in line with the general strategy for the synthesis of trifluoroethylamine peptidomimetics.

Among the resulting peptidomimetics, **LEU-3** displayed a binding affinity in the nanomolar range, with a δ versus μ affinity ratio quite similar to the ratio of natural **Leu-enkephalin**. The epimer **LEU-4** showed lower affinity, with a K_i in the micromolar range. The corresponding

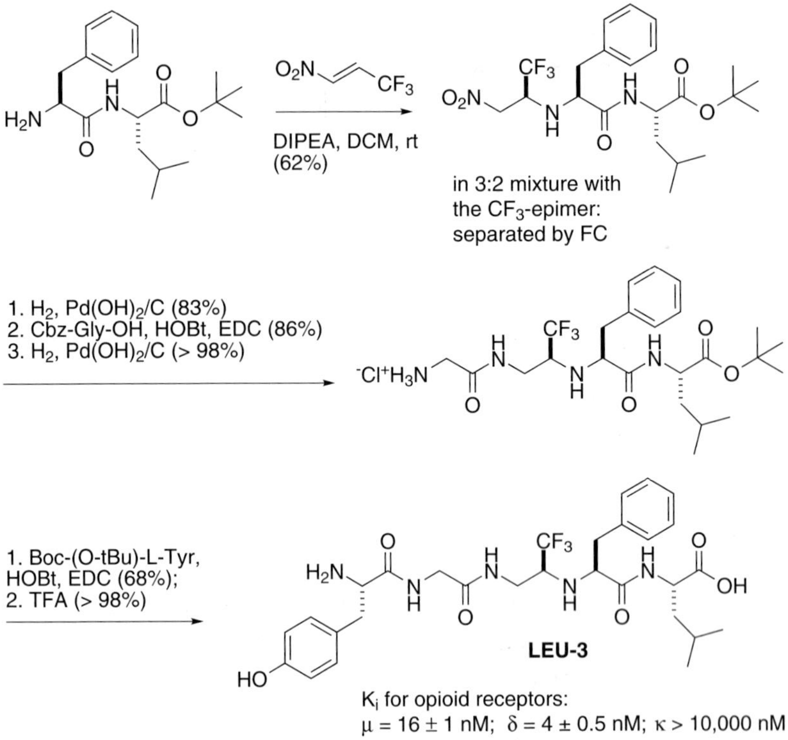

Figure 3.7. Synthesis of LEU-3.

Met-enkephalin analogues **MET-3** and **MET-4** displayed a K_i twofold higher than **LEU-3** and **LEU-4**, respectively. Rather weak affinity for both the μ and δ receptors was shown by all of the enkephalin analogues having the trifluoroethylamine replacement in the second position, namely **LEU-1,2** and **MET-1,2**. However, **LEU-3** did not display anti-nociceptive activity *in vivo*, clearly indicating that the introduction of just one trifluoroethylamine peptide-bond replacement, albeit in an appropriate position of the backbone, is not sufficient to impart drug-like properties to enkephalin peptidomimetics.

The trifluoroethylamine strategy has recently found the first validation in drug discovery. Merck developed a trifluoroethylamine compound, Odanacatib (MK-0822), that is now in phase III clinical studies for the therapy of osteoporosis (Fig. 3.8) (Gauthier *et al.*, 2008).

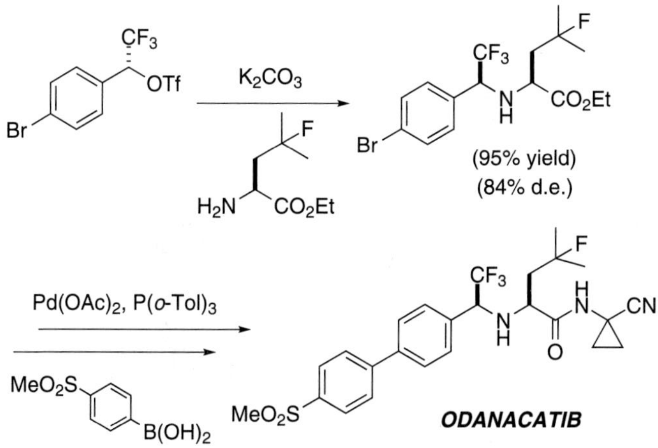

Figure 3.8. Multikilogram synthesis of Odanacatib.

Odanacatib is a highly potent and metabolically stable inhibitor of Cathepsin K, a cysteine proteinase thought to be responsible for the degradation of type I collagen in osteoclastic bone resorption, that represents a highly promising target for the therapy of osteoporosis. Encouraging results were also observed for women affected by bone metastases deriving from breast cancer.

Recently, Merck described a synthetic methodology suitable for preparing kilogram quantities of Odanacatib (Fig. 3.8) (O'Shea *et al.*, 2009).

A structurally related family of trifluoroethylhydrazino peptide mimics was also reported to inhibit the rabbit 20S proteasome at micromolar concentrations (Formicola *et al.*, 2009).

The body of experimental results listed above suggests important considerations for the successful use of the trifluoroethylamine function as a peptide/retropeptide bond mimic (see Fig. 3.9).

- When the amide or peptide bond to be replaced by the trifluoroethylamine unit is one of the reasons for the low bioavailability of the parent unfluorinated molecule, the strategy can be highly successful. Indeed the trifluoroethylamine unit seems to have high metabolic stability.

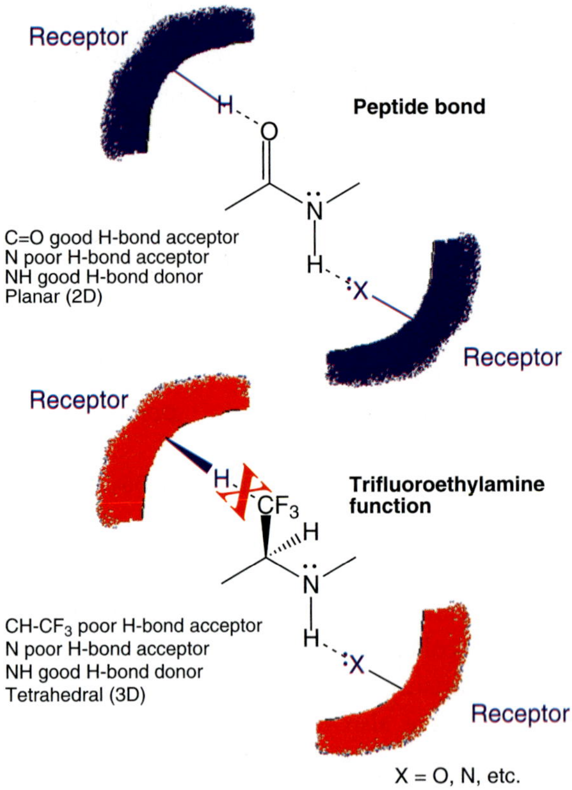

Figure 3.9. Comparison between the peptide (amide) bond and the trifluoroethylamine function.

- The trifluoromethyl group, contrarily to the carbonyl oxygen, is a weak hydrogen-bond acceptor (Dunitz and Taylor, 1997). The trifluoroethylamine function can therefore only be an effective peptide bond replacement if the carbonyl group of the original ligand's amide/peptide-bond is not involved in essential hydrogen-bonding with the receptor.
- The NH of the trifluoroethylamine unit is a good hydrogen-bond donor, due to the strong electron-withdrawing effect exerted by the CF_3 group, and can always be considered a good mimic of a peptidic NH.
- The sp^3 N atom of the trifluoroethylamine function is a poor hydrogen bond acceptor and has very little Lewis basicity, in close analogy with the peptide bond.

- The sp^3 tetrahedral configuration of the trifluoroethylamine unit is conformationally more flexible than the relatively rigid planar amide/peptide group and thus may allow better adaptation to the spatial requirements of a target binding site.

Therefore, the trifluoroethylamine function is an increasingly promising peptide- and amide-bond replacement that is expected to find important applications in medicinal chemistry and drug discovery.

3.4 Trifluoromethyl-Peptidomimetics as Protease Inhibitors

3.4.1 *MMP inhibitors*

Matrix metalloproteinases (MMPs) are a family of highly homologous Zn(II)-endopeptidases which are responsible for the cleavage of many of the constituents of the extracellular matrix (Whittaker *et al.*, 1999; Bode and Huber, 2000; Verma and Hansch, 2007). More than 20 human MMPs are known, among them are the collagenases (such as MMP-1) which can degrade fibrillar collagens that are the major components of bone and cartilage; the gelatinases (MMP-2 and MMP-9) whose main substrates are denatured collagens (gelatins); the stromelysins (such as MMP-3) which have a broad spectrum of matrix components as substrates, except for those of collagenases.

MMPs play a key role in a number of physiological processes, such as degradation of the extracellular matrix and connective tissue remodelling. They are also implicated in a number of pathological processes in humans, such as cancer cell invasion, metastasis (especially MMP-2 and MMP-9), inflammatory and autoimmune diseases, and arthritis (in particular MMP-1). Moreover, several MMPs are overexpressed in various cancers, particularly in early growth and establishment of the tumours. Selective inhibition of MMPs might, therefore, represent an attractive strategy for therapeutic intervention. Unfortunately, most rationally designed MMP inhibitors, such as broad spectrum inhibitors Marimastat and Batimastat, as well as selective inhibitors such as Trocade (MMP-1), have performed poorly in clinical trials (Hu *et al.*, 2007; Jacobsen *et al.*, 2007; Nuti *et al.*, 2007). The failure of these MMP inhibitors has been largely due to toxicity, which in

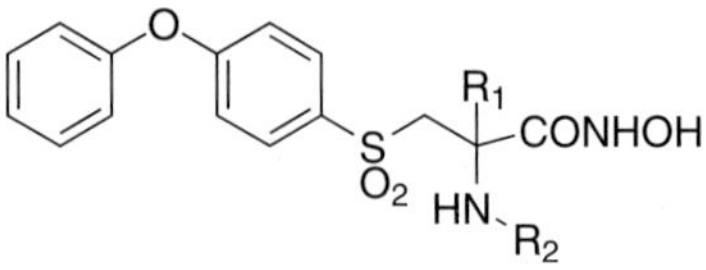

R^1 = H, CH_3, Ph (**A**)
R^1 = CF_3; R^2 = H (**1**)
R^1 = CF_3; R^2 = Me (**2**)
R^1 = CF_3; R^2 = Et (**3**)
R^1 = CF_3; R^2 = Pr (**4**)

Figure 3.10. α-Amino hydroxamic acid MMPs inhibitors.

turn is suspected to arise from insufficient specificity and selectivity (Coussens *et al.*, 2002).

Our group became interested in studying the 'fluorine effect' in selective fluorinated inhibitors of MMPs (Zanda, 2004), developing several nanomolar and selective inhibitors incorporating a trifluoromethyl group as a backbone substituent. In particular, we found that α-trifluoromethyl-α-amino-β-sulfone hydroxamates (see Fig. 3.10) represent an interesting class of MMP inhibitors. The compatibility of an electron-withdrawing CF_3 group in α-position to the hydroxamic function, which is the zinc(II)-binding group, was considered a key point to investigate. The CF_3 was therefore incorporated as R^1 substituent in structures **1–4** (Fig. 3.10) (Sinisi *et al.*, 2005), analogues of molecules **A** which have been previously reported by Becker *et al.*, (2001) as potent inhibitors of MMP-2, MMP-9 and MMP-13.

The synthesis of the racemic hydroxamic acid **1**, having a free quaternary amino group, is summarized in Fig. 3.11. The key step was the addition of a metalated methyl-sulfone to an imine obtained from trifluoropyruvate, and the resulting amino-sulfone was subsequently elaborated into the target hydroxamate **1** in a satisfactory overall yield.

The *N*-alkylated analogues **2–4** (Fig. 3.10) were prepared according to a modified procedure, in reasonable overall yields.

Inhibition tests on racemic compounds **1–4** were performed with MMP-1, MMP-3, and MMP-9.

The primary α-amino hydroxamate **1** was the most potent compound (Table 3.1), but all of the α-CF_3-hydroxamic acids **1–4** are nanomolar inhibitors of MMP-3 and MMP-9. Furthermore, **1** showed excellent selectivity versus MMP-1 (>5000-fold). This demonstrates that a CF_3 group in α-position to the Zn(II) chelating function can be successfully used as a substituent in MMP inhibitors and is very well tolerated by the enzymes.

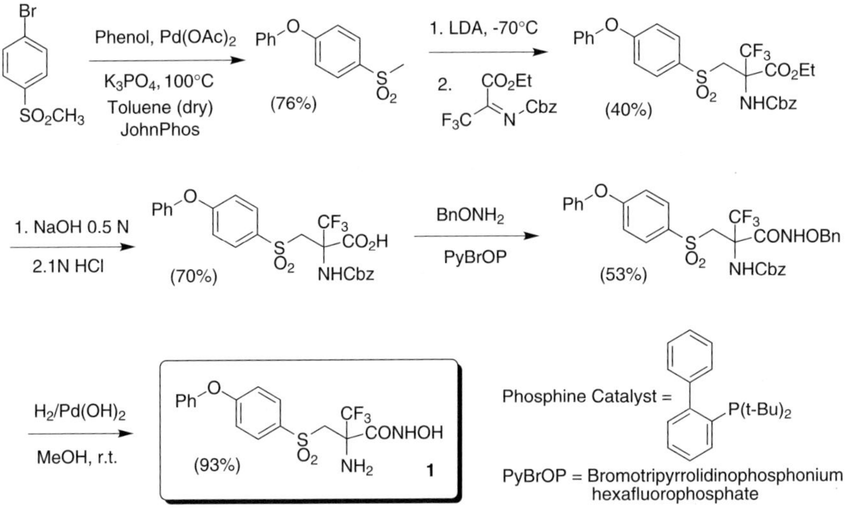

Figure 3.11. Synthesis of 1.

Table 3.1. Effect of compounds 1–4 on different MMPs proteolytic activity. Key: n.a. = not available. (Structures 1–4 in Fig. 3.10)

Compound	IC_{50}/MMP-3 (nM)	IC_{50}/MMP-9 (nM)	IC_{50}/MMP-1 (nM)
1	14	1	>5000
2	32	ca. 20	n.a.
3	28	63	n.a.
4	53	59	n.a.

3.4.2 *β-Fluoroalkyl β-sulfonyl hydroxamates*

We also explored the effect of the unique stereoelectronic features of fluoroalkyl functions in an alternative position of an MMP inhibitor. Thus, the fluoroalkyl group, including CF_3, was installed further away from the zinc(II) binding hydroxamic function (Sani *et al.*, 2005). For this purpose we selected a structurally very simple class of hydroxamate inhibitors bearing an arylsulfone moiety in the β-position, namely **B** (Fig. 3.12), which showed nanomolar inhibitory potency against MMP-2, 3 and 13 (Freskos *et al.*, 1999; Groneberg *et al.*, 1999; Salvino *et al.*, 2000).

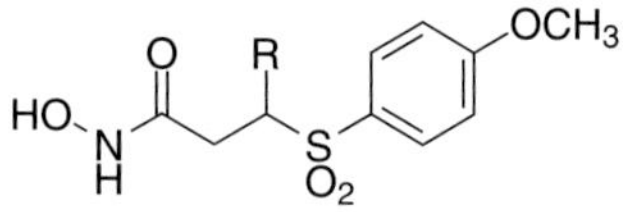

R = H, alkyl, cycloalkyl, etc. **(B)**
R = CF_3 **(5)**
R = CHF_2 **(6)**
R = C_2F_5 **(7)**

Figure 3.12. β-Sulfonyl hydroxamic MMP inhibitors.

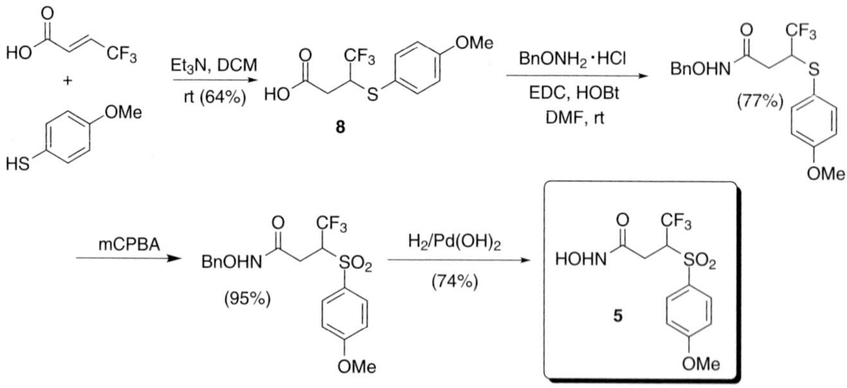

Figure 3.13. Synthesis of 5.

In the reference molecules **B**, the R side chain was critical both for the potency and the selectivity profile of enzyme inhibition. More specifically, large hydrophobic groups R (such as alkyl, cycloalkyl and arylalkyl groups) showed low nanomolar, and even subnanomolar affinity for MMP-2, 3 and 13, but very low affinity for MMP-1. The synthesis of the CF_3-compound **5** was performed as shown in Fig. 3.13. The key reaction was the Michael addition of *p*-thioanisol to trifluorocrotonic acid (Jagodzinska *et al.*, 2007), which took place in a reasonable yield of 64%. The intermediate β-trifluoromethyl-β-thio-carboxylic acid **8** was then elaborated to the target racemic hydroxamic acid **5** in good overall yields.

Analogous reaction sequences from the corresponding fluorocrotonic acids were used to furnish the fluoro-hydroxamic derivatives **6** and **7** (Fig. 3.12).

Table 3.2. IC_{50} values (nM) for the inhibition of different MMPs by β-fluoroalkyl-hydroxamates 5–7.

Substrate	MMP-1	MMP-2	MMP-3	MMP-9
5	4.0×10^3	78	8.0	52
6	1.5×10^4	734	2	6
7	947	32	93	1.7×10^3

The CF_3-compound **5** showed low nanomolar IC_{50} on MMP-3 (Table 3.2). Modest selectivity was observed versus MMP-9 and MMP-2 (about tenfold), but good selectivity versus MMP-1 (about 1000-fold). The pure enantiomers of **5** and the racemic compound showed nearly identical inhibitory profiles. Interestingly, the difluoro compound **6** was even more potent, showing a rather impressive inhibitory activity on both MMP-3 and 9, and much better selectivity versus MMP-2 (>100-fold) and MMP-1 (ca. 10^4-fold). Finally, the C_2F_5-hydroxamate **7** was less selective, displaying a remarkable loss of activity on MMP-9 (1000-fold) and to some extent on MMP-3 (50-fold), but bringing about a higher potency on MMP-2 (200-fold).

These results on fluoroalkyl-hydroxamates **1–4** and **5–7** show that (a) a fluoroalkyl group can be successfully used as a backbone substituent in metalloprotease inhibitors and is very well tolerated by the enzymes, (b) an electron-withdrawing CF_3 group in α-position to the hydroxamic function has little effect on the zinc chelating capacity of the latter.

3.4.3 *Dual ACE/NEP inhibitors*

Angiotensin converting enzyme (ACE) and neutral endopeptidase (NEP) are zinc metallopeptidases located on the outer membrane of various cell types (Skidgel and Erdos, 2004). They have an important role in the metabolism of a number of regulatory peptides of human nervous, cardiovascular, inflammatory, and immune systems. ACE is a dipeptidyl carboxypeptidase which converts angiotensin I into angiotensin II and degrades kinins, while NEP catalyses the degradation of a variety of renal and CNS-active peptides including substance P, bradykinin, enkephalins and atrial natriuretic factor.

Several inhibitors of ACE are known as effective antihypertensive drugs, such as Captopril (Kubo and Cody, 1985 and references cited therein) and its variants (Song and White, 2002). On the basis of these positive results, in a range of cardiovascular diseases, further research started to focus on the potentially beneficial effects of blocking other endogenous enzymes involved in peptide activation or degradation. In various experimental models, combined ACE/NEP inhibition led to more potent and synergistic haemodynamic and renal effects than selective inhibitors of the individual enzymes (Jandeleit-Dahm, 2006).

We became interested in pseudodipeptides and tripeptides designed from the structure of thiorphan, having general formulae HS–CH(R^1)–CH(R^2)–CONH–CH(R^3)–COOH and HS–CH(R^1)–CH(R^2)–CONH–CH(R^3)–CONH–CH(R^4)–COOH, respectively (Inguimbert *et al.*, 2002), interacting with S1′ and S2′ subsites of the enzymes and bearing a thiol group as zinc-chelating moiety. The group in α-position to the thiol moiety occupies the S1 binding subsite, and probably accounts for the subnanomolar inhibitory potencies observed for the ACE enzyme. A hypothetical model of the ACE and NEP active sites is portrayed in Fig. 3.14 (Bohacek *et al.*, 1996).

We therefore designed novel thiol-containing pseudodi- and tri-peptide analogues **8a–e** and **9**, respectively, in which the isobutyl substituent of known inhibitors **C** and **D** is replaced by a fluoroalkyl group (Fig. 3.15).

The fluoroalkyl groups were incorporated in order to modify significantly the physicochemical properties of the inhibitors, such as

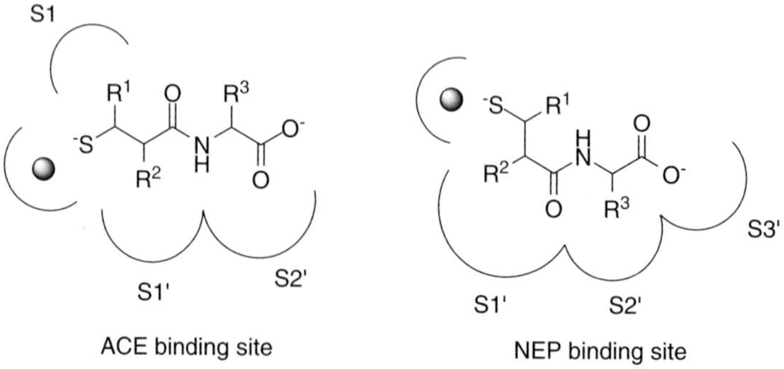

Figure 3.14. The ACE/NEP binding sites.

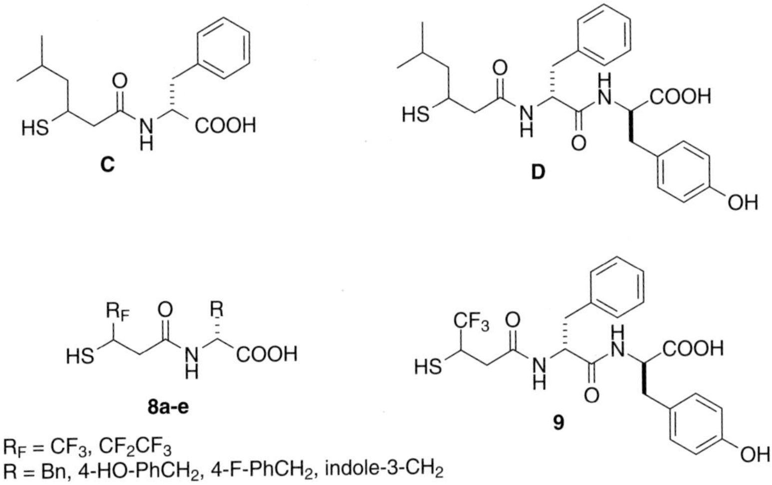

Figure 3.15. Fluorinated analogues of dual ACE/NEP inhibitors.

local hydrophobicity, acidity and nucleophilicity of the neighbouring thiol group, and also the preferred conformation, by inducing conformational constrictions due to the sterically demanding fluoroalkyl groups.

The synthesis of tripeptide inhibitor **9** is shown in Fig. 3.16. The key reaction was a Michael addition of thiolacetic acid to the trifluoromethyl acceptor **10**.

All of the partially fluorinated pseudopeptides above were assayed for their capacity to inhibit ACE and NEP (Table 3.3), showing inhibitory activity in the mid-nanomolar range towards both enzymes. Compound **8b** (see Table 3.3), bearing a tyrosine terminal group, was the most potent, showing an inhibitory activity towards NEP similar to those of the known reference compounds **C** and **D**.

In general, replacement of the isobutyl group with a trifluoromethyl group as the R^1 substituent (see Fig. 3.15) produced a significant increase of the ACE inhibitor activity of the thiol compounds **8** and **9**, as compared with the previously reported data on the non-fluorinated analogues **C** and **D** (Gomez-Monterrey *et al.*, 1993). This effect was also accompanied by a clear-cut reduction of selectivity, because these CF_3-compounds also showed a slight decrease of potency towards NEP. Replacement of

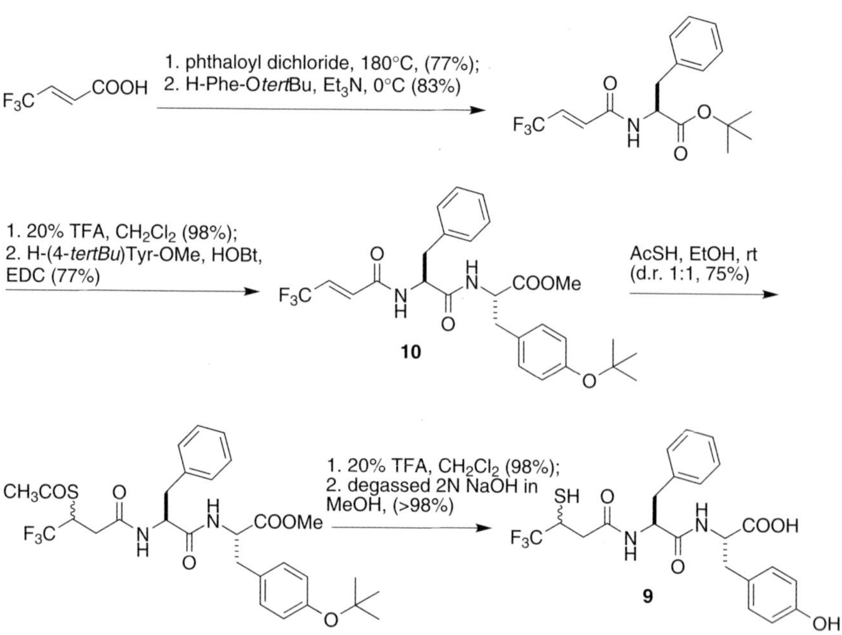

Figure 3.16. Synthesis of pseudo-tripeptide **9**.

trifluoromethyl by a pentafluoroethyl group (as in **8e**) brought about a reduction of the inhibitory capacity towards both NEP and ACE.

These results confirm that a trifluoromethyl group can be used as a replacement of an isobutyl group in protease inhibitors (Binkert *et al.*, 2006), without affecting the affinity for the active site.

3.4.4 *Crystallographic analysis of the role of the CF_3-group in the binding process to enzyme active sites*

An interesting aspect of those trifluoromethyl-peptidomimetics having strong inhibitory potency concerns the actual role of the trifluoromethyl group in the binding to the protease active site, i.e. whether it is directly involved in the binding process, making relevant interactions with the receptor, or has an indirect role, for example contributing to the binding conformation of the inhibitor. This aspect is particularly intriguing, considering the xenobiotic nature of the trifluoromethyl group and its other

Table 3.3. IC_{50} values for compounds **C**, **D**, **8a–e**, and **9**.

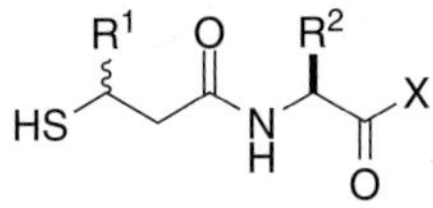

				IC_{50} (nM)	
Compounds	**R¹**	**R²**	**X**	**ACE**	**NEP**
C	isobutyl	benzyl	OH	>10000	41
D	isobutyl	benzyl	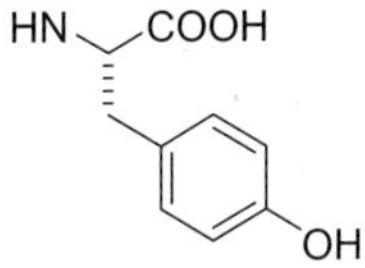	2800	50
8a	CF_3	benzyl	OH	830	300
8b	CF_3	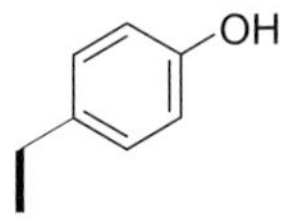	OH	160	80
8c	CF_3	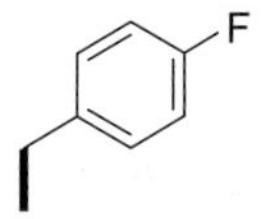	OH	300	230
8d	CF_3	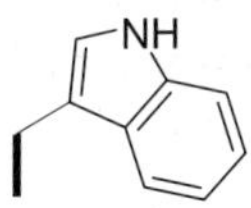	OH	400	580
8e	CF_2CF_3	benzyl	OH	1010	1150
9	CF_3	benzyl	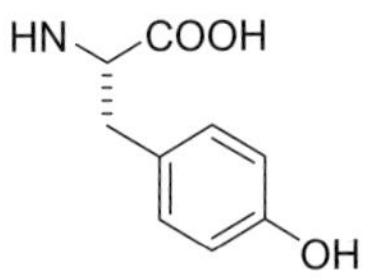	280	430

peculiar features (see Section 3.1 for a more detailed discussion). For example, its rare capacity to participate in hydrogen bonds partially compensated by its proclivity to engage in polar interactions with positively charged residues, and its capacity for hydrophobic contacts.

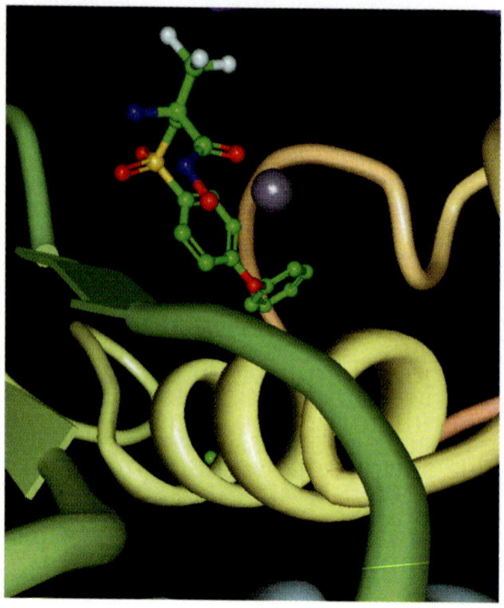

Figure 3.17. (***R***)-1 in the MMP-9 active site. Key: Grey ball = Zn(II), cyan = F, red = O, blue = N, yellow = S, green = C.

One interesting example comes from the X-ray crystallographic structure of the complex of the hydroxamate **1** with the truncated catalytic domain of MMP-9 (2.2 Å resolution, PDB code: 2OW1) (Tochowicz *et al.*, 2007). Although the compound was co-crystallized using the inhibitor in racemic form, the (*R*)-enantiomer of **1** was found to bind preferentially to the enzyme. In this case, the CF_3 group does not make significant interactions with the active site residues of the protease, but is essentially water-exposed (Fig. 3.17). This might seem surprising given the strongly hydrophobic nature of the CF_3, but apparently the overall binding energy of (*R*)-**1** to MMP-9 compensates for the low affinity of the CF_3 group for water.

A similar situation can be found for the trifluoroethylamine compound **11** (Fig. 3.18), a potent and selective inhibitor of the cysteine protease Cathepsin K, structurally related to Odanacatib (Li *et al.* 2006).

In this case, too, the trifluoromethyl group does not make any relevant interaction with the enzyme, as demonstrated by an X-ray structure (2.0 Å resolution, PDB code: 1vsn) of a complex between **11** and Cathepsin K (Fig. 3.19), but rather is directed away from the active site, into water. In this case, the CF_3 group seems to bias the conformation of the entire molecule, which can establish optimal interactions with the

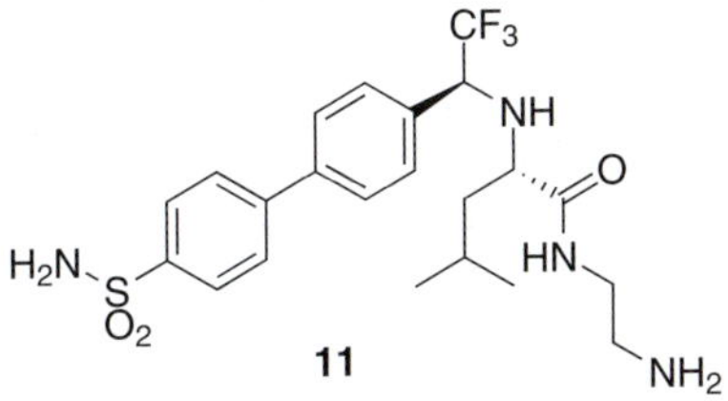

11

Figure 3.18. Merck's trifluoroethylamine compound **11**.

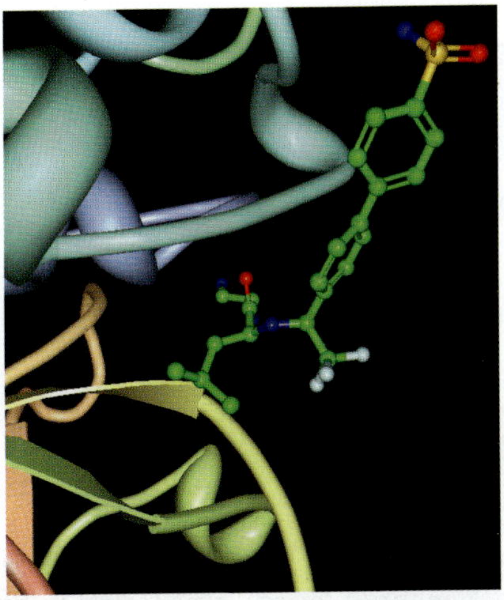

Figure 3.19. Trifluoroethylamine inhibitor **11** complexed with Cathepsin K (cyan = F, red = O, blue = N, green = C).

residues of the active site. Thus, the CF_3 group contributes indirectly to the binding process and to the high affinity of the peptidomimetic inhibitor.

A CF_3 group placed on the backbone of a peptide mimic can also directly interact with the protease active site, as demonstrated in the case of the bis-trifluoromethyl Pepstatin A analogue **12** (Fig. 3.20). Compound **12** was found to be a low nanomolar inhibitor (IC_{50} = 1 nM) of Plasmepsin II, an aspartic protease of *Plasmodium falciparum*, which is the protozoon that causes the most serious forms of malaria, and showed nearly the same potency as Pepstatin A (Binkert *et al.*, 2006).

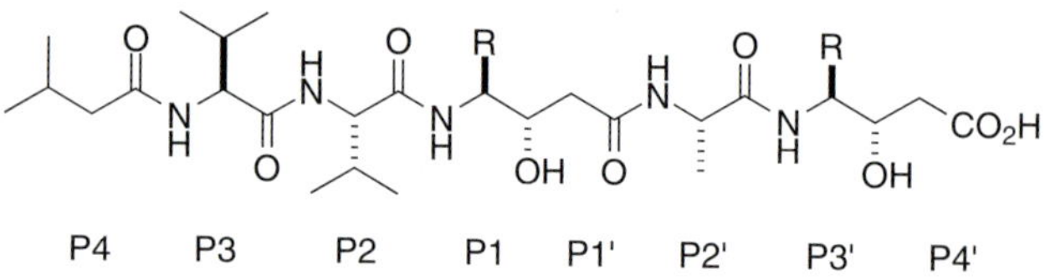

Figure 3.20. Pepstatin A and its bis-trifluoromethyl analogue 12.

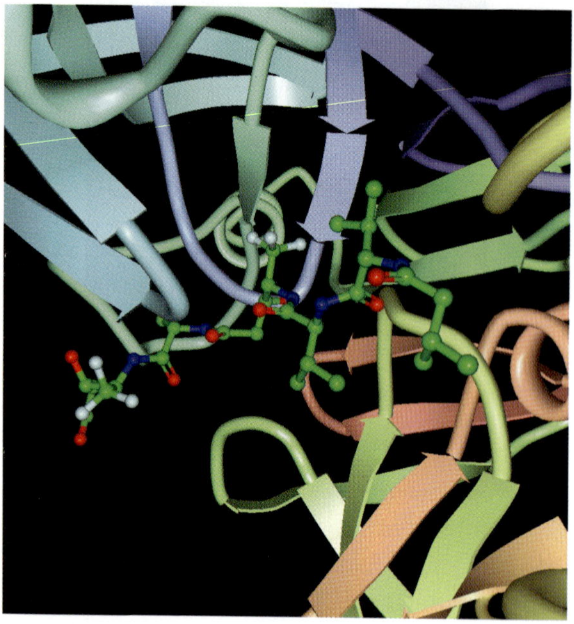

Figure 3.21. Bis-CF_3 pepstatin analogue 12 complexed with Plasmepsin II (cyan = F, red = O, blue = N, green = C).

Crystallographic analysis of bis-CF_3-pepstatin 12 complexed with Plasmepsin II (2.8 Å resolution, PDB code: 1XE6) showed that the CF_3 group occupying the P1 position is well accommodated in the S1 pocket of the active site (Fig. 3.21), and is involved in important hydrophobic interactions. In contrast, the second CF_3 group in P3′ does not make very significant interactions with the enzyme.

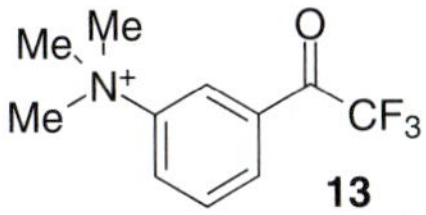

13

Figure 3.22. Structure of the trifluoroketone inhibitor of acetylcholinesterase 13.

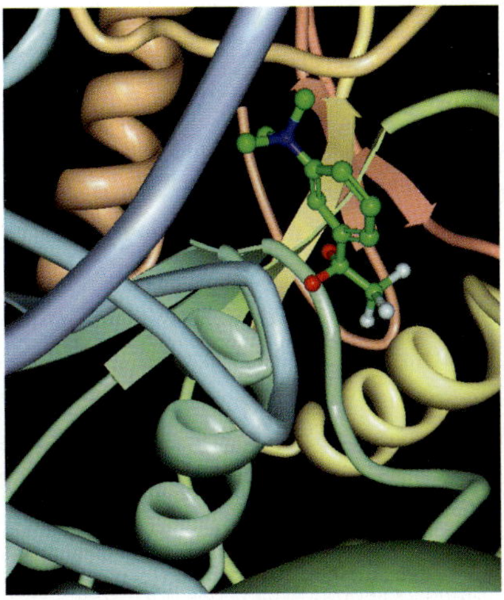

Figure 3.23. Trifluoroacetophenone **13** complexed with acetylcholinesterase (cyan = F, red = O, blue = N, green = C).

Trifluoro-ketone inhibitors are also able to accommodate the CF_3 group in an appropriate binding pocket of the target enzyme, as shown in the case of the trifluoroacetophenone **13** (Fig. 3.22) bound to acetylcholinesterase (Fig. 3.23), a serine esterase (2.8 Å resolution, PDB code: 1amn) (Harel *et al.*, 1996). The carbonyl group is hydrated, and the CF_3 fits very tightly into a concave acyl binding pocket, interacting with the hydrophobic residues (three phenylalanines, one glycine and one tryptophan) forming this pocket.

Overall, the crystallographic data confirms that a CF_3 group as a backbone substituent has a key role in imparting and modulating physicochemical and conformational properties of peptidomimetic inhibitors. In general, these results show that the final outcome of the incorporation of a trifluoromethyl group in peptide mimics is strongly dependent on the whole structure of the inhibitor, and the effect of the trifluoromethyl group on the entire structural conformation is a main factor in determining the biological activity and inhibitory potency of the molecule.

References

Becker, D.P., DeCrescenzo, G.A., Freskos, J. *et al.* (2001) α-Alkyl-α-amino-β-sulphone hydroxamates as potent MMP inhibitors that spare MMP-1. *Bioorg Med Chem Lett* 11, 2723–2725.

Bigotti, S., Volonterio, A. & Zanda, M. (2008a) The influence of fluoroalkyl-group electronegativity on stereocontrol in the synthesis of Ψ[CH(R_F)NH]Gly-peptides. *Synlett* 7, 958–962.

Bigotti, S., Meille, S.V., Volonterio, A. *et al.* (2008b) Synthesis of Ψ[CH(R_F)NH]Gly-peptides: the dramatic effect of a single fluorine atom on the diastereocontrol of the key aza-Michael reaction. *J Fluorine Chem* 129, 767–774.

Binkert, C., Frigerio, M., Jones, A. *et al.* (2006) Replacement of isobutyl by trifluoromethyl in pepstatin a selectively affects inhibition of aspartic proteinases. *ChemBioChem* 7, 181–186.

Black, W.C. & Percival, M.D. (2006) The consequences of lysosomotropism on the design of selective cathepsin K inhibitors. *ChemBioChem* 7, 1525–1535.

Bode, W. & Huber, R. (2000) Structural basis of the endoproteinase-protein inhibitor interaction. *Biochim Biophys Acta* 1477, 241–252.

Bohacek, R., De Lombaert, S., McMartin, C. *et al.* (1996) Three-dimensional models of ACE and NEP inhibitors and their use in the design of potent dual ACE/NEP inhibitors. *J Am Chem Soc* 118, 8231–8249.

Coussens, L.M., Fingleton, B. & Matrisian, L.M. (2002) Matrix metalloproteinase inhibitors and cancer-trials and tribulations. *Science* 295, 2387–2392.

Dhawan, B.N., Cesselin, R.R., Reisine, T. *et al.* (1996) International Union of Pharmacology. XII. Classification of opioid receptors. *Pharmacol Rev* 48, 567–592.

Dunitz, J.D. & Taylor, R. (1997) Organic fluorine hardly ever accepts hydrogen bonds. *Chem Eur J* 3, 89–98.

Fletcher, M.D. & Campbell, M.M. (1998) Partially modified retro-inverso peptides: development, synthesis, and conformational behavior. *Chem Rev* 98, 763–795.

Formicola, L., Maréchal, X., Basse, N. *et al.* (2009) Novel fluorinated pseudopeptides as proteasome inhibitors. *Bioorg Med Chem Lett* 19, 83–86.

Freskos, J.N., Mischke, B.V., DeCrescenzo, G.A. *et al.* (1999) Discovery of a novel series of selective MMP inhibitors: identification of the γ-sulfone-thiols. *Bioorg Med Chem Lett* 9, 943–948.

Gauthier, J.Y., Chauret, N., Cromlish, W. *et al.* (2008) The discovery of odanacatib (MK-0822), a selective inhibitor of cathepsin K. *Bioorg Med Chem Lett* 18, 923–928.

Gelb, M.H., Svaren, J.P. & Abeles, R.H. (1985) Fluoro ketone inhibitors of hydrolytic enzymes. *Biochemistry* 24, 1813–1817.

Gomez-Monterrey, I., Turcaud, S., Lucas, E. *et al.* (1993) Exploration of neutral endopeptidase active site by a series of new thiol-containing inhibitors. *J Med Chem* 36, 87–94.

Groneberg, R.D., Burns, C.J., Morrissette, M.M. *et al.* (1999) Dual inhibition of phosphodiesterase 4 and matrix metalloproteinases by an (arylsulfonyl) hydroxamic acid template. *J Med Chem* 42, 541–544.

Harel, M., Quinn, D.M., Nair, H.K. *et al.* (1996) The x-ray structure of a transition state analog complex reveals the molecular origins of the catalytic power and substrate specificity of acetylcholinesterase. *J Am Chem Soc* 118, 2340–2346.

Hu, J., Van den Steen, P.E., Sang, Q-X.A. *et al.* (2007) Matrix metalloproteinase inhibitors as therapy for inflammatory and vascular diseases. *Nat Rev Drug Discovery* 6, 480–498.

Hughes, J., Smith, T.W., Kosterlitz, H.W. *et al.* (1975) Identification of two related pentapeptides from the brain with potent opiate agonist activity. *Nature* 258, 577–579.

Inguimbert, N., Coric, P., Poras, H. *et al.* (2002) Toward an optimal joint recognition of the *S*-1′ subsites of endothelin converting enzyme-1 (ECE-1), angiotensin converting enzyme (ACE), and neutral endopeptidase (NEP). *J Med Chem* 45, 1477–1486.

Jäckel, C. & Koksch, B. (2005) Fluorine in peptide design and protein engineering. *Eur J Org Chem* 21, 4483–4503.

Jacobsen, F.E., Lewis, J.A. & Cohen, S.M. (2007) The design of inhibitors for medicinally relevant metalloproteins. *ChemMedChem* 2, 152–171.

Jagodzinska, M., Huguenot, F. & Zanda, M. (2007) Studies on a three-step preparation of β-fluoroalkyl acrylates from fluoroacetic esters. *Tetrahedron* 63, 2042–2046.

Jagodzinska, M., Huguenot, F., Candiani, G. *et al.* (2009) Assessing the bioisosterism of the trifluoromethyl group with a protease probe. *ChemMedChem* 4, 49–51.

Jandeleit-Dahm, K.A.M. (2006) Dual ACE/NEP inhibitors — more than playing the ACE card. *J Hum Hypertens* 20, 478–481.

Janecka, A., Staniszewska, R., Gach, K. *et al.* (2008) Enzymatic degradation of endomorphins. *Peptides* 29, 2066–2073.

Kirk, K.L. (2006) Fluorine in medicinal chemistry: recent therapeutic applications of fluorinated small molecules. *J Fluorine Chem* 127 1013–1029.

Kubo, S.H. & Cody, R.J. (1985) Clinical pharmacokinetics of the angiotensin converting enzyme inhibitors. A review. *Clin Pharmacokinet* 10, 377–391.
Leroux, F. (2004) Atropisomerism, biphenyls and fluorine: a comparison of rotational barriers and twist angles. *ChemBioChem* 5, 644–649.
Li, C.S., Deschenes, D., Desmarais, S. *et al.* (2006) Identification of a potent and selective non-basic cathepsin K inhibitor. *Bioorg Med Chem Lett* 16, 1985–1989.
Loffet, A. (2002) Peptides as drugs: is there a market? *J Peptide Science* 8, 1–7.
Ma, J.-A. & Cahard, D. (2008) Update 1 of: Asymmetric fluorination, trifluoromethylation, and perfluoroalkylation reactions. *Chem Rev* 108, PR1–PR43 and references therein.
Mikami, K., Itoh, Y. & Yamanaka, M. (2004) Fluorinated carbonyl and olefinic compounds: basic character and asymmetric catalytic reactions. *Chem Rev* 104, 1–16.
Molteni, M., Volonterio, A. & Zanda, M. (2003) Stereocontrolled synthesis of $\Psi[CH(CF_3)NH]$Gly-peptides. *Org Lett* 5, 3887–3890.
Molteni, M., Bellucci, M.C., Bigotti, S. *et al.* (2009) $\Psi[CH(CF_3)NH]$Gly-peptides: synthesis and conformation analysis. *Org Biomol Chem* 7, 2286–2296.
Müller, K., Faeh, C. & Diederich, F. (2007) Fluorine in pharmaceuticals: looking beyond intuition. *Science* 317, 1881–1886.
Nuti, E., Tuccinardi, T. & Rossello, A. (2007) Matrix metalloproteinase inhibitors: new challenges in the era of post broad-spectrum inhibitors. *Curr Pharm Des* 13, 2087–2100.
O'Shea, P.D., Chen, C-Y., Gauvreau, D. *et al.* (2009) A practical enantioselective - synthesis of odanacatib, a potent cathepsin K inhibitor, via triflate displacement of an alpha-trifluoromethylbenzyl triflate. *J Org Chem* 74, 1605–1610.
Regnier, T., Sarma, D., Hidaka, K. *et al.* (2009) New developments for the design, synthesis and biological evaluation of potent SARS-CoV $3CL^{pro}$ inhibitors. *Bioorg Med Chem Lett* 19, 2722–2727.
Salvino, J.M., Mathew, R., Kiesow, T. *et al.* (2000) Solid-phase synthesis of an arylsulfone hydroxamate library. *Bioorg Med Chem Lett* 10, 1637–1640.
Sani, M., Bravo, P., Volonterio, A. *et al.* (2002) Parallel solid-phase synthesis of partially modified retro and retro-inverso $\Psi[NHCH(CF_3)]$Gly-peptides. *Collect Czech Chem Commun* 67, 1305–1319.
Sani, M., Candiani, G., Pecker, F. *et al.* (2005) Novel highly potent, structurally simple γ-trifluoromethyl γ-sulfone hydroxamate inhibitor of stromelysin-1 (MMP-3). *Tetrahedron Lett* 46, 2393–2396.

Sani, M., Sinisi, R. & Viani, F. (2006) Peptidyl fluoro-ketones as proteolytic enzyme inhibitors. *Curr Top Med Chem* 6, 1545–1566.

Sani, M., Volonterio, A. & Zanda, M. (2007) The trifluoroethylamine function as peptide bond replacement. *ChemMedChem* 2, 1693–1700.

Shao, Y.-M., Yang, W.-B., Kuo, T.-H. *et al.* (2008) Design, synthesis, and evaluation of trifluoromethyl ketones as inhibitors of SARS-CoV 3CL protease. *Bioorg Med Chem* 16, 4652–4660.

Sinisi, R., Sani, M., Candiani, G. *et al.* (2005) Synthesis of α-trifluoromethyl-α-amino-β-sulfone hydroxamates: novel nanomolar inhibitors of matrix metalloproteinases. *Tetrahedron Lett* 46, 6515–6518.

Sinisi, R., Ghilardi, A., Ruiu, S. *et al.* (2009) Synthesis and in vitro evaluation of trifluoroethylamine analogues of enkephalines. *ChemMedChem* 4, 1416–1420.

Skidgel, R.A. & Erdos, E.G. (2004) Angiotensin converting enzyme (ACE) and neprilysin hydrolyze neuropeptides: a brief history, the beginning and follow-ups to early studies. *Peptides* 25, 521–525.

Smart, B.E. (1994) 'Characteristics of C–F Systems' in Banks, R.E., Smart, B.E., Tatlow, J.C. (Eds), *Organofluorine Chemistry*, Plenum, New York, pp. 57–88.

Song, J.C. & White, C.M. (2002) Clinical pharmacokinetics and selective pharmacodynamics of new angiotensin converting enzyme inhibitors — an update. *Clin Pharmacokinet* 41, 207–224.

Sydnes, M.O., Hayashi, Y., Sharma, V.K. *et al.* (2006) Synthesis of glutamic acid and glutamine peptides possessing a trifluoromethyl ketone group as SARS-CoV 3CL protease inhibitors. *Tetrahedron* 62, 8601–8609.

Tochowicz, A., Maskos, K., Huber, R. *et al.* (2007) Crystal structures of MMP-9 complexes with five inhibitors: contribution of the flexible Arg424 side chain to selectivity. *J Mol Biol* 371, 989–1006. Corrigendum: (2009) *J Mol Biol* 388, 917–918.

Verma, R.P. & Hansch, C. (2007) Matrix metalloproteinases (MMPs): chemical–biological functions and (Q)SARs. *Bioorg Med Chem* 15, 2223–2268.

Volonterio, A., Bravo, P., Moussier, N. *et al.* (2000a) Solid phase synthesis of partially-modified retro- and retro-inverso Ψ[NHCH(CF_3)]-peptides. *Tetrahedron Lett* 41, 6517–6521.

Volonterio, A., Bravo, P. & Zanda, M. (2000b) Synthesis of partially-modified retro and retroinverso Ψ[NHCH(CF_3)]-peptides. *Org Lett* 2, 1827–1830.

Volonterio, A., Bravo, P. & Zanda, M. (2001) Solution/solid phase synthesis of partially-modified retro-Ψ[NHCH(CF_3)]-peptidyl hydroxamates. *Tetrahedron Lett* 42, 3141–3144.

Volonterio, A., Bellosta, S., Bravo, P. *et al.* (2002) Solution/solid-phase synthesis of partially modified retro- and retro-inverso Ψ[NHCH(CF_3)]-peptidyl hydroxamates and their evaluation as MMP-9 inhibitors. *Eur J Org Chem* 3, 428–438.

Volonterio A., Bellosta S., Bravin F. *et al.* (2003) Synthesis, structure and conformation of partially-modified retro and retroinverso Ψ[NHCH(CF_3)]Gly peptides. *Chem Eur J* 9, 4510–4522.

Whittaker, M., Floyd, C.D., Brown, P. *et al.* (1999) Design and therapeutic application of matrix metalloproteinase inhibitors. *Chem Rev* 99, 2735–2776.

Yin, Z., Patel, S.J., Wang, W.-L. *et al.* (2006) Peptide inhibitors of Dengue virus NS3 protease. Part 1: Warhead. *Bioorg Med Chem Lett* 16, 36–39.

Zanda, M. (2004) Trifluoromethyl group: an effective xenobiotic function for peptide backbone modification. *New J Chem* 28, 1401–1411.

4

Trifluoromethyl-Substituted α-Amino Acids as Solid-State ^{19}F NMR Labels for Structural Studies of Membrane-Bound Peptides

Vladimir S. Kubyshkin,[†] *Igor V. Komarov,*[*,‡,§,¶] *Sergii Afonin,*[||] *Pavel K. Mykhailiuk,*[‡,¶] *Stephan L. Grage*[||] *and Anne S. Ulrich*[*,†,||]

4.1 Introduction

Among the plethora of native polypeptides, membrane proteins and membrane-active peptides represent one of the greatest challenges for structural biology. Despite the fact that they comprise about 30% of the

* Correspondence to: IVK: ik214@yahoo.com and ASU: anne.ulrich@kit.edu.
† Karlsruhe Institute of Technology, DFG-Center for Functional Nanostructures (CFN) and Institute of Organic Chemistry, Fritz-Haber-Weg 6, 76131 Karlsruhe, Germany.
‡ Department of Chemistry, Kyiv National Taras Shevchenko University, Vul. Volodymyrska 64, 01033 Kyiv, Ukraine.
§ Institute of High Technologies, Kyiv National Taras Shevchenko University, Vul. Volodymyrska 60, 01601, Kyiv, Ukraine.
¶ Enamine Ltd., Vul. O. Matrosova 23, 01103 Kyiv, Ukraine.
|| Karlsruhe Institute of Technology, Institute of Biological Interfaces (IBG-2), POB 3640, 76021 Karlsruhe, Germany.

human genome, the number of 3D-structures that have been determined of membrane-associated proteins is still orders of magnitude lower than for soluble proteins (Raman *et al.*, 2006). The difficulties in structure analysis, for example in the case of G-protein-coupled receptors, are related to both the production and handling of the material as well as to the limited applicability of conventional methods for structure analysis (Lagerström and Schiöth, 2008; Scheerer *et al.*, 2008; Kolb *et al.*, 2009; Rosenbaum *et al.*, 2009). Yet, there is a great demand for efficient ways to gain insight into the structures of membrane-associated peptides and proteins (Arora and Tamm, 2001; Torres *et al.*, 2003; Lacapère *et al.*, 2007). Large transmembrane receptors constitute the most important targets for modern drug and molecular therapies (Drews, 2000; Lundstrom, 2006), and small membrane-active peptides are highly promising candidates for antimicrobial therapy and intracellular drug delivery. Of special interest in the latter group are numerous antibiotic, cell-penetrating and fusogenic peptides (Epand, 2003; Sang and Blecha, 2008; Heitz *et al.*, 2009). For both groups, detailed knowledge of their structures in the native membrane-associated state is essential for rational drug design, and — more fundamentally — to understand the architectural principles and reveal functional mechanisms. In the case of complex membrane proteins, a strategy often pursued is to focus on shorter peptide segments such as transmembrane helices or peripheral amphipathic loops, for example, in the hope of gaining insight into the structure and function of the parent protein. In such an approach size is not a problem, but other difficulties arise instead — all being related to the impact of molecular motions. Dynamic interconversions, which are increasingly being described for short peptides, are evidently relevant to the membrane activities of naturally occurring and artificially designed peptide sequences. Such peptides can act specifically at the cell membranes as hormones, toxins, ionophores, or as modulators of the lipid bilayer curvature or fluidity. Understanding the mechanism of interaction with the lipid bilayer is therefore a real challenge and has been a major focus of membrane biophysics over the past two decades. Research into the effects of solid-state ^{19}F NMR on oriented membranes has been a particularly sensitive and powerful tool in revealing the structural and dynamic behaviour of such peptides (Salgado *et al.*, 2001; Sachse, 2003; Afonin *et al.*, 2004, 2007, 2008a,b; Glaser *et al.*, 2004, 2005; Kanithasen, 2005; Ulrich, 2005, 2007;

Grasnick, 2006; Strandberg *et al.*, 2006, 2008; Ulrich *et al.*, 2006; Maisch, 2008; Wadhwani *et al.*, 2008; Ehni, 2009; Maisch *et al.*, 2009; Grage *et al.*, 2010). Here, this strategy will be reviewed with particular emphasis on new designs for ^{19}F-labelled amino acids.

4.2 Solid-State NMR for Structure Analysis of Membrane-Associated Polypeptides

Membrane-bound peptides and proteins often impose multiple difficulties for structural studies using conventional techniques, such as NMR in solution or single-crystal diffraction, for various reasons: (i) these peptides often possess a high tendency to aggregate — many of them are hydrophobic and are therefore difficult to handle in aqueous solution; (ii) since their structure and function strongly depend on the native membrane environment, the presence of a model bilayer or at least a membrane-mimetic environment is essential for proper folding; (iii) due to their small size and sensitivity to their local environment, many membrane-interacting peptides are endowed with a high degree of conformational plasticity. Additionally, they often have a tendency to oligomerize functionally or to aggregate non-productively, which imposes further challenge for functionally relevant reconstitution and structure analysis.

Several methodological approaches aimed at gaining structural information from membrane-associated polypeptides have been developed in recent years. These include advanced X-ray and electron diffraction techniques, electron crystallography, liquid-state NMR; as well as circular dichroism- and infrared-based strategies, various surface analysis techniques such as AFM or TEM, and modern *in silico* methods (Wiener, 2004; Arkin, 2006; Miles and Wallace, 2006; Raman *et al.*, 2006; Schmidt-Krey, 2007; Carpenter *et al.*, 2008; Cowieson *et al.*, 2008; Engel and Gaub, 2008; Fujiyoshi and Unwin, 2008; Lindahl and Sansom, 2008; Wang, 2008). A powerful technique for detailed studies of peptides in a membrane environment is solid-state NMR. This method is one of the most informative due to its ability to determine polypeptide structures with quasi-atomic resolution, its general non-invasive nature, and — most importantly — its ability to study peptides in genuine lipid membranes, i.e. very close to their native state (Nielsen *et al.*, 2004; Ulrich, 2005; Gong *et al.*, 2007; Naito and Kawamura, 2007).

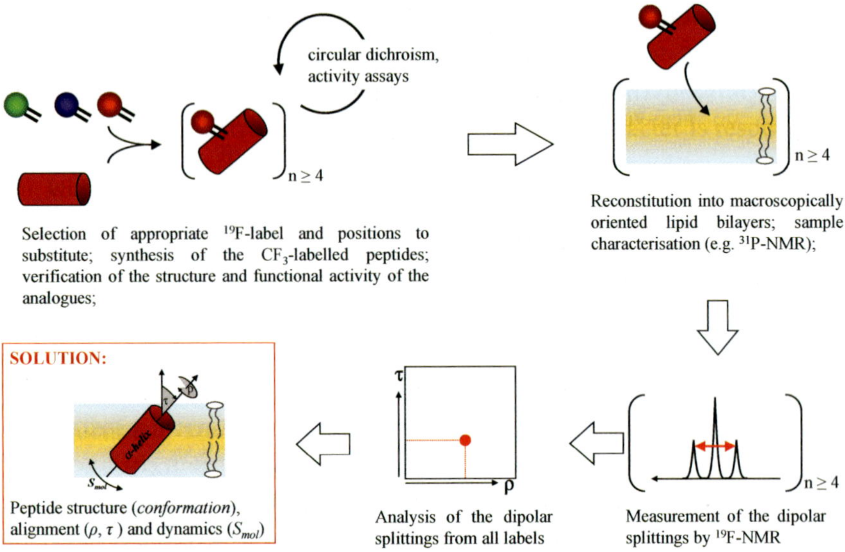

Figure 4.1. General strategy for structure analysis of membrane-active polypeptides in oriented lipid bilayers by solid-state ^{19}F NMR (see text for details).

The general strategy for determining the structure, orientation and mobility of a peptide in the lipid bilayer using solid-state NMR in oriented membrane samples consists of several steps (Fig. 4.1 outlines the approach for the case of solid-state ^{19}F NMR using CF_3-substituted amino acids as labels; see Section 4.3). First, a series of differently labelled analogues of the peptide under study is produced by chemical or recombinant methods. These analogues typically contain one reporter group that is selectively labelled by an NMR-active isotope (^{2}H, ^{13}C, ^{15}N, ^{19}F), most of which are characterized by low natural abundance. Next, those analogues that do not compromise the conformation and biological activity of the parent peptide have to be identified using qualitative structure analysis (e.g. by circular dichroism) and functional assays (this step is critical in the case of ^{19}F-labelled peptides). Then the selected labelled peptides are reconstituted, one at a time, in macroscopically oriented model membranes, usually on glass, mica or polymer slides utilizing the property of lamellar lipid bilayers to align spontaneously on a solid support. Mechanically aligned bilayers in which all peptides are uniformly oriented with respect to the sample normal are obtained in this way. Lipid composition, concentration, temperature, and other factors may have to be varied to ensure proper reconstitution of the peptide in the closest possible state to the native situation (Grage *et al.*, 2010). Typically two sets of solid-state NMR spectra are then measured for each sample. The first one to check the state of the lipids and verify a proper bilayer alignment (e.g. a ^{31}P NMR spectrum of the phospholipid headgroups, or alternatively ^{2}H or ^{13}C NMR spectra of the other moieties in the lipid). The second, truly informative set of spectra is measured on the nuclei of the corresponding label in the peptide molecule. The parameters typically obtained from these NMR data — chemical shifts and dipolar (or quadrupolar) couplings — are anisotropic, i.e. they intimately depend on the local orientation of the labelled reporter group with respect to the magnetic field. It is due to this anisotropy that solid-state NMR spectra contain information about the alignment of the molecular segment to which the NMR label has been attached. As the final step of the analysis, a structural model for the peptide is assumed (e.g. a regular α-helix, β-strand, 3_{10}-helix, π-helix, etc., in a rigid-body approximation) and a systematic grid search through all possible orientations in space is performed. By comparing the calculated segmental spectral parameters at each peptide orientation with the ones determined experimentally, the statistical goodness-of-fit (χ^2 distribution or root mean square deviation) is evaluated. Once a self-consistent solution is found, the postulated structure model is confirmed. Otherwise another structure is assumed and the analysis is repeated. Clearly, the more peptide analogues are used in the analysis, the more reliable is the output. The final result is then reported as the 3D-structure of the peptide backbone (i.e. the appropriate conformational model), its alignment in the lipid bilayer (in terms of a tilt angle τ and an azimuthal rotation angle ρ), and its mobility (in terms of uniaxial rotation and rigid-body wobbling, as characterized by an order parameter S_{mol} or more advanced fluctuation models).

The above strategy dictates several obligatory requirements on the NMR label, which will be outlined in the following sections. First, the reporter group bearing the isotope label must be rigidly connected to the peptide backbone in a geometrically well-defined fashion or be part of it (*criterion 1*). This criterion is mandatory to be able to translate the measured orientation of the reporter into the orientation of the whole peptide segment.

4.3 Choice of the CF_3 Group as a Label for ^{19}F NMR of Peptides in Membranes

The general NMR strategy outlined in the previous section can utilize peptides labelled by various isotopes: ^2H, ^{13}C, ^{15}N, or ^{19}F. Both ^{15}N and ^{13}C can be incorporated non-perturbingly in the peptide backbone simply by using the corresponding isotope labelled natural amino acid. Likewise, methyl-deuterated Ala-d_3 is perfectly suited for ^2H NMR analysis. In all these cases, the labels satisfy criterion 1; however, they have several drawbacks. Apart from the high cost of selectively ^{13}C- and ^{15}N-labelled amino acids (an important factor when peptide production on the 100 milligram scale is concerned), all three nuclei possess relatively low gyromagnetic ratios (γ) and thus very low NMR sensitivities (especially ^{15}N). Analysis of ^{13}C NMR spectra of a labelled peptide in a membrane additionally suffers from the overlap with the natural abundance ^{13}C signals from the lipids. Similar complications occur in the case of ^2H labelling when peptides are to be observed at low peptide/lipid ratios. An excellent compromise is offered by ^{19}F-labelling of peptides, since fluorine is absent in natural proteins and lipids and there is no natural background in the ^{19}F NMR spectra of the labelled peptides.

Without going into much detail (various aspects of biological ^{19}F NMR applications have been reviewed before: Gerig, 2004; Lanza *et al.*, 2005; Yu *et al.*, 2005; Malet-Martino *et al.*, 2006; Ulrich *et al.*, 2006; Murphy, 2007; Ulrich, 2007; Cobb and Murphy, 2009), in this section we will present the ^{19}F properties relevant to the solid-state NMR studies of membrane-associated peptides. Since the spin of the ^{19}F nucleus is ½, it gives rise to simple NMR spectra — only chemical shifts and dipolar interactions determine the lineshapes. The gyromagnetic ratio of the ^{19}F nucleus is high ($\gamma^{19F}/\gamma^{1H} = 0.94$) and the natural abundance of the ^{19}F isotope is 100%. These two factors ensure an exquisitely high sensitivity of ^{19}F NMR, which is an obvious advantage. The strong dipolar couplings of ^{19}F can provide intra- or intermolecular distances of up to 12 Å in solids and large chemical shift ranges (ca. 500 ppm) with high spectral resolution. The latter fact also reflects another peculiarity of the ^{19}F nucleus: its chemical shift is highly sensitive to the local environment of the label. The solvent-, concentration-, pH- and temperature-dependent variations in chemical shift can reach several ppm, which makes ^{19}F a good sensor of the surrounding conditions. However, this high environmental sensitivity makes accurate chemical shift referencing quite complicated (Ulrich *et al.*, 2003), an essential operation when analysing the chemical shift anisotropy (CSA) in a macroscopically oriented sample. Therefore, the CSA is not an optimal parameter to determine the orientation of a ^{19}F-labelled peptide in a membrane. On the other hand, dipolar ^{19}F–^{19}F couplings are absolute values that need no referencing; they depend only on distance but not on the environment and hence can be readily utilized to report anisotropic information. The strong intragroup dipolar couplings between the three magnetically equivalent nuclei of a (rapidly rotating) CF_3 group are ideally suited for NMR analysis. This fact, together with the three-fold intensity of the corresponding ^{19}F NMR signal, makes the CF_3 group an optimal reporter group for structure analysis (*criterion 2*).

Incorporation of a CF_3 reporter (Smits *et al.*, 2008) into a peptide can be achieved by replacing certain proteinogenic amino acids with their CF_3-substituted analogues. However, while non-perturbing ^{13}C and ^{15}N labels (or substitution of Ala with Ala-d_3) do not affect the properties of a peptide, when replacing a methyl group by a non-natural CF_3 reporter, the conformational and biological behaviour can be severely altered. These perturbations can occur due to several factors: the first is steric, as the volume of a CF_3 group lays in between the tert-butyl and methyl groups (Leroux, 2004, Müller *et al.*, 2007, Jagodzinska *et al.*, 2009; Chapters 2 and 4 of this book). It is, therefore, safest to use a CF_3 group as a mimic of relatively bulky hydrocarbon residues, i.e. ethyl groups. The second factor that must be taken into account when designing ^{19}F-labelled α-amino acids is the electronegativity of fluorine, as the electron-withdrawing effect of a CF_3 group is close to that of an oxygen atom (Huheey, 1965). The CF_3 group has even been used as an oxygen mimic, for example, when incorporated into a peptide backbone (Zanda, 2004). As a consequence, the acidity of the neighbouring hydrogen atoms is greatly increased (Smart, 1995). A CF_3 group will also increase the acidity and hydrogen-donating ability of neighbouring carboxylic groups and respectively decrease the basicity and nucleophilicity of adjacent amino groups (Abraham *et al.*, 1990; Koppel *et al.*, 1994; Castejon and Wiberg, 1998). This may explain the fact that all known ^{19}F-labels are used to mimic amino acids with non-polar side chains. Finally, there is also evidence of the CF_3 group participating in the formation of hydrogen bonds (Koksch *et al.*, 1997; Jäckel and Koksch, 2005). Although fluorine is a poor acceptor of hydrogen bonds (Caminati *et al.*, 2006), its effect could be significant in an intramolecular context (Banks *et al.*, 1994; Smart, 2001).

The considerable risk of potential steric and electronic perturbation of a peptide imposed by CF_3 labelling defines two further criteria to the ^{19}F-label when choosing or designing a useful CF_3-substituted amino acid. Most importantly (*criterion 3*), the conformation and chemical properties of the labelled peptide should ideally be identical to those of the native wild-type peptide. Practically, this means that for each ^{19}F-labelled peptide preservation of its global conformation and its functional intactness has to be verified. The last, pragmatic *criterion 4* for ^{19}F-labelling requires that incorporation of the label in the polypeptide chain should be easily carried out by conventional chemical or biochemical methods. This means that no racemization, decomposition or formation of other by-products should occur during the synthesis of the labelled peptides.

4.4 Suitable CF_3-Labelled Amino Acids for ^{19}F NMR Analysis

Although the repertoire of known CF_3-substituted amino acids is nowadays very large (Kukhar and Soloshonok, 1995), only few compounds have been used as labels in structural solid-state ^{19}F NMR studies, namely, (*S*) and (*R*)-trifluoromethylalanine **1** (TfmAla), (2*S*)-4′-trifluoromethylphenylglycine 2 (4-TfmPhg), and (2*S*)-3′-(trifluoromethyl)bicyclopent[1.1.1]-1′-ylglycine 3 (TfmBpg). This is not surprising, as the criteria for an amino acid to be a suitable ^{19}F NMR label are very strict and they rule out most of the CF_3-substituted amino acids synthesized to date.

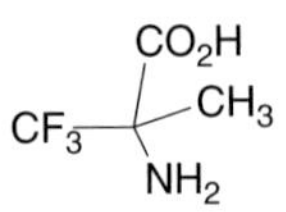

1

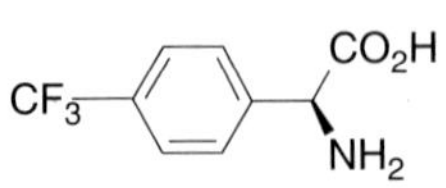

2

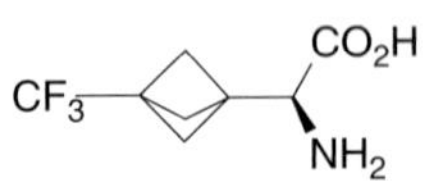

3

Many of the potential candidates for ^{19}F-labelling, for which the syntheses are well developed and documented in the literature, fail to satisfy some or all of the criteria 1–4 outlined above. For instance, let us consider the amino acids with an α-CF_3 group (α-TfmXaa). As mentioned above, the CF_3 group may be regarded as a structural analogue of a small aliphatic group, like Me, Et, *i*-Pr or *i*-Bu (Zanda, 2004). From such a perspective, the simplest representative of the series — trifluoroalanine (trifluoromethylglycine, TfmGly) — which fulfils criteria 1 and 2, might be a suitable ^{19}F-label to substitute Ala, Val, Leu or Ile in a given peptide. However, TfmGly has never been used in experimental practice, since it fails to satisfy criterion 4. Due to increased acidity of the α-proton, TfmGly and its derivatives decompose at pH ≥ 7 (Burger *et al.*, 1993; Smits *et al.*, 2008); the basic conditions usually employed in standard solid-phase peptide synthesis (SPPS) cannot be used because of the elimination of free hydrogen fluoride. The other known α-TfmXaa also have to be excluded from the list of potential ^{19}F-labels, because the α-CF_3 group in them causes severe conformational alterations of the peptide backbone. It is known that the steric effect is not linearly proportional to the size of an added moiety — it abruptly increases when both α-substituents become larger than a methyl (Toniolo and Benedetti, 1991). The only exception among α-TfmXaa is TfmAla. From the point of view of size, this amino acid may still be an appropriate substituent of the sterically demanding and β-branched proteinogenic amino acids (Val, Leu, Ile, Thr, Met, non-protonated His). Additionally, it can be placed in positions where the intrinsic conformational propensity of the TfmAla is not so critical by substituting its natural (non-proteinogenic) analogue Aib in cyclic sequences, or at the termini of a linear peptides. However, fulfilment of criterion 3 has to be carefully proven experimentally in every case.

As can be seen from the molecular structures **1–3**, fulfilment of criterion 1 is achieved here in two different ways: by attaching the CF_3 reporter directly to the peptide backbone by the α-substitution (**1**), or by inserting a rigid spacer between the CF_3 group and the backbone (compounds **2** and **3**). In the following sections, we will discuss these classes of ^{19}F-labels separately, due to certain differences in their synthesis, incorporation in peptides and applications.

4.5 α-CF_3-Substituted Amino Acids: TfmAla

4.5.1 *Synthesis*

The amino acid TfmAla represents a family of α-substituted quaternary α-amino acids, the synthesis of which — especially stereoselective — is usually a challenge (Konno *et al.*, 2005). The very first racemic synthesis of

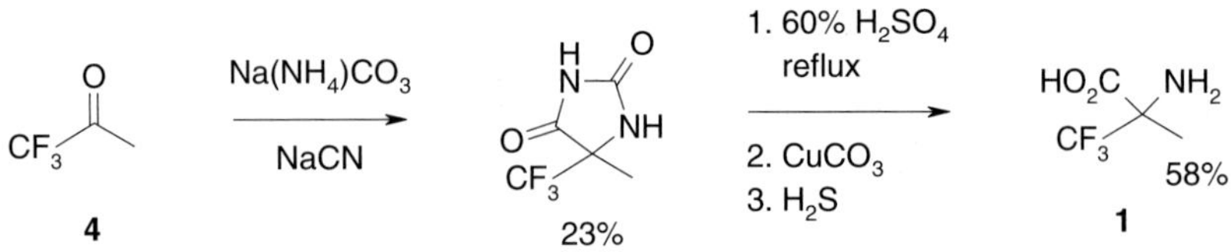

Scheme 4.1. Bucherer–Berg synthesis of racemic TfmAla.

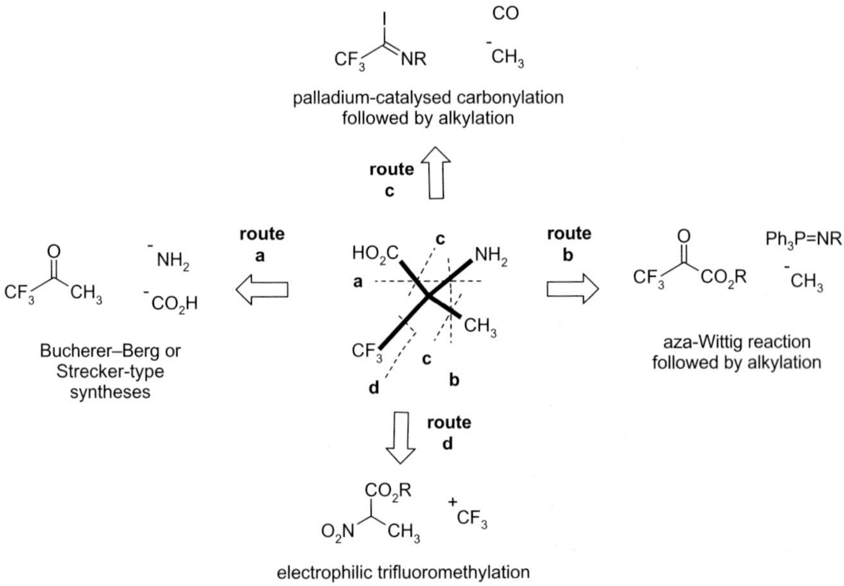

Figure 4.2. Retrosynthetic analysis of TfmAla, showing the disconnections (dashed lines) and corresponding synthons/synthetic equivalents.

this amino acid was reported in 1953 (Lontz and Raasch, 1953; Scheme 4.1). However, decades elapsed before a reliable enantioselective synthesis was published (Bravo *et al.*, 1994). Extensive research in this area, stimulated by the biological application of TfmAla and its derivatives, resulted in a number of new approaches, illustrated by the retrosynthetic disconnections in Fig. 4.2.

All the initial approaches were elaborated in non-stereoselective version and went along the routes **a**–**c**. Ten years after the initial procedure based on the Bucherer–Berg reaction (Scheme 4.1) was patented, the

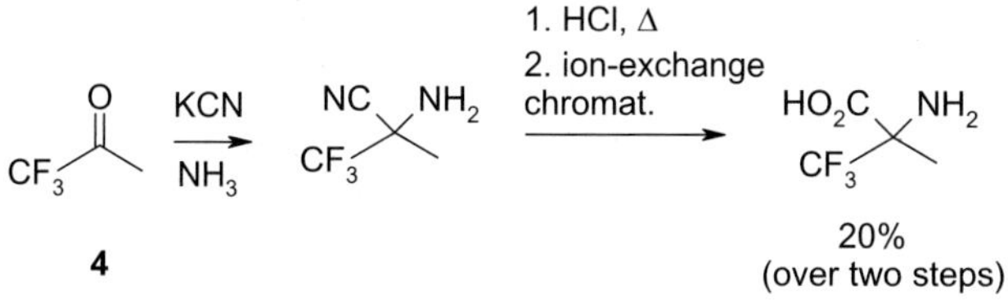

Scheme 4.2. Strecker synthesis of racemic TfmAla.

Strecker synthesis of TfmAla was reported (Christensen and Oxender, 1963, Scheme 4.2). Both syntheses were technically difficult due to the volatility of the starting compound, trifluoroacetone (**4**). In addition, the yields were far from being ideal.

Therefore, it is not surprising that other approaches towards TfmAla were explored. Success was first achieved in synthesis along route **b** (Fig. 4.2). It is based on the use of an imine, prepared from trifluoropyruvate (**5**), which is sufficiently reactive to be alkylated by organometallic reagents. It should be noted, however, that in contrast to imines prepared from alkyl/aryl-substituted aldehydes and ketones that had previously been used for the synthesis of a wide variety of amino acids, the trifluoromethyl-substituted imines were notoriously difficult to obtain from CF_3-substituted aldehydes or ketones and amines due to instability of the corresponding hem-amino alcohols (Fokin *et al.*, 1984; Fustero *et al.*, 2009). Other reactions leading to the desired imines proved to be more efficient; one of them was the aza-Wittig (Staudinger) reaction (Soloshonok *et al.*, 1987, Scheme 4.3). Methyl-trifluoropyruvate **5** was reacted with *N*-methoxycarbonyltriphenylphosphazene to give the aza-Wittig reaction product **6** in high yield. The C=N double bond in **6** was reactive enough to form **7** upon treatment with dimethylcadmium (reactions with methylmagnesium halides also proceeded similarly, but at low temperature, –78°C; the reaction with organocadmium

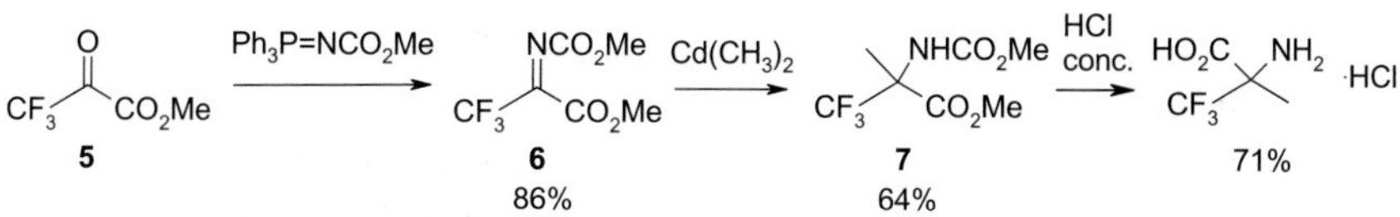

Scheme 4.3. Synthesis of racemic TfmAla based on the aza-Wittig (Staudinger) reaction.

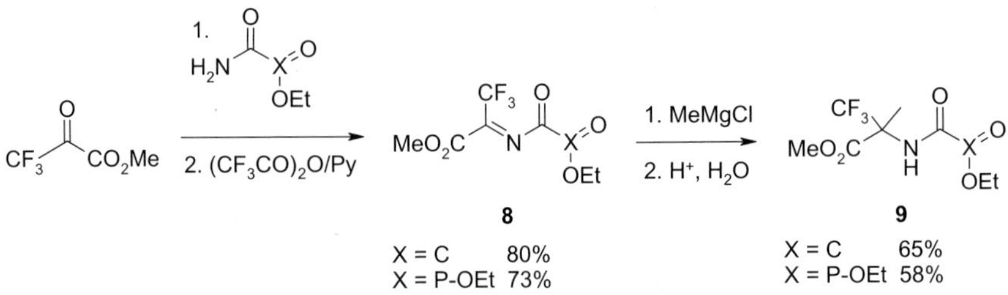

Scheme 4.4. Synthesis of imines, TfmAla precursors, and their alkylation.

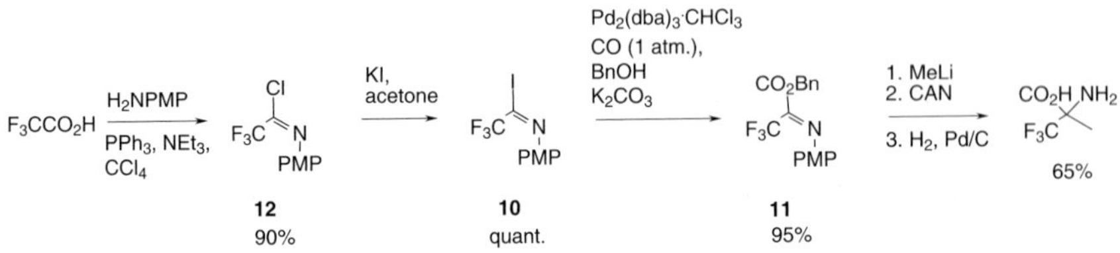

dba - dibenzylideneacetone
CAN - cerium ammonium nitrate
PMP - *p*-methoxyphenyl

Scheme 4.5. Synthesis of racemic TfmAla through palladium-catalysed imidoyliodide carbonylation.

reagent was carried out at 4°C). The intermediate **7** was hydrolysed to TfmAla hydrochloride in high yield.

An alternative route to electrophilic imines (**8**), precursors to TfmAla, was recently reported (Skarpos *et al.*, 2006). The imines **8** were alkylated using MeMgCl; however, the TfmAla derivatives **9** were not converted into the free amino acid (Scheme 4.4).

Another approach to racemic TfmAla is outlined schematically in Fig. 4.2 as route **c**. Carbonylation of trifluoroacetimidoyl iodide **10** provided imine **11**, which was reacted with MeLi to give the protected TfmAla. Deprotection was performed in two steps, using a CAN oxidation and Pd-catalysed hydrogenolysis (Scheme 4.5, Watanabe *et al.*, 1992).

Finally, route **d** (Fig. 4.2) was realized only recently, with the discovery of efficient electrophilic trifluoromethylating reagents. The

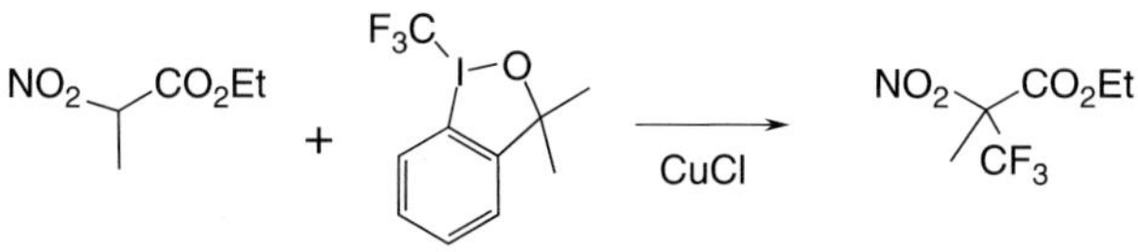

13 **14** **15**

31%

Scheme 4.6. Trifluoromethylation of an α-nitroester leading to a TfmAla precursor.

trifluoromethylation of α-nitroester **13** was achieved with the Togni reagent **14** (Kieltsch *et al.*, 2007), the preparation of which was described in the same paper (Scheme 4.6). Although the isolated yield of the TfmAla precursor **15** was low, this reaction is the first published example of an electrophilic trifluoromethylation of α-nitroesters.

The challenging problem of stereoselective synthesis of TfmAla was first solved in 1994 (Bravo *et al.*, 1994). This might be explained by the same reasons that caused the difficulties in synthesis of racemic TfmAla above, namely, by the electronic and steric effects of the CF_3 group. Classical methods for the synthesis of homochiral amino acids that had been previously developed were not successful in the case of TfmAla. For example, stereoselective alkylation of 2,5-diketopiperazine analogues of trifluoropyruvate turned out to be inapplicable for the synthesis of TfmAla (Sewald *et al.*, 1994). There was an obvious need for a new methodology, and it was developed in 1994, based on a stereoselective variant of route **b** (Fig. 4.2), where the sulfoxide functional group was used as a stereocontrolling moiety (Scheme 4.7). The anion, generated from **16**, commercially available in both enantiomeric forms, reacted with the trifluoro-*N*-benzyloxycarbonylimino derivative **17** to give **18** in moderate yield as an approximately 1:1 diastereomeric mixture. The diastereomers could then be separated by column chromatography. This allowed the synthesis of both enantiomers of TfmAla in high optical purity, after deoxygenation, desulfurization with concomitant hydrogenolysis by Raney-Ni, and hydrolysis.

A modification of the above methodology was developed later (Bravo *et al.*, 1998; Asensio *et al.*, 2001). The same chiral auxiliary, a chiral sulfoxide, was chosen. In this case, the sulfoxide group was present in the

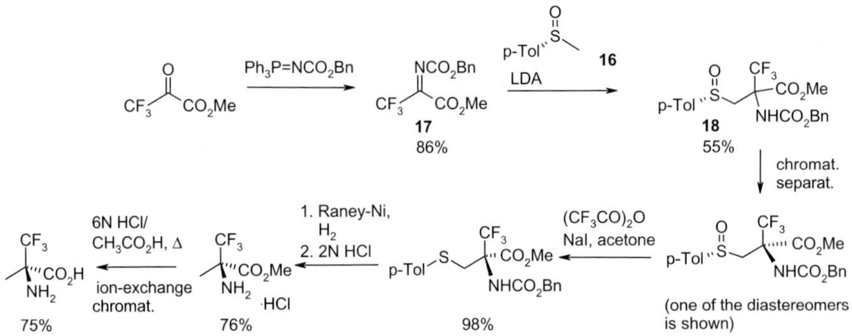

Scheme 4.7. First stereoselective synthesis of optically pure TfmAla.

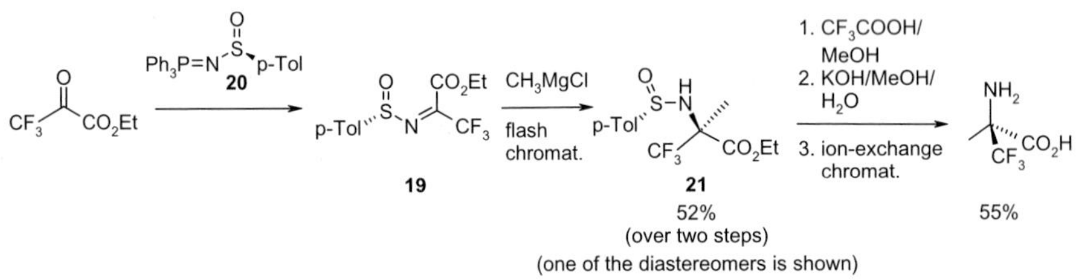

Scheme 4.8. Synthesis of homochiral TfmAla using an optically pure Staudinger reagent.

Staudinger reagent, rather than in the alkylating agent (Scheme 4.8). The chiral sulfinimine **19** was prepared by an aza-Wittig reaction of ethyl trifluoropyruvate with the optically pure Staudinger reagent **20**. The addition of methylmagnesium chloride to sulfinimine **19** proceeded with low diastereoselectivity. However, the diastereomers of **21** could easily be separated by flash chromatography and both could then be converted into the optically pure TfmAla by desulfurization, hydrolysis, and purification using ion-exchange column chromatography.

Finally, the stereoselective Strecker-type syntheses of TfmAla were sucessfully implemented. In one of them (Huguenot and Brigaud, 2006, Scheme 4.9), the oxazolidine **22** bearing a chiral auxiliary served as a precursor of the imine which was formed by treatment of **22** with a Lewis acid. The Strecker-type reaction proceeded with low diastereoselectivity, the *dr* for the aminonitrile **23** was 54:46. Nevertheless, the aminonitriles could easily be separated by column chromatography. Both diastereomers

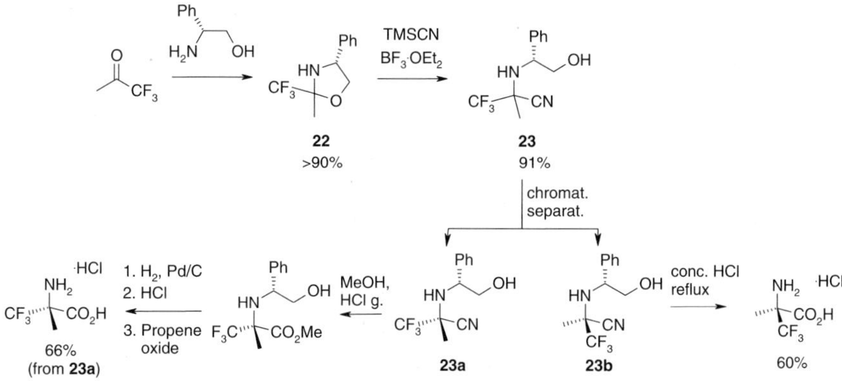

Scheme 4.9. Strecker-type synthesis of optically pure enantiomers of TfmAla.

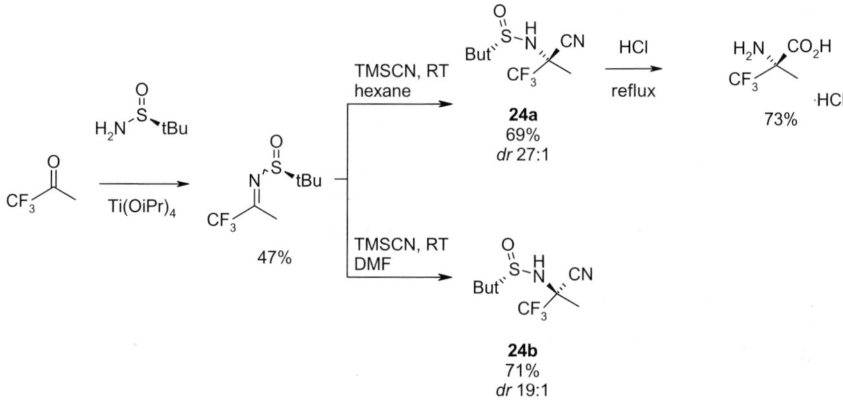

Scheme 4.10. A diastereoselective solvent-controlled Strecker-type reaction leading to non-racemic TfmAla.

23a and **23b** were transformed into their corresponding enantiomers of TfmAla hydrochlorides using different synthetic sequences.

Another Strecker-type synthesis, again employing a chiral sulfoxide as the chiral auxiliary and leading to enantiomerically enriched (*S*)-TfmAla, was based on a diastereoselective reaction rather than on separation of diastereomeric derivatives (Wang *et al.*, 2006). Remarkably, the diastereoselectivity of the Strecker reaction here was controlled by the solvent — different diastereomers **24a** and **24b** were formed in hexane and DMF respectively (Scheme 4.10).

4.5.2 *Separation of the TfmAla enantiomers*

A racemic mixture of TfmAla can be directly separated by chiral HPLC, either by using columns with chiral stationary phases or by ligand-exchange chromatography on achiral stationary phases with eluents containing chiral co-additives (Galushko, 1995). Alternatively, the stereospecificity of natural enzymes can be employed (Miyazawa, 1995). For instance, kinetic resolution by enzymatic hydrolysis of racemic *N*-trifluoroacetyl-TfmAla was reported in 1986 (Keller and Hamilton, 1986). Only the (*R*)-isomer of the derivative was hydrolysed in the presence of hog kidney aminoacylase to give (*R*)-(+)-TfmAla (40% yield); the other enantiomer was obtained in 17% yield by hydrolysis of the remaining amide by refluxing in 2M HCl. The racemic TfmAla amide was later resolved using amidase from *Mycobacterium neoaurum* (Koksch *et al.*, 2004). The (*R*)-amino acid was obtained with high enantioselectivity. Heat-stable amidase from *Klebsiella oxytoca* was also successfully used for the chiral resolution of racemic TfmAla amide; it also showed a preference to catalyse hydrolysis of the (*R*)-enantiomer (Shaw and Naughton, 2004). Notably, since TfmAla is a non-natural amino acid, enzymatic reactions may not be fully selective; thus, for complete separation, chromatography should be preferred.

Peptide synthesis is usually performed with pure enantiomers of amino acids. However, in selective ^{19}F-labelling strategies using only a single non-natural amino acid, the optical purity of the latter is not so critical. The peptide epimers that arise from the use of a racemic amino acid mixture can often be separated chromatographically. Various diastereomers containing TfmAla elute with sufficiently different retention times under carefully optimized flash-chromatography conditions, as shown in Schemes 4.8–4.10. The epimers of di- or tripeptides containing TfmAla are no exception in this respect, and their separation is usually possible. Assignment of the obtained epimers can be performed *post factum* by NMR, chiral HPLC, or by non-chiral HPLC of the diastereomeric derivatives of the amino acids obtained after peptide hydrolysis of the separated peptide epimers (see e.g. Afonin *et al.*, 2003; Glaser *et al.*, 2004; Maisch *et al.*, 2009).

4.5.3 *Incorporation of TfmAla into peptides*

Chemical methods of incorporation of α-trifluoromethylated amino acids into peptides (reviewed by Sewald and Burger 1995; Sewald *et al.* 1995) were often explored in parallel with the amino acid synthesis. Incorporation of α-trifluoromethylated amino acids possessing a quaternary α-carbon atom, including TfmAla, in peptides was accompanied by a serious problem: the α-CF_3 group deactivates the adjacent amino group towards electrophiles which, together with the steric hindrance, severely decreases the coupling rates in solid-phase peptide synthesis (SPPS) (Kobzev *et al.*, 1989). A practical solution to this problem appears to be liquid-phase peptide synthesis (LPPS) under carefully optimized conditions (for protocols see Koksch *et al.*, 1997). A representative example is shown in the Scheme 4.11. Here various urethane-protected TfmAla were transformed into the activated oxazolidinones **25** or oxazolidinediones **26**, which subsequently reacted with amino acid esters or peptide esters (Schierlinger and Burger, 1992; Burger *et al.*, 1993; Hollweck and Burger; 1995) providing the corresponding protected peptides in moderate yields. The same LPPS strategy can be used to attach the TfmAla residue to the C-terminus of a peptide (Burger *et al.*, 1998).

Several original approaches towards peptides containing TfmAla residues are based on the idea of creating the residue on a preformed peptide precursor. One example of this kind, which utilized acylimine intermediate 27 as the precursor, is shown in Scheme 4.12 (Höss *et al.*, 1993).

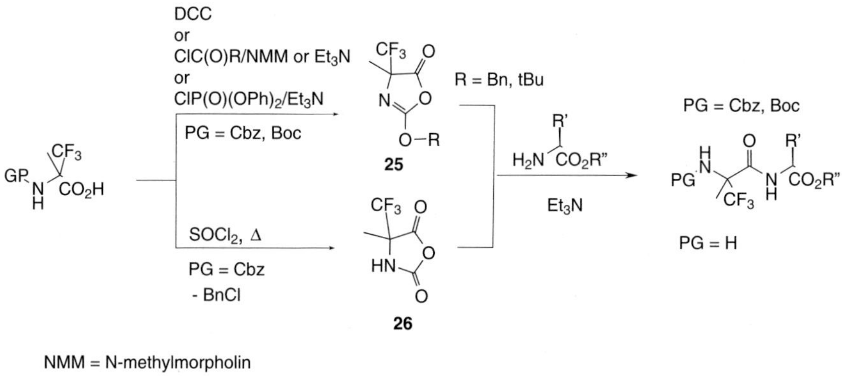

Scheme 4.11. Introduction of urethane-protected TfmAla residue into the N-terminus of peptides as a basic step in LPPS.

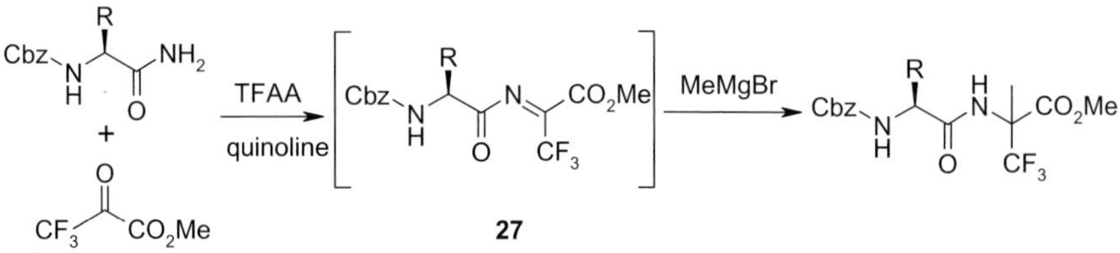

Scheme 4.12. Incorporation of TfmAla residue at the C-terminus of peptides.

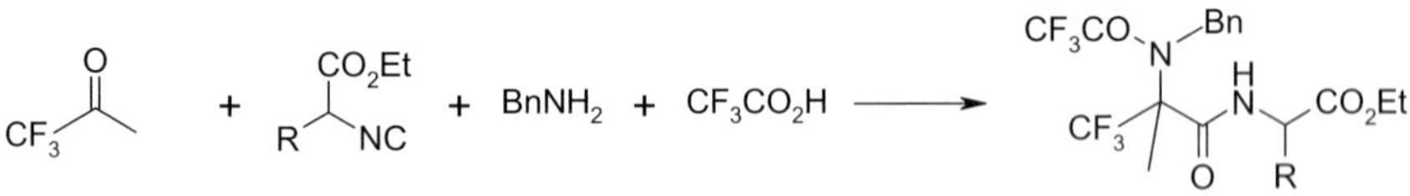

Scheme 4.13. Synthesis of dipeptides containing TfmAla by Ugi-multicomponent reaction.

Ugi-multicomponent synthesis of TfmAla-containing dipeptides (Scheme 4.13) was recently published by Gulevich *et al.* (2008). This method allowed the synthesis of protected dipeptides with the TfmAla residue at the N-terminus in medium yields of 48–56%.

Recent developments in the field of microwave-assisted solid-phase chemistry and commercialization of the corresponding equipment opened the way to yet another methodology for incorporating amino acids with sterically hindered or inactive carboxylic groups into peptides (Sabatino and Papini, 2008; Santagada *et al.*, 2009). Details of the microwave-assisted synthesis of Aib-containing peptides were recently published (Hjørringgaard *et al.*, 2009). However, the use of this approach for the synthesis of TfmAla-containing peptides is yet to be tested.

Enzymatic synthesis of peptides with an N-terminally positioned TfmAla should be briefly mentioned here (see review by Jäckel and Koksch, 2005). With the 4-guanidinophenol ester of racemic TfmAla as a substrate mimic for trypsin and α-chymotrypsin, the N-terminal incorporation of TfmAla in peptides was demonstrated (Thust and Koksch, 2003). Under optimized conditions, when the equilibrium was shifted towards peptide bond formation, yields of up to 72% were reported. Later, the same authors reported coupling of Cbz-TfmAla-OMe and Cbz-Phe-TfmAla-OMe to Leu, Val, Ala, Gly catalysed by carboxypeptidase Y in

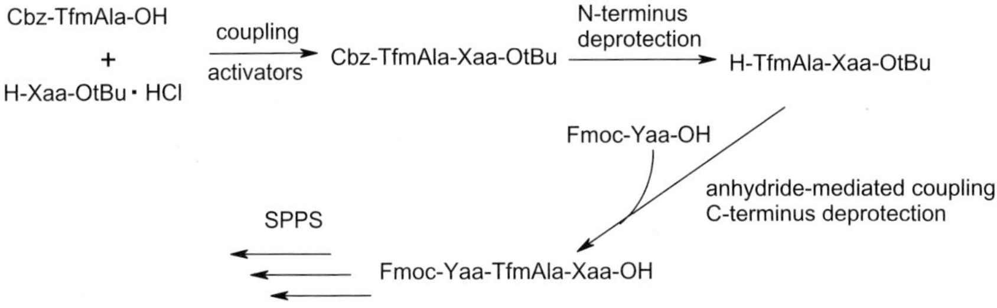

Scheme 4.14. LPPS construction of a tripeptide fragment (containing TfmAla), to be used as a building block in SPPS.

20–75% yields. Strong substrate preferences were demonstrated; the configuration of the TfmAla was suggested to be one of the factors influencing the reaction output (Thust and Koksch, 2004).

Despite the synthetic advances described above, introduction of TfmAla into long peptide sequences is achieved most economically by a combination of LPPS and SPPS (Scheme 4.14). For example, Fmoc-protected di- and tripeptides containing TfmAla can be synthesized by LPPS. Subsequently, the whole fragment can be coupled as a single building block to a polypeptide chain using standard SPPS. In this way, TfmAla was incorporated in the peptaibols harzianin HB I (10-mer) and alamethicin (20-mer) (Fanghänel, 2007; Maisch, 2008; Maisch *et al.*, 2009).

4.5.4 *^{19}F NMR structure analysis of peptides with TfmAla*

The fungal peptaibol alamethicin (F30/3, Ac-Aib-Pro-Aib-Ala-Aib-Ala-Gln-Aib-Val-Aib-Gly-Leu-Aib-Pro-Val-Aib-Aib-Glu-Gln-Pheol), which contains numerous native Aib residues, was the first example where TfmAla was applied for ^{19}F NMR structure analysis (Maisch *et al.*, 2009). According to the strategy described above (Fig. 4.1), six TfmAla-substituted analogues were produced by replacing Aib in positions 5, 10, or 16 with (*R*-) and (*S*-) TfmAla. Circular dichroism spectra and functional tests (antimicrobial and haemolytic assays) confirmed that these single substitutions were essentially non-perturbing. Solid-state ^{19}F NMR analysis demonstrated that alamethicin has an upright transmembrane alignment in liquid-crystalline lipid bilayers. The N-terminus was shown

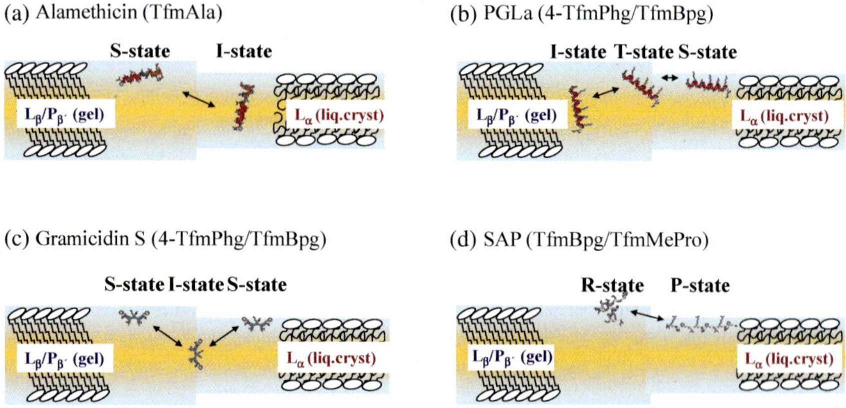

Figure 4.3. Representative examples where CF_3-substituted amino acids have been used as ^{19}F-labels in membrane-active peptides for solid-state ^{19}F NMR structure analysis, revealing their conformation and alignment in the lipid bilayer. The different peptide states are displayed as they are found below and above the lipid phase transition temperature, maintaining the proper molecular dimensions. (a) The peptaibol alamethicin has a helix–kink–helix structure, and it changes its alignment between a surface-bound S-state and an oligomeric inserted I-state. (b) The antimicrobial peptide PGLa possesses a regular α-helix, which realigns sequentially from an inserted I-state via a tilted T-state to an S-state with increasing temperature, when lipid goes from the gel to the liquid-crystalline phase. (c) The antimicrobial peptide gramicidin S is a cyclic β-sheet, which assumes an S-state on the bilayer surface in both the gel and liquid-crystalline phase, but it oligomerises and inserts into the membrane as an I-state when bilayer defects are present. (d) The cell-penetrating peptide SAP undergoes a transition from an isotropically averaged random coil (R-state) to a surfacially aligned polyproline II helix (P-state) upon the lipid phase transition.

to fold in an α-helical conformation (rather than a 3_{10}-helix), and the proline-induced kink was preserved in the membrane-inserted state (Fig. 4.3(a)).

It should be noted that the recommended exchange of Aib by TfmAla applies only under rather unique circumstances, as Aib is a rare amino acid in nature. Its α-disubstituted character strongly promotes α-helical and 3_{10}-helical secondary structures in peptides (Prasad *et al.*, 1984) due to the steric effects of the geminal methyl groups on the backbone. The steric and electronic impact of the ^{19}F-containing TfmAla is expected to be even stronger (see above).

4.6 Amino Acids with a Rigid Spacer Between Cα and the CF_3 Group: 4-TfmPhg and TfmBpg

4.6.1 *Synthesis of 4-TfmPhg*

Synthetic approaches towards 4-TfmPhg are analogous to those of most other phenylglycines carrying an electron-withdrawing substituent (for a review see Williams and Hendrix, 1992). Numerous syntheses published in the literature can be classified according to the reactions used as the key step: amination, aminohydroxylation, arylation, carboxylation, imine reduction or amino- and amido-carbonylation. We describe below the reported examples of 4-TfmPhg and note that many synthetic routes described for phenylglycines with electron withdrawing substituents other than the CF_3 group may well be applicable to the synthesis of 4-TfmPhg.

Insertion of an NH_2 group into the corresponding precursors is a widely used reaction towards arylglycines; it was exploited several times for the synthesis of 4-TfmPhg. Enantioselective approaches are especially valuable for the synthesis of an amino acid that is to be further incorporated into a peptide. For example, reaction of tert-butylcarbamate or benzylcarbamate with 4-CF_3-phenyldiazoacetate **28**, catalysed by the chiral ferrocenyl complex bpy*, gave the corresponding tert-butyl esters of Boc- and Cbz-protected 4-TfmPhg in good yields and excellent enantioselectivity (Lee and Fu, 2007, Scheme 4.15).

Another enantioselective approach was used for amination of the silylketene acetal **29**. The reaction was catalysed by a chiral rhodium complex Rh_2(*S*-TCPTTL)$_4$. The 4-TfmPhg derivative **30** was obtained in 95% yield and 97% *ee.* (Tanaka *et al.*, 2007, Scheme 4.16).

Asymmetric catalytic aminohydroxylation of styrenes has been developed as a method to synthesize different phenylglycines by Reddy and Sharpless (1998). This reaction was performed with 1-(trifluoromethyl)-4-vinylbenzene **31** using $K_2OsO_2(OH)_4$ as an oxidizing reagent, and the Cbz-chloroamine salt formed *in situ* as an amine source. Two different enantiomers of the aminoalcohol **32b** were obtained using different cinchona alkaloid catalysts: the (*S*)-isomer was obtained in the presence of $(DHQ)_2$PHAL, and the (*R*)-isomer in the presence of $(DHQD)_2$PHAL. Another regioisomer, the aminoalcohol **32a**, was also formed in both cases. The mixture of **32a** and **32b** was oxidized using sodium hypochlorite

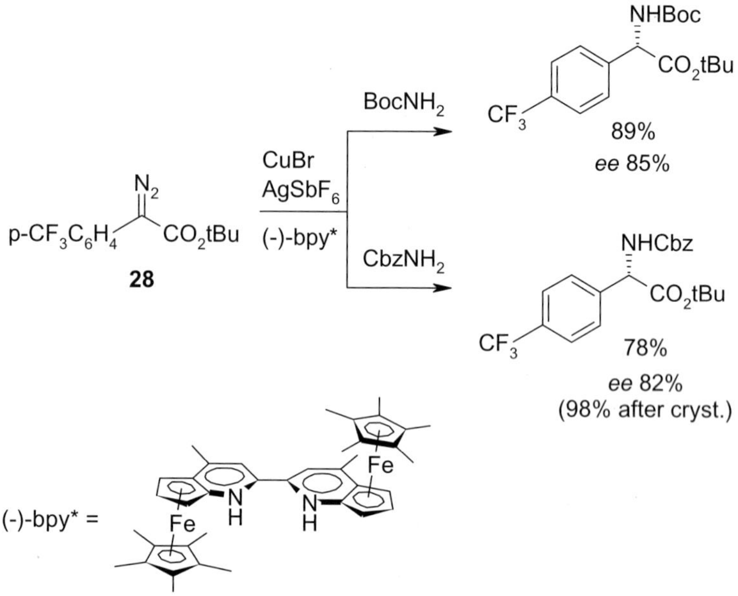

Scheme 4.15. Enantioselective synthesis of protected 4-TfmPhg via amination of a catalytically generated carbene.

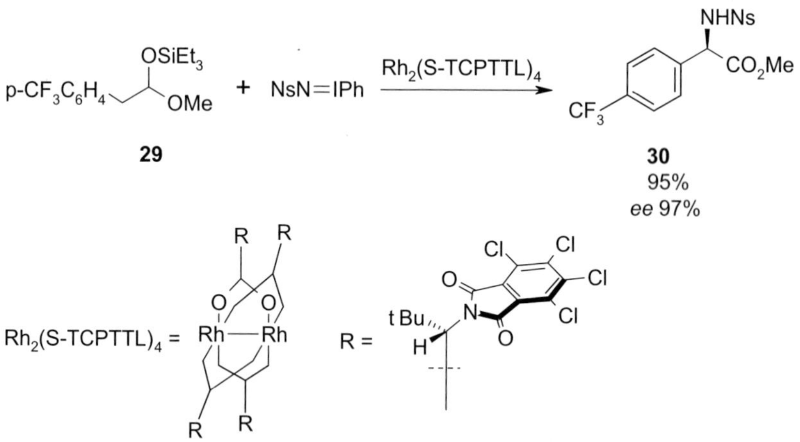

Scheme 4.16. Catalytic enantioselective synthesis of protected 4-TfmPhg using amination of a silylketene.

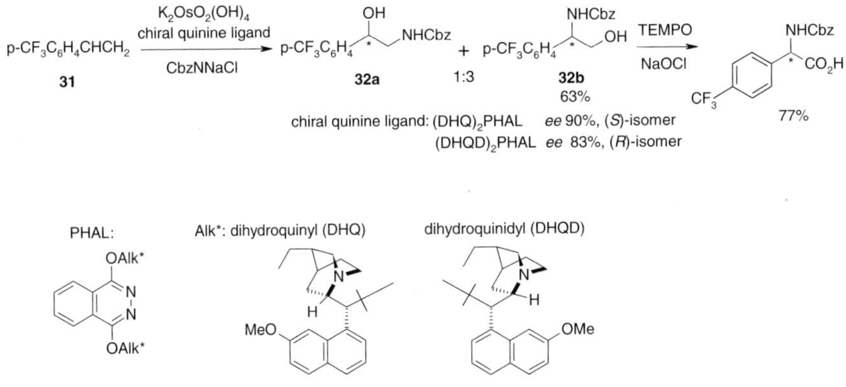

Scheme 4.17. Stereoselective synthesis of protected 4-TfmPhg starting with catalytic aminohydroxylation of the 1-(trifluoromethyl)-4-vinylbenzene (the asterisk here and in the following schemes denotes that stereoisomers with both (*R*)- and (*S*)-configurations at the corresponding centre were obtained).

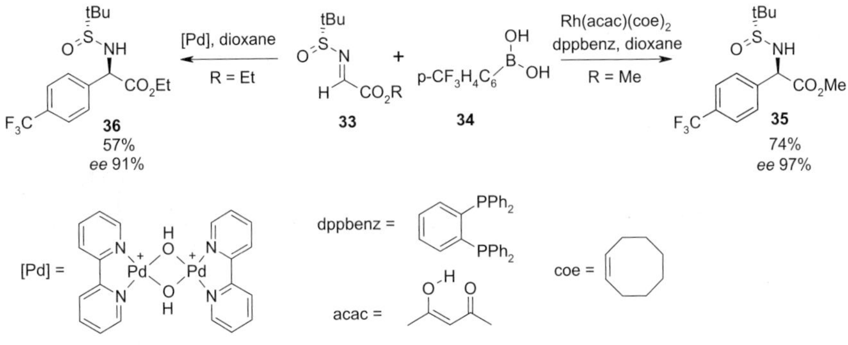

Scheme 4.18. Arylation of an azomethyne leading to protected 4-TfmPhg.

(catalysed by TEMPO) to give the Cbz-protected 4-TfmPhg in 77% yield (calculated on pure regioisomer **32b**) (Scheme 4.17).

Two methods of catalytic arylation of a precursor bearing a chiral auxiliary **33** by arylboronic acid **34** are shown in Scheme 4.18. One of the synthetic routes used a rhodium-based complex with 1,2-bis(diphenylphosphinyl)benzene as the ligand; the 4-TfmPhg derivative **35** was obtained in 74% yield and 97% *ee* (Beenen *et al.*, 2006). Another route used a

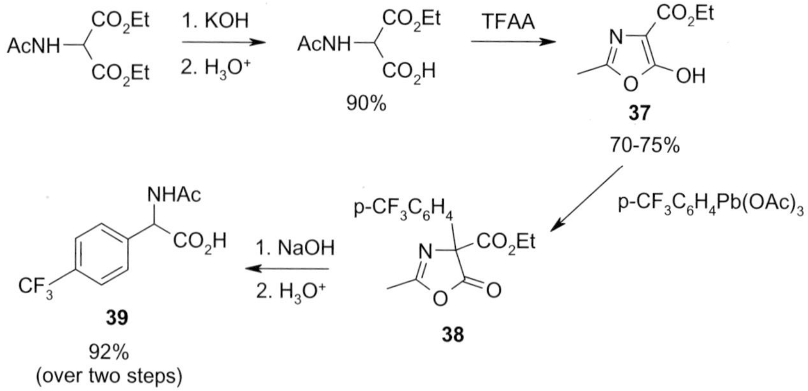

Scheme 4.19. An approach to 4-TfmPhg via arylation reaction.

palladium–bipyridine complex to give the substituted 4-TfmPhg **36** in 57% yield and 91% *ee* (Dai and Lu, 2007).

An original non-stereoselective approach to 4-TfmPhg based on an arylation reaction is illustrated in Scheme 4.19. Diethyl acetamidomalonate was hydrolysed by one equivalent of alkali, and cyclized, in the presence of trifluoroacetic acid, to form the oxazole **37**. Compound **37** was arylated using 4-$CF_3C_6H_4Pb(OAc)_3$ in a pyridine–chloroform mixture and the product **38** was hydrolysed to form a racemic mixture of the arylglycine **39** in 92% yield (Morgan *et al.*, 1997). A similar method of arylation, also starting from diethyl benzoylamidomalonate, was described by Koen *et al.* (1997).

A method of inserting the carboxylic moiety into the azomethyne **40** was described by Nemoto *et al.* (2007) (Scheme 4.20). The dicyanomethyl *tert*-butyldimethylsilyl ether **41** was suggested to be a synthetic equivalent of the carboxylic group. Dicyanide **41** reacted with the azomethyne in 88% yield to form **42**, which was then deprotected and hydrolysed. In order to measure the optical purity (40%), the amino acid was reacted with the chiral acid **43** to give the acylated product **44**. The authors argued that such a low *de* value was caused by partial epimerization during the final coupling.

Direct reduction of the C=N double bond in N-PMP-protected α-imino ester **45** by the Hantzsch ester employing chiral phosphoric acid

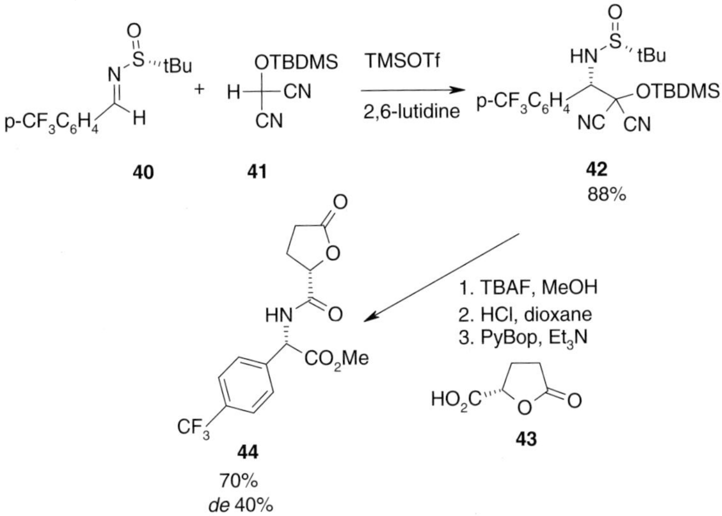

Scheme 4.20. Carboxylation of azomethyne as a route to 4-TfmPhg.

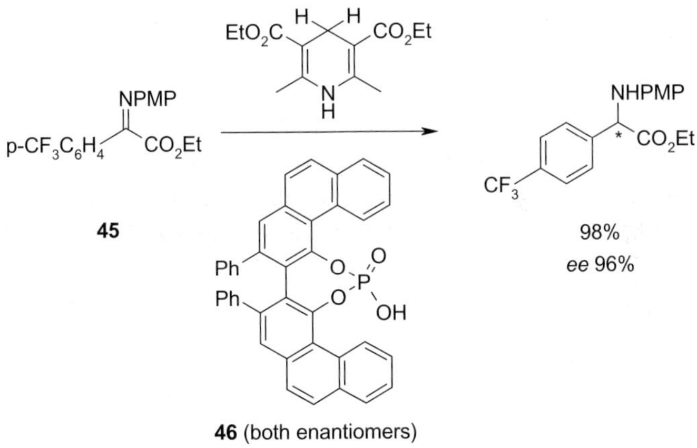

Scheme 4.21. Enantioselective reduction of the C=N double bond in an imino ester leading to protected 4-TfmPhg. Stereoisomers with both (R)- and (S)-configurations at the corresponding centre were prepared.

46 as a catalyst (Scheme 4.21) gave PMP-4-TfmPhg-OEt in 98% yield and 96% *ee* (Li *et al.*, 2007).

The first synthesis of 4-TfmPhg based on aminocarbonylation of the corresponding carbanion involved the use of chromium–carbonyl complexes

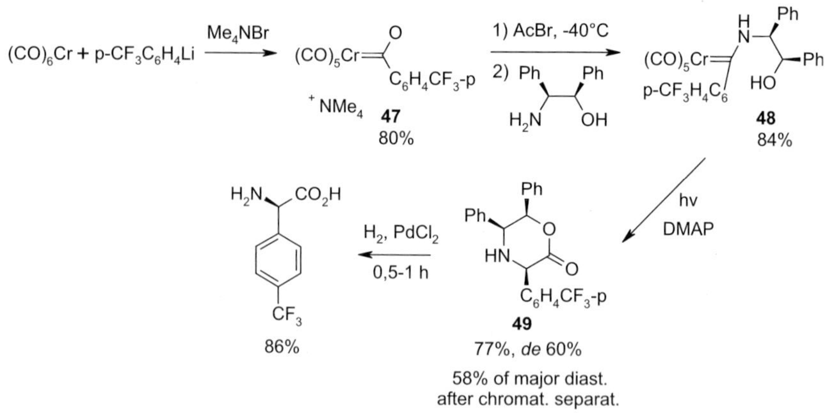

Scheme 4.22. Asymmetric synthesis of 4-TfmPhg utilizing an aminocarbonylation reaction.

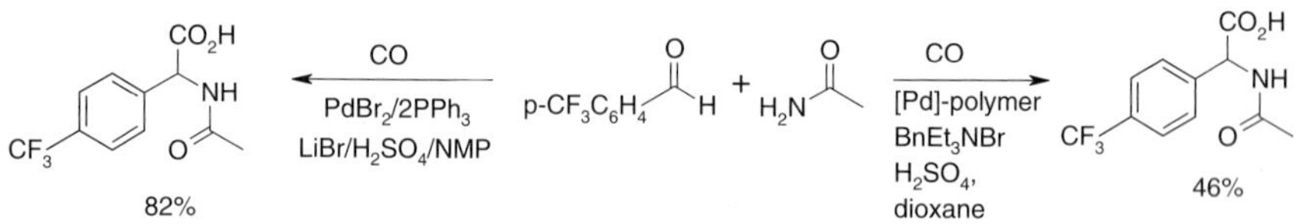

Scheme 4.23. Non-stereoselective syntheses of *N*-acylated 4-TfmPhg.

(Scheme 4.22). 4-$CF_3C_6H_4Li$ was reacted with chromium hexacarbonyl, then the oxygen atom in **47** was substituted by the chiral aminoalcohol, and the resulting complex **48** decomposed by photolysis to form the arylglycine derivative **49.** Compound **49** was deprotected under mild acidic conditions to give 4-TfmPhg in 42% overall yield and 60% *ee* (Vernier *et al.*, 1992; Hegedus *et al.*, 1990).

Two non-stereoselective syntheses of 4-TfmPhg using an amidocarbonylation reaction have also been described (Scheme 4.23). The palladium-phosphine catalysed reaction gave acylated 4-TfmPhg in 82% yield under optimized conditions (Beller *et al.*, 1998). The palladium complex with a polymer ligand was studied as a catalyst for amidocarbonylation of a wide range of aldehydes; 4-CF_3-benzaldehyde was included in this survey and gave acylated 4-TfmPhg in 46% yield (Akiyama *et al.*, 2004).

4.6.2 *Separation of the 4-TfmPhg enantiomers*

As mentioned above, the enantiomeric purity of Tfm-substituted amino acids is in many cases not critical for peptide synthesis. Nevertheless, as the yields of the peptide synthesis are much higher when optically pure amino acids are used, it is worth mentioning here the published practical methods to resolve the 4-TfmPhg enantiomers and their derivatives. Generally, chiral chromatography (Galushko, 1995) or natural enzymes (Miyazawa, 1995) are applied for this purpose. Two examples of enzymatic resolution of racemic 4-TfmPhg derivatives are illustrated in Scheme 4.24. The first was reported by Morgan *et al.* (1997): only the (*S*)-enantiomer of compound **39** has been converted into the corresponding (*S*)-amino acid (4-Tfm-L-Phg) under direct catalysis by porcine kidney acylase. The second (Arcuri *et al.*, 2003) described the dynamic kinetic resolution of enantiomeric hydantoins **50** by D-hydantoinase. This reaction gave the *N*-carbamoyl (*R*)-amino acid **51** in >98% *ee.* In this example, 100% conversion was achieved since the non-hydrolysed enantiomer of **50** was constantly racemized in the course of the enzymatic process due to the alkaline pH optimal for the catalytic activity of the D-hydantoinase.

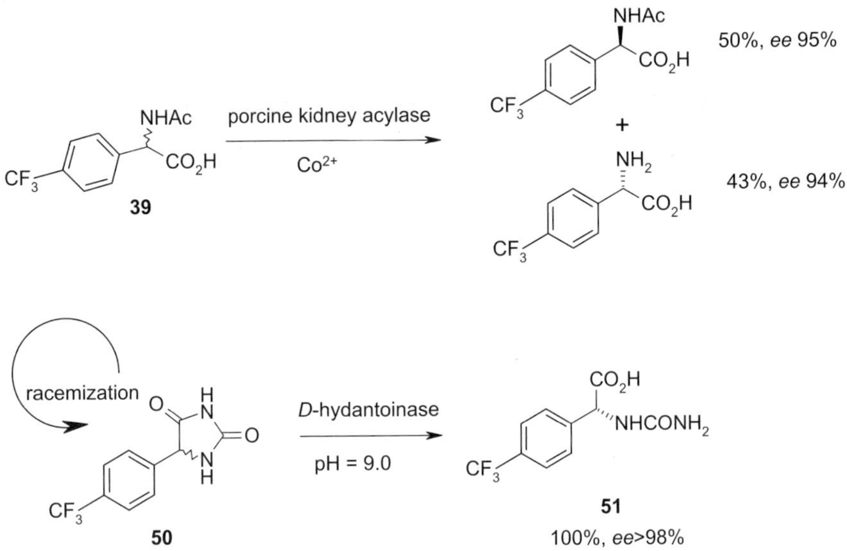

Scheme 4.24. Enzymatic approaches towards enantiomerically pure 4-TfmPhg.

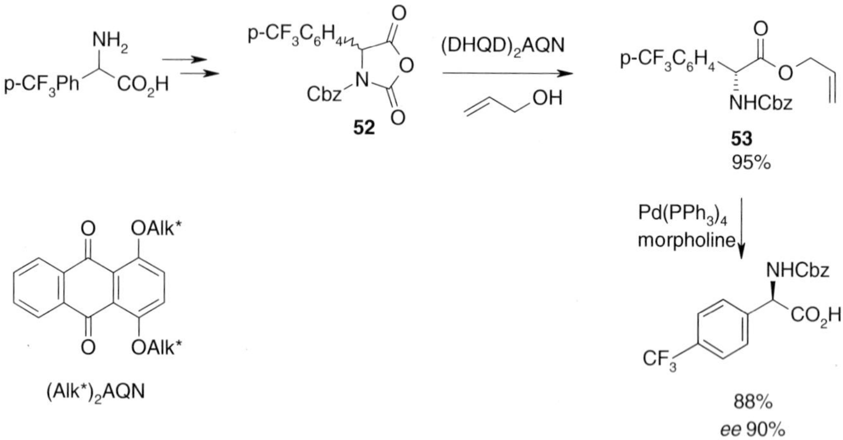

Scheme 4.25. Chemical resolution of the enantiomers of a 4-TfmPhg derivative.

Non-enzymatic resolution of 4-TfmPhg is also possible (Hang and Deng, 2002). An example is illustrated in Scheme 4.25. Alcoholysis of the urethane-protected derivative 52 in the presence of a modified cinchona alkaloid $(DHQD)_2AQN$ gave the ester 53 in 95% yield and 90% *ee*. The allyl protection was subsequently removed in the presence of a Pd(0) catalyst.

4.6.3 *Incorporation of 4-TfmPhg into peptides*

Structurally related to 4-TfmPhg, the monofluorinated 4-fluorophenylglycine (4F-Phg) was historically the first amino acid used as a ^{19}F-label for structural studies of membrane-active peptides by solid-state ^{19}F NMR (Salgado *et al.*, 2001; Afonin *et al.*, 2003, 2004). The use of 4F-Phg with the standard Fmoc SPPS protocols was applied to the fusogenic peptide B18 (Leu-Gly-Leu-Leu-Lue-Arg-His-Leu-Arg-His-His-Ser-Asn-Leu-Leu-Ala-Asn-Ile, nine monosubstituted and one disubstituted analogue were produced), the antimicrobial peptides gramicidin S ($_{cyclo}$[Pro-Val-Orn-Leu-D-Phe]$_2$, two doubly substituted analogues were synthesized), and PGLa (Gly-Met-Ala-Ser-Lys-Ala-Gly-Ala-Ile-Ala-Gly-Lys-Ile-Ala-Lys-Val-Ala-Leu-Lys-Ala-Leu-NH_2, three monosubstituted analogues were prepared). We refer to these studies here because all general observations about the synthetic behaviour of 4F-Phg, synthetic problems and proposed solutions, turned out to be the same when using 4-TfmPhg in SPPS (Sachse, 2003; Glaser *et al.*, 2004; Wadhwani *et al.*, 2006). Both amino acids are compatible with SPPS protocols, and unlike TfmAla they are sufficiently reactive to use the standard Fmoc procedures. However, due to the increased acidity of the α-proton, the phenylglycine derivatives tend to racemize completely during the synthesis. The basic conditions employed during Fmoc cleavage cause racemization. Fortunately, the diastereomers of the peptides mentioned above could be separated chromatographically (Afonin *et al.*, 2003; Glaser *et al.*, 2004). More recently it has been shown that it is possible to suppress the racemization of phenylglycine derivatives in peptide synthesis (e.g. Dettner *et al.*, 2009). As HPLC purification is always needed before the peptides can be used in NMR studies, the epimeric peptides containing different enantiomers of 4-TfmPhg (and 4F-Phg) could be conveniently separated on a preparative scale (C_{18} HPLC column with 22 mm internal diameter to prepare some tens of milligrams of the pure peptides). Even disubstituted analogues of the cyclic decapeptide gramicidin S, where the use of racemic 4-TfmPhg caused the formation of three diastereomers, were eluted with significantly different retention times under optimized chromatographic conditions (Wadhwani *et al.*, 2006).

Since the pioneering studies reported in 2003 (Sachse, 2003), 4-TfmPhg has been introduced in a range of membrane-active peptides having various length and conformational preferences. These include several antimicrobial peptides such as gramicidin S (Wadhwani *et al.*, 2006), MSI-103 ([Lys-Ile-Ala-Gly-Lys-Ile-Ala]$_3$-NH$_2$: see Kanithasen, 2005; Strandberg *et al.*, 2008), and temporin A (Phe-Leu-Pro-Leu-Ile-Gly-Arg-Val-Leu-Ser-Gly-Ile-Leu-NH$_2$: see Tiltak, 2009); several cell-penetrating peptides such as MAP (Lys-Leu-Ala-Leu-Lys-Leu-Ala-Leu-Lys-Ala-Leu-Lys-Ala-Ala-Leu-Lys-Leu-Ala-NH$_2$: see Eisele, 2007; Wadhwani *et al.*, 2008), PEP-1 (Lys-Glu-Thr-Trp-Trp-Glu-The-Trp-Trp-Thr-Glu-Trp-Ser-Gln-Pro-Lys-Lys-Lys-Arg-Lys-Val: see Eisele, 2007), penetratin (Arg-Gln-Ile-Lys-Ile-Trp-Phe-Gln-Asn-Arg-Arg-Met-Lys-Trp-Lys-Lys: see Eisele, 2007), and SAP ([Val-Arg-Leu-Pro-Pro-Pro]$_3$: see Eisele, 2007), as well as the fusogenic sequence FP23 (Ala-Val-Gly-Ile-Gly-Ala-Leu-Phe-Leu-Gly-Phe-Leu-Gly-Ala-Ala-Gly-Ser-Thr-Met-Gly-Ala-Arg-Ser: see Grasnick, 2006;

Reichert *et al.*, 2007). In all these studies, TfmPhg was utilized to substitute a proteinogenic amino acid, such as an aliphatic Leu, Ile, Val, Ala, or an aromatic Trp, Phe, or even the small Gly. Special attention was paid to ascertain that the ^{19}F-labelling would not disturb the conformation and function of the corresponding wild-type peptide. These studies revealed that the membrane-active peptides remained essentially unperturbed when the 4-TfmPhg residue was in place of Leu, Ile or Val. On the other hand, Trp and Gly could not so readily be substituted by 4-TfmPhg, as the ^{19}F-labelled analogues often had structures and/or functions significantly different from those of their unlabelled parents. Phe and Ala seem to represent intermediate cases, as in some instances the structure and/or functions were affected by the labelling, while in the others no changes were observed. These observations can be explained, at least in the case of Gly and Ala substitutions, by the considerable steric impact that is imposed by the relatively bulky side chain of 4-TfmPhg. In addition, when replacing Gly with 4-TfmPhg the local backbone torsion angles φ and ψ may be altered upon labelling. It should also be noted that 4-TfmPhg possesses an aromatic side chain and hence can undergo specific π–π or cation–π interactions. Such interactions could affect the proper folding, and their impact on the structure may be strong when a sequence is rich in aromatic and cationic amino acids — not uncommon among membrane-active peptides.

Yet another problem was recently observed when the SAP peptide was labelled with racemic 4-TfmPhg (Eisele, 2007). The SAP analogues with the 4-TfmPhg residue in place of Val7 and Leu9 were not separable by HPLC, which indicates that the resolution of the peptide epimers carrying 4-TfmPhg is sequence dependent. We may thus conclude that, despite otherwise fulfilling criteria 1–4, a significant disadvantage of 4-TfmPhg as a ^{19}F-label is its propensity to racemize extensively.

4.6.4 *Synthesis of TfmBpg*

The amino acid TfmBpg (3) was designed to overcome the above mentioned problems encountered with the use of 4-TfmPhg and TfmAla for ^{19}F NMR (Mikhailiuk *et al.*, 2006, 2010; Mykhailiuk, 2008). The idea behind the design is the following: the bicyclopentane-derived spacer between the CF_3 group and the aminocarboxylate moiety assures the

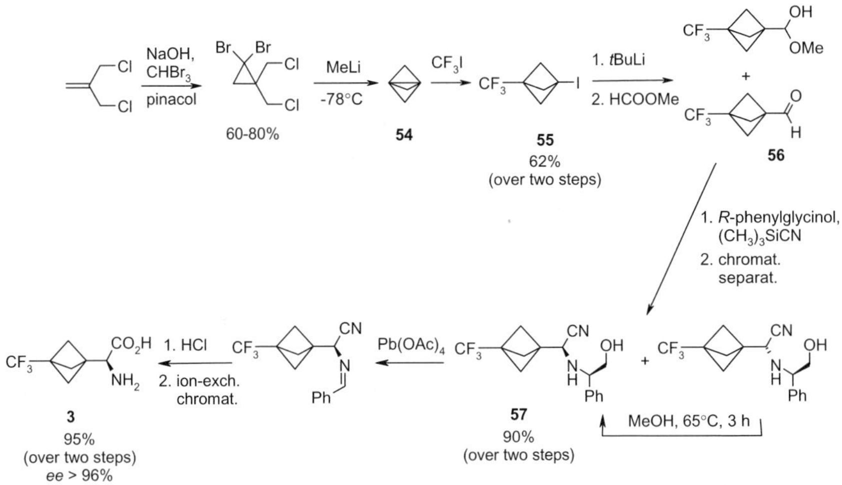

Scheme 4.26. Synthesis of TfmBpg.

rigidity of the side chain (criterion 1); at the same time, this moiety eliminates the influence of the electron-withdrawing CF_3 group on the α-proton acidity, therefore alleviating the racemization problem of 4-TfmPhg. Furthermore, by possessing an aliphatic side chain, TfmBpg is better suited for replacing aliphatic amino acids with no additional risk of interfering with aromatic interactions.

The synthesis of TfmBpg is illustrated in Scheme 4.26. Propellane **54** was transformed to the intermediate iodide **55** by addition of trifluoroiodomethane. Aldehyde **56** (together with its hemiacetal) was synthesized from **55** in good yield. A Strecker reaction with the amine bearing chiral auxiliary allowed separation of diastereomeric **57** and isolation of both enantiomers of **3**.

4.6.5 *Synthesis of peptides containing TfmBpg*

Like 4-TfmPhg, the amino acid TfmBpg was fully compatible with standard Fmoc-based SPPS, and — in contrast to the former — no racemization was observed, as desired. This was demonstrated by the synthesis of four analogues of the antimicrobial peptide PGLa (see above), labelled with TfmBpg at the very same positions that had previously been substituted by 4-TfmPhg (i.e. Ile9, Ala10, Ile13, or Ala14: see Afonin *et al.*, 2007; Mykhailiuk, 2008). All these peptide analogues synthesized with the (*S*)-enantiomer of the TfmBpg were obtained as single diastereomers, as shown by HPLC/MS analysis. The labelled peptides gave the same circular dichroism spectra in membrane-mimicking environments, and they demonstrated the same levels of biological activity in the functional antimicrobial assays as PGLa itself and the 4-TfmPhg-substituted analogues. The SAP sequence mentioned above was also labelled with TfmBpg (substitution of Leu3, Val7, Leu9, Val13, or Leu15). Here again, based on circular dichroism analysis, all the TfmBpg-substituted SAP analogues were structurally indistinguishable from the parent peptide (Mykhailiuk, 2008; Mykhailiuk *et al.*, 2008). Other peptides which have meanwhile been labelled by TfmBpg are the previously mentioned temporin A (substitution of Leu4, Ile5, Gly6, Val8, or Leu9), FP23 (substitution of Leu7, Phe8, Leu9, Phe11, or Leu12) and transportan 10 (Ala-Gly-Tyr-Leu-Leu-Gly-Lys-Ile-Asn-Leu-Lys-Ala-Leu-Ala-Ala-Leu-Ala-Lys-Lys-Ile-Leu-NH_2, substitution of Leu5, Ile8, Leu10 and Leu13: see Ehni, 2009). In the synthesis of all these peptides the Fmoc-TfmBpg behaved like the other Fmoc-protected aliphatic amino acids Leu, Ile, or Val.

4.6.6 ^{19}F NMR structure analysis of peptides with 4-TfmPhg and TfmBpg

Both 4-TfmPhg **2** and TfmBpg **3** have been widely used as ^{19}F-labels in solid-state NMR studies of peptides. From the perspective of structure analysis, both compounds have proven to be excellent reporter groups in macroscopically oriented samples. They fulfil criterion 1 as they carry the CF_3 group in a well-defined geometry on the peptide backbone: the Cα–Cβ and C–CF_3 vectors are collinear, and the side chain can rotate only around these bonds. Both side chains in **2** and **3** are effectively non-polar with a medium-sized steric volume. These properties should make them good substitutes of Phe, Trp, Tyr, Val, Leu, Ile, Ala, Met and His (uncharged). However, practical experience (see above) has corrected these expectations and suggests a more limited applicability to Leu, Ile, Val, and to a lesser extent Phe and Ala.

All the membrane-active peptides mentioned in the previous sections have been studied or are under current investigation by solid-state

^{19}F NMR as outlined in Fig. 4.1 and had been labelled for this particular purpose. Without going into details about the functional relevance of the structural and orientational results found for each peptide (these are described in the relevant publications), here we will illustrate briefly the major findings that were made possible using 4-TfmPhg and/or TfmBpg.

For the functionally similar amphiphilic peptides PGLa, MAP and MSI-103, preservation of their α-helical conformation in lipid bilayers of various compositions was confirmed by ^{19}F NMR (Sachse, 2003; Glaser *et al.*, 2004, 2005; Kanithasen, 2005; Afonin *et al.*, 2008b; Strandberg *et al.*, 2008; Wadhwani *et al.*, 2008), as expected from previous studies in other membrane(-mimicking) environments. Remarkably, these peptides were found to change their alignment in the lipid bilayer depending on concentration, lipid composition and lipid phase state. Namely, they could switch between a monomeric surface-bound (S-) state, a dimeric obliquely tilted (T-) state, and a presumably oligomeric inserted (I-) state (Fig. 4.3(b)). Very similar realignments were also demonstrated for the β-stranded gramicidin S, which exists in the S-state above and below the lipid phase transition, while flipping into an upright oligomeric I-state when a sufficient number of defects were present in the bilayer (Mykhailiuk, 2008; Afonin *et al.*, 2008a; Fig. 4.3(c)). For FP23, temporin A and transportan 10, some distorted helical conformations were determined, which again undergo alignment changes as in response to changes in the lipid composition and peptide concentration (Grasnick, 2006; Ehni, 2009; Tiltak, 2009).

MAP and MSI-103 were investigated using only 4-TfmPhg, while gramicidin S, PGLa, temporin A and FP23 were explored utilizing both 4-TfmPhg and TfmBpg. The mutual exchangeability of the two amino acids was confirmed in these studies, as both labels showed the spectra with the same ^{19}F splittings when placed in the corresponding positions along the peptide sequence (Grasnick, 2006; Afonin *et al.*, 2007; Mykhailiuk, 2008; Tiltak, 2009). The ^{19}F NMR data on 4-TfmPhg in the PGLa peptide were also compared with a corresponding ^{2}H NMR analysis using Ala-d_3 as a reporter group, showing that both labels deliver essentially the same structural results (Strandberg *et al.*, 2006). Naturally, a substitution of Ala by Ala-d_3 is less perturbing than any 4-TfmPhg label. On the other hand, the main advantage of the CF_3 group over the CD_3 group is the fact that a ^{19}F NMR spectrum yields the signed ^{19}F–^{19}F dipolar coupling (due to the CSA contribution), in contrast to the absolute value of a ^{2}H quadrupolar splitting. Hence, only about four separate ^{19}F-labels are required to fully characterize the alignment and dynamics of a rigid secondary structure element such as an α-helix, whereas twice as many ^{2}H labels are needed to obtain the same level of information.

The cell-penetrating carrier transportan 10 is the first example of a peptide of which structure and alignment in lipid membranes have been determined exclusively with TfmBpg (Ehni, 2009). The SAP peptide, where racemic 4-TfmPhg turned out to be difficult to use (see above), is the second example emerging (Mykhailiuk, 2008; Fig. 4.3(d)).

4.7 Conclusions and Perspectives

The majority of membrane-active peptides and transmembrane proteins contain hydrophobic amino acids which dominate the interactions with the intrinsically apolar interior of the lipid bilayer. These positions are highly abundant and very suitable for targeting with apolar ^{19}F-labels. As demonstrated above, hydrophobic TfmAla, 4-TfmPhg and TfmBpg are readily available for structural studies by solid-state ^{19}F NMR. Their use has already contributed to a better understanding of the functional mechanisms for several different kinds of membrane-active peptides. However, polar or even charged amino acids are also important for the specific functioning of many peptides. Most of the antimicrobial, cell-penetrating and fusogenic peptides are actually amphiphilic, with part of their surface in the folded molecule carrying polar amino acid residues. Moreover, some short homo-oligomeric stretches of cationic and anionic residues are also known to display significant levels of membrane activity. For instance, oligo-arginines represent a distinct class of cell-penetrating peptides, while short sequences rich in oligo-aspartate motifs are known to have antimicrobial function and medium-sized polyglutamines can form ion channels in lipid bilayers (Brogden *et al.*, 1997; Hirakura *et al.*, 2000; Monoi *et al.*, 2000; Futaki *et al.*, 2007). Likewise, many native polypeptides (and membrane peptides are no exception in this respect) contain amino acids known as 'breakers of a regular secondary structure', namely glycine and proline. The glycine residue is an extremely flexible backbone element; proline, by contrast, provides considerable conformational restriction to the backbone and is not able to act as a hydrogen bond donor. These different properties may lead to the formation of kinks, or to the capping of secondary structure elements in the peptides. Glycine is furthermore known to mediate specific helix–helix contacts in the G-Xaa-Xaa-Gly motifs of coiled coils, and it contributes to the conformational flexibility of the viral fusion peptides. The proline residue (often found in those secondary structure elements of peptides which are responsible for binding to other proteins) is a necessary constituent in various turns and forms the PPII-structured backbones of many 'intrinsically unstructured' proteins and sequences (Choi *et al.*, 1994; Gunasekaran *et al.*, 1997; Dyson and Wright, 2005). Obviously, limiting the repertoire of the ^{19}F-labels to only hydrophobic amino acids would exclude important areas and functionally relevant locations of many membrane peptides from being directly observable in ^{19}F NMR experiments.

The lack of suitable ^{19}F NMR reporters that are compatible with further types of side chains has prompted the development of new CF_3-substituted amino acids in recent years. Up to now, these efforts have resulted in the synthesis and application to structural studies of the proline analogue TfmMePro (**58**) (Mykhailiuk *et al.*, 2008) (Fig. 4.4). Here the design of the label was initially based on the idea that the already conformationally restricted proline molecule may serve as a scaffold to connect the CF_3 group to the peptide backbone in a rigid fashion. All known CF_3-substituted prolines (**59**, **60** and **61**: see Del Valle and Goodman, 2002; Qiu and Qing 2002a, b; Chaume *et al.*, 2006; Nadano *et al.*, 2006; Caupène *et al.*, 2009) were considered first, but all three were ruled out according to the ^{19}F-labelling criteria above. In **59** and **60** the carbon skeleton of the proline itself would retain some residual flexibility; therefore, they are still too flexible even if specific puckered conformations are favoured. The recently published 2-CF_3-proline **61**, when incorporated in peptides, would place the CF_3 group directly on the Cα-position of the peptide backbone. In this case, as in TfmAla, the presence of the bulky electron-withdrawing CF_3 moiety would severely alter the steric and electronic environment around it (Burger *et al.*, 1998).

Similarly it would also reduce the chemical reactivity of the amino acid **61**, thereby making its use in peptide synthesis difficult.

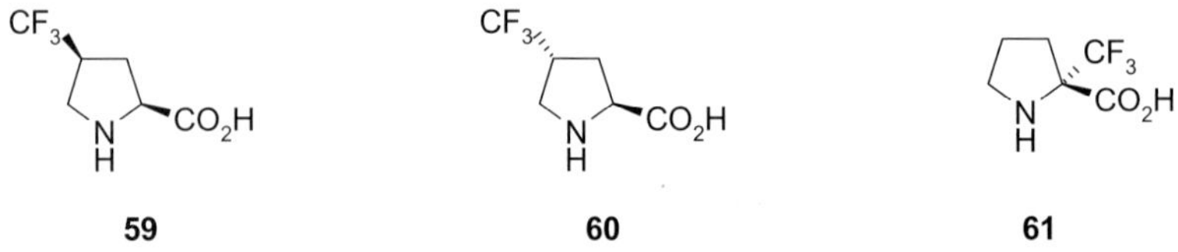

CF_3 trifluoromethylated proline analogues.

In contrast to **59** and **60**, bi- and polycyclic analogues of proline are sufficiently rigid (Komarov *et al.*, 2004) and the design of a good ^{19}F-label can exploit this fact. This idea was realized in compound **58**, which was synthesized from the known precursor **62** by reaction with the corresponding catalytically generated carbene (Scheme 4.27). The intermediate **63a** was formed along with other diastereomers **63b** and **63c** which were separated. Compound **58** was obtained in a moderate yield (ca. 20% calculated from **62**) and, compared to the rest of the isomers, was found to be the most stable in acidic media (though, in principle, all four diastereomers might have been used for ^{19}F NMR studies).

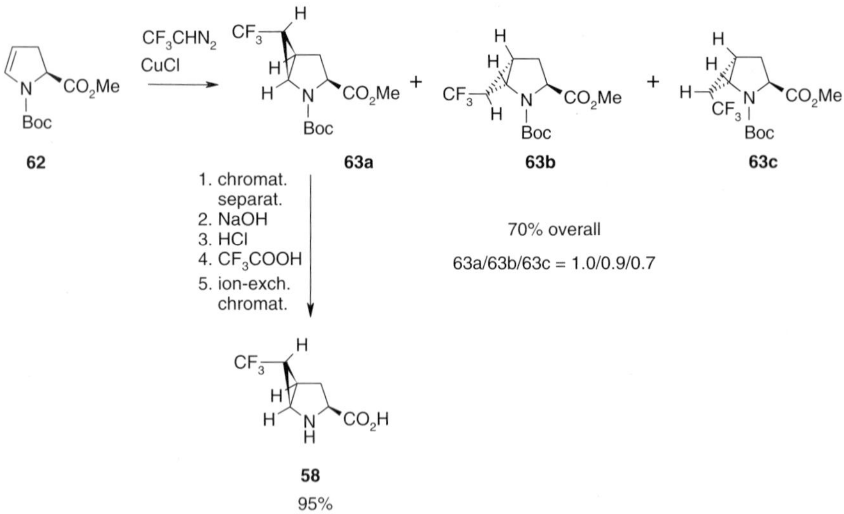

Scheme 4.27. Synthesis of TfmMePro.

The amino acid **58** turned out to be compatible with standard Fmoc SPPS (Mykhailiuk *et al.*, 2008). The proline-rich peptide SAP (see above) was synthesized with a TfmMePro substitution in place of Pro11. Neither degradation, low reactivity, nor racemization of Fmoc-**58** was observed. The TfmMePro-containing SAP was studied by ^{19}F NMR in lipid membranes under the same conditions as the TfmBpg labelled analogues (see above). Though a single label is not sufficient to perform a full structural analysis, the measured anisotropic dipolar couplings are fully consistent with the conformational and orientational behaviour of SAP as determined from the five TfmBpg-substituted analogues.

According to a circular dichroism analysis of the ^{19}F-labelled SAP, TfmMePro was also found to slightly stabilize the PPII conformation. This fact is interesting *per se*, though it could also mean that TfmMePro may turn out to be a poor ^{19}F-label. The utility of the TfmMePro is currently under investigation and still leaves much room for improvement in label design for membrane-active peptides. Starting from some proof-of-concept experiments towards the full application of trifluoromethyl-substituted α-amino acids in solid-state ^{19}F NMR (Sachse, 2003; Glaser *et al.*, 2004), the methodology of structure analysis has been constantly improving. Major effort has gone into the development of suitable labels, which is especially critical in the case of ^{19}F-labelling. A subtle balance between efficiency of the label and its non-perturbing nature must be optimized in this case.

Despite the advances already outlined, the repertoire of known 'good' ^{19}F-labels is still scarce. All the compounds which were used so far have mimicked amino acids with unpolar side chains. The challenging problem of the design, synthesis and application of new ^{19}F-labels suitable for examining polar side chains remains open. However, the experience gained from the use of unpolar Tfm-substituted amino acids suggests some basic design principles for the future polar ^{19}F-labels. The CF_3 group should be placed on a rigid (poly)cyclic spacer separating it from the peptide backbone. Simultaneously, the proper distance from the functional group in the side chain has to be maintained. These rules, when implemented, would minimize the danger of peptide properties being modified after labelling. New insights into the structure and peptide–lipid interactions will thus be accessible by ^{19}F NMR once a suitable chemical route to such labels has been paved.

Acknowledgements

The authors are grateful to all former and current members of the ASU and IVK research groups who are involved in the work on ^{19}F-labelled amino acids and acknowledge the Alexander von Humboldt Foundation and the DFG-Center for Functional Nanostructures (TP E1.2) for financial support. V.S.K. expresses special thanks to M. Tepluchin for her help with the references.

References

Abraham, M.H., Grellier, P.L., Prior D.V. *et al.* (1990) Hydrogen bonding. Part 10. A scale of solute hydrogen-bond basicity using log *K* values for complexation in tetrachloromethane. *J Chem Soc Perkin Trans 2* 4, 521–529.

Afonin, S., Glaser, R.W., Berditchevskaia, M. *et al.* (2003) 4-Fluorophenylglycine as a label for ^{19}F NMR structure analysis of membrane-associated peptides *ChemBioChem* 4, 1151–1163.

Afonin, S., Dürr, U.H.N., Glaser, R.W. *et al.* (2004) 'Boomerang'-like insertion of a fusogenic peptide in a lipid membrane revealed by solid-state ^{19}F NMR. *Mag Res Chem* 42, 195–203.

Afonin, S., Mikhailiuk, P.K., Komarov, I.V. *et al.* (2007) Evaluating the amino acid CF_3-bicyclopentylglycine as a new label for solid-state ^{19}F-NMR structure analysis of membrane-bound peptides. *J Pept Sci* 13, 614–623.

Afonin, S., Dürr, U.H.N., Wadhwani, P. *et al.* (2008a) Solid state NMR structure analysis of the antimicrobial peptide Gramicidin S in lipid membranes: concentration-dependent re-alignment and self-assembly as a β-barrel. *Topics Curr Chem* 273,139–154.

Afonin, S., Grage, S.L., Ieronimo, M. *et al.* (2008b) Temperature-dependent transmembrane insertion of the amphiphilic peptide PGLa in lipid bilayers observed by solid state ^{19}F NMR spectroscopy. *J Am Chem Soc* 130, 16512–16514.

Akiyama, R., Sagae, T., Sugiura, M. *et al.* (2004) Efficient synthesis of N-acyl-α-amino acids via polymer incarcerated palladium-catalyzed amidocarbonylation. *J Organomet Chem* 689, 3806–3809.

Arcuri, M.B., Sabino, S.J., Antunes, O.A.C. *et al.* (2003) Enantiospecific enzymatic synthesis of *N*-carbamoyl-D-*p*-fluorophenylglycine and *N*-carbamoyl-D-*p*-trifluoromethylphenylglycine. *J Fluor Chem* 121, 55–56.

Arkin, I.T. (2006) Isotope-edited IR spectroscopy for the study of membrane proteins. *Curr Opin Chem Biol* 10, 394–401.

Arora, A. & Tamm, L.K. (2001) Biophysical approaches to membrane protein structure determination. *Curr Opin Struct Biol* 11, 540–547.

Asensio, A., Bravo P., Crucianelli, M. *et al.* (2001) Synthesis of nonracemic α-trifluoromethyl α-amino acids from sulfinimines of trifluoropyruvate. *Eur J Org Chem*, 1449–1458.

Banks, R.E., Smart, B.E. & Tatlow, J.C. (Eds) (1994) *Organofluorine Chemistry: Principles and Commercial Applications,* Plenum Press, New York.

Beenen, M.A., Weix, D.J. & Ellman, J.A. (2006) Asymmetric synthesis of protected arylglycines by rhodium-catalyzed addition of arylboronic acids to *N-tert*-butanesulfinyl imino esters. *J Am Chem Soc* 128, 6304–6305.

Beller, M., Eckert, M. & Holla, E.W. (1998) Efficient synthesis of *N*-acyl-α-arylglycines via palladium-catalyzed amidocarbonylation: application to the central amino acid of chloropeptin. *J Org Chem* 63, 5658–5661.

Bravo, P., Capelli, S., Meille, S.V. *et al.* (1994) Synthesis of optically pure (*R*)- and (*S*)-α-trifluoromethyl-alanine. *Tetrahedron: Asymm* 5, 2009–2018.

Bravo, P., Crucianelli, M., Vergani, B. *et al.* (1998) Sulfinimines of trifluoropyruvate: novel intermediates for chiral non racemic α-trifluoromethyl α-amino acids. *Tetrahedron Lett* 39, 7771–7774.

Brogden, K.A., Ackermann, M. & Huttner, K.M. (1997) Small, anionic, and charge-neutralizing propeptide fragments of zymogens are antimicrobial. *Antimicrob Agents Chemother* 41, 1615–1617.

Burger, K., Mütze, K., Hollweck, W. *et al.* (1993) Untersuchungen zur proteasekatalysierten und chemischen Peptidbindungsknüpfung mit α-trifluormethylsubstituierten α-Aminosäuren. *J Prakt Chem* 335, 321–331.

Burger, K., Mütze, K., Hollweck, W. *et al.* (1998) Incorporation of α-trifluoromethyl substituted α-amino acids into C- and N-terminal position of peptides and peptide mimetics using multicomponent reactions. *Tetrahedron* 54, 5915–5928.

Caminati, W., Melandri, S., Maris, A. *et al.* (2006) Relative strengths of the O–H···Cl and O–H···F hydrogen bonds. *Angew Chem Int Edit* 45, 2438–2442.

Carpenter, E.P., Beis, K., Cameron, A.D. *et al.* (2008) Overcoming the challenges of membrane protein crystallography. *Curr Opin Struct Biol* 18, 581–586.

Castejon, H.J. & Wiberg, K.B. (1998) Effect of fluorine substitution on the carbon acidity of methane, methyl isocyanide, acetonitrile, acetaldehyde, and nitromethane. *J Org Chem* 63, 3937–3942.

Caupène, C., Chaume, G., Ricard, L. *et al.* (2009) Iodocyclization of chiral CF_3-allylmorpholinones: a versatile strategy for the synthesis of enantiopure α-Tfm-prolines and α-Tfm-dihydroxyprolines. *Org Lett* 11, 209–212.

Chaume, G., Van Severen, M.-C., Marinkovic, S. *et al.* (2006) Straightforward synthesis of (*S*)- and (*R*)-α-trifluoromethyl proline from chiral oxazolidines derived from ethyl trifluoropyruvate. *Org Lett* 8, 6123–6126.

Choi, S.H., Yu, Y.J., Shin, J.K. *et al.* (1994) Molecular dynamics simulations of *trans*- and *cis*-*N*-acetyl-*N*′-methylamides of Xaa---Pro dipeptides. *J Mol Struct* 323, 233–242.

Christensen, H.N. & Oxender, D.L. (1963) The acid strength of the amino group as a factor in the transport of amino acids. *Biochim Biophys Acta* 74, 386–391.

Cobb, S.L. & Murphy, C.D. (2009) ^{19}F NMR applications in chemical biology. *J Fluor Chem* 130, 132–143.

Cowieson, N.P., Kobe, B. & Martin, J.L. (2008) United we stand: combining structural methods. *Curr Opin Struct Biol* 18, 617–622.

Dai, H. & Lu, X. (2007) Diastereoselective synthesis of arylglycine derivatives by cationic palladium(II)-catalyzed addition of arylboronic acids to *N-tert*-butanesulfinyl imino esters. *Org Lett* 9, 3077–3080.

Del Valle, J.R. & Goodman, M. (2002) Stereoselective synthesis of Boc-protected *cis* and *trans*-4-trifluoromethylprolines by asymmetric hydrogenation reactions. *Angew Chem Int Edit* 41, 1600–1602.

Dettner, F., Hänchen, A., Schols, D. *et al.* (2009) Total synthesis of the antiviral peptide antibiotic feglymicin. *Angew Chem Int Edit* 48, 1856–1861.

Drews, J. (2000) Drug discovery: a historical perspective. *Science* 287, 1960–1964.

Dyson, H.J. & Wright, P.E. (2005) Intrinsically unstructured proteins and their functions. *Nat Rev Mol Cell Biol* 6, 197–208.

Ehni, S. (2009) *Die Charakterisierung von Transportan 10 in der Biomembran*, Diploma Thesis, University of Karlsruhe.

Eisele, A. (2007) *Aggregationsverhalten von trojanischen Peptiden*, Diploma Thesis, University of Karlsruhe.

Engel, A. & Gaub, H.E. (2008) Structure and mechanics of membrane proteins. *Annu Rev Biochem* 77, 127–148.

Epand, R.M. (2003) Fusion peptides and the mechanism of viral fusion. *Biochim Biophys Acta* 1614, 116–121

Fanghänel, S. (2007) *Synthese von ^{19}F-markierten Analoga des antibiotischen Peptaibols Harzianin für NMR-Strukturuntersuchungen*, Diploma Thesis, University of Karlsruhe.

Fokin, A.V., Kolomiets, A.F. & Vasil'ev, N.V. (1984) Fluorine-containing imines. *Russ Chem Rev* 53, 238–255.

Fujiyoshi, Y. & Unwin, N. (2008) Electron crystallography of proteins in membranes. *Curr Opin Struct Biol* 18, 587–592.

Fustero, S., Sanz-Cervera, J.F., Aceña, J.L. *et al.* (2009) Nitrogen-containing organofluorine derivatives: an overview. *Synlett* 4, 525–549.

Futaki, S., Nakase, I., Tadokoro, A. *et al.* (2007) Arginine-rich peptides and their internalization mechanisms. *Biochem Soc Trans* 35(Pt 4), 784–787.

Galushko, S. (1995) 'High-performance liquid chromatography of fluorine-containing amino acids', in Kukhar, V.P. & Soloshonok, V.A. (Eds), *Fluorine-Containing Amino Acids, Synthesis and Properties*, Wiley, Chichester, pp. 295–310.

Gerig, J.T. (2004) Fluorine NMR, in *Biophysics Textbook Online*, http://www.biophysics.org/portals/1/PDFs/Education/gerig.pdf.

Glaser, R.W., Sachse, C., Dürr, U.H.N. *et al.* (2004) Orientation of the antimicrobial peptide PGLa in lipid membranes determined from ^{19}F-NMR dipolar couplings of 4-CF_3-phenylglycine labels. *J Magn Res* 168, 153–163.

Glaser, R.W., Sachse, C., Dürr, U.H.N. *et al.* (2005) Concentration-dependent realignment of the antimicrobial peptide PGLa in lipid membranes observed by solid-state ^{19}F-NMR. *Biophys J* 88, 3392–3397.

Gong, X.-M., Franzin, C.M., Thai, K. *et al.* (2007) Nuclear magnetic resonance structural studies of membrane proteins in micelles and bilayers. *Methods Mol Biol* 400, 515–529.

Grage, S.L., Afonin, S. & Ulrich, A.S. (2010) Dynamic transitions of membrane-active peptides. *Methods Mol Biol* 618, 183–207.

Grasnick, D. (2006) *Funktionelle Untersuchungen und Festkörper-NMR Strukturanalyse am Fusionspeptid des HIV-1*, PhD Thesis, University of Karlsruhe.

Gulevich, A.V., Shevchenko, N.E., Balenkova, E.S. *et al.* (2008) The Ugi reaction with CF_3-carbonyl compounds: effective synthesis of α-trifluoromethyl amino acid derivatives. *Tetrahedron* 64, 11706–11712.

Gunasekaran, K., Ramakrishnan, C. & Balaram, P. (1997) β-Hairpins in proteins revisited: lessons for *de novo* design. *Protein Eng* 10, 1131–1141.

Hang, J., Li, H. & Deng, L. (2002) Development of a rapid, room-temperature dynamic kinetic resolution for efficient asymmetric synthesis of α-aryl amino acids. *Org Lett* 4, 3321–3324.

Hegedus, L.S., Schwindt, M.A., De Lombaert, S. *et al.* (1990) Photolytic reactions of chromium aminocarbene complexes. Conversion of amides to α-amino acids. *J Am Chem Soc* 112, 2264–2273.

Heitz, F., Morris, M.C. & Divita, G. (2009) Twenty years of cell-penetrating peptides: from molecular mechanisms to therapeutics. *Br J Pharmacol* 157, 195–206.

Hirakura, Y., Azimov, R., Azimova, R. *et al.* (2000) Polyglutamine-induced ion channels: a possible mechanism for the neurotoxicity of Huntington and other CAG repeat diseases. *J Neurosci Res* 60, 490–494.

Hjørringgaard, C.U., Pedersen, J.M., Vosegaard, T. *et al.* (2009) An automatic solid-phase synthesis of peptaibols. *J Org Chem* 74, 1329–1332.

Hollweck, W. & Burger, K. (1995) Untersuchungen zur Knüpfung der Peptidbindung mit α-trifluoromethyl substituierten α-Aminosäuren. *J Prakt Chem* 337, 391–396.

Höss, E., Rudolph, M., Seymour, L. *et al.* (1993) Peptide modification by incorporation of α-trifluoromethyl α-amino acids via trifluoromethyl-substituted acylimines. *J Fluor Chem* 61, 163–170.

Huguenot, F. & Brigaud, T. (2006) Concise synthesis of enantiopure α-trifluoromethyl alanines, diamines, and amino alcohols via the Strecker-type reaction. *J Org Chem* 71, 7075–7078.

Huheey, J.E. (1965) The electronegativity of groups. *J Phys Chem* 69, 3284–3291.

Jäckel, C. & Koksch, B. (2005) Fluorine in peptide design and protein engineering. *Eur J Org Chem*, 4483–4503.

Jagodzinska, M., Heguenot, F., Candiani, G. *et al.* (2009) Assessing the bioisomerism of the trifluoromethyl group with a protease probe. *ChemMedChem* 4, 49–51.

Kanithasen, N. (2005) *^{2}H- and ^{19}F- solid State NMR Studies of the Antimicrobial Peptide (KIAGKIA)$_3$ in Phospholipid Bilayers*, Diploma Thesis, University of Karlsruhe.

Keller, J.W. & Hamilton, B.J. (1986) Enzymatic resolution of 2-trifluoromethylalanine. *Tetrahedron Lett* 27, 1249–1250.

Kieltsch, I., Eisenberger, P. & Togni, A. (2007) Mild electrophilic trifluoromethylation of carbon- and sulfur-centered nucleophiles by a hypervalent iodine(III)–CF_3 reagent. *Angew Chem Int Edit* 46, 754–757.

Kobzev, S.P., Soloshonok, V.A., Galushko, S.V. *et al.* (1989) Fluorine-containing amino acids. VI. Acid-base properties of α-trifluoromethyl-α-amino acids. *Zhurnal Obshchei Khimii* 59, 909–912.

Koen, M.J., Morgan, J., Pinhey, J.T. *et al.* (1997) Reaction of 4-ethoxycarbonyl-2-phenyl-4,5-dihydrooxazol-5-one with organolead(IV) triacetates. A route to some α-arylglycine and α-vinylglycine derivatives. *J Chem Soc Perkin Trans* 1, 487–491.

Koksch, B., Sewald, N., Hofmann, H.-J. *et al.* (1997) Proteolytically stable peptides by incorporation of α-Tfm amino acids. *J Pept Sci* 3, 157–167.

Koksch, B., Quaedflieg, P.J.L.M., Michel, T. *et al.* (2004) Enzymatic resolution of C^{α}-fluoroalkyl substituted amino acids. *Tetrahedron: Asymm* 15, 1401–1407.

Kolb, P., Rosenbaum, D.M., Irwin, J.J. *et al.* (2009) Structure-based discovery of β_2-adrenergic receptor ligands. *Proc Natl Acad Sci USA* 106, 6843–6848.

Komarov, I.V., Grigorenko, A.O., Turov, A.V. *et al.* (2004) Conformationally rigid cyclic α-amino acids in the design of peptidomimetics, peptide models and biologically active compounds. *Russ Chem Rev* 73, 785–810.

Konno, T., Kanda, M., Ishihara, T. *et al.* (2005) A novel synthesis of fluorine-containing quaternary amino acid derivatives via palladium-catalyzed allylation reaction. *J Fluorine Chem* 126, 1517–1523.

Koppel, I.A., Taft, R.W., Anvia, F. *et al.* (1994) The gas-phase acidities of very strong neutral Brønsted acids. *J Am Chem Soc* 116, 3047–3057.

Kukhar, V.P. & Soloshonok, V.A. (Eds) (1995) *Fluorine-Containing Amino Acids: Synthesis and Properties*, Wiley, Chichester.

Lacapère, J.-J., Pebay-Peyroula, E., Neumann, J.-M. *et al.* (2007) Determining membrane protein structures: still a challenge! *Trends Biochem Sci* 32, 259–270.

Lagerström, M.C. & Schiöth, H.B. (2008) Structural diversity of G protein-coupled receptors and significance for drug discovery. *Nat Rev Drug Discov* 7, 339–357.

Lanza, G.M., Winter, P.M., Neubauer, A.M. *et al.* (2005) ^{1}H/^{19}F magnetic resonance molecular imaging with perfluorocarbon nanoparticles. *Curr Top Dev Biol* 70, 57–76.

Lee, E.C. & Fu, G.C. (2007) Copper-catalyzed asymmetric N–H insertion reactions: couplings of diazo compounds with carbamates to generate α-amino acids. *J Am Chem Soc* 129, 12066–12067.

Leroux, F. (2004) Atropisomerism, biphenyls, and fluorine: a comparison of rotational barriers and twist angles. *ChemBioChem* 5, 644–649.

Li, G., Liang, Y. & Antilla, J.C. (2007) A vaulted biaryl phosphoric acid-catalyzed reduction of α-imino esters: the highly enantioselective preparation of α-amino esters. *J Am Chem Soc* 129, 5830–5831.

Lindahl, E. & Sansom, M.S.P. (2008) Membrane proteins: molecular dynamics simulations. *Curr Opin Struct Biol* 18, 425–431.

Lontz, J.F. & Raasch, M.S. (1953) Alpha-amino acids containing alpha-fluorinated-alkyl or cycloalkyl groups. *US Pat* 2662915, *Chem Abstr* 48, 71849.

Lundstrom, K. (2006) Structural genomics for membrane proteins. *Cell Mol Life Sci* 63, 2597–2607.

Maisch, D. (2008) *Synthese und Strukturuntersuchungen des membranaktiven Peptaibols Alamethicin mittels Festkörper ^{19}F-NMR*, PhD Thesis, University of Karlsruhe.

Maisch, D., Wadhwani, P., Afonin, S. *et al.* (2009) Chemical labeling strategy with (*R*)- and (*S*)-trifluoromethylalanine for solid state ^{19}F-NMR analysis of peptaibols in membranes. *J Am Chem Soc* 131, 15596–15597.

Malet-Martino, M., Gilard, V., Desmoulin, F. *et al.* (2006) Fluorine nuclear magnetic resonance spectroscopy of human biofluids in the field of metabolic studies of anticancer and antifungal fluoropyrimidine drugs. *Clin Chim Acta* 366, 61–73.

Miles, A.J. & Wallace, B.A. (2006) Synchrotron radiation circular dichroism spectroscopy of proteins and applications in structural and functional genomics. *Chem Soc Rev* 35, 39–51.

Miyazawa, T. (1995) 'Enzymatic resolution of racemic fluorine-containing amino acids', in Kukhar, V.P. & Soloshonok, V.A. (Eds), *Fluorine-Containing Amino Acids, Synthesis and Properties*, Wiley, Chichester, pp. 267–295.

Monoi, H., Futaki, S., Kugimiya, S.-I. *et al.* (2000) Poly-*l*-glutamine forms cation channels: relevance to the pathogenesis of the polyglutamine diseases. *Biophys J* 78, 2892–2999.

Morgan, J., Pinhey, J.T. & Sherry, C.J. (1997) Reaction of organolead triacetates with 4-ethoxycarbonyl-2-methyl-4,5-dihydro-1,3-oxazol-5-one. The synthesis of α-aryl- and α-vinyl-*N*-acetylglycines and their ethyl esters and their enzymic resolution. *J Chem Soc Perkin Trans 1* 5, 613–619.

Müller, K., Faeh, C. & Diedrich, F. (2007) Fluorine in pharmaceuticals: looking beyond intuition. *Science* 317, 1881–1886.

Murphy, C.D. (2007) The application of ^{19}F nuclear magnetic resonance to investigate microbial biotransformations of organofluorine compounds. *OMICS* 11, 314–324.

Mykhailiuk, P.K., Afonin, S., Chernega, A.N. *et al.* (2006) Conformationally rigid trifluoromethyl-substituted α-amino acid designed for peptide structure analysis by solid-state ^{19}F NMR spectroscopy. *Angew Chem Int Edit* 45, 5659–5661.

Mykhailiuk, P.K. (2008) *New Fluorine-Labelled Amino Acids as ^{19}F-NMR Reporters for Structural Peptide Studies: Design, Synthesis and Applications*, PhD Thesis, University of Karlsruhe.

Mykhailiuk, P.K., Afonin, S., Palamarchuk, G.V. *et al.* (2008) Synthesis of trifluoromethyl-substituted proline analogues as ^{19}F NMR labels for peptides in the polyproline II conformation. *Angew Chem Int Edit* 47, 5765–5767.

Mykhailiuk, P.K., Voievoda, N.M., Afonin, S. *et al.* (2010) An optimized protocol for the multigram synthesis of 3-(trifluoromethyl)bicyclo[1.1.1]pent-1-ylglycine (CF_3-Bpg) *J Fluor Chem* 131, 217–220.

Nadano, R., Iwai, Y., Mori, T. *et al.* (2006) Divergent chemical synthesis of prolines bearing fluorinated one-carbon units at the 4-position via nucleophilic 5-endo-trig cyclizations. *J Org Chem* 71, 8748–8754.

Naito, A. & Kawamura, I. (2007) Solid-state NMR as a method to reveal structure and membrane-interaction of amyloidogenic proteins and peptides. *Biochim Biophys Acta* 1768, 1900–1912.

Nemoto, H., Moriguchi, H., Ma, R. *et al.* (2007) Highly diastereoselective nucleophilic addition reactions of masked acyl cyanide reagents to *tert*-butanesulfiminides. *Tetrahedron: Asymm* 18, 383–389.

Nielsen, N.C., Malmendal, A. & Vosegaard, T. (2004) Techniques and applications of NMR to membrane proteins. *Mol Membr Biol* 21, 129–141.

Prasad, B.V., Balaram, P. & Benedetti, E. (1984) The stereochemistry of peptides containing α-aminoisobutyric acid. *CRC Crit Rev Biochem* 16, 307–348.

Qiu, X.-L. & Qing, F.-L. (2002a) Synthesis of Boc-protected *cis*- and *trans*-4-trifluoromethyl-D-prolines. *J Chem Soc Perkin Trans 1*, 2052–2057.

Qiu, X.-L. & Qing, F.-L. (2002b) Practical synthesis of Boc-protected *cis*-4-trifluoromethyl and *cis*-4-difluoromethyl-l- prolines. *J Org Chem* 67, 7162–7164.

Raman, P., Cherezov, V. & Caffrey, M. (2006) The membrane protein data bank. *Cell Mol Life Sci* 63, 36–51.

Reddy, K.L. & Sharpless, K.B. (1998) From styrenes to enantiopure α-arylglycines in two steps. *J Am Chem Soc* 120, 1207–1217.

Reichert, J., Grasnick, D., Afonin, S. *et al.* (2007) A critical evaluation of the conformational requirements of fusogenic peptides in membranes. *Eur Biophys J* 36, 405–413.

Rosenbaum, D.M., Rasmussen, S.G.F. & Kobilka, B.K. (2009) The structure and function of G-protein-coupled receptors. *Nature* 459, 356–363.

Sabatino, G. & Papini, A.M. (2008) Advances in automatic, manual and microwave-assisted solid-phase peptide synthesis. *Curr Opin Drug Discov Devel* 11, 762–770.

Sachse, C. (2003) *Festkörper ¹⁹F-NMR Untersuchungen zur Orientierung und Dynamik des antimikrobiellen Peptides PGLa in Lipidmembranen*, Diploma Thesis, University of Jena.

Salgado, J., Grage, S.L., Kondejewski, L.H. *et al.* (2001) Membrane-bound structure and alignment of the antimicrobial β-sheet peptide gramicidin S derived from angular and distance constraints by solid state ¹⁹F-NMR. *J Biomol NMR* 21, 191–208.

Sang, Y. & Blecha, F. (2008) Antimicrobial peptides and bacteriocins: alternatives to traditional antibiotics. *Anim Health Res Rev* 9, 227–235.

Santagada, V., Frecentese, F., Perissutti, E. *et al.* (2009) Microwave assisted synthesis: a new technology in drug discovery. *Mini-Rev Med Chem* 9, 340–358.

Scheerer, P., Park, J.H., Hildebrand, P.W. *et al.* (2008) Crystal structure of opsin in its G-protein-interacting conformation. *Nature* 455, 497–502.

Schierlinger, C. & Burger, K. (1992) Peptide modification by introduction of α-trifluoromethyl α-amino acids via 4-trifluoromethyl-1,3-oxazolidin-2,5-diones. *Tetrahedron Lett* 33, 193–194.

Schmidt-Krey, I. (2007) Electron crystallography of membrane proteins: two-dimensional crystallization and screening by electron microscopy. *Methods* 41, 417–426.

Sewald, N., Seymour, L.C., Burger, K. *et al.* (1994) Asymmetric synthesis of α-trifluoromethyl substituted aminoacids *via* 3-hydroxy-3-trifluoromethyl-2,5-diketopiperazines. *Tetrahedron: Asymm* 5, 1051–1060.

Sewald, N. & Burger, K. (1995) 'Synthesis of β-fluorine-containing amino acids', in Kukhar, V.P. & Soloshonok, V.A. (Eds), *Fluorine-Containing Amino Acids, Synthesis and Properties*, Wiley, Chichester , pp. 139–221.

Sewald, N., Hollweck, W., Mütze, K. *et al.* (1995) Peptide modification by introduction of α-trifluoromethyl substituted amino acids. *Amino Acids* 8, 187–194.

Shaw, N.M. & Naughton, A.B. (2004) The substrate specificity of the heat-stable stereospecific amidase from *Klebsiella oxytoca*. *Tetrahedron* 60, 747–752.

Skarpos, H., Vorob`eva, D.V., Osipov, S.N. *et al.* (2006) Methyltrifluoropyruvate imines possessing *N*-oxalyl and *N*-phosphonoformyl groups — precursors to a variety of α-CF_3-α-amino acid derivatives. *Org Biomol Chem* 4, 3669–3674.

Smart, B.E. (1995) *Properties of Fluorinated Organic Compounds*, ACS Monograph 187, Vol. 1, ACS, Washington DC.

Smart, B.E. (2001) Fluorine substituent effects (on bioactivity). *J Fluorine Chem* 109, 3–11.

Smits, R., Cadicamo, C.D., Burger, K. *et al.* (2008) Synthetic strategies to α-trifluoromethyl and α-difluoromethyl substituted α-amino acids. *Chem Soc Rev* 8, 1727–1739.

Soloshonok, V.A., Gerus, I.I., Yagupolskii, Y.L. *et al.* (1987) Fluorine-containing amino acids. III. α-Trifluoromethyl-α-amino acids. *Zh Org Khim* 23, 2308–2313; *Chem Abstr* 109, 55185.

Strandberg, E., Wadhwani, P., Tremouilhac, P. *et al.* (2006) Solid state NMR analysis of the PGLa peptide orientation in DMPC bilayers: structural fidelity of ^{2}H-labels versus high sensitivity ^{19}F-NMR. *Biophys J* 90, 1676–1686.

Strandberg, E., Kanithasen, N., Tiltak, D. *et al.* (2008) Solid-state NMR analysis comparing the designer-made antibiotic MSI-103 with its parent peptide PGLa in lipid bilayers. *Biochem* 47, 2601–2616.

Tanaka, M., Kurosaki, Y., Washio, T. *et al.* (2007) Enantioselective amination of silylketene acetals with (*N*-arylsulfonylimino)phenyliodinanes catalyzed by chiral dirhodium(II) carboxylates: asymmetric synthesis of phenylglycine derivatives. *Tetrahedron Lett* 48, 8799–8802.

Thust, S. & Koksch, B. (2003) Protease-catalyzed peptide synthesis for the site-specific incorporation of α-fluoroalkyl amino acids into peptides. *J Org Chem* 68, 2290–2296.

Thust, S. & Koksch, B. (2004) Discovery of carboxypeptidase Y as a catalyst for the incorporation of sterically demanding α-fluoroalkyl amino acids into peptides. *Tetrahedron Lett* 45, 1163–1165.

Tiltak, D. (2009) *Strukturelle und funktionelle Untersuchungen der antimikrobiellen Peptide MSI-103 und Temporin A*, PhD Thesis, University of Karlsruhe.

Toniolo, C. & Benedetti, E. (1991) Structure of polypeptides from α-amino acids disubstituted at the α-carbon. *Macromolecules* 24, 4004–4009.

Torres, J., Stevens, T.J. & Samsó, M. (2003) Membrane proteins: the 'wild west' of structural biology. *Trends Biochem Sci* 28, 137–144 (Erratum in: *Trends Biochem Sci* 2003 28, 174).

Ulrich, R., Glaser, R.W. & Ulrich, A.S. (2003) Susceptibility corrections in solid state NMR experiments with oriented membrane samples. Part II: theory. *J Magn Reson* 164, 115–127.

Ulrich, A.S. (2005) Solid state ^{19}F NMR methods for studying biomembranes. *Prog NMR Spect* 46, 1–21.

Ulrich, A.S., Wadhwani, P., Dürr, U.H.N. *et al.* (2006) 'Solid-state ^{19}F-nuclear magnetic resonance analysis of membrane-active peptides', in Ramamoorthy, A. (Ed.), *NMR Spectroscopy of Biological Solids*, Taylor & Francis, Boca Raton, pp. 215–236.

Ulrich, A.S. (2007) 'Solid state NMR analysis of oriented biomembranes', in Webb, G.A. (Ed.), *Modern Magnetic Resonance*, Springer, Dordrecht, pp. 261–263.

Vernier, J.-M., Hegedus, L.S. & Miller, D.B. (1992) Synthesis of optically active arylglycines by photolysis of optically active (β-hydroxamino) carbene-chromium(0) complexes. *J Org Chem* 57, 6914–6920.

Wadhwani, P., Afonin, S., Ieronimo, M. *et al.* (2006) Optimized protocol for synthesis of cyclic gramicidin S: starting amino acid is key to high yield. *J Org Chem* 71, 55–61.

Wadhwani, P., Bürck, J., Strandberg, E. *et al.* (2008) Using a sterically restrictive amino acid as a ^{19}F NMR label to monitor and to control peptide aggregation in membranes. *J Am Chem Soc* 130, 16515–16517.

Wang, H., Zhao, X., Li, Y. *et al.* (2006) Solvent-controlled asymmetric Strecker reaction: stereoselective synthesis of α-trifluoromethylated α-amino acids. *Org Lett* 8, 1379–1381.

Wang, G. (2008) NMR of membrane-associated peptides and proteins. *Curr Protein Pept Sci* 9, 50–69.

Watanabe, H., Hashizume, Y. & Uneyama, K. (1992) Homologation of trifluoracetimidoyl iodides by palladium-catalyzed carbonylation: an approach to α-amino perfluoroalkanoic acids. *Tetrahedron Lett* 33, 4333–4336.

Wiener, M.C. (2004) A pedestrian guide to membrane protein crystallization. *Methods* 34, 364–372.

Williams, R.M. & Hendrix, J.A. (1992) Asymmetric synthesis of arylglycines. *Chem Rev* 92, 889–917.

Yu, J.-X., Kodibagkar, V.D., Cui, W. *et al.* (2005) ^{19}F: a versatile reporter for non-invasive physiology and pharmacology using magnetic resonance. *Curr Med Chem* 12, 819–848.

Zanda, M. (2004) Trifluoromethyl group: an effective xenobiotic function for peptide backbone modification. *New J Chem* 28, 1401–1411.

[1] http://www.drorlist.com/nmr/MPNMR.html

[2] http://blanco.biomol.uci.edu/Membrane_Proteins_xtal.html

[3] http://www.pdb.org

Section 2

5
Fluorine-Containing Pharmaceuticals

*Steve Swallow**

5.1 Introduction

The presence of fluorine in drug candidates and marketed drugs is now commonplace, with the small electronegative fluorine atom being an integral part of the chemist's repertoire of substitutions to modulate all aspects of molecular properties, including drug potency, physical chemistry, and pharmacokinetics. The impact of fluorine on molecular properties has been well described in the literature and others have highlighted the common occurrence of fluorine in marketed pharmaceuticals. There are numerous thorough reviews that survey the use of fluorine in marketed drugs and development compounds (Kirk, 2006; Bégué and Bonnet-Delpon, 2008; Hagmann, 2008); however, one limitation is that frequently there is limited information in the public domain on the rationale for inclusion of particular fluorine substitutions and a similar dearth of data to indicate what the impact has been.

Rather than describe comprehensively all pharmaceutical examples for different fluorine substituent patterns, this chapter will therefore aim at focusing on examples where the influence of fluorine substitution is clear or unusual and where significant advance to therapeutic intervention has been observed.

* AstraZeneca, Safety Assessment UK, Mereside, Alderley Park, Macclesfield, Cheshire, SK10 4TG, UK. E-mail: steve.swallow@astrazeneca.com

The introduction of fluorine has found great utility in many drug discovery projects that have not progressed to market; for these the reader is directed to other literature reviews and articles (Isanbor and O'Hagan, 2006; Kirk 2006; Jeschke *et al.*, 2007; Müller *et al.*, 2007; Shah and Westwell, 2007; Bégué and Bonnet-Delpon, 2008; Bohm *et al.*, 2008; Hagmann, 2008; Liu *et al.*, 2008; Purser *et al.*, 2008).

5.1.1 *Survey of fluorine-containing pharmaceuticals*

A search within the Prous Science Integrity database for launched pharmaceuticals reveals that more than 200 pharmaceuticals containing at least one fluorine substituent have been launched since the introduction of dexamethasone by Merck Sharp and Dohme in 1958. Similar literature analyses highlight the presence of fluorine in 5–15% of launched drugs in the 50 years to 2007 (Hagmann, 2008). Fluorine appears in pharmaceuticals in a variety of structural presentations as highlighted in Fig. 5.1 with

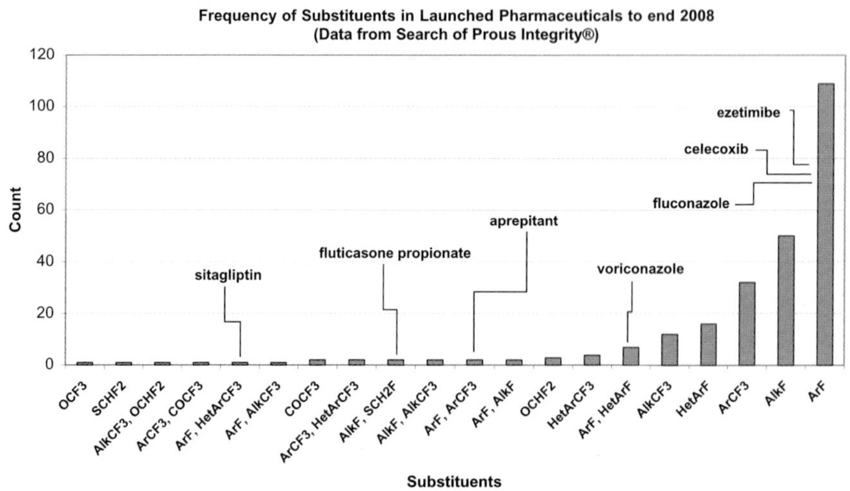

Figure 5.1. Distribution of fluorine substructural types in launched pharmaceuticals in the Prous Science Integrity database to end of 2008 (now the Thomson Reuters Integrity database). Note multiple occurrences of the same substructure only count once and are recorded under that substructure type, e.g. ezetimibe that contains two ArF substituents is counted once under ArF. Examples discussed in the text are labelled.

compounds containing between 1 and 21 fluorine atoms having found broad application in all therapeutic areas. A number of examples will be discussed in detail below.

It is perhaps not surprising that fluoroaryl, fluoroalkyl and trifluoromethylaryl substituents are the most commonly found; however, it is clear that there are other interesting fluorine containing substitutents in use. Unfortunately there is limited information on the impacts on properties for many of these, although one can speculate on them.

By far the most common fluorine substitution in pharmaceuticals is the aromatic fluorine, most commonly a 4-fluorinated phenyl ring. The substitution of an unsubstituted phenyl ring for a 4-fluorinated version is frequently introduced by medicinal chemists to enhance metabolic stability by preventing phenyl oxidation, although such substitution can also have positive impacts on potency through the introduction of additional specific interactions. Substitution of Ar–H for Ar–F has minimal impact on the other properties of the molecule and is arguably often incorporated into discovery project chemistry pre-emptively and in the absence of data highlighting specific metabolic liability because of this minimal impact on other properties. It is interesting that a study by Pfizer chemists (Lewis and Cucurull-Sanchez, 2009) examining the impact of simple substitutions on the metabolic stability of compounds, through an analysis of 'matched pairs' (Leach *et al.*, 2006), highlighted that a 4-F substituent leads to a decrease in metabolism in human liver microsomal clearance assays in only 9.2% of 491 pairs examined, while increasing clearance in 4.5% of pairs. Arguably, the value in this observation is that 4-F substitution has minimal impact on other properties when compared with higher performing groups (with respect to metabolic stability) that modulate physical properties more radically, and perhaps also highlights that maximum value from this strategy is derived when the metabolic pathway is known to involve phenyl oxidation. Aromatic fluorine substitution can also be used to block access to metabolic pathways leading to reactive metabolites (Wu *et al.*, 2003) that may be of concern from a toxicological perspective. Indeed there are a number of well-described examples where metabolism data have indicated the impact of Ar–F substitution on modulating the metabolism of a phenyl substituent, some of which are described below.

5.2 Case Studies

5.2.1 *Ezetimibe (Zetia)*

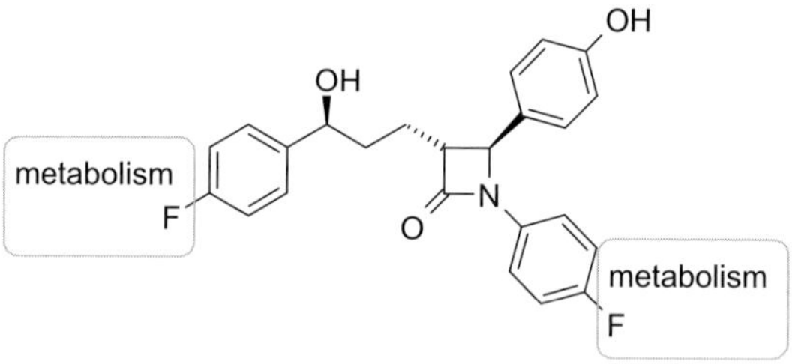

The discovery of ezetimibe (Zetia), a first in class compound that inhibits the absorption of cholesterol from the intestine, is an excellent example highlighting the role of fluorine substitution in metabolism-based drug optimization.

Following the success of the statin class in reducing serum lipoprotein cholesterol (LDL-C) and total cholesterol through inhibiting cholesterol biosynthesis, other cholesterol-lowering approaches have been pursued. Reduction in the absorption of dietary cholesterol was considered an attractive approach owing to its potential as a stand-alone or additive combination therapy and pursuit of this objective by Schering Plough led to the launch in 2002 of ezetimibe, leading to combined sales of $4.6bn in 2008 from ezetimibe monotherapy and a single pill combination of ezetimibe with simvastatin (Vytorin).

The whole discovery story is fascinating (Rosenblum *et al.*, 1998; Earl and Kirkpatrick, 2003), but focusing on the early structure–activity relationship that led to the initial probe compound is instructive, highlighting the value in this case of a switch from optimization against acyl-CoA cholesteryl acyl transferase (ACAT; the initial molecular target) to a classical pharmacology-based optimization in a rat disease model, following the observation of a disconnect between *in vitro* ACAT inhibitory activity and *in vivo* biological effects.

Following identification of the azetidinone **1**, initial optimization led to SCH48461 (Fig. 5.2). Among the structure–activity relationships described in the literature, it was observed that introduction of the 4-methoxy substituent in the azetidine C-4 phenyl substituent had a

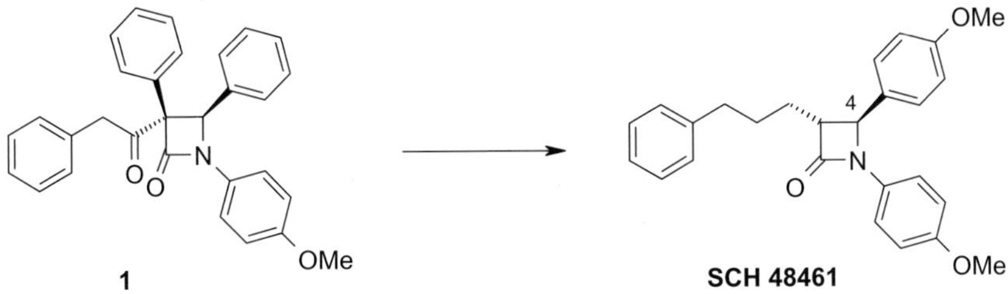

1 SCH 48461

Figure 5.2. Structures of initial lead and first clinical candidate to show proof-of-concept.

profound impact on enhancing compound activity *in vivo* and the tolerance of other polar substituents in this ring that are of particular importance to the later metabolism-based optimization.

Completion of phase II clinical trials with SCH48461 provided the initial proof-of-concept for the inhibition of cholesterol absorption and SCH48461 showed positive effects, reducing serum LDL levels by 15% at 25mg/day dose. However, the modest activity in humans coupled with a complex metabolite profile led to a programme of further optimization involving the use of aromatic fluorine substitution to minimize the formation of inactive metabolites.

In an elegant approach using metabolite identification, isolation and testing, in combination with metabolite structure hypothesis generation and an understanding of prior structure–activity relationship, a targeted set of compounds was designed that combined metabolite features expected to have positive effects on potency with features expected to block metabolism to inactive structures (Fig. 5.3) (Van Heek *et al.*, 1997; Rosenblum *et al.*, 1998).

Synthesis and testing critically demonstrated that the C-4 aryl *O*-demethylated derivative **2** retained activity (consistent with the structure–activity relationship noted above) while demethylation of the N-4-methoxyphenyl group **3** was shown to reduce activity as was the *para* hydroxylation of the pendant C-2 phenyl group **4** (Fig. 5.4). Retention of the C-4 phenol group, blocking *para* hydroxylation through fluorination, replacing the OMe with fluorine to prevent formation of polar (inactive) metabolites in this ring, and concomitant introduction of a benzylic alcohol led to SCH58235, a compound that had a 50-fold increase in potency compared with SCH48461 in the *in vivo* efficacy model. As a consequence, in addition to the improved *in vivo* efficacy, a much simplified metabolite

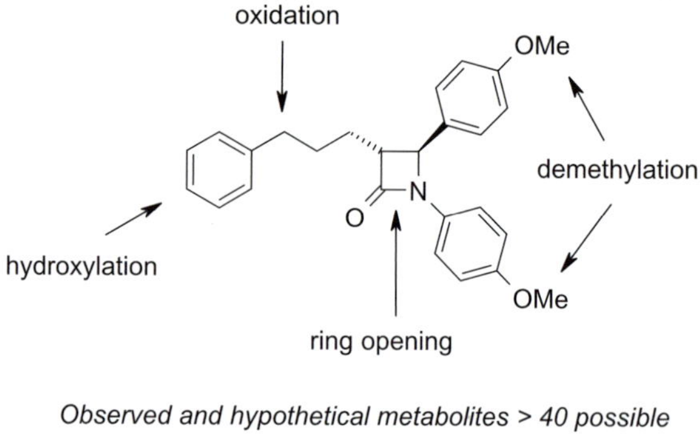

Figure 5.3. Putative sites of metabolism of SCH48461.

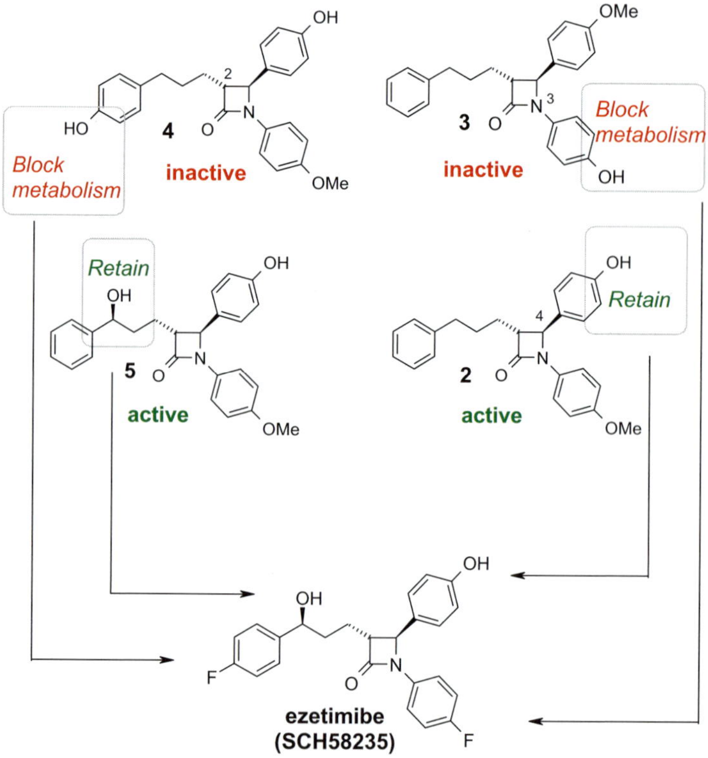

Figure 5.4. Key compounds and outcomes: blocking metabolism while retaining polar features in **2** and **5** leads to ezetimibe.

profile was also obtained in which the phenol glucuronide of SCH48461 was observed as the major metabolite in rats and subsequently in man (Sweeney and Johnson, 2007). Interestingly this glucuronide has been shown to be biologically active and is recycled to the site of action in the intestine via enterohepatic recirculation.

5.2.2 *Celecoxib (Celebrex)*

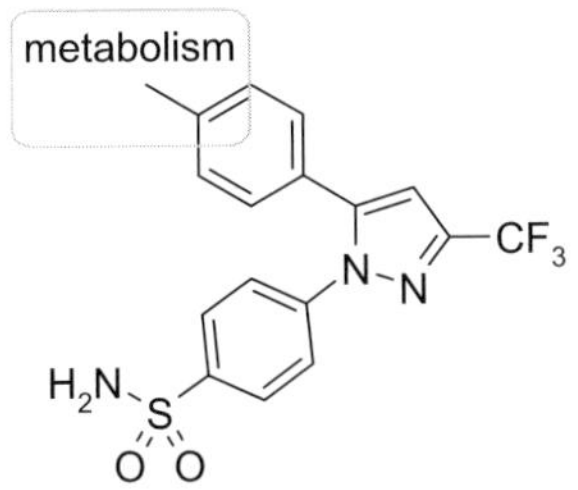

The propensity for aromatic fluorination to lead to high metabolic stability can potentially be problematic as exemplified in the discovery of celecoxib by researchers at Searle (Penning *et al.*, 1997). The early lead compounds **6** and **7** proved to be metabolically stable leading to a very long biological half-life (Fig. 5.5). Replacement with metabolically more labile groups to provide a potential site for metabolism led to compounds with a much more acceptable half-life from which celecoxib was selected.

5.2.3 *Sitagliptin (Januvia)*

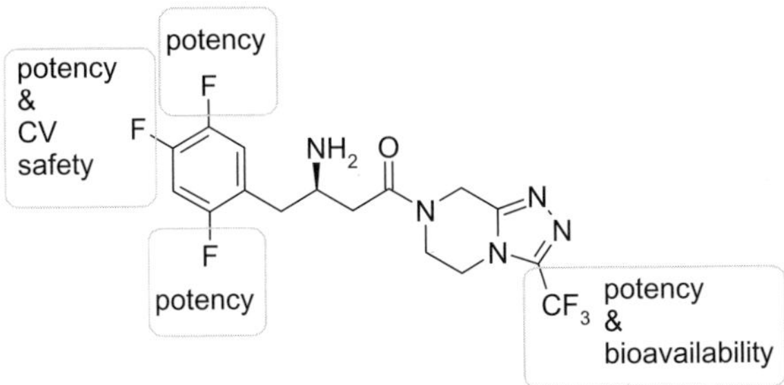

The discovery of sitagliptin, an oral dipeptidylpeptidase IV (DPP-IV) inhibitor for the treatment of type II diabetes, is an excellent example that

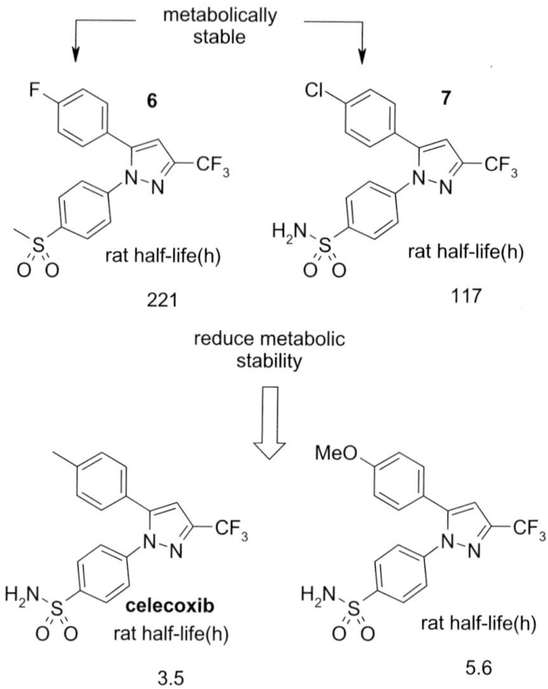

Figure 5.5. Changes to early Cox-2 inhibitors to reduce metabolic stability conferred by halogen substituents.

highlights the multiple impacts of fluorine on a compound's pharmacological and pharmacokinetic profile, although interestingly in this case it is oral absorption rather than metabolism that is impacted in the pharmacokinetic profile.

The search for oral DPP-IV inhibitors was based on the laboratory observation that DPP-IV inhibitors caused an increase in insulin secretion, a process believed to be mediated primarily by inhibiting the proteolytic cleavage of incretin hormone glucagon-like peptide-1 (GLP-1), a DPP-IV substrate, which has a clearly established role in glucose-dependent insulin biosynthesis and secretion (Drucker and Nauck 2006; Weber and Thornberry 2007).

Many companies embarked on oral DPP-IV inhibitor programmes and, while not first to enter this arena, Merck were first to market with sitagliptin following a rapid development that was, in part, the result of an innovative use of biomarkers that led to a faster than average development programme. Sitagliptin was approved by the US Food and Drug Administration (FDA) in 2006 and in 2008 achieved annual sales of $1.4bn.

5.2.3.1 *Impact of fluorine on pharmacology profile*

In contrast to many companies, Merck's DPP-IV drug discovery programme concentrated on compounds that did not rely on covalent binding to a nitrile electrophile because of concerns about potential compound instability associated with prototypic inhibitors at the time (Weber and Thornberry, 2007). Multiple approaches were pursued, but it was elaboration of high-throughput screening (HTS) hits that ultimately led to a successful outcome.

As part of an HTS programme, compound **8** (Fig. 5.6) was identified as a low micromolar potency hit and early structure–activity relationship studies and simplification of the structure highlighted thiazolidine **9**.

It was at this point that further structure–activity relationships started to indicate the significant impact of fluorine substitution on compound potency in this series as highlighted in Fig. 5.7: addition of a 2-fluoro substituent giving compound **11** a threefold enhancement of potency and additional fluorine substitutions enhanced potency a further eightfold. The third fluorine in the 4-position (compound **13**) arguably

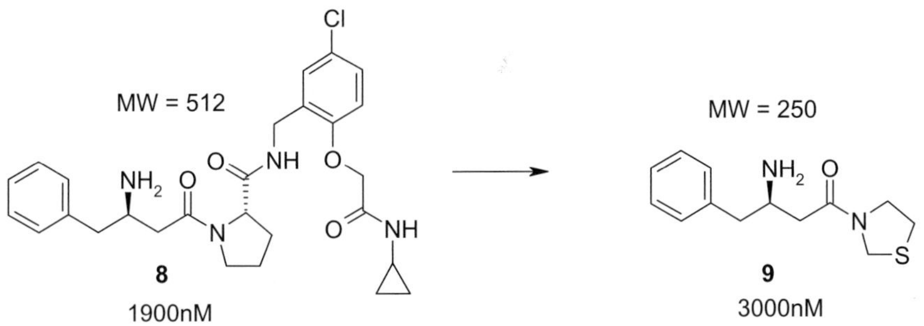

Figure 5.6. Initial screening hit and early lead.

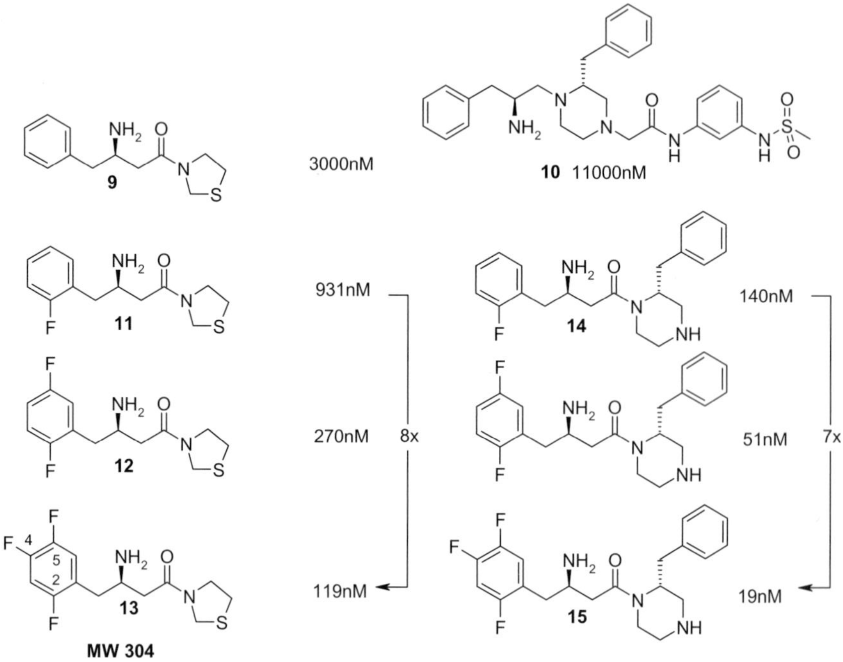

Figure 5.7. Impact of fluorine substitution on DPP-IV potency.

has limited impact on potency in this series; however, this subsequently became an important entity having significant impact on the off-target pharmacology despite its small size. It is also noteworthy that compound **13** is a highly efficient ligand compared with the initial lead **8**, with more than tenfold improvement in potency and a concomitant 40% reduction in molecular weight, and as such represented a highly attractive lead molecule. Unfortunately pharmacokinetics of **12** and **13** were poor owing to high clearance and low bioavailability (Xu *et al.*, 2004).

A parallel optimization that started from HTS hit **10** highlighted the cross-series applicability of this fluorine substitution and is illustrated for the matched pair **14** and **15** showing similar sevenfold enhancement in potency for the addition of the second and third fluorine atoms (Brockunier *et al.*, 2004). In common with the thiazolidines these compounds uniformly showed poor pharmacokinetics with high clearance and low bioavailability.

5.2.3.2 *Contribution of fluorine to pharmacokinetic profile*

In vitro metabolite identification studies implicated piperazine metabolism as problematic in this second, more potent, series and attempts to address this through appending heterocycles ultimately proved successful in this respect (Kim *et al.*, 2005). A number of piperazines containing fused heterocycles were studied; however, it is the triazole fused compounds that are of most interest.

Structure–activity relationships with the 3,4-difluoro-substituted phenyl series highlighted that activity lost by removal of the piperazine benzyl-substituent **16** could be partly restored through introduction of fused 5-membered ring heterocycles such as **17** and **18**. Whilst incorporation of an ethyl substituent **19** had a modest impact on potency and the oral

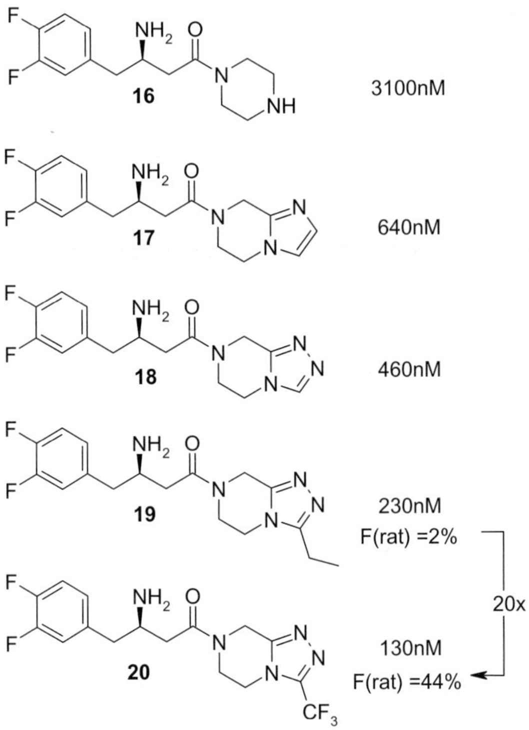

Figure 5.8. Impact of fused heterocycles on potency and bioavalability.

bioavailability remained low, increased metabolic stability appeared to have been achieved as the hepatic extraction ratio in rats was low (10–20%). *In vitro* permeability, intestinal loop, and rat portal vein cannulation studies also indicated that poor permeability was causing the low bioavailability.

Surprisingly, replacement of the ethyl side chain with a trifluoromethyl substituent **20** had a profound impact on bioavailability (F = 44%) while slightly increasing potency relative to **19**. An explanation for this phenomenon has not been put forward; however, the strong electron withdrawing effect of the trifluoromethyl group will likely reduce the strong hydrogen bonding potential of the triazole, impacting the desolvation required for membrane permeation.

Adjusting the fluorine substitution pattern to the more potent 2,4,5- and 2,5-substitution arrangements increased potency further, providing sitagliptin and desfluorositagliptin (Fig. 5.9), both with excellent potency and cross-species pharmacokinetic profiles. The high apparent metabolic stability and bioavailability of sitagliptin were ultimately demonstrated *in vivo* in healthy volunteers (CL_P = 416 ml/min, F = 87%) leading to a 100 mg once a day dosing regimen (Bergman and Krishna, 2007).

Follow-up studies on close analogues of sitagliptin further highlighted the consistent effect of other fluorosubstituted triazoles on oral absorption as indicated in Fig. 5.10 (Kim *et al.*, 2007).

desfluorositagliptin

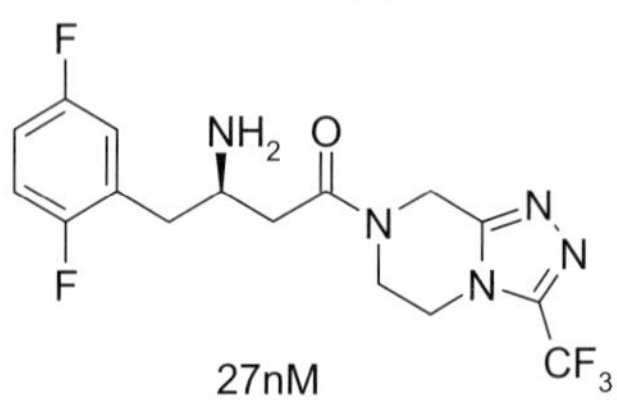

27nM

	Clp (ml/min/kg)	T1/2 (h)	F (%)
rat	43	1.6	51

sitagliptin

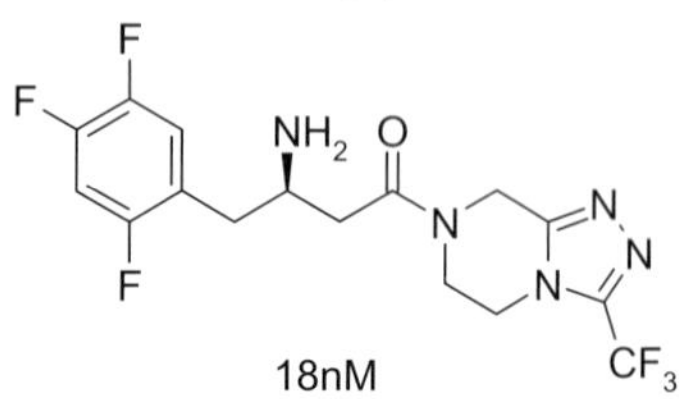

18nM

	Clp (ml/min/kg)	T1/2 (h)	F (%)
rat	60	1.7	76
dog	6	4.9	100
monkey	28	3.7	68

Figure 5.9. Similarity of desfluorositagliptin and sitagliptin.

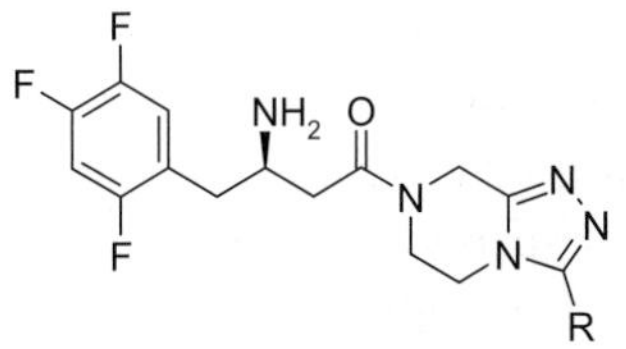

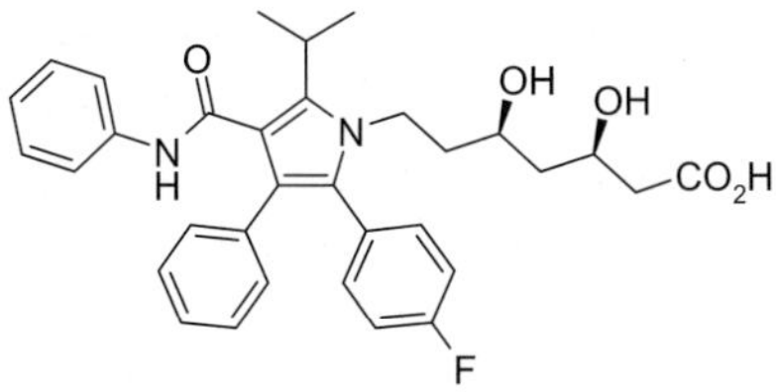

atorvastatin

Rat pharmacokinetics

R	Clp (ml/min/kg)	T1/2 (h)	F (%)
H	40	1.0	3.0
Et	70	1.7	2.0
cyclopropyl	68	1.8	3.0
CHF_2	66	1.3	39
CF_3	60	1.7	76
CF_2CF_3	58	2.3	61

Figure 5.10. Fluoroalkyl-substituted triazoles showing the impact on bioavailability (F%), structure of atorvastatin.

5.2.3.3 *Structural aspects*

The potency enhancements provided by S-1 aryl group fluorination are arguably more than one would expect based on lipophilicity arguments; an analysis of the X-ray structure of sitagliptin and other DPP-IV inhibitors bound to DPP-IV has suggested that the 2-fluoro substituent $C^{\delta+}$–$F^{\delta-}$ dipole makes a favourable electrostatic interaction with the positively charged Arg125 (Fig. 5.11a) (Kuhn *et al.*, 2007). A survey of protein data bank (PDB) structures has highlighted this as a common favourable fluorine interaction (Müller *et al.*, 2007), one other notable example being in the structure of atorvastatin (Lipitor) (see structure in Fig. 5.10) (Istvan and Deisenhofer, 2001). Early structure–activity relationships in the development of this compound highlighted a fivefold improvement in potency for the 4-F analogue relative to the unsubstituted ring (Roth *et al.*, 1990), this benefit being highlighted by the presence of this motif in the other statins (rosuvastatin, pitavastatin and cerivastatin).

The 4- and 5-F substituents in sitagliptin presumably contribute to potency owing to other close contacts and optimal fitting of the pocket.

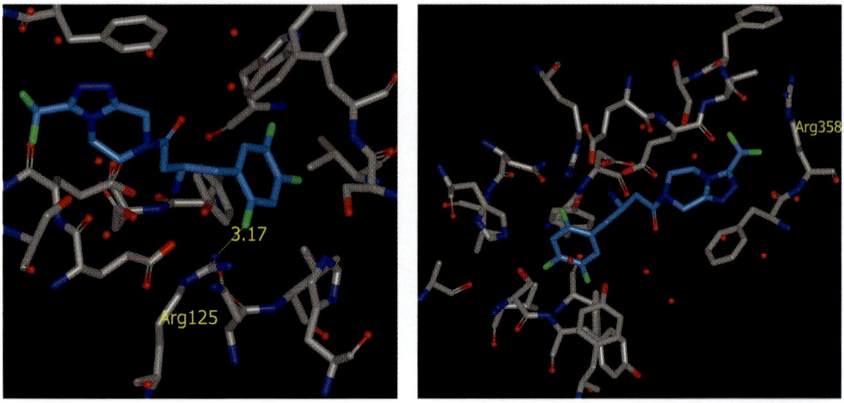

Figure 5.11. Sitagliptin bound to DPP-IV (PDB code1X70, selected residues removed for clarity): (a) showing S1 pocket with Arg125–F interaction highlighted in the foreground; (b) showing the CF_3 group which also appears to make favourable interactions with an arginine residue (here Arg358), though arguably in this case the potency contribution is limited (see structure–activity relationship above).

5.2.3.4 *Contribution of fluorine to safety profile*

As a final comment on the discovery of sitagliptin, it is interesting that desfluorositagliptin was progressed ahead of sitagliptin and preferential differentiation of sitagliptin only came following assessment of cardiovascular safety in which sitagliptin was demonstrated to have a much cleaner profile with a no observable effect level (NOEL) of 10 mg/kg with maximal plasma concentrations of 59 μM compared to 1 mg/kg and 6.5 μM for desfluorositagliptin, respectively. The mechanism of the cardiovascular effect has not been described, but this again serves to highlight the significant impact that a single fluorine substituent can have on the pharmacological profile (Weber and Thornberry, 2007).

5.2.4 *Fluconazole (Diflucan) and Voriconazole (Vfend)*

Before the discovery of the azole antifungals by Pfizer scientists, the treatment of serious fungal infections was limited and the discovery of the highly successful compounds fluconazole and voriconazole were due, in part, to the favourable properties imparted by aryl fluorine substitution as discussed below. Before its patent expiry in 2004, fluconazole generated annual sales in excess of \$1bn (2008 sales >\$300 m). Voriconazole has become the new standard of care in the treatment of invasive aspergillosis and generated sales in excess of \$700 m in 2008.

5.2.4.1 *Fluconazole (Diflucan)*

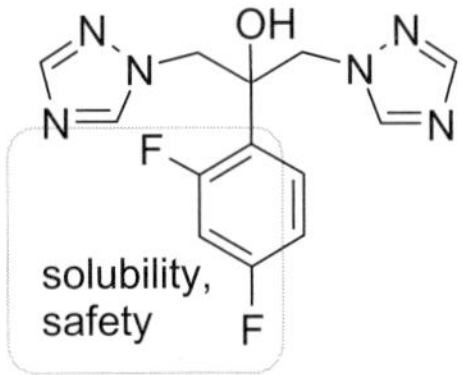

In the early 1970s, a programme was initiated at Pfizer with the objective of identifying compounds to treat serious systemic fungal infection by exploiting the known propensity of imidazole compounds to possess potent and selective *in vitro* activity against a wide range of fungal pathogens. They have the ability to inhibit the function of 14α-demethylase, a cytochrome P450-containing enzyme that is essential for the production of the principal fungal sterol ergosterol that is required to maintain the viability of fungi through its effects on fungal membrane fluidity.

The evolution of the programme is highlighted below and in Fig. 5.12 (Richardson, 1996). An initial compound of interest, tioconazole, was effective when administered topically; however, its high lipophilicity rendered it a poor compound for oral and intravenous administration owing to low oral bioavailability, high plasma protein binding and high metabolic clearance. Targeting compounds with lower lipophilicity led to ketoconazole, a compound with improved metabolic stability and good oral bioavailability, but still with high plasma-protein binding. This was an important advance but further improvements were sought. Extensive structure–activity relationship studies eventually traced the poor metabolic stability to the imidazole moiety, and replacement with alternative heterocycles identified UK-46,245 which was twofold more potent in an *in vivo* model than the corresponding imidazole despite sixfold lower *in vitro* activity, suggesting that improved metabolic stability had been achieved. Further attempts to reduce lipophilicity and metabolic lability led to the replacement of the *n*-hexyl group with a

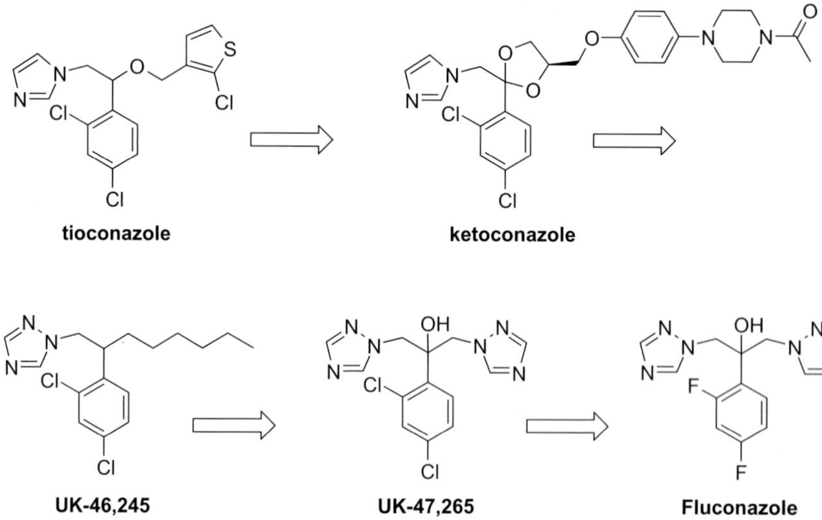

Figure 5.12. Evolution of fluconazole from tioconazole.

second 1,2,4-triazole moiety leading to UK-47,265, a compound with unprecedented levels of activity being 100-fold more potent than ketoconazole by the oral or intravenous route. Pharmacokinetic evaluation showed a promising profile with excellent bioavailability and half-life; however, preclinical safety studies were disappointing as UK-47,265 proved to be hepatotoxic in mice and dogs, and teratogenic in rats. A back-up programme had identified a series of additional promising compounds, among them the 2,4-difluoro analogue, fluconazole, which showed striking improvements compared to UK-47,265. In particular, it showed good aqueous solubility, was excreted largely unchanged in urine in animal studies, and it was devoid of teratogenicity and hepatotoxicity.

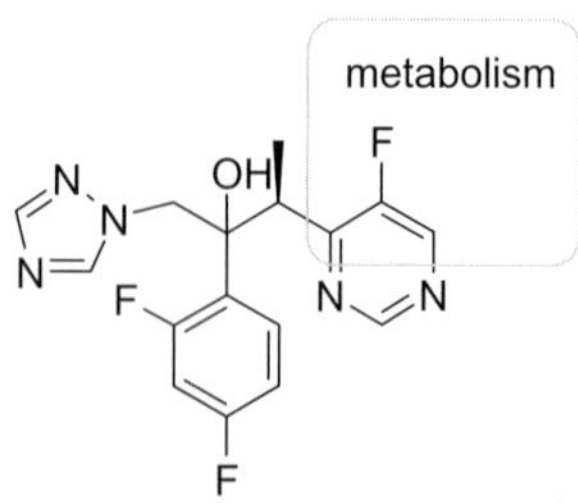

5.2.4.2 *Voriconazole (Vfend)*

Fluconazole became the agent of choice for treatment of infections due to *Candida albicans* and *Cryptococcus neoformans*; however, it is poorly effective against *Aspergillus* infections and compounds combining the favourable profile of fluconazole with a broader spectrum of action were sought (Dickinson *et al.*, 1996).

Introduction of a methyl group alpha to one of the triazole groups **21** resulted in increased potency against *Aspergillus fumigatus* and paved the way for further optimization through replacement of the proximal triazole with other heterocycles (Fig. 5.13). Replacement of the 1,2,4-triazole with 4-pyridinyl **22** or 4-pyrimidinyl groups **23** gave improved *in vitro* potency; however, introduction of a fluorine atom into the 5-position of the pyrimidine had a profound impact on *in vivo* potency despite the slight reduction in *in vitro* potency. This outcome has been ascribed to the reduced metabolic clearance imparted to these heterocycles (Dickinson *et al.*, 1996). Comparative pharmacokinetic data have

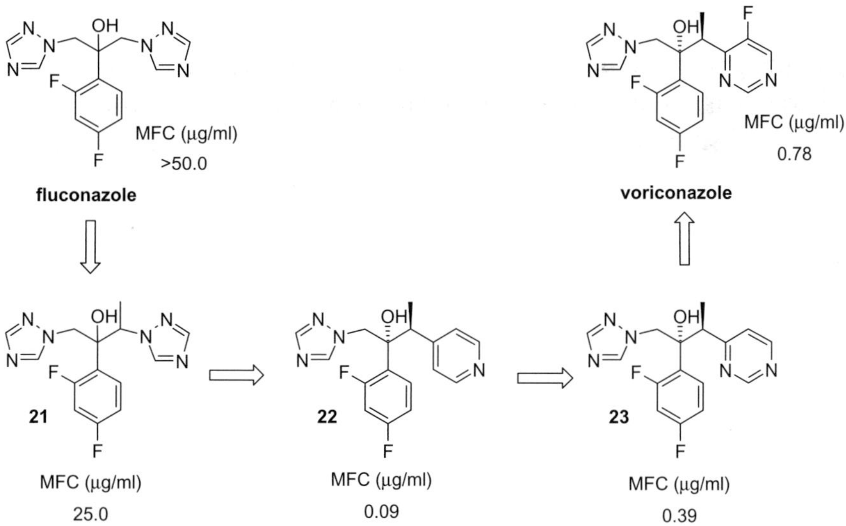

Figure 5.13. Improvements in *in vitro* potency against *Aspergillus fumigatus*. (MFC is the minimum fungicidal concentration giving at least 90% reduction in colony formation compared with drug free control.)

not been described, although the fact that voriconazole is extensively metabolized in multiple species, including man, to major metabolites derived from oxidation of the heterocycle, suggests that attenuation of the extent and/or rate of metabolism in this region of the molecule is a reasonable explanation for the impact of fluorination (Roffey *et al.*, 2003). Voriconazole, while active against *Aspergillus*, has also increased potency against *Candida* spp. compared to fluconazole (Barry and Brown, 1996).

5.2.5 *Fluoroquinolones*

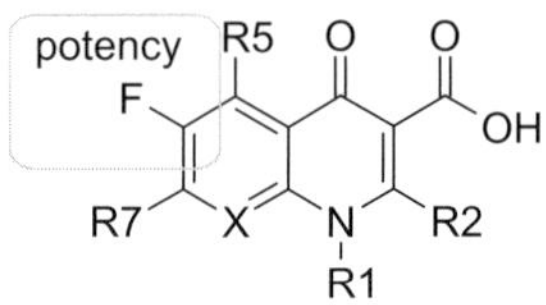

The fluoroquinolone antibiotics are a large and diverse series of more than 20 pharmaceuticals that have evolved from the initial discovery of nalidixic acid (Fig. 5.14), a compound discovered as an impurity in the manufacture of quinine in the 1960s (Andersson and MacGowan, 2003). Activity of the early compounds was limited to the treatment of urinary tract infections (UTIs), but has since grown to encompass activity across a broad spectrum of Gram positive and Gram negative bacteria, and clinical utilization in many areas from the original UTIs to systemic and respiratory infections. Several compounds in this class, for example levofloxacin (Levaquin)/ofloxacin (Floxin and moxifloxacin (Avelox)), achieved sales in excess of \$0.5bn in 2008.

The structural class can be broadly encompassed by the generic structure shown in Fig. 5.14 which clearly highlights the requirement for fluorine in the 6-position. It is interesting to note that since its introduction into the core structure of this class of molecules, in early drugs such as pefloxacin (the first quinolone antibiotic with systemic activity), this fluorine substituent has remained in all but one of the subsequent drugs to reach the market.

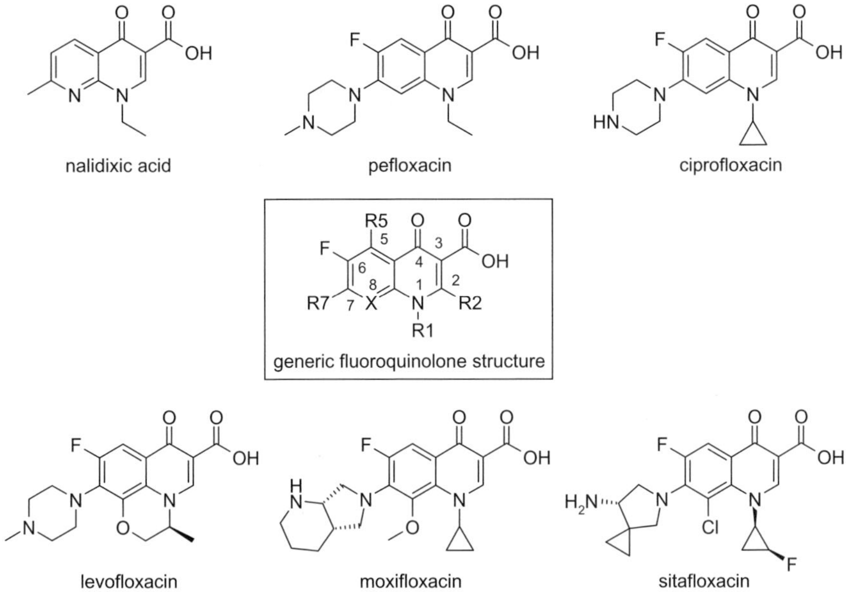

Figure 5.14. Generic and specific fluoroquinolone structures.

Despite the early introduction of this feature and extensive work in the area it took 25 years before a good understanding of the impact of fluorination was established through the careful analysis of the activities of a range of compounds in which the impact of single changes in molecular structure, a matched pair analysis, were assessed (Domagala *et al.*, 1986). For example, comparison of enzyme inhibitory activities and antibacterial potencies for the two matched pairs — norfloxacin and desfluoronorfloxacin **25**, and enoxacin and desfluoroenoxacin **26** — illustrates the dramatic impact of this substitution (Fig. 5.15).

The incorporation of a fluorine into the 6-position of **25** to give norfloxcin led to a 17.5-fold increase in DNA gyrase-inhibitory potency and a 63-fold increase in potency in the minimum inhibition concentration (MIC) against *Escherichia coli* H560, with similar increases in antibacterial activity being observed in other strains. A similar enhancement was observed with enoxacin when compared with its desfluoro analogue **26**. Such dramatic effects have been rationalized in terms of combined effects on compound binding and cellular penetration but are not seen with all

Compound	Minimum inhibition conc. of gyrase cleavage µg/ml	Antibacterial activity (MICs) µg/ml *E.coli* H560	*Klebsiela pneum.* MGH-2	*Pseudo. aerug* UI-18
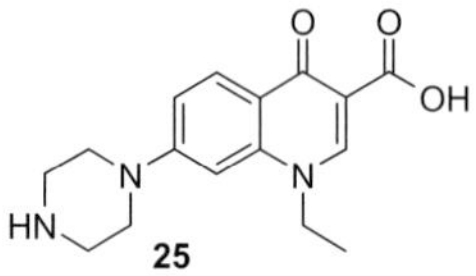 25	18	6.3	6.3	25
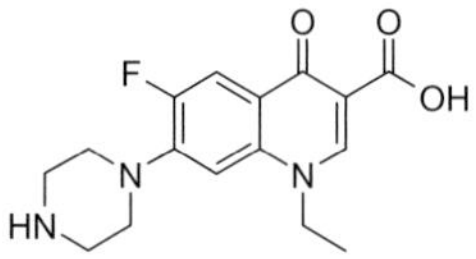 norfloxacin	1	0.1	0.05	0.2
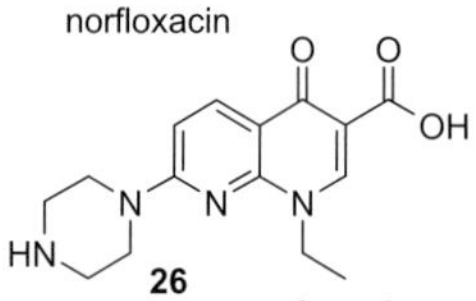 26	75	3.1	12.5	1.5
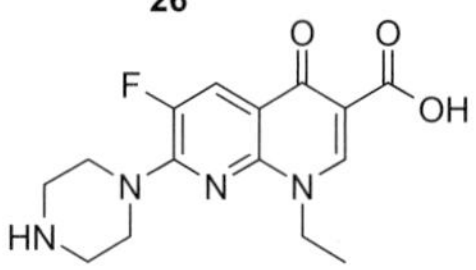 enoxacin	5	0.1	0.1	0.8

Figure 5.15. Matched pairs showing the impact of 6-fluorosubstitution on enzyme inhibition and antibacterial potency.

examples (gyrase-inhibitory potency enhancements range from 2- to 17-fold and cellular potency enhancements range from 2- to 100-fold), but appear to be particularly effective when combined with a piperazine at C-7 and may reflect synergistic, but as yet undefined, effects on this substituent, for example through pK_a modulation. The most recently approved fluoroquinolone, sitafloxacin (Fig. 5.14), contains an unusual fluorinated cyclopropane that is unique amongst marketed pharmaceuticals and which has been described to lower overall lipophilicity and improve selectivity against mammalian topoisomerase II (Kimura *et al.*, 2002).

5.2.6 *Fluticasone propionate (Flovent, Flixotide)*

Fluticasone propionate is a trifluorinated glucocorticoid receptor ligand that has been extensively used as an inhaled pharmaceutical for the treatment

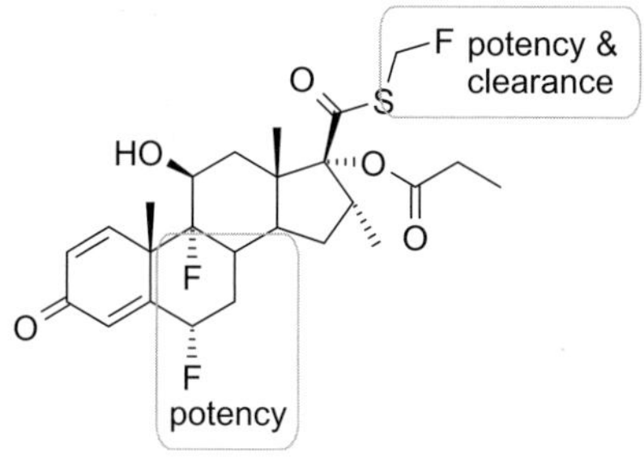

of asthma. In combination with salmeterol, a long acting β2-adrenergic agonist, it has become one of the world's leading pharmaceuticals (Advair, Serotide) generating sales in excess of £4bn in 2008, and is widely used in the treatment of asthma and chronic obstructive pulmonary disease. The impacts of fluorine on the development of this compound are several-fold as discussed below and build on many years of research in the steroid field.

The utility of fluorine in the discovery of glucocorticoids has a long history, with compounds such as dexamethasone, registered in the late 1950s, being amongst the first fluorine containing pharmaceuticals to be marketed. The early development of such compounds was based on improving the properties of the natural steroid cortisol (Fig. 5.16), specifically attempting to increase its topical potency while removing undesired effects through improving selectivity. The increase in potency, typically measured in *in vivo* systems, was achieved by one of several approaches: (i) by insertion of a double bond at the 1,2-position in the steroid nucleus, (ii) by the introduction of 6α-fluoro, 6α-methyl or 9α-fluoro substituents, or (iii) through combinations of these changes (Phillipps, 2009). Although anti-inflammatory potency was increased through these changes, undesired mineralocorticoid activity was increased to an even greater extent. However, this latter activity could be attenuated with substitutions in the 16-position as exemplified in the structures of dexamethasone (Fig. 5.16). It was subsequently found that masked alcohols at the 16- and 17-positions were preferred, giving rise to, for example, fluocinolone 16,17-acetonide. Inhalation provides topical delivery of such masked compounds, thus minimizing the systemic side effects for the treatment of airway diseases. This has led to compounds like beclometasone dipropionate having proven value in the treatment of bronchial asthma and rhinitis (Fig. 5.16).

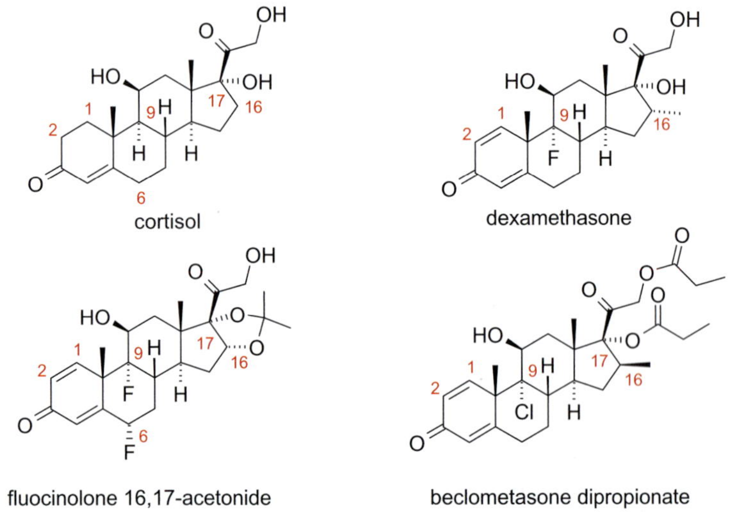

Figure 5.16. Structures of early glucocorticoids showing steroid numbering.

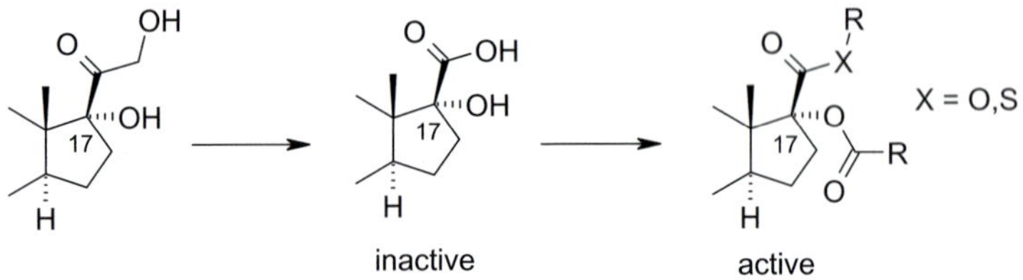

Figure 5.17. Core structure change leading to fluticasone discovery.

Despite these advances, there was a desire to improve further on systemic side effects, and workers at Glaxo embarked on a programme of research that delivered fluticasone propionate through exploitation of a novel series of 17β-carboxylates which marked a departure from the normal core structure containing the two-carbon 17β-side chain found in cortisol and derivatives (Fig. 5.17) (Bain *et al.*, 1974; Phillipps, 1990). Esterification of the inactive 17β-carboxylic acid provided both high potency and a pharmocokinetic deactivation handle that led to excellent selectivity combined with negligible systemic exposure.

A wide range of analogues were tested using a combination of assays to assess their potential relative to fluocinolone acetonide (Phillipps *et al.*, 1994). A vasoconstriction assay based on topical administration to the skin of human volunteers was used to assess anti-inflammatory activity and tissue penetration. Anti-inflammatory activity was further assessed in a topical rat model that allowed simultaneous assessment of systemic exposure on the undesired hypothalamic–pituitary–adrenal (HPA) function. Some key compounds are highlighted in Fig. 5.18 with potencies referenced to fluocinolone acetonide: in particular introduction of fluorine in the thioester **28** has a profound eightfold impact on the vasoconstriction potency relative to the desfluoro compound **27**. Introduction of the 6α-fluorine and saturation of the 16-exo double bond to give the 16α-methyl group improves the anti-inflammatory potency in rats while maintaining low HPA inhibitory potency, providing fluticasone proprionate, a compound with improved therapeutic index over fluocinolone acetonide.

The lack of effect on HPA function has been ascribed to the potential for metabolic inactivation; comparison of HPA effects following oral and subcutaneous dosing seemed to confirm this hypothesis with high first-pass turnover to the acid being observed in rats and zero bioavailability being subsequently observed in man (Harding, 1990). The weak glucocorticoid activity after oral administration is particularly valuable in the treatment of airway diseases where a high proportion of the inhaled dose is swallowed.

5.2.6.1 *Structural aspects*

The contribution of 9α-fluorination to glucocorticoid potency has been highlighted by a number of reports to be in the order of seven- to tenfold (Phillipps, 1990). A study in which the binding of cortisol and 9α-fluorocortisol to the glucocorticoid receptor were compared, shows a fourfold difference (Berger *et al.*, 1992), while another report comparing binding to the mineralocorticoid receptor suggests a difference of more than tenfold for the introduction of a fluorine (Genard and Palem-Vliers, 1983).

The X-ray structures of several 9α-fluorinated compounds, including fluticasone propionate, have now been solved, highlighting the presence of a hydrogen bond between the 11β-hydroxyl group and an amide H-bond acceptor in the protein (Bledsoe *et al.*, 2002; Kauppi *et al.*, 2003;

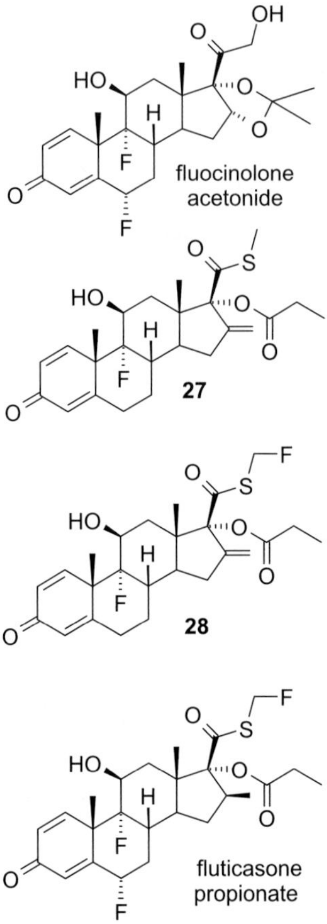

Compound	Human V	Rat AIT	Rat HPA
fluocinolone acetonide	100	100	100
27	187	19	<0.2
28	1497	20	2.5
fluticasone propionate	945	85	1.5

(27 → 28: 8x)

Figure 5.18. Activities of selected glucorticoid agonists relative to flucinolone acetonide (100-top) as described by Phillipps *et al.* (1994), highlighting the impact of fluorine substitution. V is topical human vasoconstrictor activity, AIT is topical anti-inflammatory activity and HPA is the undesired hypothalamic–pituitary–adrenal activity assessed by measuring circulating corticosterone levels following topical administration of compound.

Biggadike *et al.*, 2008). Introduction of the 9α-fluoro substituent might be expected to increase the H-bond strength through enhancing the hydroxyl group acidity (Mock and Zhang, 1990) while also making favourable interactions with other residues. The structure of dexamethasone-bound glucocorticoid receptor is highlighted in Fig. 5.19.

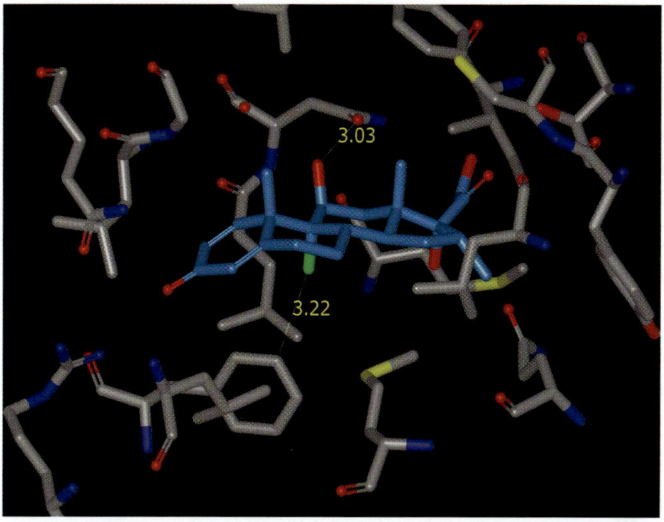

Figure 5.19. Structure of dexamethasone bound to the glutocorticoid receptor (PDB 1M2Z) (Bledsoe *et al.*, 2002) showing close contacts of 11β-OH and 9α-F.

A structural understanding of the contribution to binding of the fluorine in the fluoromethylthio group seems less clear, though favourable electrostatic interactions with an asparagine residue have been highlighted (Biggadike *et al.*, 2008) and inductive effects on the polarity of the thioester should not be ignored.

5.2.7 *Aprepitant (Emend)*

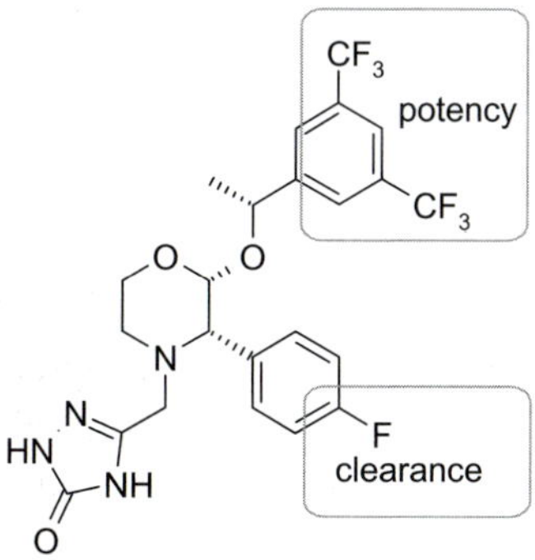

Aprepitant was the first NK1 receptor antagonist to be launched and was approved by the FDA in 2003 for use as an antiemetic for the prevention of acute and delayed nausea and vomiting associated with cancer treatment using emetogenic chemotherapeutics such as high dose cisplatin. In 2008 this compound generated sales of $264m for Merck & Co.

The discovery of this compound originates from early work on the NK1 receptor and one of its natural substrates, the undecapeptide substance P, implicating NK1 in a variety of biological responses including pain transmission and the emetic response. Early NK1 antagonists were peptidic in nature and had value only as research tools owing to undesirable properties such as poor solubility, pharmacokinetics and selectivity. It was the discovery of the first selective, non-peptidic NK1 antagonist, CP-96345 (Fig. 5.20), by workers at Pfizer that provoked a growth in the area with many groups pursuing this molecular target (Humphrey, 2003).

In an early systematic exploration of this series by workers at Merck, it was shown that the benzylamine substituent in CP-96345 could be modified to an ether substituent and potency maintained through the introduction of 3,5-substituents (Swain *et al.*, 1993). The 3,5-dimethyl analogue **31** proved to be 20- and 100-fold more potent than the methyl-substituted analogue **30** and the unsubstituted parent compound **29** (Fig. 5.20). In related work on an acyclic scaffold aimed at minimizing cardiovascular effects considered to be associated with the quinuclidine scaffold, the 3,5-bistrifluoromethyl substituent pattern was shown to be at least as potent as the 3,5-dimethyl (compare, for example, **34** and **32** in Fig. 5.21) (Williams *et al.*, 1994). Further minimization of the pharmacophore additionally highlighted the benefit of this group from a potency perspective, with the

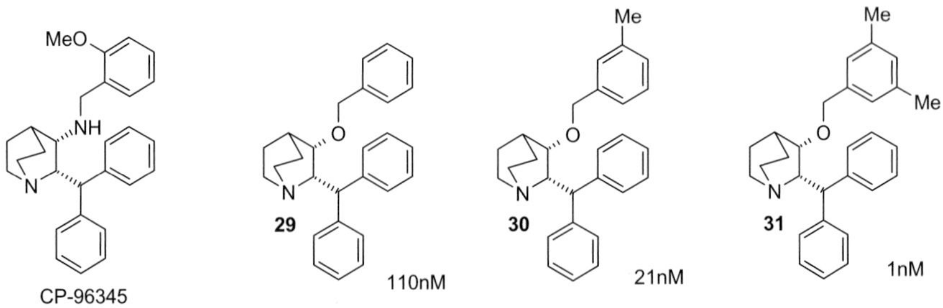

Figure 5.20. Early NK1 antagonists.

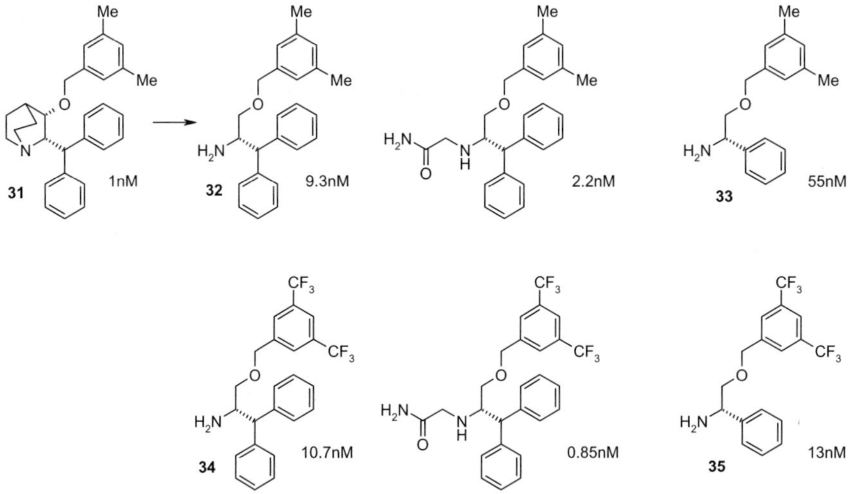

Figure 5.21. Showing comparable or improved IC_{50} values of representative bistrifluoromethylphenyl substituted NK1 antagonists relative to dimethylphenyl substituted analogues.

3,5-bistrifluoromethyl pattern **35** being fourfold more potent than the 3,5-dimethyl analogue **33** (Owens *et al.*, 1995). It is clear from the literature that this group was preferred, having been widely adopted in later work by this and other groups, presumably because of its high potency and metabolic stability compared with other options (Humphrey 2003).

In the absence of crystal structures the nature of the molecular interactions in this region are unclear. However, a number of independent publications, using for example point mutation studies, suggest a favourable interaction between His265 of transmembrane region 6 and the bistrifluoromethyl benzyl. Increased lipophilicity of the bistrifluoromethyl aryl group relative to the mesityl group may improve both receptor-binding and metabolic stability.

Recyclization of **35** (Fig. 5.22) gave among other analogues the piperidine L-733,060 in which activity resided only in the (2*S*,3*S*) enantiomer (Harrison *et al.*, 1994). However, this compound was shown to have modest affinity for the L-type Ca^{2+} channel, a cardiovascular ion channel activity that has been suggested to be responsible for the adverse cardiovascular effects of quinuclidine based NK1 antagonsists such as CP-96345.

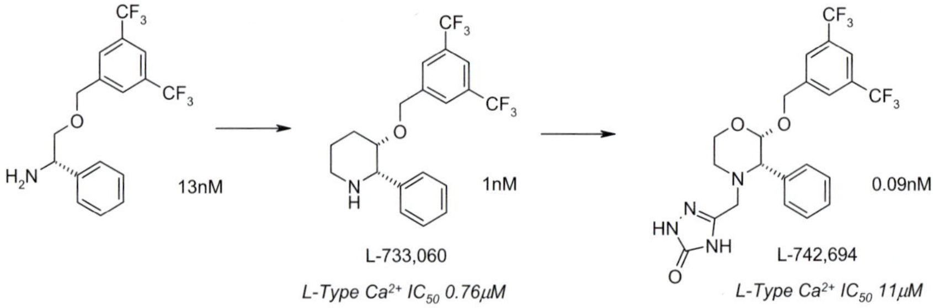

Figure 5.22. Showing development of NK1 antagonists with reduced cardiovascular risk.

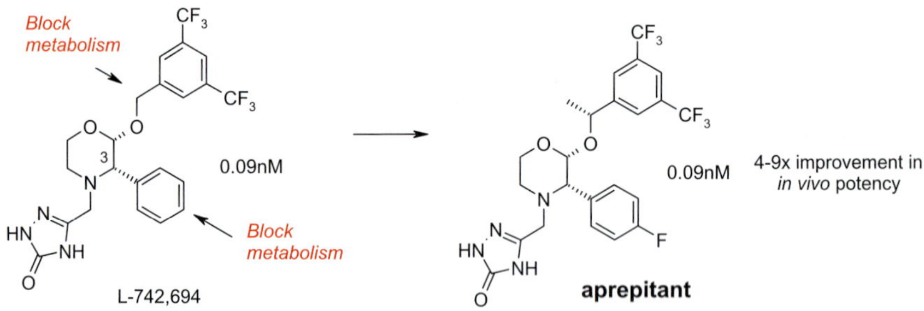

Figure 5.23. Strategy to minimize metabolic clearance of L-742694 leading to aprepitant.

To remove this activity heterocycles were added via a methylene spacer and basicity was reduced via conversion of the piperidine to a morpholine, ultimately providing some improvement in L-type calcium channel activity along with a significant improvement in potency L-742,694 (Harrison *et al.*, 1995; Hale *et al.*, 1996).

Aprepitant was finally delivered through efforts to minimize potential metabolic weaknesses in L-742,694 identified as benzylic cleavage through oxidation and *para* hydroxylation of the 3-phenyl substituent. Introduction of a 4-fluoro subtituent had no impact on receptor binding while giving a twofold improvement in potency in two *in vivo* efficacy models, a peripherally mediated NK-1 inflammation assay and a centrally mediated NK-1 assay that provides a convenient assessment of central nervous system (CNS) penetration. In combination with a benzylic methyl group the *in vivo* potency improvements are four- and ninefold, respectively, and are consistent with potential effects on metabolism (Hale *et al.*, 1998).

A later publication on the disposition of aprepitant in rats and dogs highlights the stability of the fluorinated regions of the molecule, with metabolism occurring on the morpholine ring and nitrogen substituent (Huskey *et al.*, 2004). The human metabolism is also apparently similar (Chavez-Eng *et al.*, 2004).

5.3 Summary and Future Outlook

Returning to Fig. 5.1, it is clear that only a small fraction of compounds which have reached the market have been covered in this review, with the omission of some interesting substructural classes as well as some important pharmaceuticals. For many of these compounds the precise contribution of fluorine to the story of their development is either less clear, for example, tafluprost (Matsumura, 2009), or exemplifies further the principles that have already been described, for example, fluoxetine (Robertson *et al.*, 1988). The use of fluoroalkylethers, a potentially interesting structural motif which is not covered here, has been reviewed recently (Jeschke *et al.*, 2007). For additional information on other compounds or

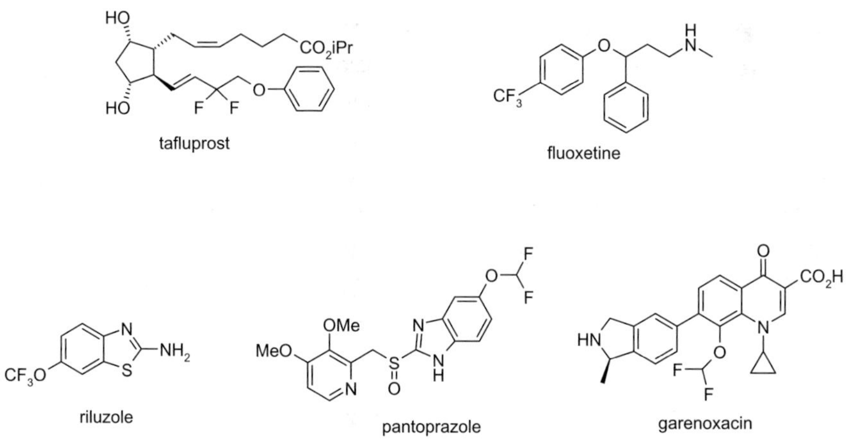

Figure 5.24. Structures of tafluprost, fluoxetine and fluoroalkylether-containing pharmaceuticals riluzole, patoprazole, and garenoxacin.

structural types, the interested reader should consult the reviews cited in the introduction to this chapter or see the primary literature.

Looking forward it seems likely, as has been suggested previously (Hagmann, 2008; Muller and Bohm, 2009), that the continued interest in the development of improved synthetic methods for the introduction of fluorine, for example Watson *et al.* (2009), will lead to even more effective utilization of the opportunities that fluorine substitution can bring and further increases in the number and structural types of fluorine-containing pharmaceuticals reaching patients.

References

Andersson, M.I. & Macgowan, A.P. (2003) Development of the quinolones. *J Antimicrob Chemoth* 51, S1–S11.

Bain, B.M., May, P.J., Phillipps, G.H. *et al.* (1974) Anti-inflammatory esters of steroidal carboxylic acids. *J Steroid Biochem* 5, 299.

Barry, A.L. & Brown, S.D. (1996) *In vitro* studies of two triazole antifungal agents (Voriconazole [UK-109,496] and Fluconazole) against *Candida* species. *Antimicrob Agents Ch* 40, 1948–1949.

Bégué, J.-P. & Bonnet-Delpon, D. (2008) 'Fluorinated drugs', in *Bioorganic and Medicinal Chemistry of Fluorine*, John Wiley & Sons, Chichester, pp. 279–351.

Berger, T.S., Parandoosh, Z., Perry, B.W. *et al.* (1992) Interaction of glucocorticoid analogues with the human glucocorticoid receptor. *J Steroid Biochem* 41, 733–738.

Bergman, A., Ebel, D., Liu, F. *et al.* (2007). Absolute bioavailability of sitagliptin, an oral dipeptidyl peptidase-4 inhibitor, in healthy volunteers. *Biopharm Drug Dispos* 28, 315–322.

Biggadike, K., Bledsoe, R.K., Hassell, A.M. *et al.* (2008) X-ray crystal structure of the novel enhanced-affinity glucocorticoid agonist fluticasone furoate in the glucocorticoid receptor ligand binding domain. *J Med Chem* 51, 3349–3352.

Bledsoe, R.K., Montana, V.G., Stanley, T.B. *et al.* (2002) Crystal structure of the glucocorticoid receptor ligand binding domain reveals a novel mode of receptor dimerization and coactivator recognition. *Cell* 110, 93–105.

Bohm, H.-J., Banner, D., Bendels, S. *et al.* (2004) Fluorine in medicinal chemistry. *ChemBioChem* 5, 637–643.

Brockunier, L.L., He, J., Colwell, L.F. *et al.* (2004) Substituted piperazines as novel dipeptidyl peptidase IV inhibitors. *Bioorg Med Chem Lett* 14, 4763–4766.

Chavez-Eng, C.M., Constanzer, M.L. & Matuszewski, B.K. (2004) Simultaneous determination of aprepitant and two metabolites in human plasma by high-performance liquid chromatography with tandem mass spectrometric detection. *J Pharmaceut Biomed* 35, 1213–1229.

Dickinson, R.P., Bell, A.S., Hitchcock, C.A. *et al.* (1996) Novel antifungal 2-aryl-1-(1H-1,2,4-triazol-1-yl)butan-2-ol derivatives with high activity against *Aspergillus fumigatus. Bioorg Med Chem Lett* 6, 2031–2036.

Domagala, J.M., Hanna, L.D., Heifetz, C.L. *et al.* (1986) New structure–activity relationships of the quinolone antibacterials using the target enzyme. The development and application of a DNA gyrase assay. *J Med Chem* 29, 394–404.

Drucker, D.J. & Nauck, M.A. (2006) The incretin system: glucagon-like peptide-1 receptor agonists and dipeptidyl peptidase-4 inhibitors in type 2 diabetes. *Lancet* 368, 1696–1705.

Earl, J. & Kirkpatrick, P. (2003) Ezetimibe. *Nat Rev Drug Discov* 2, 97–98.

Genard, P. & Palem-Vliers, M. (1983) Role of hydrophobic effects and polar groups in steroid–mineralocorticoid receptor interactions. *J Steroid Biochem* 19, 1639–1645.

Hagmann, W.K. (2008) The many roles for fluorine in medicinal chemistry. *J Med Chem* 51, 4359–4369.

Hale, J.J., Mills, S.G., Maccoss, M. *et al.* (1996) 2(*S*)-((3, 5-bis(trifluoromethyl) benzyl)oxy)-3(*S*)-phenyl-4-((3-oxo-1,2,4-triazol-5-yl)methyl)morpholine (1): A potent, orally active, morpholine-based human neurokinin-1 receptor antagonist. *J Med Chem* 39, 1760–1762.

Hale, J.J., Mills, S.G., Maccoss, M. *et al.* (1998) Structural optimization affording 2-(*R*)-(1-(*R*)-3,5-bis(trifluoromethyl)phenylethoxy)-3-(*S*)-(4-fluoro)phenyl-4-(3-oxo-1,2,4-triazol-5-yl)methylmorpholine, a potent, orally active, long-acting morpholine acetal human NK-1 receptor antagonist. *J Med Chem* 41, 4607–4614.

Harding, S.M. (1990) The human pharmacology of fluticasone propionate. *Resp Med* 84, S25–S29.

Harrison, T., Williams, B.J., Swain, C.J. *et al.* (1994) Piperidine-ether based HNK1 antagonists 1: determination of the relative and absolute stereochemical requirements. *Bioorg Med Chem Lett* 4, 2545–2550.

Harrison, T., Owens, A.P., Williams, B.J. *et al.* (1995) Piperidine-ether based HNK1 antagonists 2: investigation of the effect of N-substitution. *Bioorg Med Chem Lett* 5, 209–212.

Humphrey, J.M. (2003) Medicinal chemistry of selective neurokinin-1 antagonists. *Curr Top Med Chem* 3, 1423–1435.

Huskey, S.W., Dean, B.J., Doss, G.A. *et al.* (2004) The metabolic disposition of aprepitant, a substance P receptor antagonist, in rats and dogs. *Drug Metab Dispos* 32, 246–258.

Isanbor, C. & O'Hagan, D. (2006) Fluorine in medicinal chemistry: a review of anti-cancer agents. *J Fluorine Chem* 127, 303–319.

Istvan, E.S. & Deisenhofer, J. (2001) Structural mechanism for statin inhibition of HMG–COA reductase. *Science* 292, 1160–1164.

Jeschke, P., Baston, E. & Leroux, F.R. (2007) α-Fluorinated ethers as exotic entity in medicinal chemistry. *Mini Rev Med Chem* 7, 1027–1034.

Kauppi, B., Jakob, C., Farnegardh, M. *et al.* (2003) The three-dimensional structures of antagonistic and agonistic forms of the glucocorticoid receptor ligand-binding domain: RU-486 induces a transconformation that leads to active antagonism. *J Biol Chem* 278, 22748–22754.

Kim, D., Wang, L., Beconi, M. *et al.* (2005) (2*R*)-4-Oxo-4-[3-(trifluoromethyl)-5,6-dihydro[1,2,4]triazolo[4,3-a]pyrazin-7(8*H*)-yl]-1-(2,4,5-trifluorophenyl)butan-2-amine: a potent, orally active dipeptidyl peptidase IV inhibitor for the treatment of type 2 diabetes. *J Med Chem* 48, 141–151.

Kim, D., Kowalchick, J.E., Edmondson, S.D. *et al.* (2007) Triazolopiperazine-amides as dipeptidyl peptidase IV inhibitors: close analogs of Januvia® (sitagliptin phosphate). *Bioorg Med Chem Lett* 17, 3373–3377.

Kimura, Y., Atarashi, S., Kawakami, K. *et al.* (1994) (Fluorocyclopropyl) quinolones. 2. Synthesis and stereochemical structure–activity relationships of chiral 7-(7-amino-5-azaspiro[2.4]heptan-5-yl)-1-(2-fluorocyclopropyl) quinolone antibacterial agents. *J Med Chem* 37, 3344–3352.

Kirk, K.L. (2006) Fluorine in medicinal chemistry: recent therapeutic applications of fluorinated small molecules. *J Fluorine Chem* 127, 1013–1029.

Kuhn, B., Hennig, M. & Mattei, P. (2007) Molecular recognition of ligands in dipeptidyl peptidase IV. *Curr Top Med Chem* 7, 609–619.

Leach, A.G., Jones, H.D., Cosgrove, D.A. *et al.* (2006) Matched molecular pairs as a guide in the optimization of pharmaceutical properties: a study of aqueous solubility, plasma protein binding and oral exposure. *J Med Chem* 49, 6672–6682.

Lewis, M. & Cucurull-Sanchez, L. (2009) Structural pairwise comparisons of HLM stability of phenyl derivatives: introduction of the Pfizer metabolism index (PMI) and metabolism-lipophilicity efficiency (MLE). *J Comput-Aid Mol Des* 23, 97–103.

Liu, P., Sharon, A. & Chu, C.K. (2008) Fluorinated nucleosides: synthesis and biological implication. *J Fluorine Chem* 129, 743–766.

Matsumura, Y. (2009) 'Fluorinated prostanoids: development of tafluprost, a new anti-glaucoma agent', in Ojima, I. (Ed.), *Fluorine in Medicinal Chemistry and Chemical Biology*, John Wiley & Sons, Chichester, pp. 47–66.

Mock, W.L. & Zhang, J.Z. (1990) Concerning the relative acidities of simple alcohols. *Tetrahedron Lett* 31, 5687–5688.

Müller, K., Faeh, C. & Diederich, F. (2007) Fluorine in pharmaceuticals: looking beyond intuition. *Science* 317, 1881–1886.

Müller, K. & Bohm, H.-J. (2009) Facilitating the design of fluorinated drugs. *Chem Biol* 16, 1130–1131.

Owens, A.P., Williams, B.J., Harrison, T. *et al.* (1995) Phenyl-glycinol based NK1 receptor antagonists: towards the minimum pharmacophore. *Bioorg Med Chem Lett* 5, 2761–2766.

Penning, T.D., Talley, J.J., Bertenshaw, S.R. *et al.* (1997) Synthesis and biological evaluation of the 1,5-diarylpyrazole class of cyclooxygenase-2 inhibitors: identification of 4-[5-(4-methylphenyl)-3-(trifluoromethyl)-1H-pyrazol-1-yl]benzenesulfonamide (SC-58635, celecoxib). *J Med Chem* 40, 1347–1365.

Phillipps, G.H. (1990) Structure–activity relationships of topically active steroids: the selection of fluticasone propionate. *Resp Med* 84, S19–S23.

Phillipps, G.H., Bailey, E.J., Bain, B.M. *et al.* (1994) Synthesis and structure–activity relationships in a series of antiinflammatory corticosteroid analogues, halomethyl androstane-17.beta-carbothioates and -17.beta-carboselenoates. *J Med Chem* 37, 3717–3729.

Phillipps, G. H. (2009) 'Locally active corticosteroids: structure activity relationships', in Wilson, L. & Markes, R. (Eds), *Mechanisms of Topical Corticosteroid Activity*, Churchill Livingstone, London, pp. 1–18.

Purser, S., Moore, P.R., Swallow, S. *et al.* (2008) Fluorine in medicinal chemistry. *Chem Soc Rev* 37, 320–330.

Richardson, K. (1996) The discovery of fluconazole. *Contemporary Org Syn* 3, 125–132.

Robertson, D.W., Jones, N.D., Swartzendruber, J.K. *et al.* (1988) Molecular structure of fluoxetine hydrochloride: a highly selective serotonin-uptake inhibitor. *J Med Chem* 31, 185–189.

Roffey, S.J., Cole, S., Comby, P. *et al.* (2003) The disposition of voriconazole in mouse, rat, rabbit, guinea pig, dog, and human. *Drug Metab Dispos* 31, 731–741.

Rosenblum, S.B., Huynh, T., Afonso, A. *et al.* (1998) Discovery of 1-(4-fluorophenyl)-(3*R*)-[3-(4-fluorophenyl)-(3*S*)-hydroxypropyl]-(4*S*)-(4-hydr oxyphenyl)-2-azetidinone (SCH58235): a designed, potent, orally active inhibitor of cholesterol absorption. *J Med Chem* 41, 973–980.

Roth, B.D., Ortwine, D.F., Hoefle, M.L. *et al.* (1990) Inhibitors of cholesterol biosynthesis. 1. *Trans*-6-(2-pyrrol-1-ylethyl)-4-hydroxypyran-2-ones: a novel series of HMG–CoA reductase inhibitors. 1. Effects of structural modifications at the 2- and 5-positions of the pyrrole nucleus. *J Med Chem* 33, 21–31.

Shah, P. & Westwell, A.D. (2007) The role of fluorine in medicinal chemistry. *J Enzym Inhib Med Ch* 22, 527–540.

Swain, C.J., Seward, E.M., Sabin, V. *et al.* (1993) Quinuclidine based NK-1 antagonists 2: determination of the absolute stereochemical requirements. *Bioorg Med Chem Lett* 3, 1703–1706.

Sweeney, M.E. & Johnson, R.R. (2007) Ezetimibe: an update on the mechanism of action, pharmacokinetics and recent clinical trials. *Expert Opin Drug Met* 3, 441–450.

Van Heek, M., France, C.F., Compton, D.S. *et al.* (1997) *In vivo* metabolism-based discovery of a potent cholesterol absorption inhibitor, SCH58235, in the rat and rhesus monkey, through the identification of the active metabolites of SCH48461. *J Pharmacol Exper Ther* 283, 157–163.

Watson, D.A., Su, M., Teverovskiy, G. *et al.* (2009) Formation of ArF from LPdAr(F): catalytic conversion of aryl triflates to aryl fluorides. *Science* 325, 1661–1664.

Weber, A.E. & Thornberry, N. (2007) 'Case history: Januvia™ (sitagliptin), a selective dipeptidyl peptidase IV inhibitor for the treatment of type 2 diabetes', in Macor, J.E. (Ed.), *Annual Reports in Medicinal Chemistry,* Vol. 42, Academic Press, London, pp. 95–109.

Williams, B.J., Teall, M., Mc Kenna, J. *et al.* (1994) Acyclic NK-1 antagonists: 2-benzhydryl-2-aminoethyl ethers. *Bioorg Med Chem Lett* 4, 1903–1908.

Wu, Y.J., Davis, C.D., Dworetzky, S. *et al.* (2003) Fluorine substitution can block cyp3a4 metabolism-dependent inhibition: identification of (*S*)-*N*-[1-(4-fluoro-3-morpholin-4-ylphenyl)ethyl]-3-(4-fluorophenyl)acrylamide as an orally bioavailable KCNQ2 opener devoid of CYP3A4 metabolism-dependent inhibition. *J Med Chem* 46, 3778–3781.

Xu, J., Ok, H.O., Gonzalez, E.J. *et al.* (2004) Discovery of potent and selective [beta]-homophenylalanine based dipeptidyl peptidase IV inhibitors. *Bioorg Med Chem Lett* 14, 4759–4762.

6

Applications of Pentafluorosulfanyl Substitution in Life Sciences Research

*John T. Welch**

6.1 Introduction

The organic chemistry of the pentafluorosulfanyl (SF_5) group, previously reviewed (Lentz *et al.*, 1999; Winter *et al.*, 2005; Lal and Syvret, 2008; Gard, 2009) and extensively developed by Gard, has only recently come under more widespread investigation with the ready availability of building blocks and reagents previously difficult to access (Kirsch, 2004). Pentafluorosulfanylarene chemistry is largely based on the development of modern synthetic reactions which make possible the commercial availability of arene building blocks. In particular, it was the availability of kilogram quantities of arylpentafluorosulfanyl compounds, such as 4-nitro-pentafluorosulfanylbenzene, by direct fluorination methods that facilitated exploration of their utility (Chambers *et al.*, 1996).

However, these building blocks would have had much less impact without the basic understanding and chemical explorations provided by Thrasher who revealed the potential of pentafluorosulfanylarene compounds (Sipyagin *et al.*, 2001, 2004). Pentafluorosulfanyl groups are

* Department of Chemistry, University at Albany, SUNY, 1400 Washington Ave., Albany, New York 12222, USA. E-mail: jwelch@uamail.albany.edu.

both relatively chemically and hydrolytically stable (Kirsch *et al.*, 1999, 2000, 2001; Bowden *et al.*, 2000; Kirsch and Bremer, 2000). Aromatic pentafluorosulfanyl groups exhibit equal or greater hydrolytic stability than trifluoromethyl groups; they withstand Brønsted acids and bases and are stable under conditions required for Ni-, Pd- or Pt-catalysed hydrogenation (Kirsch, 2004). However, the SF_5 group may react with some alkyl lithium reagents, such as n-butyllithium; but reagents such as *tert*-butyllithium are compatible (Kirsch, 2004).

The steric demand and symmetry of the SF_5 group can be compared and contrasted with both the *tert*-butyl and trifluoromethyl groups. The volume of the SF_5 group is slightly less than that of a *tert*-butyl group (Anthony 1984; Lentz *et al.*, 1999) and therefore considerably greater than that of a trifluoromethyl group (CF_3). However, the electrostatic surface presented by SF_5 is comparable to CF_3 in that it presents a highly fluorinated surface, a pyramid of electron density as opposed to the inverted cone of density associated with CF_3 group. The electron-withdrawing effects of an SF_5 and a CF_3 group appear to be similar in magnitude, as assessed by 1s-photoelectron spectra of corresponding aryl derivatives (Brant *et al.*, 1981; True *et al.*, 2003). The electronegativity of the SF_5 group has been proposed to be as high as 3.65 in comparison to a value of 3.36 for the CF_3 group (Saethre *et al.*, 2001). In electrophilic substitution reactions, the Hammet σ_p value for SF_5 was determined to be 0.68 in contrast to a σ_p value for CF_3 of 0.54 (Sheppard, 1962). This has been further refined to a σ_I value for SF_5 of 0.55 and a σ_R value of 0.11 (Sheppard, 1962), in contrast to σ_I value for CF_3 of 0.39 and a σ_R value of 0.12 (Taft and Lewis, 1959; Taft, 1960). It is important to note the decreased resonance and increased inductive contributions, a trend that is consistent with the electronic effects observed in the estimation of electronegativity (Brant *et al.*, 1981; True *et al.*, 2003).

The unique properties of the SF_5 group have shown diverse utility in biological applications. There is ample evidence that SF_5 is not simply a more expensive CF_3 analogue. With the unique octahedral geometry around sulfur and a square pyramidal array of fluorines, the SF_5 group has a reduced barrier to rotation and as such can optimize receptor interactions efficiently. The potential dehydrofluorination of a CF_3 group to form a Michael-type acceptor and hence a site for covalent attachment that would be associated with mechanism-based inhibition is not possible with an SF_5 group. In environmental degradation studies of SF_5-substituted molecules, degradation was shown to lead to environmentally benign products (Jackson and Mabury, 2008, 2009). Lastly, the remarkable hydrophobicity and steric demand of the SF_5 group can profoundly influence molecular conformation in aqueous solutions as indicated in our heptapeptide studies.

This chapter presents a synopsis of both the chemistry and biological applications of arene pentafluorosulfanyl compounds. Where available, the biological data are also presented. It is evident from the number of diverse reactions that the transformations of arene pentafluorosulfanyl groups are even more general and predictable than might have been imagined.

6.2 General Preparative Information

As mentioned above, the availability of pentafluorosulfanyl building blocks has been key to the exploration of the potential of the SF_5 group as a substituent in life sciences research. However, at present the variety of available pentafluorosulfanylated molecules is still limited to only a handful of arenes. The first pentafluorosulfanylated arenes, prepared by direct fluorination, were reported in 1960 (Sheppard, 1960), with processes for improved preparation by others (Williams and Foster, 1994; Bowden *et al.*, 1997, 2000). The exciting advance in the preparation of pentafluorosulfanylbenzene (Fig. 6.1) that avoids the necessity for the use of molecular fluorine promises to dramatically lower the cost of these starting materials (Umemoto, 2009).

The further expansion of the utility of SF_5-containing arenes was made possible by Thrasher who developed methods for the preparation of the previously inaccessible *ortho*-substituted (Sipyagin *et al.*, 2001) starting materials by recognizing the importance of the nitro-substituent on fluorination (Fig. 6.2).

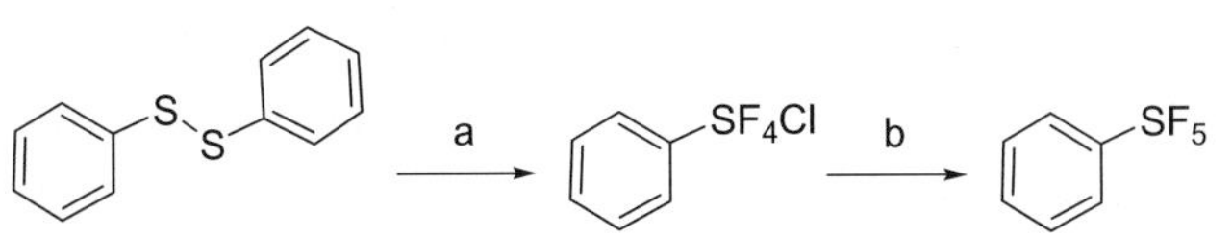

Figure 6.1. Preparation of pentafluorosulfanylbenzene: (a) KF, Cl_2; (b) ZnF_2.

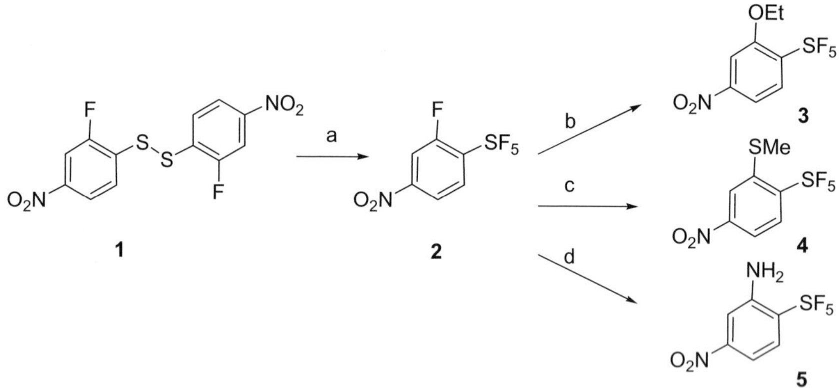

Figure 6.2. Possible transformations of a substituted nitro-disulfide: (a) AgF_2, CFC 113, 60–120°C (18%); (b) EtOH, KOH reflux (49%); (c) NaSMe (46%); (d) NH_4OH, 130–135°C (61%).

6.2.1 *Synthesis of 1-fluoro-4-nitro-2-(pentafluorosulfanyl)benzene and derivatives*

As an example of the variety of transformations possible, starting from a substituted nitro-disulfide three reactions are illustrated (Sipyagin *et al.*, 2001) (Fig. 6.2). While the yield of the fluorination step is modest, direct access to the substitution pattern as mentioned earlier is especially significant.

6.2.2 *Synthesis of 4,5-dihydroisoxazoles with allylic pentafluorosulfanyl substituents*

There are relatively few examples of heterocycles with pentafluorosulfanyl-containing aliphatic substitutents. One example of this type of molecule is derived from the addition of SF_5Cl to a diene to form **6** (Brel, 2006). This reaction was made possible by the discovery of Dolbier and coworkers of the utility of triethylborane on improving the homolytic addition process at the expense of the undesired ionic fluorobromination reaction (Aiet-Mohand and Dolbier, 2002; Dolbier and Ait-Mohand, 2004, Dolbier *et al.*, 2006). Elimination to give the alkenes **8** and **11** and isomerization thence to allylic substituents proceeded uneventfully (Fig. 6.3).

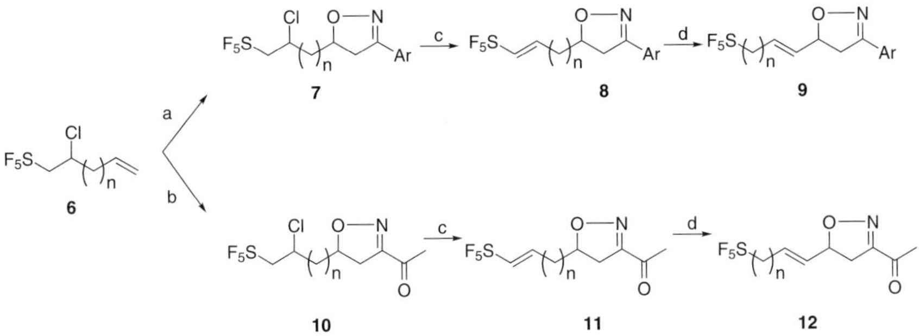

Figure 6.3. Preparation of 4,5-dihydroisoxazoles **9** and **12**: (a) *N*-hydroxy-benzenecarboximidoyl chloride Et_3N; (b) $(NH_4)_2Ce(NO_3)_6$, acetone; (c) K_2CO_3, DMF; (d) Cs_2CO_3, MeOH.

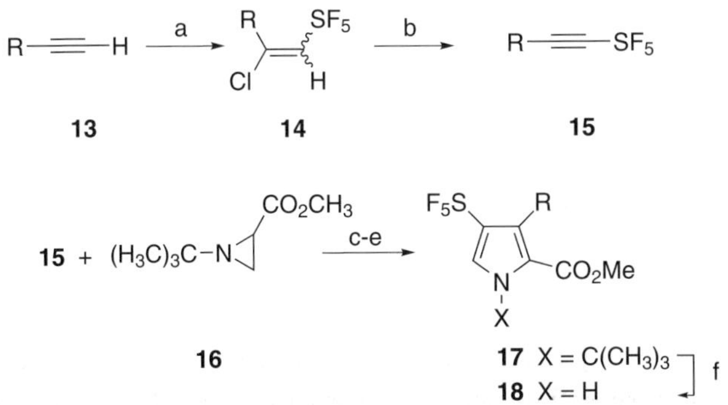

Figure 6.4. Preparation of 4-(pentafluorothio)pyrrole-2-carboxylic acid ester **18**: (a) SF_5Cl, Et_3B; (b) LiOH, DMSO; (c) xylene, 135°C; (d) DDQ; (e) TfOH; (f) DDQ.

6.2.3 *Pentafluorosulfanyl (SF_5) pyrrole carboxylic acid esters*

The direct substitution of heterocycles with the pentafluorosulfanyl group is extremely uncommon. Dolbier and Zheng (2009) developed a method for the synthesis of these useful heterocycles based on a pentafluorosulfanyl-substituted alkyne **15**, easily prepared by the aforementioned addition reaction of pentafluorosulfanyl chloride (Fig. 6.4). These authors have reported that most of the well known methods for preparing pyrroles, especially those that involve ring formation utilizing condensation reactions via carbanionic intermediates, fail with SF_5-substituted substrates. To overcome this shortcoming the cycloaddition of an azomethine ylide derived from **16** and **15** was used to prepare the corresponding 4-(pentafluorothio)pyrrole-2-carboxylic acid ester **18** in a 53–78% yield.

6.3 Agrochemical Applications

The first patent describing the utility of pentafluorosulfanylarenes in life sciences research appeared in 1963 (Raasch, 1963). A number of patents have subsequently appeared claiming the utility of these compounds generically as pesticides (Pilato and Wu, 1996; Herman *et al.*, 1998; Manning and Wu, 1998; Phillips *et al.*, 1998) and more specifically as herbicides (Kay *et al.*, 1994; Hayashizaki *et al.*, 1997), fungicides (Alt *et al.*, 1990), parasiticides (Banks, 1997, 1998a,b, 1999) and insecticides (Prichard and Stacey, 1975; Salmon, 1994; Salmon *et al.*, 1994, Howard and Stevenson, 1995; Chern *et al.*, 1999; Huber, 1999).

In addition to the previous examples, the following preparations were selected as illustrations not only of the reactivity and stability of pentafluorosulfanylarenes, but also of agrochemical utility.

6.3.1 *3-(2-Chloro-4-(pentafluorosulfanyl)phenoxy)benzoic acid*

Diphenylether herbicides **23** and **24** were prepared from 4-pentafluorosulfanyl aniline **20** (Barton and Mitchell, 1994) (Fig. 6.5). The reaction conditions described are mild and result in acceptable yields of the ethers.

At 125 g/ha post-emergence, **24** gave 90–100% control of several weeds, including *Chenopodium album, Amaranthus retroflexus, Ipomoea hederacea* and *Abutilon theophrasti.*

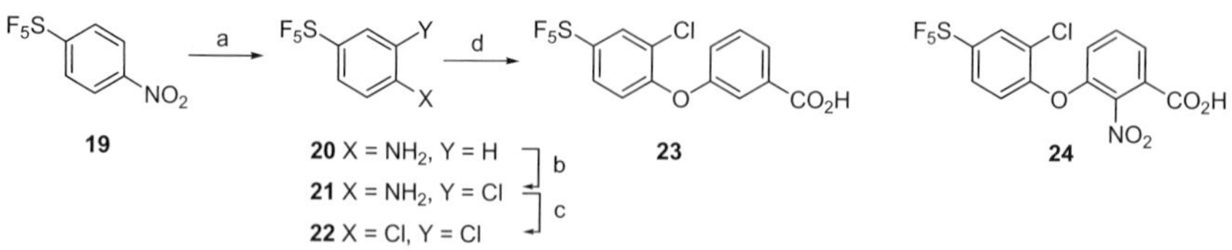

Figure 6.5. Preparation of diphenylether herbicides 23 and 24: (a) HCl (conc), Fe, *i*-PrOH (93%); (b) NCS, CH_3CN (84%); (c) CuCl, *t*-BuONO(89%); (d) *m*-hydroxybenzoic acid, KOH then K_2CO_3, DMSO (58%).

6.3.2 *Pentafluorosulfanylphenyl and benzoylisoxazoles*

The preparation of pentafluorosulfanylarene-substituted isoxazoles described by Hawkins (1997) clearly illustrates the stability of pentafluorosulfanylarenes under a variety of reaction conditions (Fig. 6.6). In particular, transmetalation of **27** with *tert*-butyllithium proceeded in 36% yield. The SF_5 group was also stable in the presence of the magnesium enolate of *tert*-butyl 3-cyclopropyl-3-oxopropanoate.

The products showed selectivity for grassy weeds in the presence of broadleaf crops such as cotton and soyabeans both pre- and post-emergence.

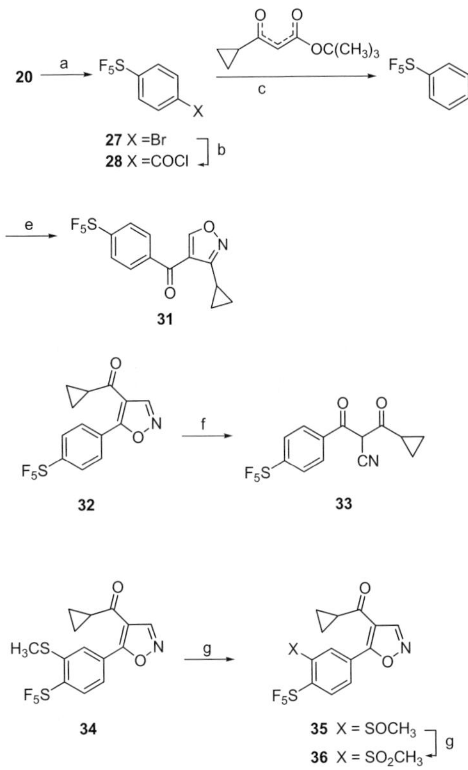

Figure 6.6. Preparation of pentafluorosulfanylarene-substituted isoxazoles: (a) $CuBr_2$, *t*-BuONO, CH_3CN (quantitative); (b) *t*-BuLi, CO_2(36%) then ClCOCOCl (quantitative); (c) Mg, $PhCH_3$ then TsOH, reflux(80%); (d) $HC(OEt)_3$ (quantitative); (e) $NH_4OH{\cdot}HCl$, EtOH, NaOAc; (f) Et_3N; (g) MCPBA.

6.3.3 *Trifluralin analogue*

Thrasher and coworkers were the first to publish the preparation of a pentafluorosulfanyl analogue (**39**) of Treflan (**40**) (Sipyagin *et al.*, 2004). However, those authors were not able to publish a comparative study of the pentafluorosulfanyl analogue with the trifluoromethyl-containing parent, Treflan. A novel synthesis and the biological data were subsequently reported several years later by others (Lim *et al.*, 2007).

A side-by-side comparison of the spectrum of herbicidal activity of **40** and **39** clearly demonstrates that in post-emergence applications, the pentafluorosulfanyl replacement of the trifluoromethyl group affords no apparent advantage (Lim *et al.*, 2007) (Fig. 6.7). In pre-emergence applications, the herbicidal performance of **39** is greater than the parent trifluralin **40** as assessed by visual injury estimation. When parent trifluralin was applied at 4 and 1 kg/ha, crop injury for maize in pre-emergence was 100% (complete death) and 20% damage, respectively. Compared to the parent trifluralin **40**, the crop injury was much less in crops treated with **39**. For example, crop injury for maize on treatment with **40** at 2.68 kg/ha resulted only in 10% growth inhibition. Therefore, the observation that crops were relatively more tolerant to **39** than trifluralin, might at least account for the differential selectivity to the parent trifluralin **40**. Although trifluralin **40** effectively controlled several weeds at 4 kg/ha application rates, the herbicidal activity of **39** toward grass weeds was greater than that of parent trifluralin. At 0.67 kg/ha application rate, the herbicidal activity of **39** afforded the same level of control as trifluralin at 4 kg/ha. Moreover, herbicidal activity in barnyard grass and crabgrass

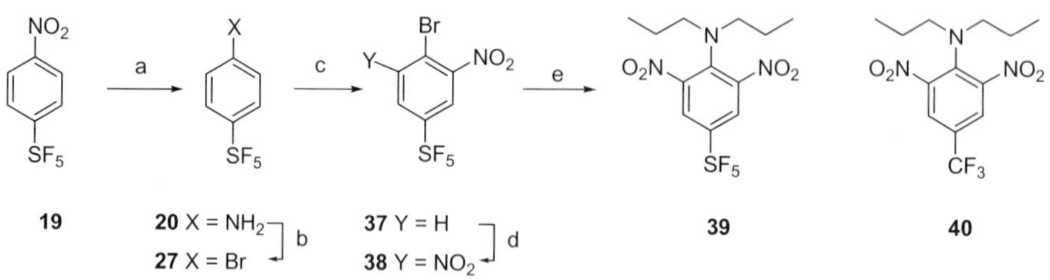

Figure 6.7. Synthesis of trifluralin: (a) Fe, HCl, EtOH (86%); (b) HBr, $NaNO_2$, CuBr (67%); (c) H_2SO_4, HNO_3, 0°C (99%); (d) H_2SO_4, HNO_3, 80°C (73%); (e) 8% NaOH, *N,N*-dipropylamine (99%).

at 0.168 kg/ha resulted in over 90% damage. In conclusion, our results indicate that herbicidal activity of **39** on pre-emergence application, in terms of crop injury and weed control, is nearly fivefold greater than the parent trifluralin **40** (Lim *et al.*, 2007).

6.3.4 *Insecticidal derivatives of substituted phosphorylated phenylalkyl iminooxazolines and iminothiazolines*

As mentioned in the introduction to this section, insecticidal compounds based on the incorporation of pentafluorosulfanyl arenes are known. The class of compounds represented by **41** is an example of the materials claimed for insectidal applications (Fig. 6.8). Unfortunately, however, as the class was not further exemplified, no specific data on preparative activity is available.

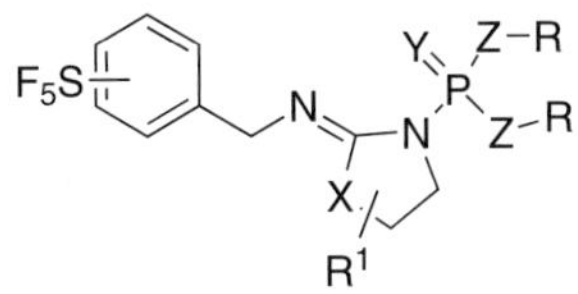

41

Figure 6.8. Phenylalkyl iminooxazolines and iminothiazolines.

6.3.5 *Fungicidal (E)-methyl 2-(2-(3-(pentafluorosulfanyl) phenoxymethyl) phenyl)-3-methoxyacrylate, 44*

3-Pentafluorosulfanylaniline was diazotized and hydrolysed in hot aqueous sulfuric acid to give 3-pentafluorophenol in 43% yield (Worthington and Streeting, 1994). Etherification with (*E*)-methyl 3-methoxy-2-[2-(bromomethyl)phenyl]propenoate with potassium carbonate yielded **44** (Fig. 6.9). Conversion of the 4-pentafluorosulfanyl aniline to the phenol proceeded similarly, albeit requiring the addition of copper nitrate trihydrate and copper oxide on dediazoniation.

In tests at 100 ppm (foliar spray), **44** gave complete protection against several fungi including *Septoria nodorum*, *Venturia inaequalis* and *Phytophthora infestans lycopersici.*

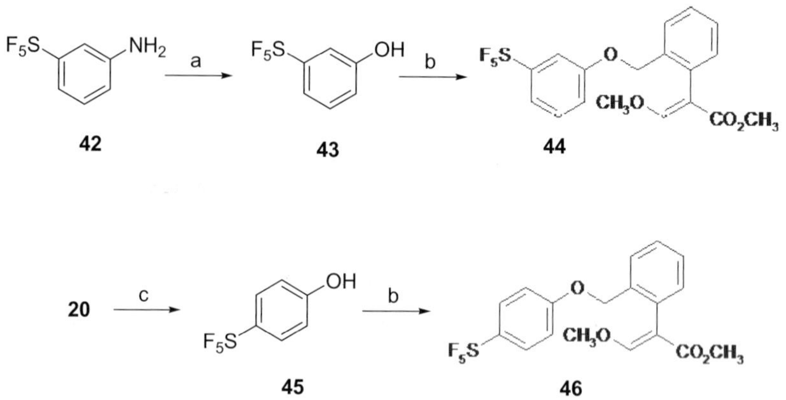

Figure 6.9. Preparation of compound **46**: (a) H_2SO_4, $NaNO_2$ and then H_2SO_4 and H_2O, 120°C (43% for both steps); (b) K_2CO_3 and (*E*)-methyl 2-(2-(bromomethyl)phenyl)-3-methoxyacrylate (45%); (c) H_2SO_4, $Cu(NO_3)_2$ and then CuO (61%).

6.3.6 *N-(3-Phenylpropyl) and (3-phenylethyl)benzamides*

In an usual series of claims, compounds with two pentafluorosulfanyl groups were claimed as fungicides (Mansfield *et al.*, 2007; Coqueron *et al.*, 2008) (Fig. 6.10). Both the propyl and ethyl series were described; however, no preparative or biological data is available.

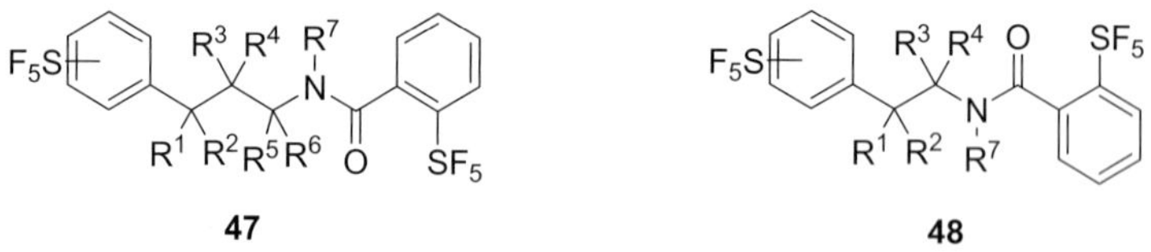

Figure 6.10. *N*-(3-Phenylpropyl) and (3-phenylethyl)benzamides.

6.4 Medicinal Chemistry

The same features of the pentafluorosulfanyl group that make it an attractive substituent in agrochemical development also suggest its utility in medicinal chemistry. The examples published in both the open and patent literature are much more recent than those describing agrochemical utility. The examples discussed are representative of both the biological activity and the stability and reactivity of pentafluorosulfanyl arenes.

6.4.1 *1-(Pentafluorosulfanylphenyl)-3-(1,2,4-triazol-3-ylthioalkyl)-3-azabicyclo[3.1.0]hexanes, dopamine D3 receptor modulators*

Diazotization of 4-pentafluorosulfanylaniline **20** formed the reactive diazonium intermediate that underwent addition to maleimide (Andreotti *et al.*, 2006) (Fig. 6.11). The pentafluorosulfanyl group was unaffected by the cyclopropanation with the dimethyloxosulfonium methylide or by borane reduction.

Compound **53** had pK_i values within the range 7.0–10.5 at the dopamine D_3 receptor with selectivities >30 over the dopamine D_2 receptor in a guanosine γ-thiophosphate (*GTPγS*) scintillation proximity assay. The therapeutic effect of many antipsychotic compounds is a consequence of indiscriminate binding to the dopamine D receptors. Inhibition of D_2 has been identified with the undesirable side effects associated with this class of inhibitors. These compounds may minimize the undesired side effects yet retain activity as antipsychotic agents with utility in the treatment of

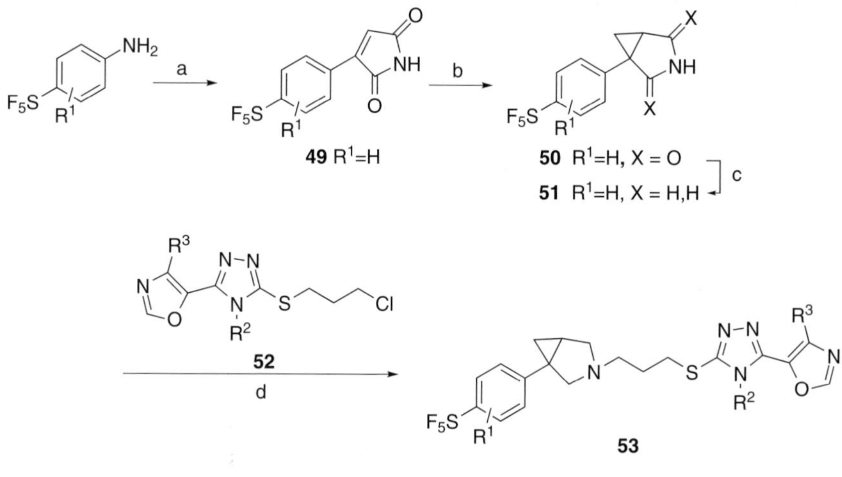

Figure 6.11. Preparation of compound 53: (a) maleimide, *t*-BuONO; (b) $(CH_3)_3SO^+I^-$, NaH (56% for two steps); (c)BH_3·THF, 65°C (51%); (d) Na_2CO_3, NaI, DMF (43%).

diseases such as psychosis, substance abuse, premature ejaculation, cognition impairment and obsessive–compulsive spectrum disorders.

6.4.2 *Pentafluorosulfur piperazinylpiperidines*

From the parent pentafluorosulfanyl benzene it was possible to introduce both a vinyl and an allyl group by palladium-promoted coupling (Cao *et al.*, 2009) (Fig. 6.12). Furthermore, the SF_5 group was unaffected during Grubbs metathetic ring closure or on epoxidation.

Compound **64** bound to CCR5 chemokine receptors with an IC_{50} of less than 500 nM. CCR5 is a member of the β-chemokine receptor family of integral membrane proteins associated with HIV-1 entry into CD4 cells.

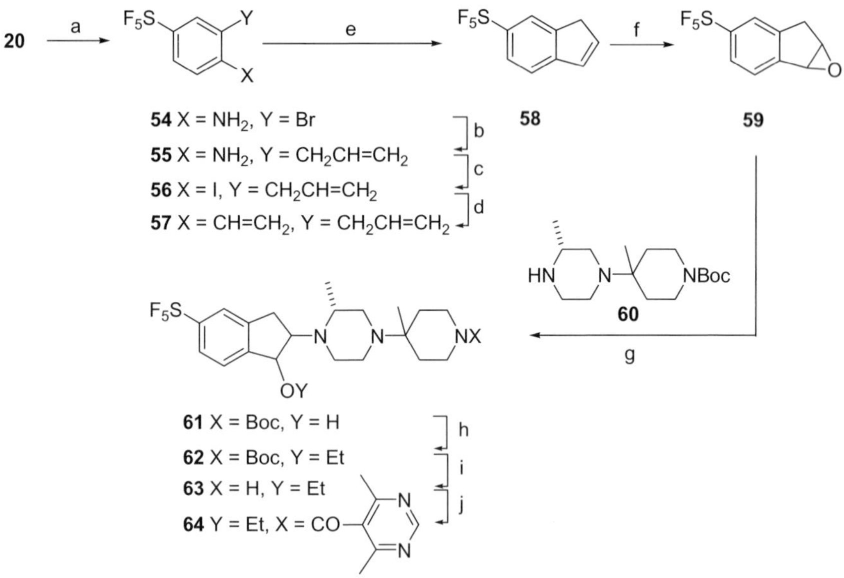

Figure 6.12. Preparation of compound **64**: (a) NBS, DMF (78%); (b) CsF, $(Ph_3P)_4Pd$, 2-allyl-4,4,5,5-tetramethyl-1,3,2-dioxaborolane (70%); (c) $NaNO_2$, HCl then KI (63%); (d) Bu_3SnH, 2,6-(*t*-Bu)-4-MePhOH, $(Ph_3P)_4Pd$ (69%); (e) $(Cy_3P)_2PhCH_2RuCl$, CH_2Cl_2 (78%); (f) MCPBA; (g) *i*-PrOH (14% for two steps); (h) NaH, EtI (63%); (i) 4.0 M HCl in dioxane; (j) 4,6-dimethylpyrimidine-5-carboxylic acid, 1-hydroxybenzotriazole, *N*-(3-dimethylaminopropyl)-*N′*-ethyl carbodiimide (35% for two steps).

6.4.3 *Pentafluorosulfanyl arene containing pyrazoles*

The pentafluorosulfanyl group was found to be effective in the development of therapeutic cannabinoid receptor ligands (Carroll *et al.*, 2009) (Fig. 6.13).

The major psychoactive constituent of marijuana interacts with two receptor subtypes, one highly expressed in the central nervous system (CB_1) and the other abundantly associated with the cells of the immune system (CB_2). The effects associated with CB_1 inhibition (euphoria, sedation, catalepsy and anxiety) have prevented the development of broad spectrum inhibitors and thereby obviated the use of CB_2 inhibition associated with reduction in neuropathic pain. Compounds such as **68** bind to CB_2 receptors with a K_i of less than 1000 nM.

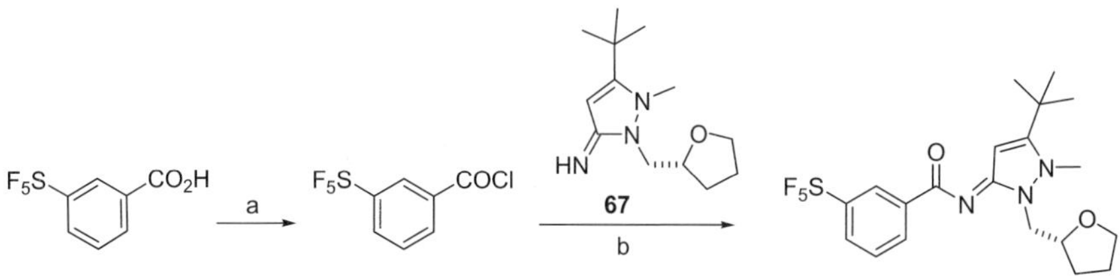

65 66 68

Figure 6.13. Preparation of compound **68**: (a) $SOCl_2$, reflux; (b) Et_3N, 16 h (56% for two steps).

6.4.4 *N-(phenoxycyanomethylethyl)(pentafluorosulfanyl) benzamide*

This preparation is distinguished by the use of an oxidative transformation to form the necessary benzoic acid **71** from 4-pentafluorosulfanyl styrene **70** (Comlay *et al.*, 2008) (Fig. 6.14). The styrene was prepared by Pd-promoted vinylation using vinyl tributyltin.

Haemonchus contortus is a parasitic nematode with a remarkably great range, but is more prevalent in warm moist regions rather than cold dry ones. *H. contortus* frequently inhabits the abomasum (fourth stomach) of ruminent animals and has been found in humans in Brazil and Australia. The compounds described in this patent were evaluated in *H. contortus* L3 (HcL3) assays (*in vitro*) and **74** demonstrated a minimum effective dose of 3μg/mL.

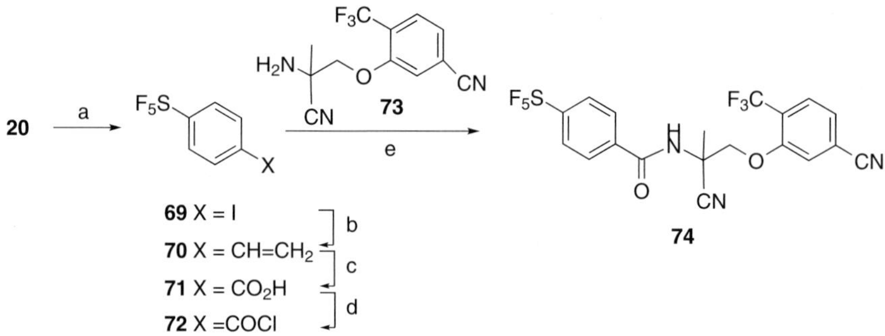

Figure 6.14. Preparation of compound 74: (a) HCl, $NaNO_2$ then KI (74%); (b) $Bu_3SnCH{=}CH_2$, $(Ph_3P)_4Pd$, DMF (quantitative); (c) $NaIO_4$, $RuCl_3$, CH_3CN, H_2O, CCl_4 (33%); (d) $SOCl_2$, 65°C (83%); (e) Hünig's base (49%).

6.4.5 *Preparation of pentafluorosulfanyl-substituted compounds for use as vanilloid receptor VR1 ligands*

The utility of the pentafluorosulfanylated isothiocyanates in formation of thioureas has been demonstrated (Frank *et al.*, 2006) (Fig. 6.15). These building blocks were reported in the course of a programme to prepare pharmaceutically active pentafluorosulfanylarene-containing compounds including vanilloid receptor inhibitors.

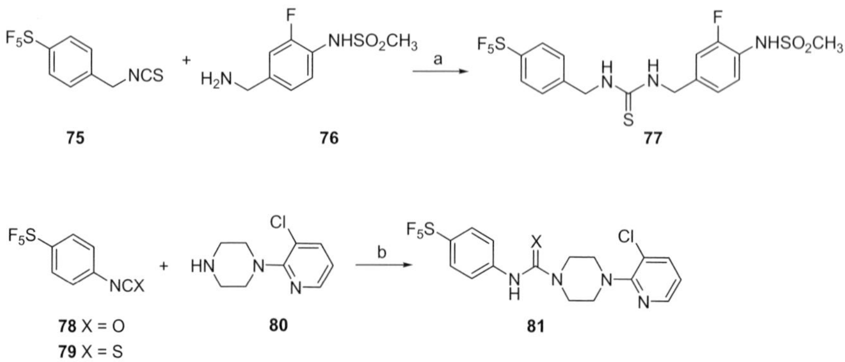

Figure 6.15. Preparation of compound **81**: (a) DMF, Et_3N (57%); (b) DMF, Et_3N, X=O (75%), X=S (60%).

Drugs acting at vanilloid receptors could potentially be used to treat neuropathic pain associated with multiple sclerosis, chemotherapy or amputation, as well as pain associated with the inflammatory response of damaged tissue, such as in osteoarthritis. Closely related benzamide analogues are claimed in a capsaicin receptor inhibition assay to have K_i values ranging from 11.9 to 69.1nM (Frank *et al.*, 2008). However, no data on the reported pentafluorosulfanylated compounds has been reported.

6.4.6 *4-Fluoro-N-(4-pentafluorosulfanylphenyl)-4-(3-fluoropyridin-2-yl)cyclohexanecarboxamide 84*

The preparation of compound **84** is noteworthy as a description of the reactivity of the 4-pentafluorosulfanylaniline in a palladium-mediated carbonylation and coupling reaction (Gomtsyan *et al.*, 2009) (Fig. 6.16). While the yield of the reaction was not reported in the patent literature, several examples utilizing different enol triflates are described.

Nociceptors are primary sensory afferent neurons that are activated by noxious stimuli. The vanilloid capsaicin activates these neurons by antagonism of the transient receptor potential cation channel, subfamily V, member 1 receptor (TRPV1 receptor). The analgesic component of TRPV1 activation mediates capsaicin desensitization and thus has prompted the clinical use of capsaicin analogues as analgesic agents.

The functional activity of the compounds was determined with a Ca^{2+} influx assay and measurement of the intracellular Ca^{2+} levels. The inhibitors

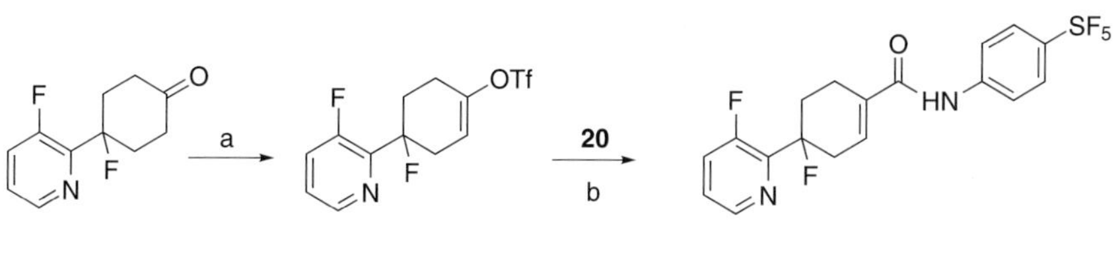

82 **83** **84**

Figure 6.16. Synthesis of compound **84**: (a) 1,1,1-trifluoro-*N*-phenyl-*N*-(trifluoromethylsulfonyl)methane sulfonamide, THF, LiHMDS; (b) Et_3N, Pd $(OAc)_2$, Cy_2P-2′-($(CH_3)_2N$) Ph-Ph, CO.

are effective TRPV1 antagonists with IC_{50} values from about 10 μM to about 10 nM, and exhibited efficacy in relieving pain.

6.4.7 *Pentafluorosulfanylarene aminoimidazoles*

Preparation of the protected Weinreb amide **88** proceeded in good yields (Heinelt *et al.*, 2009) (Fig. 6.17). Conversion to the acetyl product **89** in 73% yield illustrated the stability of the pentafluorosulfanyl group during reactions with Grignard reagents.

Protease-activated receptor 1 is a member of the G-protein-coupled receptor superfamily. In particular, the effect of a coagulation protein (thrombin) is mediated by protease-activated receptors (PARs). Thrombin signalling contributes to clotting. Endothelial PARs participate in the regulation of vascular tone and permeability, while in vascular smooth muscle they mediate contraction and proliferation. PARs also contribute to the pro-inflammatory responses observed in atherosclerosis, as

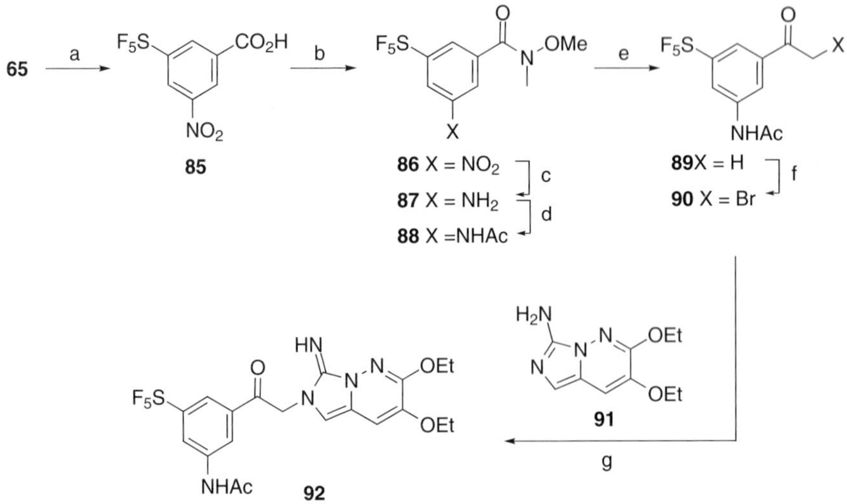

Figure 6.17. Preparation of compound **92**: (a) H_2SO_4, HNO_3, 75°C (72%), 5h; (b) $SOCl_2$ then CH_3ONHCH_3, Hünig's base (92%); (c) RaNi, H_2, CH_3OH (45%); (d) Ac_2O, Et_3N (96%); (e) LiHMDS, MeMgBr (73%); (f) $PhN(CH_3)_3{}^+Br^-$, THF (45%); (g) Hünig's base (69%).

proteinase-activated receptor (PAR-1) inhibitors. In PAR-1 assays, compounds related to **92** exhibited IC_{50} values of 0.18 and 0.005 μM.

6.4.8 *3-Phenylhydantoins*

The synthesis of the pentafluorosulfanyl-containing hydantoin proceeded uneventfully, however, yield information for the individual transformations was not provided (Jaehne *et al.*, 2008) (Fig. 6.18).

Cannabinoid receptors are cell membrane receptors of the G-protein-coupled receptor superfamily. The CB_1 receptor is expressed mainly in the central nervous system and in the lungs, liver and kidneys. Cannabinoid receptor inhibitors have both therapeutically undesirable psychotropic

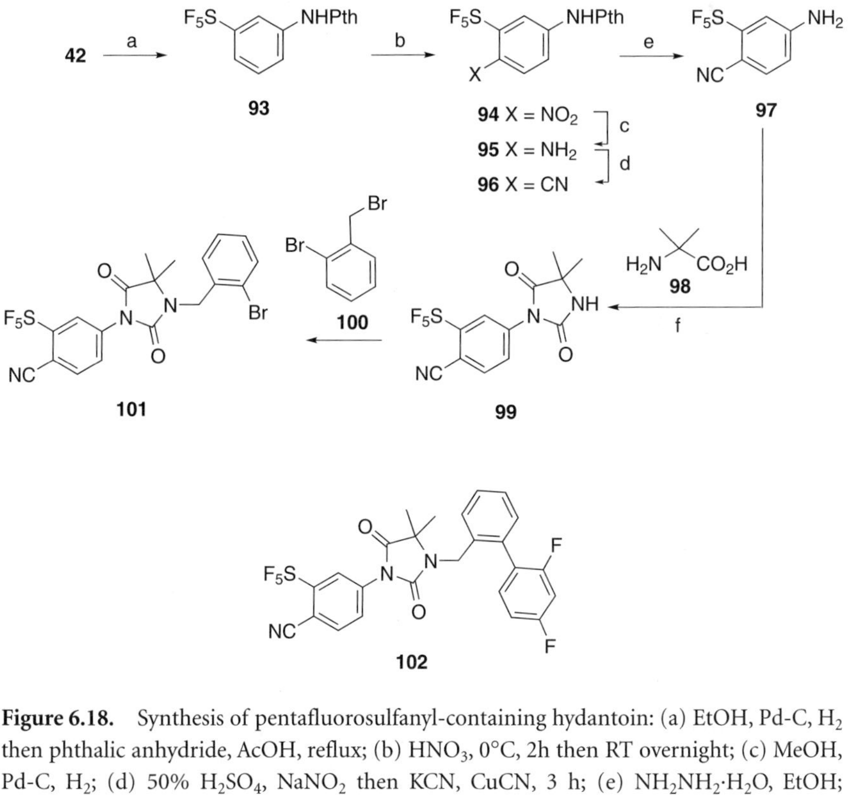

Figure 6.18. Synthesis of pentafluorosulfanyl-containing hydantoin: (a) EtOH, Pd-C, H_2 then phthalic anhydride, AcOH, reflux; (b) HNO_3, 0°C, 2h then RT overnight; (c) MeOH, Pd-C, H_2; (d) 50% H_2SO_4, $NaNO_2$ then KCN, CuCN, 3 h; (e) $NH_2NH_2{\cdot}H_2O$, EtOH; (f) triphosgene, Et_3N then 2-amino-2-methylpropanoic acid.

effects and clinically desirable effects, such as moderation of haemorrhagic or endotoxin induced hypotension. In human CB_1 receptor inhibition assays, 99 examples of phenylhydantoins related to **102** exhibited IC_{50} values ranging from 0.7 to 208 nM.

6.4.9 *Pentafluorosulfanyl benzoylguanidines*

Of particular note in the synthesis of **113** is the scale of the transformations of **19** to **106**, starting from approximately a kilogram scale to 50 g scale for the latter transformations (Kleemann, 2005a) (Fig. 6.19). The palladium-assisted methylation and cyanation reactions were apparently unaffected by the SF_5 substituent.

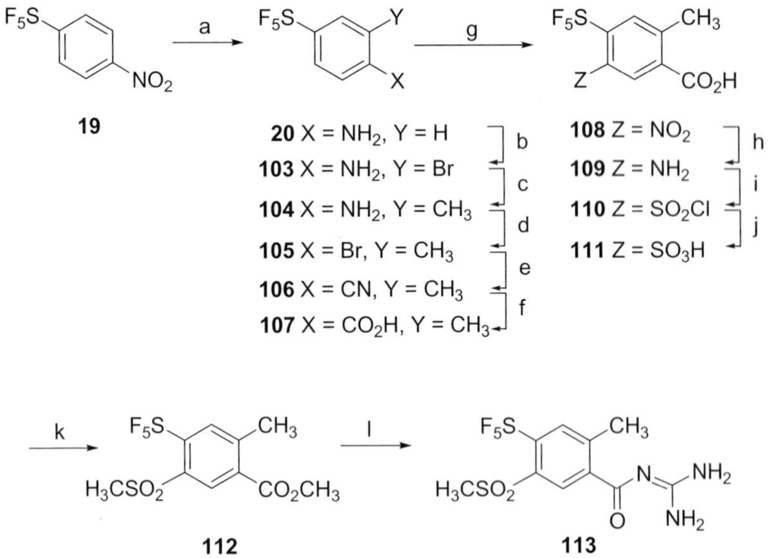

Figure 6.19. Synthesis of compound 113: (a) $SnCl_2$, conc HCl, 80–100°C (99%); (b) dibromohydantoin, CH_2Cl_2 (86%); (c) trimethylborate, $PdCl_2$ (dppf), 70°C (76%); (d) *t*-BuONO, $CuBr_2$, CH_3CN (89%); (e) $Zn(CN)_2$, Zn powder, $PdCl_2$ (dppf), DMA (80%); (f) ethylene glycol, NaOH, H_2O, 130°C (93%); (g) 90% HNO_3, 96% H_2SO_4, RT (92%); (h) AcOH, 10% Pd-C, H_2 (1 bar) (98%); (i) AcOH-HCl, $NaNO_2$ then CuCl, $CuCl_2$, SO_2 in AcOH (quantitative); (j) Na_2SO_3, H_2O 70°C (93%); (k) 2N NaOH then CH_3I, DMF (48%); (l) guanidinium hydrochloride, *t*-BuOK, DMF (40%).

NHE1 is a Na^+/H^+ antiporter (also known as sodium–hydrogen antiporter 1, SLC9A1) that is a ubiquitous membrane-bound enzyme involved in pH regulation of vertebrate cells. In an NHE1 inhibition assay, benzoylguanidines such as **113** exhibited IC_{50} values as low as 49 nM. These compounds are claimed to be useful as antiarrhythmic agents.

6.4.10 *Pentafluorosulfanylphenoxy-substituted benzoylguanidines*

The Aventis group extended their previously described NHE1 inhibitor investigation to include the ethers **116** and **120** (Kleemann, 2005b) (Fig. 6.20). It is noteworthy that phenol **43** retained sufficiently nucleophilicity so that it could be successfully employed in the required aromatic substitution reaction to form **115**.

Compounds such as **116** and **120** are claimed to inhibit processes associated with ischemia-induced damage to the heart, in particular ischemia-induced cardiac arrhythmias and heart failure. Activities as low as IC_{50} of 3.9nM were reported for inhibition of NHE1 receptors (see Section 6.4.3).

6.4.11 *Functionalization of pentafluorosulfanylphenoxy-substituted benzoylguanidines*

This work describes the further transformations of **115** (Kleemann, 2006) (Fig. 6.21).

Compounds such as **126** are described as being potentially useful as antiarrhythmic medicaments as they inhibit, even preventively, the ischemic induced damage, particularly during the cardiac arrhythmias.

6.4.12 *Pentafluorosulfanyl-containing diarylamine trypanothione reductase inhibitors*

By analogy to phenol **43** in Section 6.4.10, aniline **20** retained sufficient nucleophilicity to participate in a nucleophilic aromatic displacement reaction with 2,6-difluoro-nitrobenzene (Stump *et al.*, 2009) (Fig. 6.22). Remaining manipulations proceeded in good to excellent yield in a synthesis that readily accommodated reactions of other amines in addition to dimethylamine. Analysis of crystal structures of the pentafluorosulfanyl

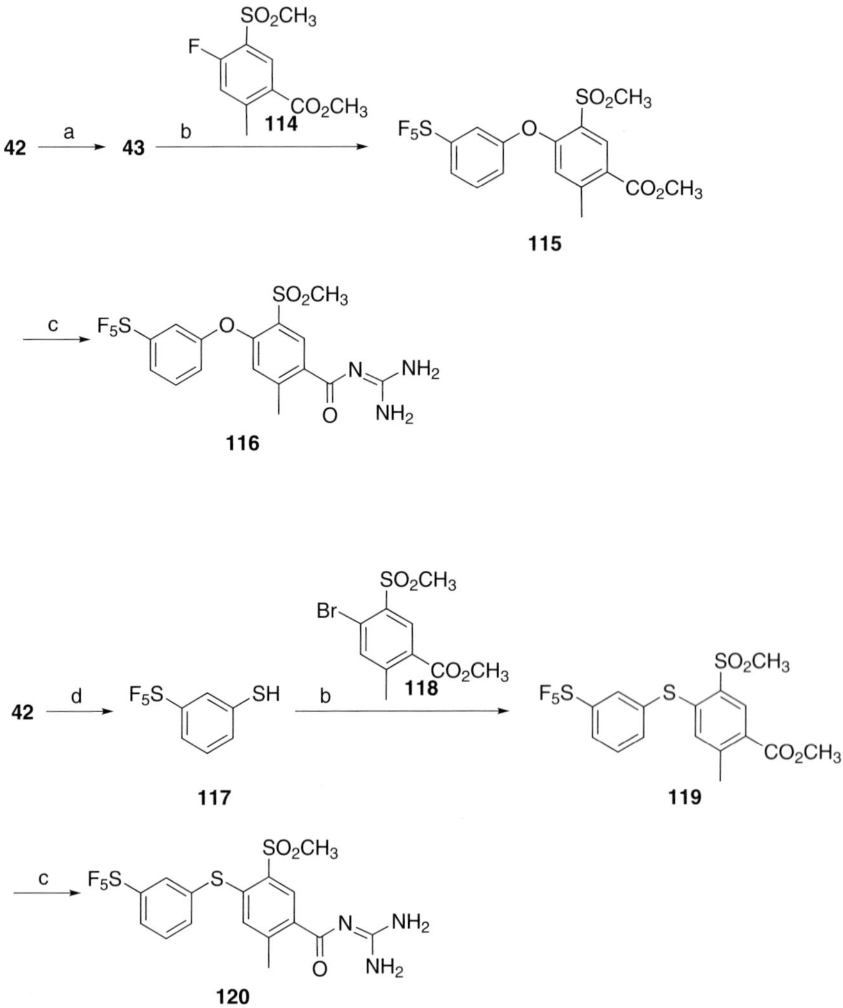

Figure 6.20. Preparation of pentafluorosulfanylphenoxy-substituted benzoylguanidines: (a) H_2SO_4, $NaNO_2$ then $Cu(NO_3)_2$, CuO (70%); (b) Cs_2CO_3, DMF (21%); (c) guanidinium hydrochloride; (d) HOAc, 35% H_2SO_4, $NaNO_2$ then Na_2S_2, 50–60°C then $LiAlH_4$ (9%).

analogues bound to trypanothione reductase facilitated comparison of the SF_5, CF_3 and *tert*-butyl groups. The steric demand of the SF_5 group lies between those of CF_3 and $C(CH_3)_3$, but the profound electron withdrawing character of the pentafluorosulfanyl group was found to promote a

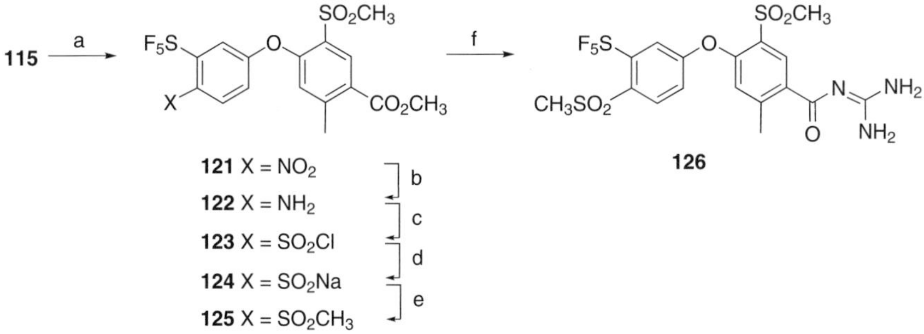

Figure 6.21. Functionalization of pentafluorosulfanylphenoxy-substituted benzoylguanidines: (a) 90% HNO_3 (quantitative); (b) AcOH, MeOH, Pd-C, H_2 6 bar, 24 h (93%); (c) AcOH-HCl, $NaNO_2$ then CuCl, $CuCl_2$, SO_2 in AcOH (87%); (d) Na_2SO_3, H_2O 70°C (87%); (e) 2N NaOH then CH_3I, DMF (17%); (f) guanidinium hydrochloride, Cs_2CO_3.

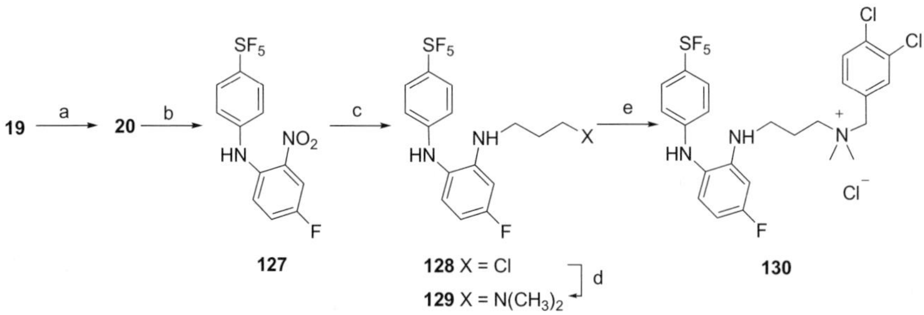

Figure 6.22. Preparation of pentafluorosulfanyl-containing diarylamine trypanothione reductase inhibitors: (a) H_2 5 bar, RaNi, MeOH, 65°C (97%); (b) 2,5-difluoro-nitrobenzene, *t*-BuOK, DMSO (25%); (c) Zn, NH_4Cl, 3-chloropropionyl chloride CH_3OH, 65°C (74%); (d) Me_2NH in H_2O, DMF, 60°C (91%); (e) 3,4-dichlorobenzyl chloride, 50°C (67%).

T-shaped interaction with nearby tryptophan. Relative to the CF_3 substituted analogue, the SF_5-containing molecules were associated more closely with Trp21. Interaction of the aniline nitrogen with Glu18 was also considerably closer with the pentafluorosulfanylated compound than with either the CF_3- or $C(CH_3)_3$-substituted compounds, presumably reflecting a strengthened hydrogen bonding interaction. The efficacy of the pentafluorosulfanyl compound was likely derived from the electronic effects of the pentafluorosulfanyl group that compensated for the greater steric obstruction of that group to binding.

Trypanosomatid parasites possess trypanothione [N1,N8-bis(glutathionyl)spermidine] and trypanothione reductase. Trypanothione reductase, a key enzyme of the trypanothione-based antioxidant defence systems of parasitic trypanosomes and *Leishmania* sp., is a promising target for the new antiparasitic drugs. *Trypanosoma brucei* is the causative agent of African sleeping sickness which threatens millions of people. *Trypanosoma cruzi* is the pathogen responsible for 14,000 deaths each year in central and southern America. Leishmaniasis is a disease resulting from infection with *Leishmania* parasites: the disease infects millions of people and leads to nearly 50,000 deaths annually (Stump *et al.*, 2009). While none of the diarylamines showed a significant effect on the growth of axenic *Leishmania donovani*, the non-quaternized dimethylamine **129** moderately inhibited the growth of *T. cruzi.*

SF_5 is a suitable substituent for novel diarylamine-based trypanothione reductase inhibitors with micromolar affinities. A preference of the mepacrine-binding site for CF_3 and SF_5 substituents over the bulkier *t*-Bu residue can be rationalized by molecular modelling. Ligands with SF_5 substituents display the lowest cytotoxicity among all compounds tested and show good membrane permeability (Stump *et al.*, 2009).

6.4.13 *A pentafluorosulfanyl-containing quinoline, a mefloquine analogue*

The first report on an SF_5–quinoline construction significantly expands the repertoire of pentafluorosulfanyl chemistry (Wipf *et al.*, 2009) (Fig. 6.23). Upon completion of the synthesis, both **141** and **142** were characterized by single crystal X-ray diffraction studies.

Mefloquine, the CF_3-containing parent quinoline, is an orally-administered antimalarial agent. Unfortunately, mefloquine is associated with adverse effects, including anxiety, depression, halucinations and seizures which limit its utility (Wipf *et al.*, 2009).

The antimalarial activities and selectivities of **141** and **142** were compared to mefloquine analogues in which the quinoline ring was substituted at the 6- and 7-positions with a trifluoromethyl group. The 50% and 90% inhibitory concentrations (IC_{50} and IC_{90}) against four drug resistant strains of *Plasmodium falciparum*, and the 50% lethal concentration

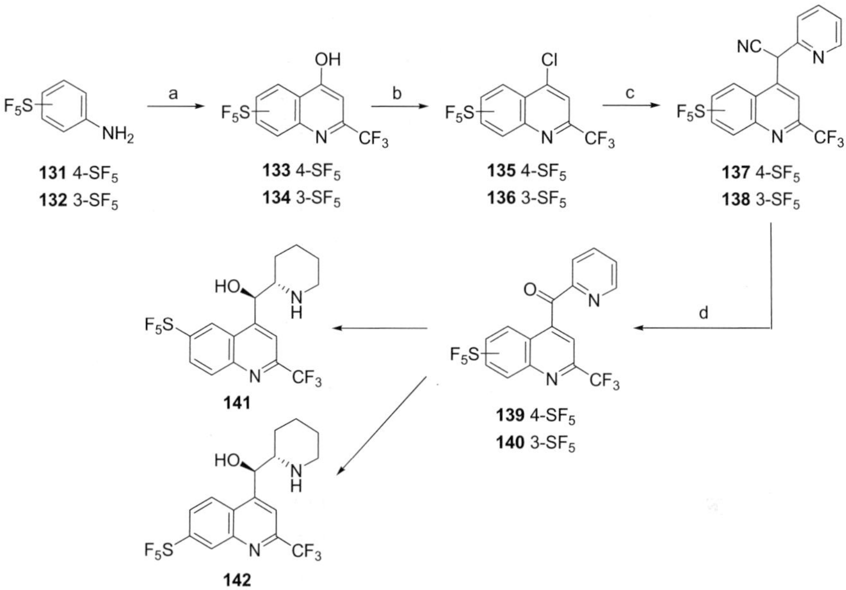

Figure 6.23. Synthesis of compounds **141** and **142**: (a) ethyl 4,4,4-trifluoro-3-oxobutanoate, PPA (4-SF_5 44%, 3-SF_5 75%); (b) $POCl_3$ (4-SF_5 77%, 3-SF_5 78%); (c) 2-(pyridin-2-yl) acetonitrile, NaH, DMF, $PhCH_3$ (4-SF_5 86%, 3-SF_5 92%); (d) H_2O_2, AcOH (4-SF_5 85%, 3-SF_5 92%).

(LC_{50}) against a mammalian cell line were determined. Compound **141** exhibited generally equivalent or lower IC_{50} and IC_{90}, and greater selectivity than its CF_3-congener and mefloquine. The IC_{50} and IC_{90} of **142** were generally equivalent to those of the CF_3-analogue and mefloquine (Wipf *et al.*, 2009).

6.4.14 *Fluoxetine analogues*

Commercially available 1-nitro-4-pentafluorosulfanylbenzene **19** was easily reduced to 1-amino-4-pentafluorosulfanylbenzene **20** under classic conditions with iron powder and concentrated hydrochloric acid (Welch and Lim, 2007) (Fig. 6.24). Subsequent diazotization and dediazoniation was also affected in an uneventful manner. In contrast to methods for the preparation of the fluoxetine **149** (Fig. 6.25), where the 4-chloro-trifluoromethylbenzene undergoes a ready nucleophilic aromatic displacement reaction with **20**, **27** did not react in our hands. However, following introduction of the sacrificial auxiliary nitro group, 1-bromo-2-nitro-4-pentafluorosulfanylbenzene **37** did undergo the desired displacement reaction, albeit in modest yield. It was found that protection of the methylamine group of **144** was necessary for successful reductive dediazoniation of easily prepared aminoarene **146**. In the absence of protection, the reduction was accompanied by the apparent intramolecular reaction of the reactive amine group. Protection of the methylamino group by benzylation, *tert*-butoxycarbonylation, benzyloxycarbonylation, trimethylsilylation or *tert*-butyldimethylsilylation all failed to lead to the desired reduction product. Fortunately benzoylation to form **145** was effective in suppressing the undesired side reactions. Diazotization and reductive dediazoniation of **146** with *tert*-butylnitrite in DMF proceeded smoothly. Deprotection of the **147** by reduction with diisobutylaluminum hydride led uneventfully to the desired pentafluorosulfanyl fluoxetine analogue **148**.

In secondary screening, the K_i values were determined for those receptors where at the original test concentration of 10μM >50% inhibition was observed. For **148**, substitution of the CF_3 by the SF_5 group diminished the affinity for 5-HT_{2a} and 5-HT_{2c} but had no effect on 5-HT_{2b} (Fig. 6.25).

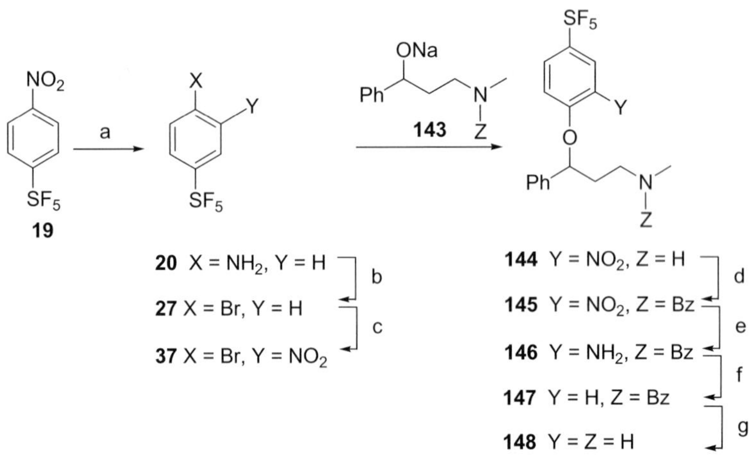

Figure 6.24. Preparation of fluoxetine analogues **144**–**148**: (a) Fe, HCl (86%); (b) HBr/$NaNO_2$ (67%), (c) H_2SO_4/HNO_3 (99%); (d) BzCl (36%); (e) Fe, HCl (76%); (f) *t*-BuONO, DMF (33%); (g) DIBAL-H (80%).

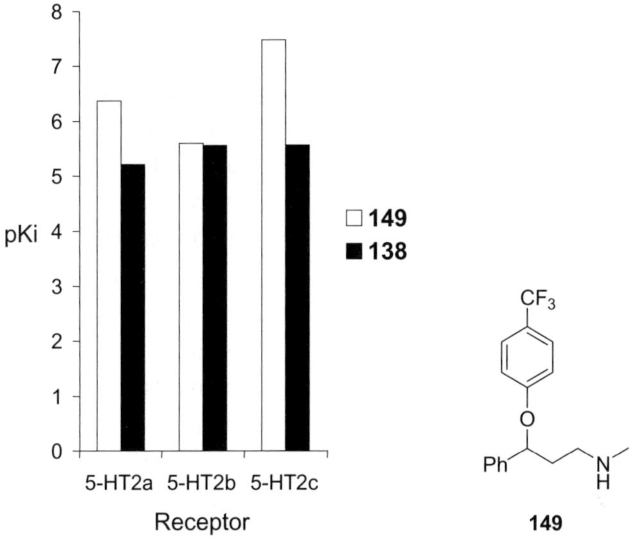

Figure 6.25. Replacement of the trifluoromethyl group of **149** by pentafluorosulfanyl group as in **148**. Influence of substitution on receptor binding. Data from the NIMH Psychoactive Drug Screening Program (Welch and Lim, 2007).

6.4.15 *Fenfluramine and norfenfluramine*

The pentafluorosulfanyl analogues of fenfluramine and norfenfluramine were prepared from readily available pentafluorosulfanylbenzene **150** (Welch and Lim, 2007) (Fig. 6.26). Bromination led to the formation of 1-bromo-3-pentafluorosulfanylbenzene **151**. Metallation with *tert*-butyllithium was followed by quenching of the resultant anion with dimethylformamide, which on workup afforded the desired aldehyde **152**. Condensation with nitroethane formed the olefin **153** which was subsequently reduced with lithium aluminium hydride. The pentafluorosulfanyl analogue of norfenfluramine **154** was reductively alkylated with acetaldehyde and sodium triacetoxyborohydride to form the fenfluramine analogue **155**.

In Fig. 6.27 it is evident that the pentafluorosulfanyl group enhances the affinity of **154** for 5-HT_{2b}, 5-HT_{2c} and 5-HT_6 relative to fenfluramine **157**. Of especial note is the increased affinity for 5-HT_{2b} and 5-HT_6, with binding increasing nearly tenfold. Binding to 5-HT_{2b} has been associated with adverse valvulopathy. Unfortunately the increase in affinity for the

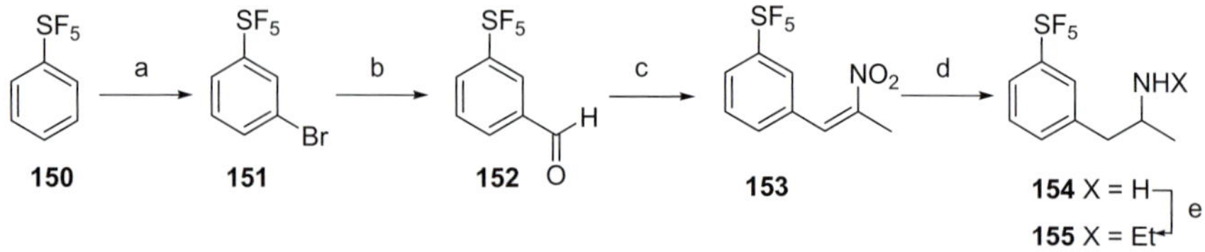

Figure 6.26. Preparation of pentafluorosulfanyl analogues of fenfluramine and norfenfluramine **154** and **155**: (a) NBS, H_2SO_4/TFA (93%); (b) *t*-BuLi/DMF, –78 to 0°C (70%); (c) $CH_3CH_2NO_2$, NH_4OAc (70%); (d) LAH/THF (59%); (e) CH_3CHO, $NaBH(OAc)_3$ (30%).

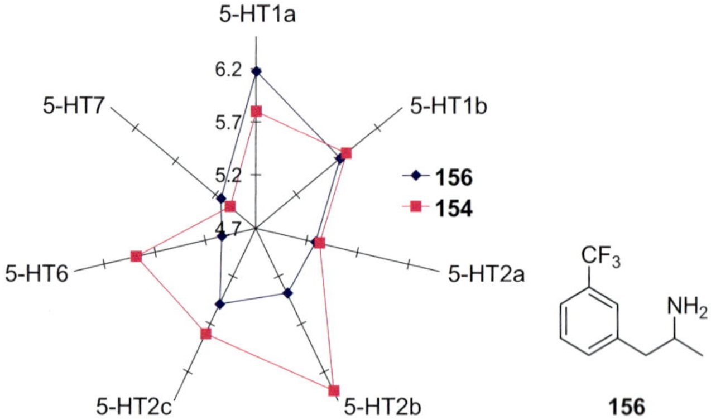

Figure 6.27. Comparison of pK_i values of **156** and **154** for a series of 5 HT receptors.

5-HT_{2c} receptor is much less, so it is likely that the analogue **154** would not be as safe as the clinical agent. By contrast, while the affinity for 5-HT_{2b} receptor of **155** relative to **157** is enhanced (Fig. 6.28), the increase is much less than in the case of **154** relative to **156** (Fig. 6.27). Perhaps more strikingly, the pentafluorosulfanyl group substitution in the norfenfluramine structure showed the same general pattern of selectivity observed with the parent compound. As determined in the primary inhibition assays there was little affinity for 5-HT_{1a}, 5-HT_{1e}, 5-HT_3 or 5-HT_{5a}, and this selectivity was unaffected by substitution.

6.4.16 *5-Hydroxytryptamine*

Formation of the diazonium salt from **20** was effected by slow addition of aqueous sodium nitrite to an acidic methanolic solution (Welch and

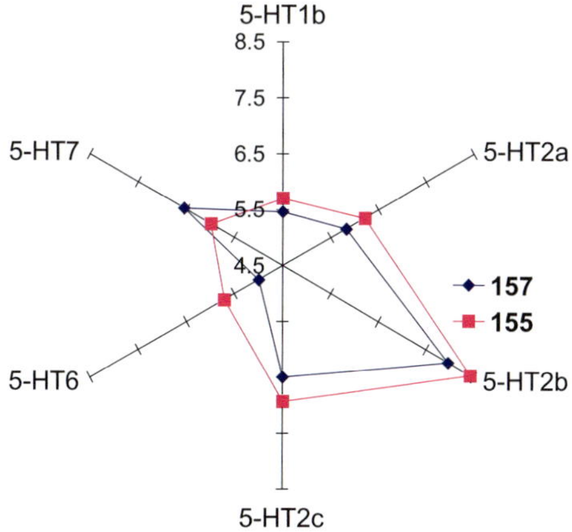

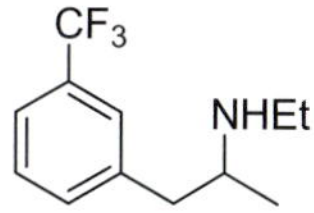

157

Figure 6.28. Comparison of pK_i values for **157** and **155** with a series of 5 HT receptors.

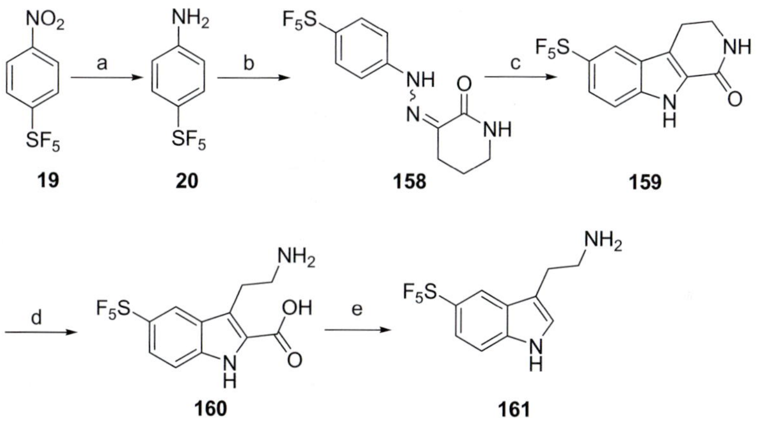

Figure 6.29. Preparation of tryptamine analogue **161**: (a) Fe, HCl (96%); (b) HCl/$NaNO_2$ then 2-oxopiperidine-3-carboxylic acid, NaOAc (88%); (c) aq.HCO_2H, reflux (27%); (d) KOH, EtOH-H_2O, reflux (58%); (e) Cu/quinoline, 190–200°C/1h (21%).

Lim, 2009) (Fig. 6.29). The addition of a solution of 2-oxopiperidine-3-carboxylic acid followed by buffering of the solution with sodium acetate produced the desired 3-(2-(4-(pentafluorosulfanyl)phenyl)hydrazono) piperidin-2-one **158** as a solid. On dissolution in aqueous formic acid (88%) and heating under reflux for 1.5 hours the cyclized lactam **159** was formed in limited but unoptimized yield. The lactam was saponified to yield the carboxylic acid **160**. Decarboxylation to the desired tryptamine analogue **161** was affected on heating to 200°C in quinoline in the presence of copper powder.

6.5 Conclusions and Outlook

The utility of the pentafluorosulfanyl group is just beginning to be demonstrated. In numerous examples it exhibits significant differences from trifluoromethyl substitution. With a growing body of experimental practice accessible in the literature, it is clear that the SF_5 group is quite compatible with the majority of common functional group transformations. None of these publications considers the important environmental observations of Jackson and Mabury (2008, 2009) indicate that pentafluorosulfanyl substitution may have environmental benefits not previously appreciated. The recent report of Umemoto (2009) on the synthesis of the SF_5 group without the need for molecular fluorine may be the most significant of all. This procedure suggests the cost of pentafluorosulfanyl substitution can be dramatically reduced, perhaps to a cost similar to that of trifluoromethyl substituents. It is likely that the coming years will see an ever expanding number of reports of the utility of pentafluorosulfanylation in biological chemistry, a suitable memorial to the pioneering work of Bill Sheppard so long ago.

Acknowledgements

This work was supported by Petroleum Research Fund of the American Chemical Society and by Air Products and Chemicals Inc.

References

2009 Insecticidal derivatives of substituted phosphorylated phenylalkyl iminooxazolines and iminothiazolines. European patent app. 2096114.

Aiet-Mohand, S. & Dolbier, W.R., Jr. (2002) New and convenient method for incorporation of pentafluorosulfanyl (SF_5) substituents into aliphatic organic compounds. *Org Lett* 4, 3013–3015.

Alt, G.H., Pratt, J.K., Phillips, W.G. *et al.* (1990) Preparation of thiazole-5-carboxanilides as agrochemical fungicides. European patent 371950.

Andreotti, D., Checchia, A., Hamprecht, D. *et al.* (2006) Preparation of 1-(pentafluorosulfanylphenyl)-3-(1,2,4-triazol-3-ylthioalkyl)-3-azabicyclo[3.1.0]hexanes as selective modulators of dopamine D_3 receptors. WO patent 2006108700.

Anthony, M. (1984) Serotonin antagonists. *Aust NZ J Med* 14, 888–895.

Banks, B.J. (1997) Parasiticidal pyrazole derivatives and their preparation and use. WO patent 9707102.

Banks, B.J. (1998a) Preparation of cyclopropylpyrazoles with parasiticidal activity. WO patent 9824767.

Banks, B.J. (1998b) Preparation of parasiticidal 3-cyano-1-phenylpyrazoles. WO patent 9804530.

Banks, B.J. (1999) Preparation of 1-aryl-4-cyclopropylpyrazoles as medical and agrochemical parasiticides and pesticides. European patent 933363.

Barton, J.E.D. & Mitchell, G. (1994) Herbicidal compounds having a pentafluorosulfanyl group. GB patent 2276379.

Bowden, R.D., Greenhall, M.P., Moilliet, J.S. *et al.* (1997) Preparation of organosulfurpentafluorides by the fluorination of organic disulfides. WO patent 9705106.

Bowden, R.D., Comina, P.J., Greenhall, M.P. *et al.* (2000) A new method for the synthesis of aromatic sulfurpentafluorides and studies of the stability of the sulfurpentafluoride group in common synthetic transformations. *Tetrahedron* 56, 3399–3408.

Brant, P., Berry, A.D., DeMarco, R.A. *et al.* (1981) Gas phase and solid state X-ray photoelectron spectra of the substituted acetylenes $(SF_5)_nC_2(CF_3)_{2-n}$($n = 0$, 1, 2): a comparison of the pentafluorosulfur and trifluoromethyl groups. *J Electron Spectrosc Relat Phenom* 22, 119–129.

Brel, V.K. (2006) Synthesis of 4,5-dihydroisoxazoles connected by short spacers to the pentafluoro-λ^6-sulfanyl group. *Synthesis* 16, 2665–2670.

Cao, G., Xue, C.-B. & Metcalf, B. (2009) Preparation of pentafluorosulfur piperazinylpiperidines as CCR5 chemokine receptor antagonists. WO patent 2009012259.

Carroll, W.A., Dart, M.J., Perez-Medrano, A. *et al.* (2009) Preparation of pyrazoles, oxazoles, and other nitrogen-containing heterocyclic compounds as therapeutic cannabinoid receptor ligands. US patent 2009105306.

Chambers, R.D., Greenhall, M.P., Hutchinson, J. *et al.* (1996) *Proceedings of the 211th National Meeting of American Chemical Society*, New Orleans, LA, Fluo11.

Chern, R.T., Zingerman, J.R., Clark, J.N. *et al.* (1999) Sulfur pentafluorophenyl pyrazoles for controlling ectoparasitic infestations. WO patent 9947139.

Comlay, S.N., Hannam, J.C., Howson, W. *et al.* (2008) Preparation of N-(phenoxycyanomethylethyl)(pentafluorothio)benzamide for use as antiparasitic agents. WO patent 2008096231.

Coqueron, P.-Y., Grosjean-Cournoyer, M.-C. Mansfield, D.G. *et al.* (2008) Preparation of N-(3-phenylpropyl)benzamide derivatives and their use as fungicides. WO patent 2008101975.

Dolbier, W.R., Jr. & Ait-Mohand, S. (2004) Stereoselective method and catalysts for incorporation of pentafluorosulfanyl substituents into aliphatic and aromatic compounds. WO patent 2004011422.

Dolbier, W.R., Aiet-Mohand, S., Schertz, T.D. *et al.* (2006) A convenient and efficient method for incorporation of pentafluorosulfanyl (SF_5) substituents into aliphatic compounds. *J. Fluorine Chem* 127, 1302–1310.

Dolbier, W.R., Jr. & Zheng, Z. (2009) Preparation of pentafluorosulfanyl (SF_5) pyrrole carboxylic acid esters. *J Org Chem* 74, 5626–5628.

Frank, R., Sundermann, B. & Schick, H. (2006) Preparation of pentafluorosulphanyl-substituted compounds and for use in producing medicaments. WO patent 2006122773.

Frank, R., Bahrenberg, G., Christoph, T. *et al.* (2008) Preparation of N-benzyl-2-phenylpropanamides as vanilloid receptor antagonists. WO patent 2008125296.

Gard, G.L. (2009) Recent milestones in SF_5-chemistry. *Chem Today* 27, 10–13.

Gomtsyan, A.R., Schmidt, R.G., Bayburt, E.K. *et al.* (2009) Preparation of 4-pyridinylcyclohexene-1-carboxamides as TRPV1 antagonists. US patent 20090124671.

Hawkins, D.W. (1997) Preparation of pentafluorosulfanylphenyl- and benzoylisoxazoles and analogs as herbicides. DE patent 19711953.

Hayashizaki, K., Usui, Y. & White, C.J. (1997) 1,3-Oxazin-4-one derivatives useful as herbicides, and a process and intermediates for their preparation. WO patent 9740041.

Heinelt, U., Wehner, V., Herrmann, M. *et al.* (2009) Preparation of aminoimidazoles as proteinase activated receptor (PAR-1) inhibitors. WO patent 2009097973.

Herman, N.D., Huber, S.K., Huang, J. *et al.* (1998) Preparation of pesticidal 1-polyarylpyrazoles. European patent 839810.

Howard, M.H., Jr. & Stevenson T. M. (1995) Preparation of arthropodicidal pentafluorothio-substituted anilides. WO patent 9516676.

Huber, S.K. (1999) Flea control in cats and dogs. European patent 963695.

Jackson, D.A. & Mabury, S.A. (2008) Breaking down the substituent of the future: environmental properties of pentafluorosulfanyl compounds. Abstracts of Papers, 236th ACS National Meeting, Philadelphia, PA, USA, August 17–21, 2008.

Jackson, D.A. & Mabury, S.A. (2009) Environmental properties of pentafluorosulfanyl compounds: physical properties and photodegradation. *Environ Toxicol Chem* 28, 1866–1873.

Jaehne, G., Stengelin, S., Gossel, M. *et al.* (2008) Preparation of 3-phenylhydantoins as cannabinoid receptor inhibitors. WO patent 2008017381.

Kay, I.T., Barton, J.E.D., Collins D.J. *et al.* (1994) Preparation of [(Phenyl)hetereocyclyl] Carbamates or Thiocarbamates as Herbicides. WO patent 9413652.

Kirsch, P., Bremer, M., Heckmeier, M. *et al.* (1999) Liquid crystals based on hypervalent sulfur fluorides: pentafluorosulfuranyl as polar terminal group. *Angew Chem Intl Edit* 38, 1989–1992.

Kirsch, P. & Bremer, M. (2000) Nematic liquid crystals for active matrix displays: molecular design and synthesis. *Angew Chem Intl Edit* 39, 4216–4235.

Kirsch, P., Bremer, M., Taugerbeck, A. *et al.* (2001) Difluorooxymethylene-bridged liquid crystals: a novel synthesis based on the oxidative alkoxydifluorodesulfuration of dithianylium salts. *Angew Chem Intl Edit* 40, 1480–1484.

Kirsch, P. (2004) *Modern Fluoroorganic Chemistry. Synthesis, Reactivity and Applications*, Wiley-VCH, Weinheim.

Kleemann, H.-W. (2005a) Preparation of pentafluorosulfanyl benzoylguanidines as antiarrhythmic agents. WO patent 2005047239.

Kleemann, H.-W. (2005b) Preparation of pentafluorosulfanylphenyl-substituted benzoylguanidines for treating or preventing diseases which are related to NHE (sodium-proton exchanger). US patent 20050043401.

Kleemann, H.-W. (2006) Pentafluorosulfanylphenyl-substituted benzoylguanidines, method for the production thereof, their use as a medicament or diagnostic agent, and a medicament containing these compounds. WO patent 2006027130.

Lal, G.S. & Syvret, R.G. (2008) Enabling chemistry innovation with new SF_5 synthons from air products. *Chemistry Today, Focus on Fluorine Chemistry*, 26.

Lentz, D., Seppelt, K. & Akiba, K.-Y. (1999) 'The SF_5, SeF_5 and TeF_5 groups in organic chemistry', in Akiba, K.-Y. (Ed.), *Chemistry of Hypervalent Compounds*, Wiley–VCH, New York.

Lim, D.S., Choi, J.S., Pak, C.S. *et al.* (2007) Synthesis and herbicidal activity of a pentafluorosulfanyl analog of trifluralin. *J Pestic Sci (Tokyo, Jpn)* 32, 255–259.

Manning, D.T. & Wu, T.-T. (1998) Pesticidal 1-aryl-3-iminopyrazoles. WO patent 9856767.

Mansfield, D., Coqueron, P.-Y., Rieck, H. *et al.* (2007) Preparation of N-(1-methyl-2-phenylethyl)benzamides as agrochemical fungicides. WO patent 2007060162.

Phillips, J., Pilato, M. & Wu, T.-T. (1998) Preparation of arylpyrazoles as pesticides. WO patent 9828277.

Pilato, M.T. & Wu, T.-T. (1996) Preparation of pesticidal 5-amino-3-cyano-1-(4-pentafluorothiophenyl)pyrazoles. WO patent 9625401.

Prichard, W.W. & Stacey, F.W. (1975) 1,5-Diphenyl-3-formazancarbonitriles and insecticidal compositions containing them. DE patent 2460255.

Raasch, M.S. (1963) Ureido-substituted arylsulfur pentafluorides. US patent 3073861.

Salmon, R. (1994) Dihydropyrazole compounds useful as insecticides. GB patent 2276382.

Salmon, R., Pearson, D.P.J., Parry, D.R. *et al.* (1994) Acaricidal and insecticidal pentafluoro(phenyl)sulfur or pentafluoro(pyridinyl)sulfur compounds. WO patent 9421606.

Sæthre, L.J., Berrah, N., Bozek, J.D. *et al.* (2001) Chemical insights from high-resolution X-ray photoelectron spectroscopy and ab initio theory: propyne, trifluoropropyne, and ethynylsulfur pentafluoride. *J Am Chem Soc* 123, 10729–10737.

Sheppard, W.A. (1960) Arylsulfur trifluorides and pentafluorides. *J Am Chem Soc* 82, 4751–4752.

Sheppard, W.A. (1962) Electrical effect of the sulfur pentafluoride group. *J Am Chem Soc* 84, 3072–3076.

Sipyagin, A.M., Bateman, C.P., Tan, Y.-T. *et al.* (2001) Preparation of the first ortho-substituted pentafluorosulfanylbenzenes. *J Fluorine Chem* 112, 287–295.

Sipyagin, A.M., Enshov, V.S., Kashtanov, S.A. *et al.* (2004) New 4-pentafluorosulfanyl and 5-perfluoroalkylthio derivatives of 1-chloro-2-nitro- and 1-chloro-2,6-dinitrobenzenes. *J Fluorine Chem* 125, 1305–1316.

Stump, B., Eberle, C., Schweizer, W.B. *et al.* (2009) Pentafluorosulfanyl as a novel building block for enzyme inhibitors: trypanothione reductase inhibition and antiprotozoal activities of diarylamines. *ChemBioChem* 10, 79–83.

Taft, R.W., Jr. & Lewis, I.C. (1959) Evaluation of resonance effects on reactivity by application of the linear inductive energy relationship. V. Concerning a σ_R scale of resonance effects. *J Am Chem Soc* 81, 5343–5352.

Taft, R.W., Jr. (1960) Sigma values from reactivities. *J Physical Chem* 64, 1805–1815.

True, J.E., Thomas, T.D., Winter, R.W. *et al.* (2003) Electronegativities from core-ionization energies: electronegativities of SF_5 and CF_3. *Inorg Chem* 42, 4437–4441.

Umemoto, T. (2009) New processes for producing arylsulfur pentafluorides. 284. 19th International Symposium on Fluorine Chemistry, Jackson Hole, WY, 23–28 August 2009.

Welch, J.T. & Lim, D.S. (2007) The synthesis and biological activity of pentafluorosulfanyl analogs of fluoxetine, fenfluramine, and norfenfluramine. *Bioorg Med Chem* 15, 6659–6666.

Welch, J.T. & Lim, D.S. (2009) 'Pentafluorosulfanyl serotonin analogs: synthesis, characterization, and biological activity', in Gakh, A. & Kirk, K.L. (Eds), *Fluorinated Heterocycles*, American Chemical Society, Washington DC.

Williams, A.G. & Foster, N.R. (1994) Process for the preparation of aryl- and heteroarylsulphurpentafluorides. WO patent 9422817.

Winter, R.W., Dodean, R.A., Gard, G.L. *et al.* (2005) 'SF_5 synthons: pathways to organic derivatives of SF_6', in Soloshonok, V.A. (Ed.), *Fluorine Containing Synthons*, American Chemical Society, Washington D.C.

Wipf, P., Mo, T., Geib, S.J. *et al.* (2009) Synthesis and biological evaluation of the first pentafluorosulfanyl analogs of mefloquine. *Org Biomol Chem* 7, 4163–4165.

Worthington, P.A. & Streeting, I.T. (1994) Fungicidal compounds containing a phenylsulfurpentafluoride group. GB patent 2276381.

7

Strategic Incorporation of Fluorine into Taxoid Anticancer Agents

Antonella Pepe,[*,†] *Liang Sun*[†] *and Iwao Ojima*[†]

7.1 Introduction

Covalently bound fluorine plays a crucial role in bioorganic and medicinal chemistry (Bégué and Bonnet-Delpon, 2006; Isanbor and O'Hagan, 2006; Ojima, 2009). A recent estimate accounts for 138 US Food and Drug Administration (FDA) approved fluorine-containing molecules for human treatment, of which 23 have been withdrawn from the market, and 33 are currently used for veterinary applications (Pepe *et al.*, 2009a). Fluorine-containing biologically active compounds, in which a C–H or C–O bond is replaced by a C–F bond, often exhibit stronger binding to target molecules, higher metabolic stability, increased lipophilicity and higher membrane permeability, as compared with the parent compound. Due to the small atomic radius of the fluorine atom and its high electronegativity, the C–F bond is short (1.35 Å) and highly polarized. The strong ionic character contributes to its unusual strength, hence the C–F bond is more resistant to metabolic oxidation. Because of the recognized

* Current address: Laboratory of Synthetic Chemistry, Developmental Therapeutics Program Support, SAIC-Frederick, Inc.; National Cancer Institute-Frederick, Frederick, MD 21702, USA.

† Institute of Chemical Biology & Drug Discovery and Department of Chemistry, State University of New York at Stony Brook, Stony Brook, NY 11794-3400, USA.

value of fluorine in medicinal chemistry, it is now common practice to investigate fluoro-analogues of lead compounds under development (Kirk, 2006; Yamazaki *et al.*, 2009). Recently, increasing numbers of experimental and computational studies have been performed systematically to better understand and predict how the replacement of H with F affects the electronic nature and conformation of small molecules in their interactions with proteins and enzymes (Müller *et al.*, 2007; Yamazaki *et al.*, 2009). These studies will eventually lead to rational design and development of fluoro-organic molecules as new and effective biochemical tools.

The importance of fluorine in medicinal chemistry is further enhanced by its paramagnetic property. The nuclear spin of ½, the natural abundance, the high NMR sensitivity along with the large ^{19}F–^{1}H coupling constants and the virtual absence of ^{19}F in living tissues, allow application of ^{19}F NMR spectroscopy to the investigation of biological processes (Martino *et al.*, 2000; O'Hagan *et al.*, 2002). Solid-state ^{19}F NMR is a particularly powerful tool for investigation of the protein-bound structures of biologically active molecules (Ojima, 2009). In addition, *in vivo* ^{19}F magnetic resonance (MR) has been applied to monitor fluorine compounds and their metabolites in preclinical and clinical settings (Schneider and Lin, 2009).

This chapter provides a concise overview of our research on the strategic incorporation of fluorine into taxoid ('taxoid' = Taxol-like compound) anticancer agents through exploitation of the unique properties of this element.

7.2 Paclitaxel, Docetaxel and New-Generation Taxoids

Paclitaxel (Taxol) and its semi-synthetic analogue docetaxel are currently two of the most widely used anticancer drugs for the treatment of ovarian cancer, breast cancer, melanoma, non-small-cell lung cancer and Kaposi's sarcoma as well as neck, prostate and cervical cancers (Rowinsky, 1997; FDA, 2004) (see Fig. 7.1).

Paclitaxel (and taxoids) binds to the β-subunit of αβ-tubulin heterodimer and accelerates the polymerization of the tubulin dimer units by stabilizing the interaction between these units (Schiff *et al.*, 1979; Jordan *et al.*, 1993). Thus, microtubules arising from paclitaxel-bound αβ-tubulin dimers are less dynamic with a growth rate higher than the disassembling rate. The paclitaxel-bound stabilized microtubules do not function normally,

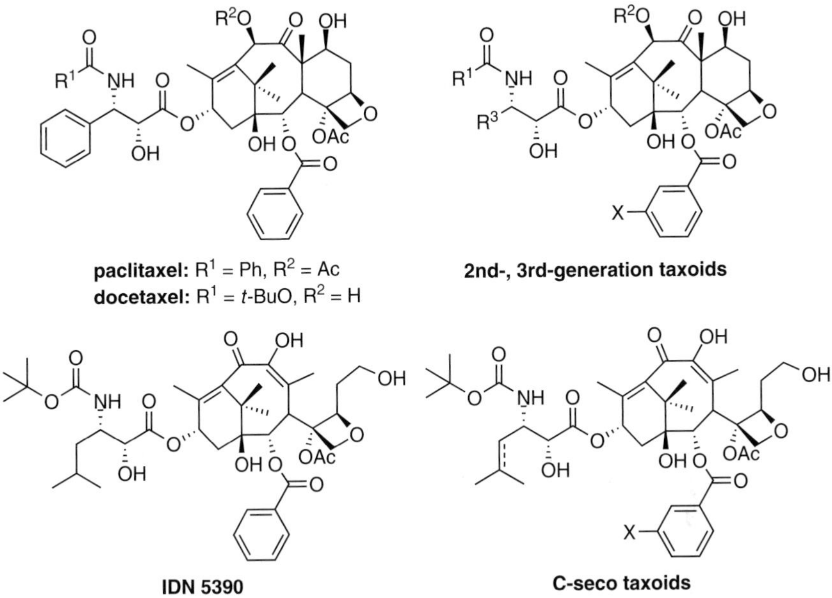

Figure 7.1. Structures of paclitaxel, docetaxel, new-generation taxoids.

leading to the inhibition of depolymerization and interruption of the cell division cycle. This event activates a cell-signalling cascade, that induces apoptosis (Schiff *et al.*, 1979; Jordan *et al.*, 1993).

Despite the remarkable success of paclitaxel and docetaxel, chemotherapy with these drugs is often associated with undesirable side effects, as well as the occurrence of drug resistance (Dumontet and Sikic, 1999). Therefore, there is a strong need for developing new taxoid anticancer drugs and their efficacious drug delivery systems with fewer side effects, superior pharmacological properties and improved activity against drug-resistant tumours and various classes of cancers.

Several mechanisms of drug resistance to paclitaxel, docetaxel and other microtuble-targeting agents have been indicated (Sève and Dumontet, 2008). In general, the development of drug resistance is a multifactor phenomenon and likely to be mediated by a combination of events (Gottesman *et al.*, 2002). The first mechanism of paclitaxel resistance reported is multidrug resistance (MDR) caused by the overexpression of ABC transporters, typically P-glycoprotein drug-efflux pumps, which can keep the intracellular drug concentration below therapeutic level (Chevillard *et al.*, 1996). Point mutation of class I β-tubulin has been reported for one-third of human non-small cell lung cancer, and is in direct correlation with cell sensitivity to paclitaxel (Seve and Dumontet, 2005). The complexity of the MDR phenotype is demonstrated by the absence of point mutations in tubulins isolated from the paclitaxel-resistant lung and ovarian tumours of cancer patients (Derry *et al.*, 1997). Another mechanism of drug resistance, which has received increased attention in recent years, is the selective overexpression of β-tubulin isotypes. Different β-tubulin isotypes form tubulin dimers with anomalous behaviours *in vitro* with regard to assembly, dynamics, conformation and ligand binding (Derry *et al.*, 1997). Microtubules with altered β-tubulin isotype composition respond differently to paclitaxel (Sullivan, 1988; Banerjee *et al.*, 1992; Panda *et al.*, 1994; Derry *et al.*, 1997; Kavallaris *et al.*, 1997; Hari *et al.*, 2003). Among β-tubulin subtypes, class III β-tubulin is less sensitive to the action of paclitaxel and other microtubule-stabilizing anticancer agents, due to higher intrinsic dynamic instability or to different binding interaction with those agents. Thus, the overexpression of class III β-tubulin in microtubules leads to paclitaxel-resistance (Dumontet *et al.*, 2009).

Our extensive studies on the design, synthesis and structure–activity relationship (SAR) of taxoid anticancer agents have led to the discovery and development of new-generation taxoids bearing non-aromatic substituents at the C3′ position and various acyl groups at the C10 position (Ojima *et al.*, 1996, 2008) as well as *meta*-substituted benzoyl groups at the C2 position (Ojima *et al.*, 1999, 2008) (see Fig. 7.1). These new-generation taxoids possess two to three orders of magnitude higher potency than paclitaxel and docetaxel against drug-resistant cancer cell lines, expressing MDR phenotype (Ojima *et al.*, 1996, 1999, 2008). Further improvement of the pharmacological properties of these taxoids has been studied through strategic incorporation of fluorine.

Our recent study on the metabolic stability of C3′-isobutyl- and C3′-isobutenyl-taxoids has disclosed that the metabolism of new-generation taxoids (e.g. SB-T-1214 and SB-T-1103) exhibits marked difference from that of docetaxel and paclitaxel (Gut *et al.*, 2006). One of the cytochrome P450 family enzymes in humans, CYP 3A4, metabolizes these taxoids

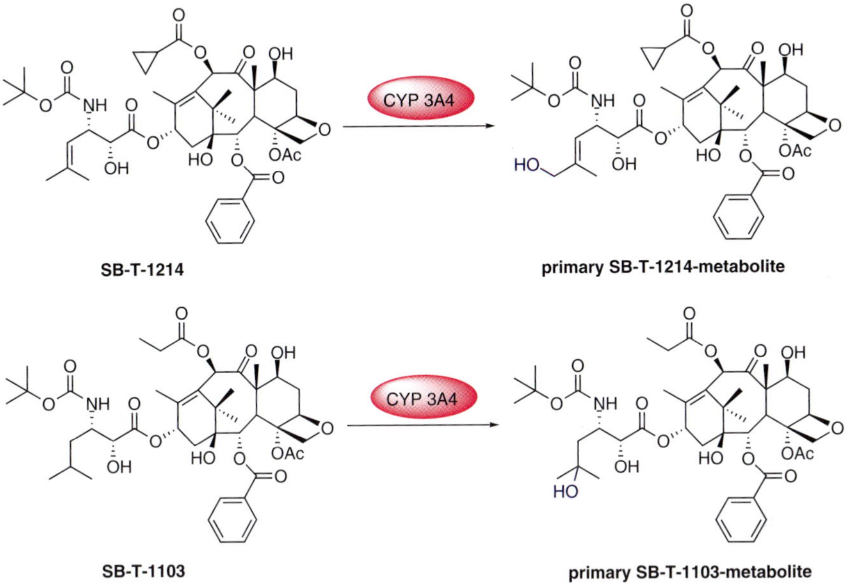

Figure 7.2. Primary sites of hydroxylation on new generation taxoids by a P450 family enzyme.

through hydroxylation primarily at the two allylic methyl groups of the C3′-isobutenyl group and the methyne moiety of the C3′-isobutyl group (Fig. 7.2). The result shows a stark contrast to the fact that the *tert*-butyl group of the C3′ *N*-Boc moiety is the single predominant metabolic site for docetaxel (Vuilhorgne *et al.*, 1995).

To address these unique metabolic profiles of the new-generation taxoids, we designed and synthesized a new series of fluorine-containing taxoids bearing C3′-difluoromethyl-, C3′-trifluoromethyl- and C3′-difluorovinyl-taxoids in order to block the oxidation by CYP 3A4 mentioned above, which should enhance the metabolic stability and activity *in vivo.*

7.3 Synthesis and Biological Evaluation of Fluorine-Containing New-Generation Taxoids

C3′-Trifluoromethyl-, C3′-difluoromethyl- and C3′-difluorovinyl-taxoids were synthesized through coupling of fluorine-containing β-lactams with

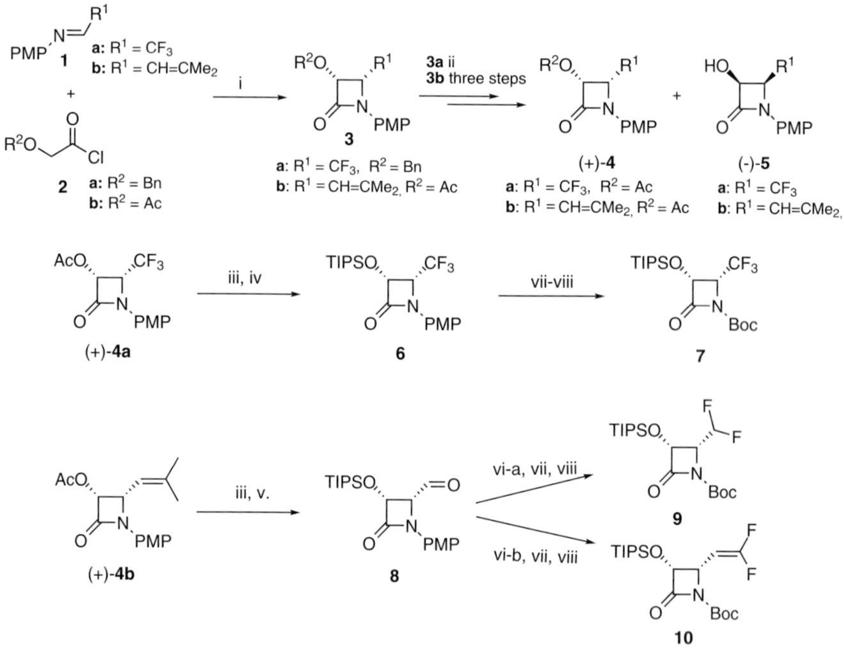

i) Et_3N, CH_2Cl_2 70-80%; ii) PS-Amano, buffer pH 7, 10% CH_3CN; iii) KOH, THF, -5 °C 94%; iv) TIPSCl, Et_3N, CH_2Cl_2 85-95%; v) O_3, MeOH/CH_2Cl_2, -78 °C; Me_2S 73%; vi-a) DAST, CH_2Cl_2 86%; vi-b) CBr_2F_2, HMPT, Zn, THF, 84%; vii) CAN, CH_3CN/H_2O, -10 °C, 68-84%; viii) *t*-Boc_2O, Et_3N, DMAP, CH_2Cl_2 80-98%.

Scheme 7.1. Synthesis of enantiopure 1-Boc-3-TIPSO-4-Rf-β-lactams.

modified baccatins (Ojima *et al.*, 1997b, 2000; Kuznetsova *et al.*, 2004, 2008). Syntheses of enantiopure coupling-ready 4-Rf-β-lactams, 7, **9** and **10**, are illustrated in Scheme 7.1.

Enantiopure β-lactams **9** and **10** bearing a CF_2H group and a difluorovinyl group at the C4 position, respectively, were prepared from a common intermediate, 4-formyl-β-lactam **8**. Racemic β-lactam **3a,b** was prepared through [2+2] ketene-imine cycloaddition of *N*-PMP-aldimine **1a,b** and acetyl chloride **2a,b**. A racemic mixture of β-lactams **3a,b** was resolved by enzymatic catalysis, using PS Amano lipase at 0–5°C for **3a** and 50°C for **3b** (Kuznetsova *et al.*, 2004). Subsequent functional group manipulations of (+)-**4a** gave (3*R*,4*S*)-1-Boc-3-TIPSO-4-CF_3-β-lactam 7. Enantiopure β-lactam (+)-**4b** was converted to 4-formyl-β-lactam **8** by ozonolysis, followed by difluoromethylation with DAST and *N*-protecting group manipulation to yield (3*R*,4*S*)-1-Boc-3-TIPSO-4-CF_2H-β-lactam **9**. (3*R*,4*S*)-1-Boc-3-TIPSO-4-difluorovinyl-β-lactam **10** was prepared in three steps from 4-formyl-β-lactam **8** using the Wittig-type reaction of the formyl moiety with difluoromethylphosphorus ylide generated *in situ* from $(Me_2N)_3P$, CF_2Br_2 and Zn (Pepe *et al.*, 2009a).

The 4-CF_3-, 4-CF_2H- and 4-difluorovinyl-β-lactams **7**, **9** and **10** thus obtained, were subjected to the Ojima–Holton coupling (Ojima, 1995) with C10- and/or C2-modified baccatins **11** (Ojima *et al.*, 1999, 2008) in the presence of LiHMDS at –40°C in THF, followed by removal of silicon protecting groups to give the corresponding novel C3′-Rf-taxoids, **15**, **16** and **17**, in good overall yields (Scheme 7.2).

Novel C3′-Rf-taxoids **15** (Rf = CF_3) and **16** (Rf = CF_2H) were evaluated for their cytotoxic potency *in vitro* against human breast cancer cell lines (MCF7-S and LCC6-WT), their corresponding drug-resistant cell lines (MCF7-R and LCC6-MDR: MCF7-R is now re-designated as NCI/ADR), as well as human non-small-cell lung cancer cell line (H460) and colon cancer cell line (H-29). The cytotoxicity of C3′-difluorovinyl-taxoids **17** were examined against MCF7-S and MCF7-R (NCI/ADR). Several of them were also assayed against HT-29 and a human pancreatic cancer cell line (PANC-1). Results are summarized in Table 7.1 (C3′-CF_3- and C3′-CF_2H-taxoids) and Table 7.2 (C3′-difluorovinyl-taxoids).

As Table 7.1 shows, all C3′-Rf-taxoids **15** and **16** exhibit substantially higher potencies than those of paclitaxel and docetaxel, with single-digit

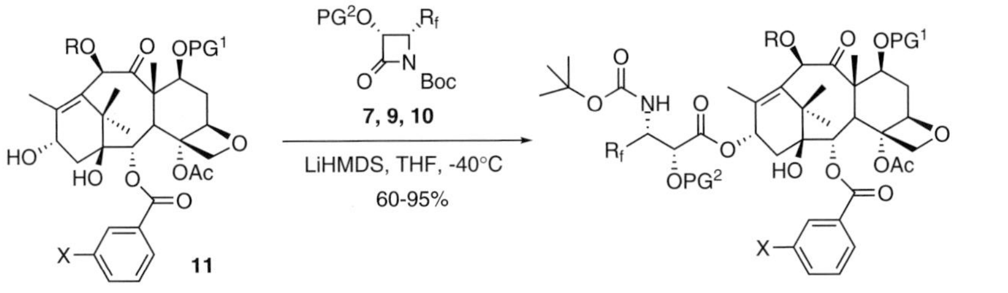

R = Ac, EtCO, c-PrCO, Me_2NCO, MeOCO
X = H, MeO, F, Cl, N_3

HF/pyridine
pyridine/CH_3CN
57-91%

12: $R_f = CF_3$, **13:** CF_2H, **14:** $CH{=}CF_2$
PG^1 = TES, PG^2 = TIPS

15: $R_f = CF_3$, **16:** CF_2H, **17:** $CH{=}CF_2$
PG^1, PG^2 = H

Scheme 7.2. Synthesis of C3′-CF_3-, C3′-CF_2H- and C3′-difluorovinyl-taxoids.

Table 7.1. *In vitro* cytotoxicity (IC_{50} nM)[a] of C3′-CF_3 and C3′-CF_2H-taxoids (15 and 16).

Taxoid	Rf	R	X	MCF7-S[b] (breast)	MCF7-R[c] (breast)	R/S[d]	LCC6-WT[b] (breast)	LCC6-MDR[e] (breast)	R/S[d]	H460[f] (lung)	HT-29[g] (colon)
Paclitaxel	Ph	Ac	H	1.7	300	176	3.1	346	112	4.9	3.6
Docetaxel	Ph	H	H	1.0	215	215	…	…	…	…	1.0
SB-T-12821-1	CF_3	Ac	MeO	0.32	8.8	28	0.33	3.99	12	0.38	0.69
SB-T-12821-2	CF_3	Ac	F	0.45	5.58	13	0.38	5.93	16	0.49	1.11
SB-T-12821-3	CF_3	Ac	Cl	0.40	5.04	13	0.22	4.96	23	0.5	0.85
SB-T-12821-4	CF_3	Ac	N_3	0.47	3.85	8.2	1.18	4.00	3.4	0.20	0.50
SB-T-12822-1	CF_3	Et-CO	MeO	0.19	2.16	11	0.45	4.24	9	0.41	0.54
SB-T-12822-2	CF_3	Et-CO	F	0.68	3.78	5.6	0.82	4.27	5.2	0.59	1.15
SB-T-12822-3	CF_3	Et-CO	Cl	0.34	3.28	9.6	0.39	2.54	6.5	0.63	1.11
SB-T-12822-4	CF_3	Et-CO	N_3	0.38	1.61	4.2	1.09	2.56	2.3	0.20	0.40
SB-T-12823-1	CF_3	Me_2NCO	MeO	0.57	1.84	3.2	0.28	4.48	16	0.35	0.68
SB-T-12823-2	CF_3	Me_2NCO	F	0.32	2.64	8.3	0.32	5.57	17	0.5	0.76
SB-T-12823-3	CF_3	Me_2NCO	Cl	0.12	1.02	8.5	0.27	2.55	9.4	0.42	0.45
SB-T-12823-4	CF_3	Me_2NCO	N_3	0.47	2.61	5.6	1.27	3.52	2.8	0.30	0.50
SB-T-12824-1	CF_3	MeOCO	MeO	0.17	2.88	17	0.27	3.99	15	0.38	0.53
SB-T-12824-2	CF_3	MeOCO	F	0.31	4.88	16	0.39	5.81	15	0.61	0.85
SB-T-12824-3	CF_3	MeOCO	Cl	0.65	4.72	7.3	0.29	5.08	18	0.43	0.68
SB-T-12824-4	CF_3	MeOCO	N_3	0.47	2.92	6.2	1.09	4.00	3.7	0.20	0.40
SB-T-12841-1	CF_2H	Ac	MeO	0.34	4.16	12	0.26	5.57	21	0.38	0.52
SB-T-12841-2	CF_2H	Ac	F	0.44	5.33	13	0.52	10.0	19	0.20	0.35

(*Continued*)

Table 7.1. *(Continued)*

Taxoid	Rf	R	X	MCF7-S[b] (breast)	MCF7-R[c] (breast)	R/S[d]	LCC6-WT[b] (breast)	LCC6-MDR[e] (breast)	R/S[d]	H460[f] (lung)	HT-29[g] (colon)
SB-T-12841-3	CF_2H	Ac	Cl	0.40	6.48	16	0.31	5.80	19	0.49	1.94
SB-T-12841-4	CF_2H	Ac	N_3	0.32	1.68	5.3	0.22	1.57	7.1	0.48	0.57
SB-T-12842-1	CF_2H	Et-CO	MeO	1.14	4.05	3.5	0.69	4.92	7.1	0.40	0.59
SB-T-12842-2	CF_2H	Et-CO	F	0.53	7.24	14	0.88	4.63	3.5	0.41	0.86
SB-T-12842-3	CF_2H	Et-CO	Cl	0.44	5.20	12	0.52	4.71	9.1	0.30	0.43
SB-T-12842-4	CF_2H	Et-CO	N_3	0.32	0.96	3.0	0.39	1.15	2.9	0.27	0.37
SB-T-12843-1	CF_2H	Me_2N-CO	MeO	0.45	4.51	10	0.69	7.06	10	0.40	0.43
SB-T-12843-2	CF_2H	Me_2N-CO	F	0.52	8.13	16	0.69	10.6	15	0.20	0.35
SB-T-12843-3	CF_2H	Me_2N-CO	Cl	0.31	2.96	9.5	0.21	3.87	18	0.36	0.58
SB-T-12843-4	CF_2H	Me_2N-CO	N_3	0.37	1.44	3.9	0.29	1.69	5.8	0.52	0.40
SB-T-12844-1	CF_2H	MeO-CO	MeO	0.81	6.59	8.1	1.03	10.2	9.9	0.30	0.44
SB-T-12844-2	CF_2H	MeO-CO	F	0.59	11.38	19	0.86	12.6	15	0.30	0.43
SB-T-12844-3	CF_2H	MeO-CO	Cl	0.26	2.08	8.0	0.13	1.82	14	0.25	0.29
SB-T-12844-4	CF_2H	MeO-CO	N_3	1.69	2.56	1.5	0.26	2.06	7.9	0.23	0.36

[a] The concentration of compound which inhibits 50% (IC_{50}, nM) of the growth of a human tumour cell line after 72 h drug exposure.

[b] Human breast carcinoma.

[c] Multidrug-resistant human breast/ovarian cancer cell line (currently renamed to NCI/ADR).

[d] Drug-resistance factor.

[e] Multidrug-resistant human breast carcinoma.

[f] Human non-small cell lung carcinoma.

[g] Human caucasian colon adenocarcinoma.

Table 7.2. *In vitro* cytotoxicity (IC_{50} nM)[a] of 3′-difluorovinyl-taxoids (**17**).

Entry	Taxoid	R	X	MCF7-S[b] (breast)	MCF7-R[c] (breast)	R/S	HT-29[d] (colon)	PANC-1[e] (pancreatic)
1	Paclitaxel			1.2	300	250	3.6	25.7
2	SB-T-12851	Ac	H	0.099	0.95	9.6	0.41	1.19
3	SB-T-12852	c-Pr-CO	H	0.12	6.0	50	0.85	5.85
4	SB-T-12853	Et-CO	H	0.12	1.2	10	0.34	0.65
5	SB-T-12854	Me_2N-CO	H	0.13	4.3	33	0.46	1.58
6	SB-T-12855	MeO-CO	H	0.14	1.29	9.2	…	…
7	SB-T-12851-1	Ac	MeO	0.25	1.5	6.0	…	…
8	SB-T-12852-1	c-Pr-CO	MeO	0.092	0.48	5.2	…	…
9	SB-T-12853-1	Et-CO	MeO	0.34	0.57	1.7	…	…
10	SB-T-12854-1	Me_2N-CO	MeO	0.11	0.5	4.5	…	…
11	SB-T-12855-1	MeO-CO	MeO	0.078	0.50	6.4	…	…
12	SB-T-12851-2	Ac	F	0.14	1.53	11	…	…
13	SB-T-12852-2	c-Pr-CO	F	0.071	1.72	24	…	…
14	SB-T-12853-2	Et-CO	F	0.22	2.54	12	…	…
15	SB-T-12854-2	Me_2N-CO	F	0.17	2.25	13	…	…

(*Continued*)

Table 7.2. (*Continued*)

Entry	Taxoid	R	X	MCF7-S[b] (breast)	MCF7-R[c] (breast)	R/S	HT-29[d] (colon)	PANC-1[e] (pancreatic)
16	SB-T-12855-2	MeO-CO	F	0.12	1.85	15	...	...
17	SB-T-12851-3	Ac	N_3	0.092	0.34	3.7	...	...
18	SB-T-12852-3	c-Pr-CO	N_3	0.092	0.45	4.9	...	...
19	SB-T-12853-3	Et-CO	N_3	0.14	0.38	2.7	...	...
20	SB-T-12854-3	Me_2N-CO	N_3	0.14	0.46	3.3	...	...
21	SB-T-12855-3	MeO-CO	N_3	0.078	0.40	5.3	...	...
22	SB-T-12851-4	Ac	Cl	0.13	0.70	5.4	...	...
23	SB-T-12852-4	c-Pr-CO	Cl	0.12	0.50	4.2	...	...
24	SB-T-12853-4	Et-CO	Cl	0.13	0.45	3.5	...	...
25	SB-T-12854-4	Me_2N-CO	Cl	0.93	2.60	2.8	...	...
26	SB-T-12855-4	MeO-CO	Cl	0.099	1.15	12	...	...

[a–d] See footnotes to Table 1.

[e] Human pancreatic carcinoma.

nanomolar IC_{50} values against drug-sensitive MCF7-S, LCC6-WT, H460 and HT-29 cancer cell lines (except for a few cases). The potency of **15** and **16** against MDR cell lines, MCF7-R (NCI/MDR) and LCC6-MDR, is remarkable, with almost all C3′-Rf-taxoids exhibiting single-digit nanomolar IC_{50} values, which are two orders of magnitude more potent than paclitaxel on average. The C3′-Rf-taxoids **15** and **16** exhibit, in general, comparable cytotoxicity against all cancer cell lines examined. It appears, however, that the potency of **15** against MCF7-S and LCC6-WT is higher and less fluctuated as compared to that of **16** (except for two cases). In contrast, **15** exhibits more uniform potency against MCF7-R (NCI/ADR) and LCC6-MDR cell lines than **15**. For **16**, potency against MDR cell lines is dependent on the nature of *meta* substituents of the C2-benzoate moiety, i.e. the order of potency is F < MeO < Cl < N_3. By contrast, such a trend is unclear for **16** against these MDR cell lines. Among these C3′-Rf-taxoids examined, SB-T-12842-4 (R = *n*-propanoyl; X = N_3) appears to be the most potent compound, showing a resistance factor (R/S ratio) of only 2.9–3.0 against MCF7-R (NCI/ADR) and LCC6-MDR.

As Table 7.2 indicates, all difluorovinyl-taxoids **17** are remarkably more potent than paclitaxel. A clear effect of C2-benzoate modification at the *meta* position is observed on the potency against drug-sensitive and drug-resistant MCF7 cell lines. In a manner similar to that observed for **16**, the potency of **17** against MCF7-R (NCI/ADR) is dependent on the nature of *meta* substituents of the C2-benzoate moiety, and the order of potency is F < Cl ≤ MeO < N_3. Some of these taxoids with 2,10-modifications possess impressive potency, i.e. their IC_{50} values are below 100 pM (range 78–92 pM) and in the subnanomolar range (0.34–0.50 nM) against MCF7-R (NCI/ADR), which is three orders of magnitude more potent than paclitaxel (entries 8, 11, 17, 18 and 21). The resistance factor for these taxoids is 3.7–6.4, while that of paclitaxel is 250. SB-T-12852-2 (R = cyclopropanecarbonyl; X = F) exhibits extremely high potency (IC_{50} 71 pM) against MCF7-S, but the resistance factor is 24. Taxoids **17** with unmodified C2-benzoate moiety (entries 2–6) also possess substantially enhanced potency against MCF7-S and MCF7-R (NCI/ADR) as compared to paclitaxel. These taxoids exhibit excellent potency against HT-29 (human colon) and PANC-1 (human pancreatic) cancer cell lines as well. It appears that SB-T-12853 is particularly promising against these gastrointestinal

(GI) cancer cell lines. Although taxoids **17** with 2,10-modifications have not been evaluated against HT-29 and PANC-1 cell lines yet, it is more than reasonable to assume that these taxoids would exhibit exceptional potency against these GI cancer cell lines.

7.4 Synthesis and Biological Evaluation of Fluorine-Containing C-Seco-Taxoids

Due to the complexity and multiplicity of drug resistance, a multidirectional approach appears to be needed to discover and develop efficacious anticancer drugs. As mentioned above, besides MDR caused by the overexpression of ABC transporters such as P-glycoprotein, another type of paclitaxel-resistance arising from the overexpression of class III β-tubulin has recently been identified as a clinically significant drug resistance mechanism (Dumontet *et al.*, 2009). Recently, a C-seco-taxoid, IDN5390 (see Fig. 7.1), has been shown to be more potent than paclitaxel against drug-resistant ovarian cancer cell lines overexpressing class III isotype (Ferlini *et al.*, 2005). Accordingly, we set out to perform a structure–activity relationship (SAR) study, including fluorine incorporation, of this C-seco-taxoid.

Microtubules play a crucial role in the mitosis stage of the cell division cycle, as well as in the locomotion of most cell types, by taking part in coordinating the direction of cell movement (Liao *et al.*, 1995; Ueda *et al.*, 1997). Accordingly, paclitaxel and taxoids are likely to possess a certain level of activity to inhibit cancer metastasis. It has been shown that tumours develop an extended network of blood vessels in order to grow and spread to different tissues and organs in the body (Folkman, 1971). The uncontrolled tumour cell proliferation generates areas of necrotic tissue, wherein the lack of oxygen (hypoxia) activates the vascular endothelial growth factor (VEGF) (Shweiki *et al.*, 1992), which initiates the process of capillary formation (angiogenesis) (Papetti and Herman, 2002). Many crucial endothelial cell activities relevant to angiogenesis require a functional cytoskeleton (Goltlieb *et al.*, 1981; Coan *et al.*, 1993). Thus, microtubule-targeting agents can inhibit the endothelial cell proliferation and migration by blocking microtubule dynamics required for the G_2–M transition (Selden *et al.*, 1981; Belotti *et al.*, 1996; Klauber *et al.*, 1997). In fact, paclitaxel exhibits some antiangiogenic activity at sub-cytotoxic doses (Taraboletti *et al.*, 2002). Recently, IDN5390, a C-seco-taxoid just mentioned above, has also shown potent antimotility and low cytotoxicity on endothelial cells, together with the ability to down-regulate the two main angiogenic factors VEGF and bFGF (Taraboletti *et al.*, 2002). Thus, we included the evaluation of antiangiogenic activity in the SAR study of C-seco-taxoids. As a part of this SAR study, we investigated two fluorine-containing analogues, SB-T-10104 and SB-T-10204.

Two C-seco-fluorotaxoids, SB-CST-10204 and SB-CST-10104, were synthesized through the Ojima–Holton coupling of protected 2-(3-fluorobenzoyl)-C-seco-baccatin **20** with β-lactams **21a** and **21b**, respectively, in the presence of LiHMDS in THF at –40°C, followed by deprotection with HF-pyridine (Scheme 7.3) (Pepe *et al.*, 2009b). C2-fluorinated C-seco-baccatin **20** was prepared from 2-(3-fluorobenzoyl)baccatin **18** (Ojima *et al.*, 1997b) following Appendino's protocol (Appendino *et al.*, 1997, 2003). C2-fluorinated baccatin **18** was oxidized at C10 with $Cu(OAc)_2$ and air. The resulting 10-oxo-baccatin **19** was treated with L-selectride at –78°C in THF, followed by protection of hydroxyl groups at C7 and C9 to

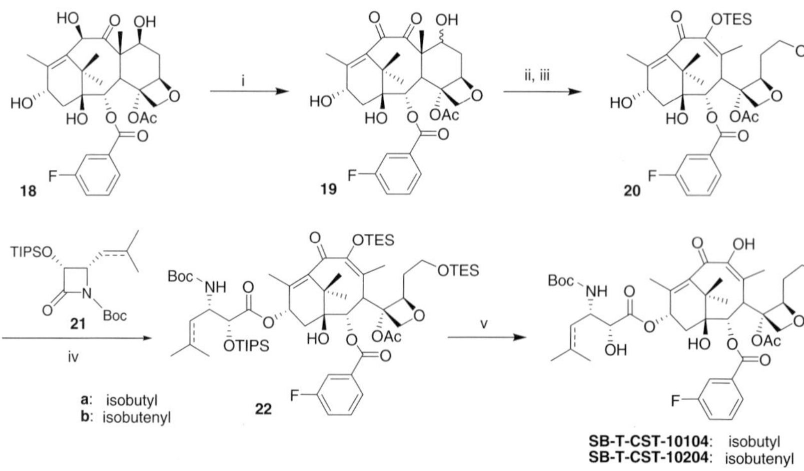

i: $Cu(OAc)_2$, MeOH, 77–86 %; ii: L-selectride, THF, –78°C 50–70%, iii: methyl imidazole, TESCl, DMF, 0°C, 50–80%; iv: LiHMDS, THF, –40°C, 70–80%; v: HF/pyridine, CH_3CN/pyridine, 0°C to RT, 52–92%.

Scheme 7.3. Synthesis of C2-fluoro-C-seco-taxoids.

afford **20**. Further transformed as described above (Scheme 7.2) then produced C-seco-fluoro-taxoids, SB-CST-10104 and SB-CST-10204, in good overall yields.

Novel C-seco-fluoro-taxoids SB-CST-10104, and SB-CST-10204 were assayed against several human ovarian adenocarcinoma cell lines, including A2780wt (drug sensitive, wild type), A2780CIS, A2780ADR, A2780TOP (resistant to cisplatin, adriamicin and topotecan/doxorubicin, respectively), A2780TC1 and A2780TC3 (resistant to both paclitaxel and cyclosporine A) (Table 7.3) (Pepe *et al.*, 2009b). Drug resistance in the A2780TC1 and A2780TC3 cell lines is caused by the overexpression of class III β-tubulin subunit and other possible mutations (Kavallaris *et al.*, 1999; Ferlini *et al.*, 2005), whereas that in the A2780ADR cell line is based on MDR. Accordingly, the activity of SB-CST-10104 and SB-CST-10204 against A2780TC1 and A2780TC3 cell lines is the critical point of interest.

As Table 7.3 shows (Pepe *et al.*, 2009b), these two C-seco-fluoro-taxoids, especially, SB-CST-10104, exhibit remarkable potency, as compared to paclitaxel, against A2780TC3 cell line, i.e. the most drug-resistant cell line for paclitaxel in this series (24–38 times more potent than paclitaxel and 3–5 times more potent than IDN5390). The resistance factor for this cell line, i.e. IC_{50} (A2780TC3)/IC_{50} (A2780wt), is 10 470 for paclitaxel, but

Table 7.3. *In vitro* cytotoxicity (IC_{50} nM)[a] of C-seco-fluorotaxoids.

C-seco-Taxoid	A2780wt[b]	A2780CIS[c]	A2780TOP[d]	A2780ADR[e]	A2780TC1[f]	A2780TC3[g]
Paclitaxel	1.7±1.2	2.2±0.2	7.2±1.5	1239±265	10027±3195	17800±5,499
IDN5390	17.4±1.5	16.8±3.1	27.5±5.1	2617±1028	2060±344	2237±471
SB-CST-10104	11.1±8.4	11.8±1.0	12.8±3.5	3726±198	1497±31	460±128
SB-CST-10204	6.1±0.6	4.9±0.2	6.9±0.8	2218±588	4454±1391	745±60

[a] The concentration of compound which inhibits 50% (IC_{50}, nM) of the growth of a human tumour cell line after 72 h drug exposure.
[b] Human ovarian carcinoma wild type.
[c] Cisplatin-resistant A2780.
[d] Topotecan-resistant A2780.
[e] Adriamycin-resistant A2780.
[f,g] Clones derived from chronic exposition of A2780 to paclitaxel and cyclosporine.

only 41 for SB-CST-10104. For comparison, IDN5390 exhibits 8.0 times higher potency than paclitaxel with a resistance factor of 129 against the same cell line. This result is intriguing considering the fact that the only structural difference between IDN5390 and SB-CST-10104 is one fluorine incorporation at the *meta* position of the C2-benzoate moiety of the C-seco-taxoid molecule. These C-seco-fluorotaxoids also possess 2.3–6.7 times higher potency than paclitaxel against the A2780TC1 cell line. The potency of these two C-seco-fluoro-taxoids, especially SB-CST-10204, against A2780CIS, A2780TOP and A2780ADR is consistently lower than paclitaxel, but higher than that of the parent IDN5390. The C3′-substitutents of C-seco-fluoro-taxoids, i.e. a 3′-isobutyl or 3′-isobutenyl group, show subtle effects on potency, which appears to be directly related to their interactions with the class III β-tubulin. SB-CST-10204 (C3′ = isobutenyl) exhibits higher potency than SB-CST-10104 (C3′ = isobutyl) against A2780wt, A2780CIS, A2780TOP and A2780ADR. However, the reversal of this SAR is observed against A2780TC1 and A2780TC3 in which the class III β-tubulin is overexpressed.

The antiangiogenic activity of SB-CST-10204 and SB-CST-10104 was evaluated based on their ability to inhibit the microcapillary tubule formation in human umbilical vein endothelial cells (HUVEC) supported in Matrigel. Microscopy images were quantified by the Bioquant Image Analysis System. The results are illustrated in Fig. 7.3 and typical phase-controlled microscopy images of the microcapillary tubules formed at drug concentrations of 10 nM and 100 nM are shown in Fig. 7.4. The length of the tubules formed (Graph A) or the number of junctions (Graph B), normalized to the control experiment, are a measure of the antiangiogenic activity (Hollingshead *et al.*, 2004).

As Fig. 7.3 shows, at the lowest concentration tested (0.1 nM), SB-CST-10204 and SB-CST-10104 show 20–30% inhibition, i.e. microcapillary tubules formed in HUVEC cells treated with these C-seco-fluoro-taxoids are 20% shorter and the number of junctions is 30% smaller than in the control experiments. IDN5390 does not show any inhibition at this concentration and even an increase in the number of junctions is observed. At 1 nM concentration, SB-CST-10204 shows substantially better activity than IDN5390. At 10 nM concentration, SB-CST-10104 starts exhibiting superior activity to those of IDN5390 and SB-CST-10204, controlling the

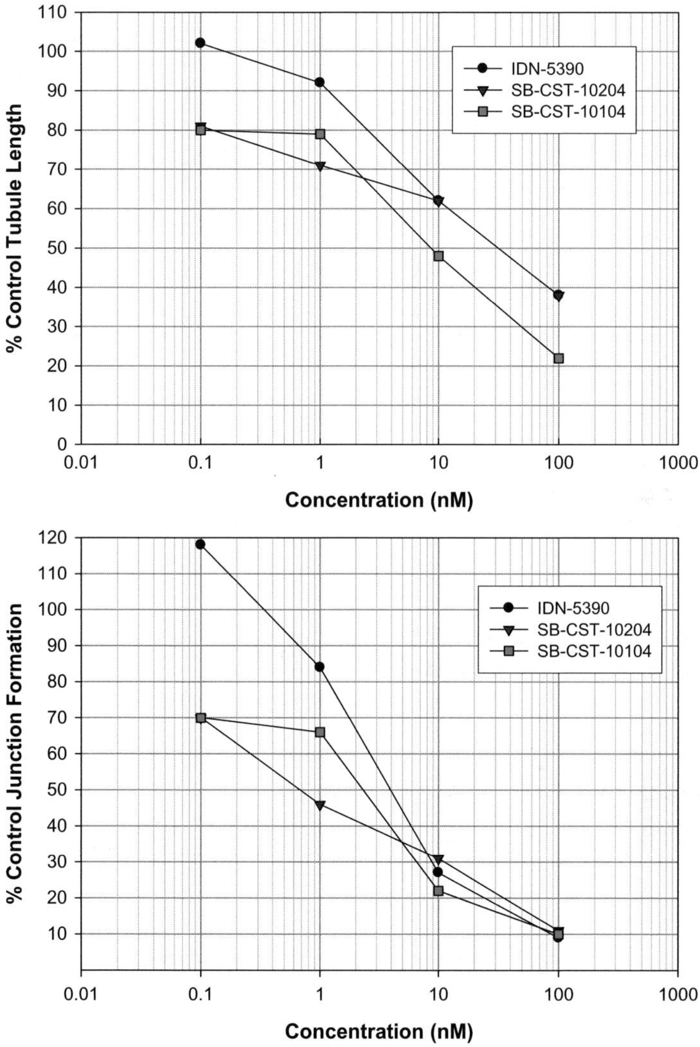

Figure 7.3. Inhibition of microcapillary tubule formation by C-seco-taxoids. Graph A: % control tubule length; Graph B: % control junction formation.

tubule growth to less than 50% and the junction formation to 22%. At 100 nM concentration, SB-CST-10104 inhibits the growth of tubules by 78%, while the growth inhibition by SB-SCT-10204 and IDN5390 is 62%. At this concentration, all C-seco-taxoids inhibit the junction formation by 90%.

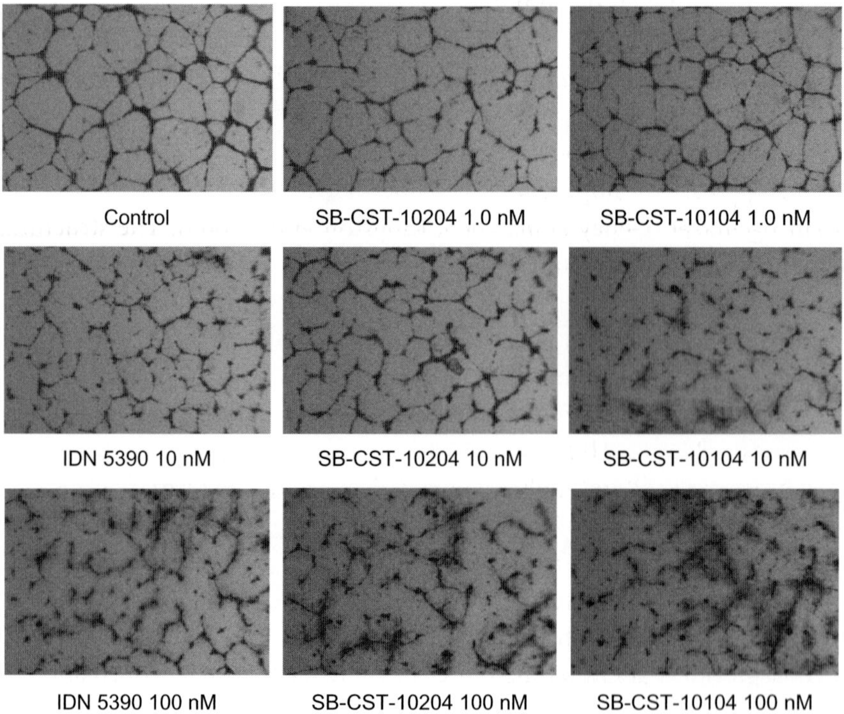

Figure 7.4. Microscopy images of microcapillary tubules in the presence of C-seco-taxoids.

These results confirm that the *meta* fluoro-substituent of the C2 benzoyl group as well as the C3′ moiety have critical effects on the inhibitory activity of microcapillary formation, with an isobutyl group being a better C3′ group than an isobutenyl group. This new class of taxoids may hold unique prospect with their dual activities against two distinct functions, i.e. cytotoxicity and antiangiogenic activity.

7.5 Use of Solid-State ^{19}F NMR and Computational Analysis for the Determination of Bioactive Conformation of Paclitaxel and Fluorinated Taxoids

Paclitaxel binds to the β-tubulin component of the α,β-tubulin dimer, promotes the polymerization of tubulins, stabilizes microtubules and blocks microtubular dynamics, which eventually leads to apoptosis (Schiff *et al.*, 1979; Jordan and Wilson, 2004). Although its mechanism of action as a microtubule-stabilizing agent was discovered almost 30 years ago, the structure of paclitaxel bound to β-tubulin has not been fully elucidated. Investigation into the bioactive conformation of paclitaxel could lead to the design and development of novel drugs with much simpler structures than paclitaxel (Geney *et al.*, 2005; Kingston *et al.*, 2005). The structural biology study of paclitaxel did not start until the first cryo-electron microscopy (cryo-EM) (or 'electron crystallography') structure of a microtubule model, i.e. Zn^{2+}-stabilized α,β-tubulin dimer with a paclitaxel molecule, was reported in 1998 at 3.7 Å resolution (PDB-code: 1TUB structure) (Nogales *et al.*, 1998). The structure was refined to 3.5 Å resolution (PDB-code: 1JFF structure) in 2001 (Löwe *et al.*, 2001), but the resolution was still not high enough to show the exact binding conformation of paclitaxel.

A combination of ^{19}F NMR and advanced 2D NMR methods provide very powerful tools to study dynamic conformational equilibria. The wide dispersion of fluorine chemical shifts simplifies the observation of molecular conformers at low temperature. To study the binding conformation of paclitaxel and other taxoids, fluorine-containing taxanes were successfully used as probes for NMR analysis in conjunction with molecular modelling (Ojima *et al.*, 1997a). The solid-state magic angle-spinning (SSMAS) ^{19}F NMR analysis with the radiofrequency-driven dipolar recoupling (RFDR) method was also used to measure the F–F distance for the microtubule-bound conformation of a F_2-docetaxel derivative (Fig. 7.5) (Ojima *et al.*, 1998). The rotational echo double resonance (REDOR) method was used to investigate the structure of the microtubule-bound paclitaxel by determining the ^{19}F–^{13}C distances and the ^{19}F–^{2}H distances of fluorinated paclitaxel derivatives (Fig. 7.5) (Li *et al.*, 2000; Paik *et al.*, 2007). Since real microtubules, i.e. not the Zn^{2+}-stabilized tubulin dimer model, were used in this experiment, the results were critically important to confirm the relevance of the cryo-EM structures (1TUB, 1JFF).

On the basis of the REDOR distances, photoaffinity labelling results and molecular modelling, a bioactive conformation of paclitaxel, 'REDOR-Taxol', was proposed as the most plausible microtubule-bound paclitaxel structure in 2005 (Geney *et al.*, 2005). In the REDOR-Taxol structure, the

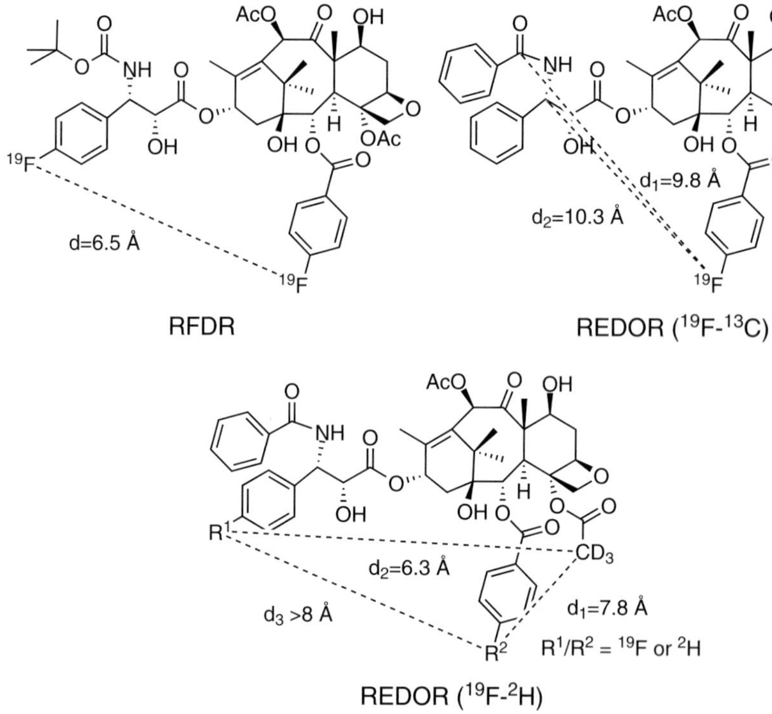

Figure 7.5. Solid-state NMR studies on microtubule-bound taxane fluorine-probes.

C2′-OH group interacts with His 229 as the H-bond donor, which is consistent with the well-established SAR studies (Kant *et al.*, 1993; Williams *et al.*, 1993). The 'REDOR-Taxol' was further refined using the 1JFF atomic coordinate (Fig. 7.6), and it has been shown that the 'REDOR-Taxol' is not only fully consistent with the new REDOR experiments, but also accommodates highly active macrocyclic paclitaxel analogues (Sun *et al.*, 2008, 2009).

The molecular modelling of C3′-Rf-taxoids was performed to predict their bioactive conformations based on the REDOR-Taxol structure. SB-T-1284, SB-T-1282 and SB-T-12853 (Fig. 7.6) were oriented according to the REDOR-Taxol conformation and docked into β-tubulin binding pocket by superimposing the baccatin cores. Energy minimization of the tubulin-bound structures (InsightII, CVFF) provided three C3′-Rf-taxoid–β-tubulin

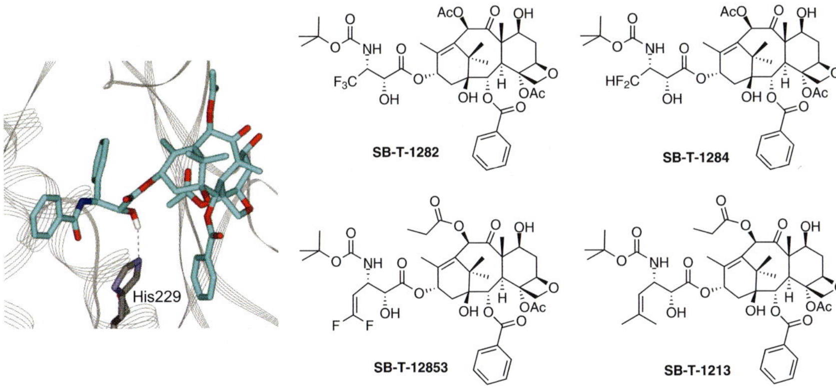

Figure 7.6. REDOR-Taxol structure and C3′-Rf-taxoids analysed by molecular modelling.

complexes. The computer-generated binding structures of these fluorotaxoids are shown in Fig. 7.7(a–c).

As Fig. 7.7 shows, the baccatin moiety occupies virtually the same space in all cases, as anticipated. Each taxoid resides comfortably in the pocket without high-energy contacts with the protein, which indicates that there is no substantial difference between the REDOR-Taxol structure and those of C3′-Rf-taxoids. There is a strong hydrogen bond between the C2′-OH of C3′-Rf-taxoids and His229, similar to that in the REDOR-Taxol structure (Geney *et al.*, 2005). The molecular modelling analysis indicates that the CF_2H and CF_3 moieties fill essentially the same space (Fig. 7.7(a), (b)). The difluorovinyl moiety of SB-T-12853 occupies more extended hydrophobic space as compared to CF_3 and CF_2H moieties (Fig. 7.7(c)). The overlay of SB-T-12853 with a representative new-generation taxoid, SB-T-1213, shows very good fit, but with appreciable difference (Fig. 7.7(d)). The result may indicate that the difluorovinyl group mimics isobutenyl group, but there is a difference in size and electronic nature between these two groups.

The difluorovinyl group is in between vinyl and isobutenyl groups in size, but two fluorine atoms may mimic two hydroxyl groups rather than two methyl groups electronically. Accordingly, the difluorovinyl group is a very unique structural component in medicinal chemistry, which can be widely explored in drug design, including its anticipated metabolic stability against P-450 family enzymes.

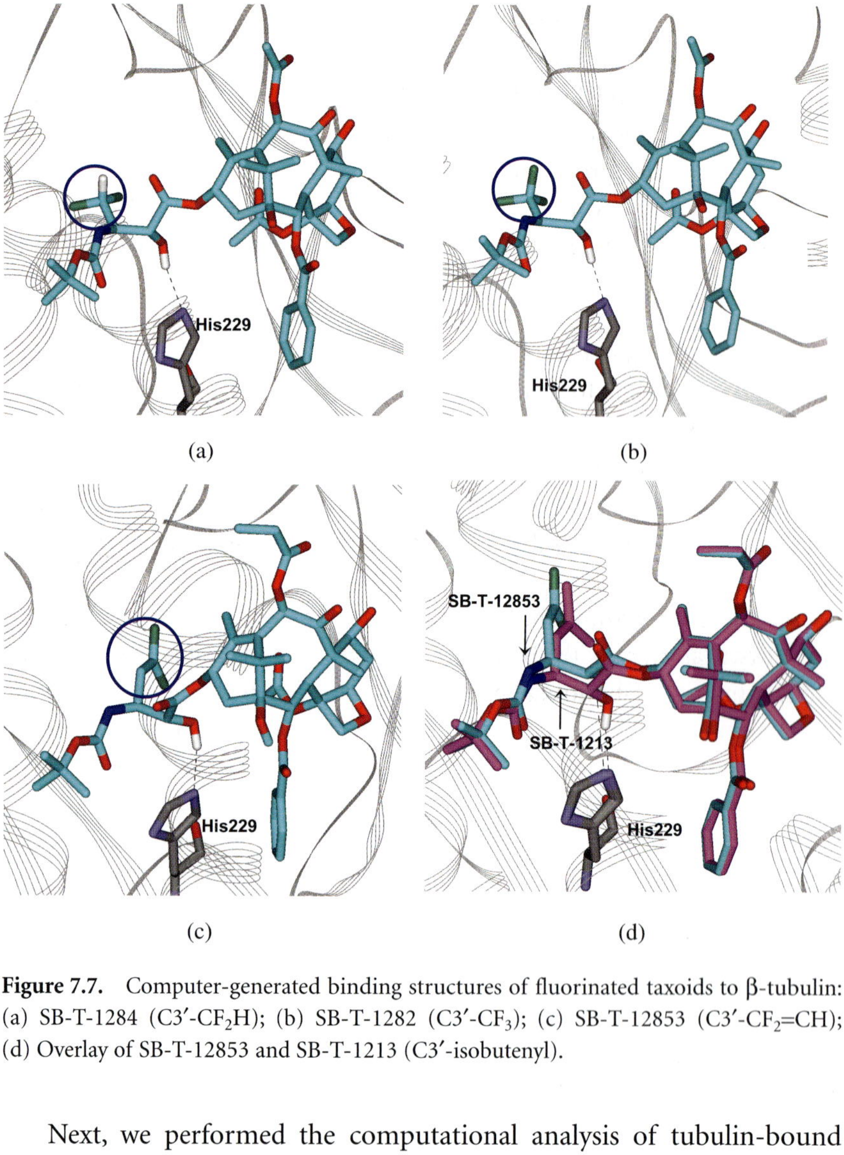

Figure 7.7. Computer-generated binding structures of fluorinated taxoids to β-tubulin: (a) SB-T-1284 (C3′-CF_2H); (b) SB-T-1282 (C3′-CF_3); (c) SB-T-12853 (C3′-CF_2=CH); (d) Overlay of SB-T-12853 and SB-T-1213 (C3′-isobutenyl).

Next, we performed the computational analysis of tubulin-bound C-seco-fluoro-taxoid structures with class I and III β-tubulins (Pepe *et al.*, 2009b). Following up on recent molecular modelling studies of paclitaxel and IDN5390 in human class I and III β-tubulins (Ferlini *et al.*, 2005; Magnani *et al.*, 2006; Pepe *et al.*, 2009b), molecular dynamics (MD) simulation was employed to predict the binding conformation of C-seco-taxoids in class I and III β-tubulins. A cryo-EM crystal structure of bovine brain tubulin (1JFF) (Löwe *et al.*, 2001) was used as the template to create 3D models of class I and III β-tubulin models, TubB1 and TubB3, respectively. The protein sequences were obtained from SwissProt and a standard comparative modelling procedure was used by replacing the side chains of the template, followed by energy minimization (Pepe *et al.*, 2009b). The C-seco-fluoro-taxoid (SB-CST-10204) molecule was manually docked into TubB1 and TubB3 (Pepe *et al.*, 2009b), based on the conformation of REDOR-Taxol in 1JFF (Sun *et al.*, 2008). A 600 ps simulation of the complexes was performed with a weak restraint on the protein backbones in a 8 Å truncated octahedron of TIP3P explicit water (AMBER9, ff03) (Case *et al.*, 2005). Snapshots of the binding conformation of SB-CST-10204 in TubB1 and TubB3 at the end of the simulations are shown in Fig. 7.8. The overall protein structures of TubB1 and TubB3 are similar, and the two mutations (Cys241Ser and Ser277Ala) in the binding site do not have direct interaction with SB-CST-10204.

The M-loop (from Thr276 to Arg284) of TubB3 (cyan) has higher flexibility than that of TubB1 (magenta), which is very likely to be caused by the loss of an H-bonding between Ser277 and Ser280 in TubB1 with the

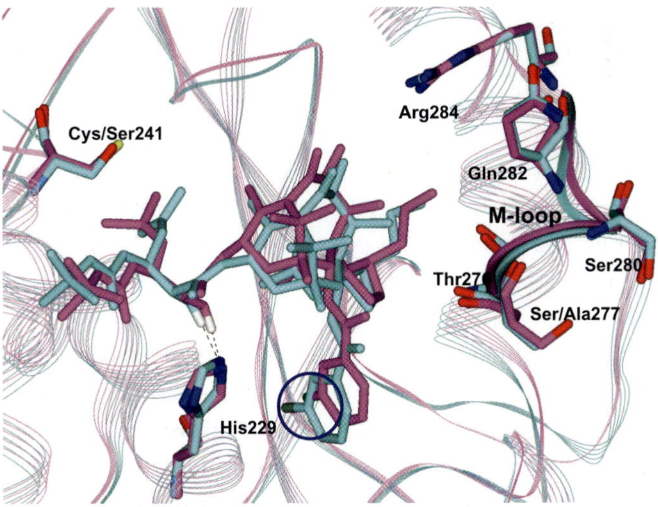

Figure 7.8. Snapshots of SB-CST-10204 in TubB1 (magenta) and TubB3 (cyan).

mutation of Ser277 to Ala277 in TubB3 (Löwe *et al.*, 2001). The flexible C7-OH of SB-CST-10204 can form an H-bond with Gln282 or Arg284, which would not be affected by the increased flexibility of the M-loop in the class III β-tubulin, allowing C-seco-taxoids to keep high affinity to class III β-tubulin. In contrast, the H-bond between the C7-OH of paclitaxel with Arg284 would be substantially weakened by increased flexibility of the M-loop in TubB3. Thus, the binding of paclitaxel to the class III β-tubulin should be less favourable than that to class I β-tubulin. The H-bond between C2′-OH and His229 is very stable throughout the MD simulation. There is also a favorable interaction between His229 and the *meta*-substituent of the C2-benzoate moiety in both TubB1 and TubB3.

Consequently, the molecular modelling study on the tubulin-bound conformations of paclitaxel, and SB-CST-10204 in TubB1 and TubB3 revealed considerable difference in the conformations between the two β-tubulin subclasses. This study also demonstrates the importance of the hydrophobic interaction between the C2-benzoyl moiety and His229.

7.6 Use of Fluorine in Tumour-Targeting Anticancer Agents

The development of highly specific anticancer agents that can target cancer cells without affecting normal cells is of critical importance in chemotherapy. Such drugs would eliminate the often severe side effects associated with current cancer chemotherapy. As anticancer drugs disrupt the cell division cycle at different stages, rapidly proliferating tumour cells should be more affected by anticancer drugs than normal cells. Unfortunately, this is not true for fast regenerating non-malignant cells, such as those of the gastrointestinal tract and bone marrow, and thus these cells cannot escape the attack of cytotoxic drugs. This causes undesirable side effects on patients. Therefore, it is very important to develop new chemotherapeutic agents with improved tumour specificity.

To reduce side effects of chemotherapeutic agents, tumour-targeting anticancer drug conjugates have been developed by connecting a potent cytotoxic drug to a tumour-targeting molecule through a mechanism-based cleavable linker. The tumour-targeting molecule has high affinity to cancer-specific biomarkers, which are often receptors necessary for cell growth, overexpressed on the tumour cell surface, and promote the internalization of the drug conjugate into tumour cells. Once inside the tumour cell, the linker should be cleaved by intracellular substances and the cytotoxic agent released. It is important that the conjugate is stable while in the blood circulation, but readily cleavable inside the tumour.

Most commonly used tumour-targeting molecules are monoclonal antibodies (mAbs), their derivatives and fragments, hyaluronic acid, folic acid, biotin, somatostatin peptide mimic and aptamers (Chen *et al.*, 2005; Jaracz *et al.*, 2005; Ojima, 2008). mAbs and their fragments are the most widely used tumour-targeting functionalities. They specifically bind to antigens that are overexpressed on the surface of tumour tissues or cells, and are much less abundant on normal tissues. Therefore, in principle, mAb–cytotoxic drug conjugates can be specifically delivered to the tumour, internalized via receptor-mediated endocytosis and release the parent drug intracellularly (Liu *et al.*, 1996; Chari, 1998; Chen *et al.*, 2005). Mylotarg (gemtuzumab-ozogamicin) (Hamann *et al.*, 2002) is the first mAb–drug immunoconjugate approved by the FDA for the treatment of acute myelogenous leukemia (AML). Several other mAb–drug conjugates have advanced to human clinical trials (Liu *et al.*, 1996; Saleh *et al.*, 1998; Gillespie *et al.*, 2000; Chan *et al.*, 2003; Lam *et al.*, 2003; Widdison *et al.*, 2006). We have developed novel mAb–taxoid conjugates as tumour-targeting anticancer agents, which exhibited extremely promising results in human cancer xenografts in SCID mice. These conjugates clearly demonstrated tumour-specific delivery of a taxoid anticancer agent without any noticeable toxicity to the animals and cured all animals tested (Ojima *et al.*, 2002). For these mAb–taxoid conjugates, we used a disulfide linker, stable in blood circulation, but efficiently cleaved by glutathione or other thiols in the tumour. In fact, it has been shown that the glutathione level is 1000 times higher in tumour tissues than blood plasma (Kigawa *et al.*, 1998). However, in this first-generation mAb–taxoid conjugates, the original taxoid molecule was not released because of the compromising modification of the taxoid molecule to attach the disulfide linker. Thus, the cytotoxicity of the taxoid released in these conjugates was 8–10 times weaker than the parent taxoid.

Accordingly, we have devised second-generation self-immolative disulfide linkers (Ojima, 2008). Once the conjugate is internalized into the tumour cells, endogenous glutathione triggers a cascade reaction that

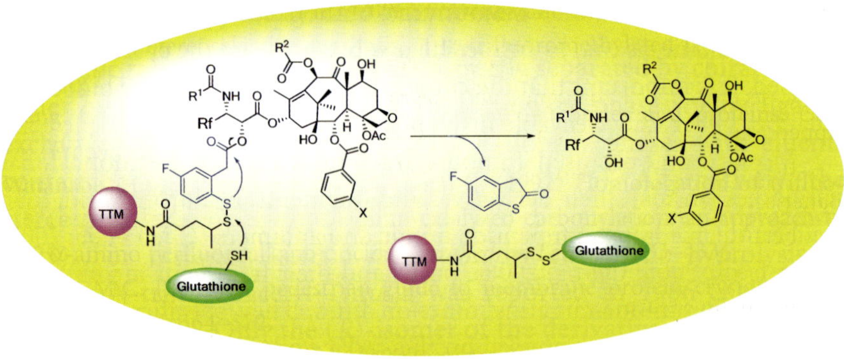

Scheme 7.4. Glutathione-triggered cascade drug release using self-immolative linker.

cleaves the linker through thiolactonization to release the parent unmodified drug molecule (Scheme 7.4). The strategic incorporation of fluorine allows us to monitor the cleavage of the linker and the release of the drug by ^{19}F NMR spectroscopy (Ojima, 2004). In addition, the presence of a fluorine substituent at the *para* position to the disulfide linkage would direct the cleavage of this linkage by a thiol to generate the desirable thiophenolate or sulfhydrylphenyl species for thiolactonization. This type of self-immolative linker is highly versatile and readily applicable to any tumour-targeting drug conjugates, including the conjugates with highly potent C3′-Rf-taxoids.

Moreover, a combination of a fluorine-containing linker and a C3′-Rf-taxoid may serve as fluorine-probes for monitoring the internalization and drug release of these conjugates in the tumour cells and tissues by ^{19}F NMR spectroscopy as an alternative method to confocal fluorescence microscopy using fluorescence labelled probes (Ojima, 2008). Further applications of the strategic incorporation of fluorine(s) into medically active substances to enhance potency and metabolic stability as well as to provide useful tools for biomedical research are actively underway in these laboratories.

Acknowledgements

This research was supported by grants from the National Cancer Institute (CA 103314 to I.O.). Generous support from Indena, SpA (to I.O.) is also gratefully acknowledged.

References

Appendino, G., Danieli, B., Jakupovic, J. *et al.* (1997) Synthesis and evaluation of C-seco paclitaxel analogues. *Tetrahedron Lett* 38, 4273–4276.

Appendino, G., Noncovich, A., Bettoni, P. *et al.* (2003) The reductive fragmentation of 7-hydroxy-9,10-dioxotaxoids. *Eur J Org Chem* 22, 4422–4431.

Banerjee, A. & Luduena, R.F. (1992) Kinetics of colchicine binding to purified beta-tubulin isotypes from bovine brain. *J Biol Chem* 267, 13335–13339.

Bégué, J.-P. & Bonnet-Delpon, D. (2006) Recent advances (1995–2005) in fluorinated pharmaceuticals based on natural products. *J Fluorine Chem* 127, 992–1012.

Belotti, D., Vergani, V., Drudis, T. *et al.* (1996) The microtubule affecting drug paclitaxel has on antiangiogenic activity. *Clin Cancer Res* 2, 1843–1849.

Case, D.A., Cheatham, T.E., Darden, T. *et al.* (2005) The Amber biomolecular simulation programs. *J Comput Chem* 26, 1668–1688.

Chan, S.Y., Gordon, A.N., Coleman, R.E. *et al.* (2003) A phase 2 study of the cytotoxic immunoconjugate CMB-401 (hCTM01-calicheamicin) in patients with platinum-sensitive recurrent epithelial ovarian carcinoma. *Cancer Immunol Immun* 52, 243–248.

Chari, R.V.J. (1998) Targeted delivery of chemotherapeutics: tumor-activated prodrug therapy. *Adv Drug Deliver Rev* 31, 89–104.

Chen, J., Jaracz, S., Zhao, X. *et al.* (2005) Antibody-cytotoxic agent conjugates for cancer therapy. *Expert Opin Drug Del* 2, 873–890.

Chevillard, S., Pouillart, P., Beldjord, C. *et al.* (1996) Sequential assessment of multidrug resistance phenotype and measurement of S-phase fraction as predictive markers of breast cancer response to neoadjuvant chemotherapy. *Cancer* 77, 292–300.

Coan, D.E., Wechezak, A.R., Viggers, R.F. *et al.* (1993) Effect of shear stress upon localization of the Golgi apparatus and microtubule organizing center in isolated cultured endothelial cells. *J Cell Sci* 104, 1145–1153.

Derry, W.B., Wilson, L., Khan, I.A. *et al.* (1997) Taxol differentially modulates the dynamics of microtubules assembled from unfractionated and purified beta-tubulin isotypes. *Biochemistry* 36, 3554–3562.

Dumontet, C. & Sikic, B.I. (1999) Mechanisms of action of and resistance to anti-tubulin agents: microtubule dynamics, drug transport, and cell death. *J Clin Oncol* 17, 1061–1070.

Dumontet, C., Jordan, M.A. & Lee, F.F.Y. (2009) Ixabepilone: targeting βIII-tubulin expression in taxane-resistant malignancies. *Mol Cancer Ther* 8, 17–25.

FDA (2004) Drug Approvals. http://www.fda.gov/cder/approval/t.htm.

Ferlini, C., Raspaglio, G., Mozzetti, S. *et al.* (2005) The seco-taxane IDN5390 is able to target class III β-tubulin and to overcome paclitaxel resistance. *Cancer Res* 65, 2397–2405.

Folkman, J., Bach, M., Rowe, J.W. *et al.* (1971) Tumor angiogenesis: therapeutic implications. *New Eng J Med,* 285.

Geney, R., Sun, L., Pera, P. *et al.* (2005) Use of the tubulin bound paclitaxel conformation for structure-based rational drug design. *Chem Biol* 12, 339–48.

Gillespie, A.M., Broadhead, T.J., Chan, S.Y. *et al.* (2000) Phase I open study of the effects of ascending doses of the cytotoxic immunoconjugate CMB-401 (hCTMO1-calicheamicin) in patients with epithelial ovarian cancer. *Ann Oncol* 11, 735–41.

Gotlieb, A., May, L., Subrahmanyan, L. *et al.* (1981) Distribution of microtubule organizing centers in migrating sheets of endothelial cells. *J Cell Biol* 91, 589–594.

Gottesman, M.M., Fojo, T. & Bates, S.E. (2002) Multidrug resistance in cancer: role of ATP-dependent transporters. *Nat Rev Cancer* 2, 48–58.

Gut, I., Ojima, I., Vaclavikova, R. *et al.* (2006) Metabolism of new-generation taxanes in human, pig, minipig and rat liver microsomes. *Xenobiotica* 36, 772–792.

Hamann, P.R., Hinman, L.M., Hollander, I. *et al.* (2002) Gemtuzumab ozogamicin, a potent and selective anti-CD33 antibody-calicheamicin conjugate for treatment of acute myeloid leukemia. *Bioconjugate Chem* 13, 47–58.

Hari, M., Yang, H., Zeng, C. *et al.* (2003) Expression of class III beta-tubulin reduces microtubule assembly and confers resistance to paclitaxel. *Cell Motil Cytoskel* 56, 45–56.

Hollingshead, M.G., Alley, M.C., Kaur, G. *et al.* (2004) 'NCI specialized procedures in preclinical drug evaluations', in Teicher, B.A. & Andrews, P.A. (Eds), *Anticancer Drug Development Guide*, 2nd edn, Humana Press, Totowa.

Isanbor, C. & O'Hagan, D. (2006) Fluorine in medicinal chemistry: a review of anti-cancer agents. *J Fluorine Chem* 127, 303–319.

Jaracz, S., Chen, J., Kuznetsova, L.V. *et al.* (2005) Recent advances in tumor-targeting anticancer drug conjugates. *Bioorg Med Chem* 13, 5043–5054.

Jordan, M.A., Toso, R.J. & Thrower, D. (1993) Mechanism of mitotic block and inhibition of cell proliferation by taxol at low concentration. *Proc Natl Acad Sci USA* 90, 9552–9556.

Jordan, M.A. & Wilson, L. (2004) Microtubules as a target for anticancer drugs. *Nat Rev Cancer* 4, 253–265.

Kant, J., Huang, S., Wong, H. *et al.* (1993) Studies toward structure–activity relationships of Taxol(R): synthesis and cytotoxicity of Taxol(R) analogues with C-2′ modified phenylisoserine side chains. *Bioorg Med Chem Lett* 3, 2471–2474.

Kavallaris, M., Kuo, D.Y.-S., Burkhart, C.A. *et al.* (1997) Taxol-resistant epithelial ovarian tumors are associated with altered expression of specific beta-tubulin isotypes. *J Clin Invest* 100, 1282–1293.

Kavallaris, M., Burkhart, C.A. & Horwitz, S.B. (1999) Antisense oligonucleotides to class III β-tubulin sensitize drug-resistant cells to Taxol. *Brit J Cancer* 80, 1020–1025.

Kigawa, J., Minagawa, Y., Kanamori, Y. *et al.* (1998) Glutathione concentration may be a useful predictor of response to second-line chemotherapy in patients with ovarian cancer. *Cancer* 82, 697–702.

Kingston, D.G.I., Bane, S. & Snyder, J.P. (2005) The taxol pharmacophore and the T-taxol bridging principle. *Cell Cycle* 4, 279–289.

Kirk, K.L. (2006) Fluorine in medicinal chemistry: recent therapeutic applications of fluorinated small molecules. *J Fluorine Chem* 127, 1013–1029.

Klauber, N., Parangi, S., Flynn, E. *et al.* (1997) Inhibition of angiogenesis and breast cancer cell in mice by the microtubule inhibitors 2-methoxyestradiol and Taxol. *Cancer Res* 57, 81–86.

Kuznetsova, L.V., Ungureanu, I.M., Pepe, A. *et al.* (2004) Trifluoromethyl- and difluoromethyl-β-lactams as useful building blocks for the synthesis of fluorinated amino acids, dipeptides, and fluoro-taxoids. *J Fluorine Chem* 125, 487–500.

Kuznetsova, L.V., Pepe, A., Ungureanu, I.M. *et al.* (2008) Syntheses and structure-activity relationships of novel 3'-difluoromethyl and 3′-trifluoromethyl-taxoids. *J Fluorine Chem* 129, 817–828.

Lam, L., Lam, C., Li, W. *et al.* (2003) Recent advances in drug–antibody immunoconjugates for the treatment of cancer. *Drug Future* 28, 905–910.

Li, Y., Poliks, B., Cegelski, L. *et al.* (2000) Conformation of microtubule-bound paclitaxel determined by fluorescence spectroscopy and REDOR NMR. *Biochemistry* 39, 281–291.

Liao, G., Nagasaki, T. & Gundersen, G.G. (1995) Low concentrations of nocodazole interfere with fibroblast locomotion without significantly affecting microtubule level: implications for the role of dynamic microtubules in cell locomotion. *J Cell Sci* 108, 3473–3483.

Liu, C., Tadayoni, B.M., Bourret, L.A. *et al.* (1996) Eradication of large colon tumor xenografts by targeted delivery of maytansinoids. *Proc Natl Acad Sci USA* 93, 8618–8623.

Löwe, J., Li, H., Downing, K.H. *et al.* (2001) Refined structure of [alpha][beta]-tubulin at 3.5 Å resolution. *J Mol Biol* 313, 1045–1057.

Magnani, M., Ortuso, F., Soro, S. *et al.* (2006) The beta I/beta III-tubulin isoforms and their complexes with antimitotic agents: docking and molecular dynamics studies. *FEBS J* 273, 3301–3310.

Martino, R., Malet-Martino, M. & Gilard, V. (2000) Fluorine nuclear magnetic resonance, a privileged tool for metabolic studies of fluoropyrimidine drugs. *Curr Drug Metab* 1, 271–303.

Müller, K., Faeh, C. & Diederich, F. (2007) Fluorine in pharmaceuticals: looking beyond intuition. *Science* 317, 1881–1886.

Nogales, E., Wolf, S. G. & Downing, K.H. (1998) Structure of the αβ tubulin dimer by electron crystallography. *Nature,* 391, 199–203.

O'Hagan, D., Schaffrath, C., Cobb, S.L. *et al.* (2002) Biochemistry: biosynthesis of an organofluorine molecule. *Nature,* 416, 279.

Ojima, I. (1995) Recent advances in the β-lactam synthon method. *Acc Chem Res* 28, 383–389.

Ojima, I., Slater, J.C., Michaud, E. *et al.* (1996) Syntheses and structure-activity relationships of the second generation antitumor taxoids. Exceptional activity against drug-resistant cancer cells. *J Med Chem* 39, 3889–3896.

Ojima, I., Kuduk, S.D., Chakravarty, S. *et al.* (1997a) A novel approach to the study of solution structures and dynamic behavior of paclitaxel and docetaxel using fluorine-containing analogs as probes. *J Am Chem Soc* 119, 5519–5527.

Ojima, I., Slater, J.C., Pera, P. *et al.* (1997b) Synthesis and biological activity of novel 3′-trifluoromethyl taxoids. *Bioorg Med Chem Lett* 7, 133–138.

Ojima, I., Kuduk, S.D. & Chakravarty, S. (1998) 'Recent advances in the medicinal chemistry of taxoid anticancer agents', in Maryanoff, B.E. & Reitz, A.B. (Eds), *Advances Medicinal Chemistry,* JAI Press, Greenwich, CT.

Ojima, I., Wang, T., Miller, M.L. *et al.* (1999) Synthesis and structure–activity relationships of new second-generation taxoids. *Bioorg Med Chem Lett* 9, 3423–3428.

Ojima, I., Lin, S., Slater, J.C. *et al.* (2000) Syntheses and biological activity of C-3′-difluoromethyl-taxoids. *Bioorg Med Chem* 8, 1619–1628.

Ojima, I., Geng, X., Wu, X. *et al.* (2002) Tumor-specific novel taxoid-monoclonal antibody conjugates. *J Med Chem* 45, 5620–5623.

Ojima, I. (2004) Use of fluorine in the medicinal chemistry and chemical biology of bioactive compounds — a case study on fluorinated taxane anticancer agents. *ChemBioChem* 5, 628–635.

Ojima, I. (2008) Guided molecular missiles for tumor-targeting chemotherapy: case studies using the 2nd-generation taxoids as warheads. *Acc Chem Res* 41, 108–119.

Ojima, I., Chen, J., Sun, L. *et al.* (2008) Design, synthesis and biological evaluation of new second-generation taxoids. *J Med Chem* 51, 3203–3221.

Ojima, I. (2009) *Fluorine in Medicinal Chemistry and Chemical Biology,* Wiley-Blackwell, Chichester.

Paik, Y., Yang, C., Metaferia, B. *et al.* (2007) Rotational-echo double-resonance NMR distance measurements for the tubulin-bound paclitaxel conformation. *J Am Chem Soc* 129, 361–370.

Panda, D., Miller, H.P., Banerjee, A. *et al.* (1994) Microtubule dynamics in vitro are regulated by the tubulin isotype composition. *Proc Natl Acad Sci USA* 91, 11358–11362.

Papetti, M. & Herman, I.M. (2002) Mechanism of normal and tumor-derived angiogenesis. *Am J Physiol-Cell Ph* 282, C947–C970.

Pepe, A., Kuznetsova, L., Sun, L. *et al.* (2009a) 'Fluoro-taxoid anticancer agents', in Ojima, I. (Ed.), *Fluorine in Medicinal Chemistry and Chemical Biology*, Wiley-Blackwell, Chichester.

Pepe, A., Sun, L., Zanardi, I. *et al.* (2009b) Novel C-seco-taxoids possessing high potency against paclitaxel-resistant cancer cell lines overexpressing class III [beta]-tubulin. *Bioorg Med Chem Lett* 19, 3300–3304.

Rowinsky, E.K. (1997) The development and clinical utility of the taxane class of antimicrotubule chemotherapy agents. *Annu Rev Med* 48, 353–374.

Saleh, M.N., Lobuglio, A.F. & Trail, P.A. (1998) 'Immunoconjugate therapy of solid tumors: studies with BR96-doxorubicin', in Grossbard, M. (Ed.), *Monoclonal Antibody-based Therapy of Cancer*, Marcel Dekker, New York, 397–416.

Schiff, P.B., Fant, J. & Horwitz, S.B. (1979) Promotion of microtubule assembly in vitro by taxol. *Nature* 277, 665–667.

Schneider, E. & Lin, R. (2009) 'Study of metabolism of fluorine-containing drugs using *in vivo* magnetic resonance spectroscopy', in Ojima, I. (Ed.), *Fluorine in Medicinal Chemistry and Chemical Biology*, Wiley-Blackwell, Chichester.

Selden, S.C.I., Rabinovitch, P.S. & Schwartz, S.M. (1981) Effect of cytoskeletal disrupting agents on replication of bovine endothelium. *J Cell Physiol* 108, 195–211.

Sève, P. & Dumontet, C. (2005) Chemoresistance in non-small cell lung cancer. *Curr Med Chem: Anti-Cancer Agent* 5, 73–88.

Sève, P. & Dumontet, C. (2008) Is class III β-tubulin a predictive factor in patients receiving tubulin-binding agents? *Lancet Oncol* 9, 168–175.

Shweiki, D., Itin, A., Soffer, D. *et al.* (1992) Vascular endothelial growth factor induced by hypoxia may mediate hypoxia-induced angiogenesis. *Nature* 359, 843–845.

Sullivan, K.F. (1988) Structure and utilization of tubulin isotypes. *Annu Rev Cell Biol* 4, 687–716.

Sun, L., Geng, X., Geney, R. *et al.* (2008) Design, synthesis, and biological evaluation of novel C14-C3' BzN-linked macrocyclic taxoids. *J Org Chem* 73, 9584–9593.

Sun, L., Simmerling, C. & Ojima, I. (2009) Recent advances in the study of the bioactive conformation of Taxol. *ChemMedChem* 4, 719–731.

Taraboletti, G., Micheletti, G., Rieppi, M. *et al.* (2002) Antiangiogenic and antitumor activity of IDN5390, a new taxane derivative. *Clin Cancer Res* 8, 1182–1188.

Ueda, M., Gräf, R., Macwilliams, H.K. *et al.* (1997) Centrosome positioning and directionality of cell movements. *Proc Natl Acad Sci USA* 94, 9674–9678.

Vuilhorgne, M., Gaillard, C., Sanderink, G.J. *et al.* (1995) 'Metabolism of taxoid drugs', in Georg, G.I., Chen, T.T., Ojima, I. *et al.* (Eds), *Taxane Anticancer Agents: Basic Science and Current Status, ACS Symp. Ser. 583*, American Chemical Society, Washington, DC.

Widdison, W.C., Wilhelm, S.D., Cavanagh, E.E. *et al.* (2006) Semisynthetic maytansine analogues for the targeted treatment of cancer. *J Med Chem* 49, 4392–4408.

Williams, H.J., Scott, A.I., Dieden, R.A. *et al.* (1993) NMR and molecular modeling study of the conformations of Taxol and of its side chain methylester in aqueous and non-aqueous solution. *Tetrahedron*, 49, 6545–6560.

Yamazaki, T., Taguchi, T. & Ojima, I. (2009) 'Unique properties of fluorine and their relevance to medicinal chemistry and chemical biology', in Ojima, I. (Ed.), *Fluorine in Medicinal Chemistry and Chemical Biology*, Wiley-Blackwell, Chichester.

8

Synthesis and Antiviral, Antitumour Activities of Fluorinated Sugar Nucleosides

Feng Zheng, Xiao-Long Qiu* and Feng-Ling Qing**

8.1 Introduction

Known subunits of deoxyribonucleic acid (DNA) and ribonucleic acid (RNA), nucleosides play key roles in neurotransmission (Baldwin *et al.*, 1999), regulation of cardiovascular activity (Shryock *et al.*, 1997), and as signalling molecules (Schachter *et al.*, 1995), in addition to their roles as intermediates for many essential cellular biosynthetic pathways. In addition, nucleosides and their analogues play an important role in medicinal chemistry as a structural basis for the development of antiviral and antitumour agents (De Clercq, 2004). Numerous modifications to the sugar ring as well as the heterocyclic nucleobase moieties have been utilized in recent years to increase chemotherapeutic activity (Agrawal *et al.*, 1998; Li *et al.*, 2008). Fluorinated nucleosides have drawn great attention due to their bioactivity profile (Pankiewicz, 2000; Meng *et al.*, 2006; Liu *et al.*, 2008).

Fluorine has been considered as a suitable bioisostere for hydrogen due to its small size and high compactness. Its electronegativity along with

* Key Laboratory of Organofluorine Chemistry, Shanghai Institute of Organic Chemistry, Chinese Academy of Science, 345 Lingling Lu, Shanghai 200032, PR China. E-mail: flq@mail.sioc.ac.cn.

its capability to engage in dipolar interactions make fluorine a close 'iso-electronic' replacement for the hydroxyl group. The replacement of a hydroxyl group or a hydrogen by a fluorine atom causes only a minor change in the steric effect of the functionality, however, such a substitution has profound effects on the chemical–physical as well as biological properties: (i) the glycosyl bond is strengthened, resisting enzymatic hydrolysis by phosphorylases (Tsuchiya, 1990). (ii) The polar hydrophobic aspect of the short C–F bond reduces the polarizability of molecules and increases their hydrophobicity (Biffinger *et al.*, 2004; DiMagno *et al.*, 2006). (iii) The strong *gauche* and antiperiplanar effects of the fluorine substituent caused by its high electronegativity, have profound stereoelectronic effects on the neighbouring groups; thereby the fluorine substituent governs the overall conformation of the sugar ring (Guschlbauer *et al.*, 1980; Cheng *et al.*, 1983; Uesugi *et al.*, 1983; Van der Boogaart *et al.*, 1994). In fact, virtually every hydrogen atom or hydroxyl group from the known endogenous nucleosides has been replaced with fluorine via either single- or multiple-atom substitutions. Many of these novel agents exhibit modified biological potency compared with their corresponding parent compound (Meng *et al.*, 2006; Liu *et al.*, 2008).

In view of the interesting discovery of carbocyclic nucleosides (Schneller, 2002; Rodriguez *et al.*, 2003), thionucleosides (Yokoyama, 2000), phosphanucleosides (Yamashita *et al.*, 1998) and azanucleosides (Yokoyama *et al.*, 1999), fluorinated analogues were prepared and used to explore the effect of the fluorine on structure–activity relationships in attempts to improve the antiviral and antitumour efficacy of these modified molecules. This chapter concentrates on the synthesis of fluorinated sugar nucleosides and their antiviral and antitumor activities, but does not cover a large group of nucleosides fluorinated at the nucleobase.

8.2 Nucleosides Fluorinated at C2′

The unique role of the substituent (hydrogen or hydroxyl) on the C2′ atom in nucleotides, as the distinguishing feature between DNA and RNA, prompted the investigation of the biological properties of nucleosides

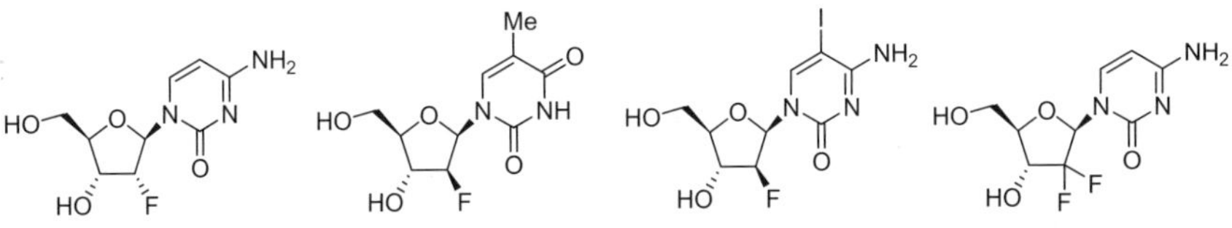

Figure 8.1. 2′-Fluoronucleosides with potent antiviral activity.

containing substituents other than hydrogen or hydroxyl at this position. Accordingly, it was interesting to study the biological properties of C2′ fluorinated nucleosides which could mimic both their hydrogenated and hydroxylated parent compounds. So far, a number of 2′-fluorinated nucleosides have been synthesized and biologically evaluated, some of which showed broad and potent biological activities (Fig. 8.1). For example, 2′-deoxy-2′-fluorocytidine (FdC) is a potent inhibitor of the hepatitis C virus (HCV) RNA replication in cell cultures and thus exhibited high anti-HCV activity (Stuyver *et al.*, 2004). 2′-Deoxy-2′-fluoro-5-methyl-1-β-D-arabinosyluracil (FMAU) (Watanabe *et al.*, 1979; Chu *et al.*, 1995) and 2′-deoxy-2′-fluoro-5-iodo-1-β-D-arabinosylcytosine (FIAC) (Watanabe *et al.*, 1979) showed not only potent activities against the herpes simplex virus (HSV), but also excellent activities against the hepatitis B virus (HBV) and other viruses such as varicella zoster virus (VZV), cytomegalovirus (CMV), and Epstein–Barr virus (EBV). Furthermore, Gemcitabine (2′-deoxy-2′,2′-difluorocytidine) has been approved by the US Food and Drug Administration (FDA) for the treatment of pancreatic cancer (Noble *et al.*, 1997). It has also shown antitumour activity against a wide spectrum of human solid tumours (Hertel *et al.*, 1990). These impressive results motivated organic chemists and pharmaceutical scientists to investigate various types of 2′-monofluorinated nucleosides bearing different functional groups.

8.2.1 *2′-α-Fluoro nucleosides*

Fox and coworkers (Codington *et al.*, 1961, 1964) synthesized a series of 2′-deoxy-2′-fluororibosyl nucleosides **4** by treatment of 2,2′-anhydro nucleoside **3** with hydrogen fluoride (Scheme 8.1). Subsequently, via base transformation procedure, they also prepared 2′-deoxy-2′-fluorocytidine (Doerr *et al.*, 1967). Later, 2′-deoxy-2′-fluoroadenosine and 2′-deoxy-2′-fluoroguanosine were synthesized in the Ranganathan's group (Ranganathan, 1977) and the Kawasaki's group (Kawasaki *et al.*, 1993), respectively. Both of them introduced the fluorine atom into the 2′-position via nucleophilic substitution of the corresponding triflate using tetrabutylammonium fluoride (TBAF).

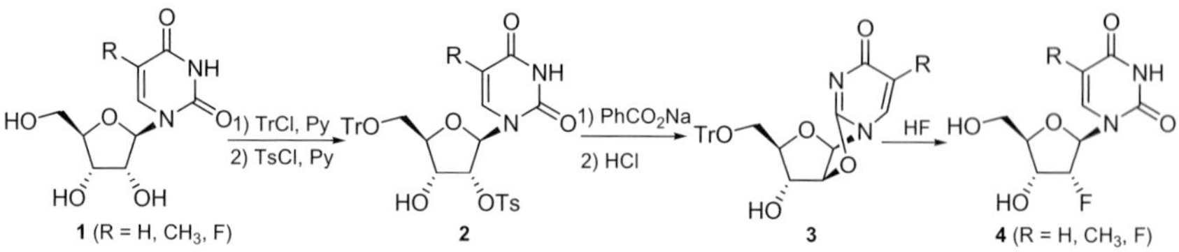

Scheme 8.1.

Fluorination using DAST (Et_2NSF_3) is one of the most important methods in nucleoside chemistry for a one-step exchange of a hydroxyl group by fluorine. Shi *et al.* (2005) found that the direct fluorination of 2,2′-anhydro-5-fluorouridine **5** with KF or Py·nHF failed. Thus they turned to a DAST fluorination approach to accomplish the synthesis. Compound **5** was hydrolysed under basic conditions and then fluorinated with DAST to deliver 2′,5-difluorouridine **7** in excellent yield (Scheme 8.2).

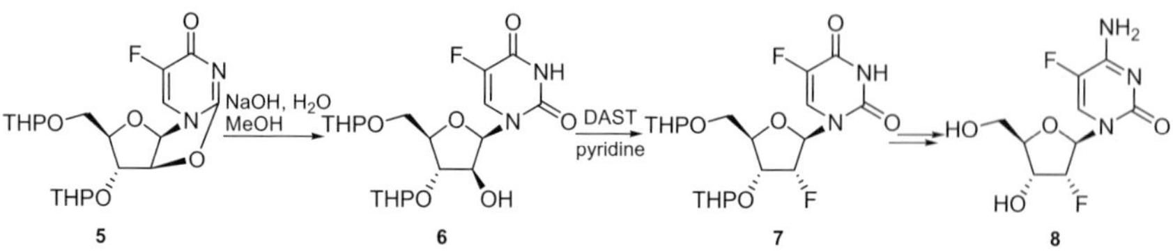

Scheme 8.2.

DAST fluorination of tertiary hydroxyl groups also proved to be successful. Using this strategy, Watanabe and coworkers designed and synthesized a series of D-2′-α-fluoro-2′-methylnucleosides (Clark *et al.*, 2005) (Scheme 8.3). Treatment of 2′-ketone **9** with methyllithium gave compound **10**. After removal of the silyl group and benzoylation, compound **10** was converted to precursor **11**, which was subsequently fluorinated with DAST and deprotected with NH_3/MeOH to furnish the target D-2′-α-fluoro-2′-methycytosine **12**. It is noteworthy that D-2′-α-fluoro-2′-methycytosine **12** demonstrated potent and selective inhibition of HCV replication with an EC_{90} value of 5.4 μM, with no apparent toxicity up to 100 μM (Clark *et al.*, 2005, 2006; Murakami *et al.*, 2007).

Besides the nucleophilic fluorination methods described above, Liotta and coworkers (1998) reported a novel fluorination method using electrophilic *N*-fluorobenzenesulfonimide (NFSI) to prepare D-2′-α-fluoro-lactone **14** (and its L-isomer) diastereoselectively. Fluoride **14** was used to synthesize a number of 2′,3′-dideoxy-2′-fluoro nucleosides **15** (McAtee *et al.*, 1998) (Scheme 8.4).

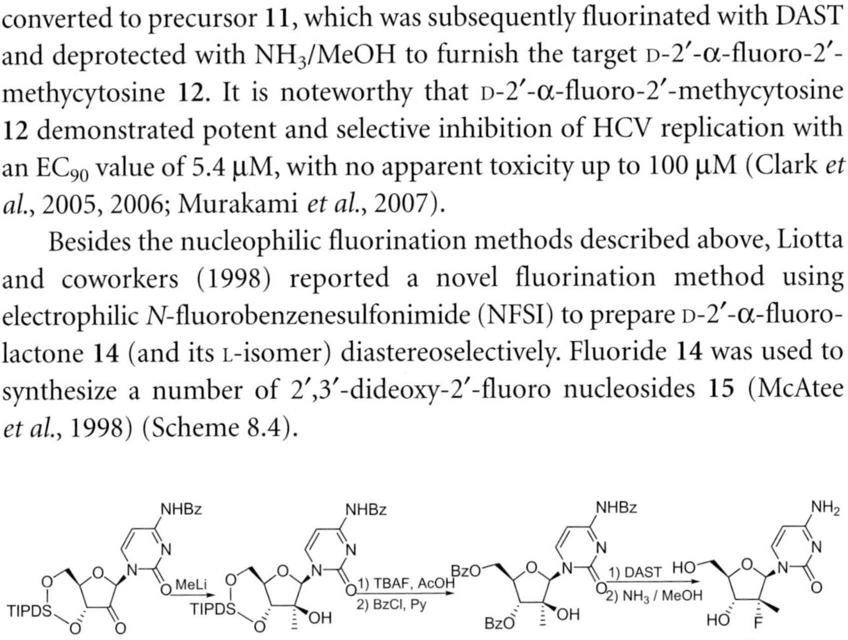

Scheme 8.3.

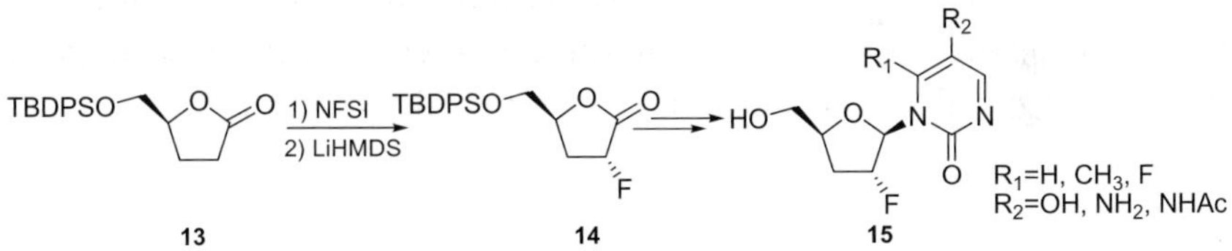

Scheme 8.4.

For the synthesis of 2′-deoxy- and 2′,3′-dideoxy-2′-α-fluorothio-carbocyclic nucleosides, nucleophilic fluorination agents, such as HF, Py·nHF, KHF_2, DAST and TBAF, were also widely utilized to replace the corresponding hydroxyl group with fluorine. Interestingly, during the synthesis of 2′,3′-deoxy-2′-fluorothionucleosides, Marquez and coworkers found that the stereochemical outcome of the DAST-mediated fluorination was largely dependent on the oxidation state of the sulfur atom (Scheme 8.5) (Jeong *et al.*, 1994a,b, 1995b).

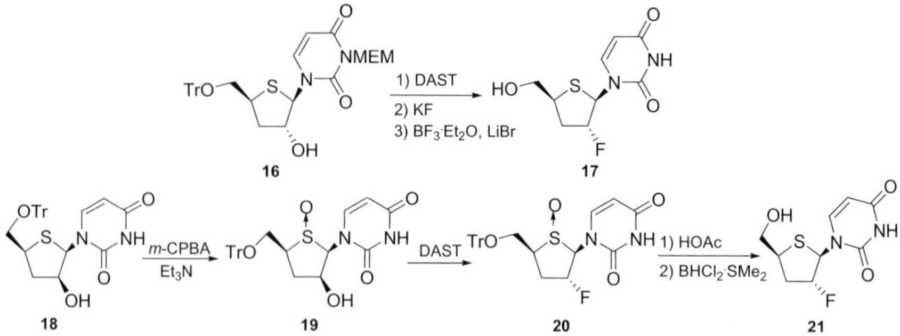

Scheme 8.5.

8.2.2 2′-β-Fluoro nucleosides

The first synthesis of 2′-β-fluoronucleosides, 2′-deoxy-2′-fluoro-β-D-arabinosyl-adenine (F-ara-A) and -cytosine (F-ara-C), was accomplished in the Fox group (Wright *et al.*, 1969, 1970) (Scheme 8.6). Treatment of the methyl 2′,3′-anhydro-5′-*O*-benzyl-α-D-riboside **22** with KHF_2 afforded a mixture of fluoro sugars **23** (as the major product) and its 3-β-F isomer. Compound **23** was converted to the corresponding glycosyl bromide **24**, which was used for coupling with adenine and cytosine to give F-ara-A and F-ara-C. 1′-(2′-Deoxy-2′-fluoro-β-D-arabinosyl)cytosine (F-ara-C) (Wright *et al.*, 1970) showed potent inhibitory activity against L1210 leukaemia and is now a clinically used anticancer agent, 1-(β-D-arabinosyl)cytosine.

Due to the low yield of 2-β-fluorosugar **23**, Watanabe's group (Reichman *et al.*, 1975) subsequently developed a practical synthetic approach to 2-deoxy-2-fluoro-D-arabinose **29** from a readily available D-glucose

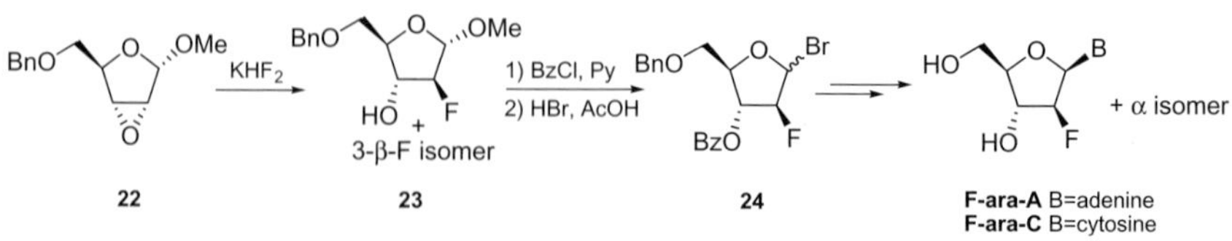

Scheme 8.6.

derivative **25** (Scheme 8.7). The key step of the synthesis is an oxidation of the 3-deoxy-3-fluoro-D-glucose **27** with $NaIO_4$, delivering exclusively the desired 2-deoxy-2-fluoro-β-D-arabinose **28**. After bromination of **28** a series of pyrimidine and purine nucleosides, including the highly bioactive FMAU, FIAC and FEAU were prepared from glycosylbromide **19**.

The potent antiviral activities of FIAC and FMAU motivated Tann *et al.* (1985) to develop an even more efficient method for the construction of the 2-β-F-arabino configuration (Scheme 8.8). The exclusively formed 1-α-glycosyl bromide **33** was used as a key building block for the preparation of 2′-deoxy-2′-β-fluoroarabinonucleosides, in particular for pyrimidine nucleosides. Similarly, Chu's group (Ma *et al.*, 1996; Du *et al.*, 1999) prepared the *L*-enantiomer of 1-α-glycosyl bromide **33** starting from L-ribose, which was used to synthesize L-FMAU and its derivates. They discovered that L-FMAU is a potent antiviral agent against HBV as well as EBV.

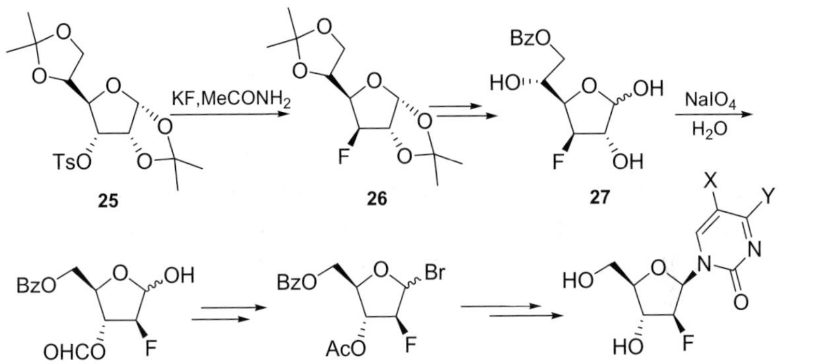

FMAU X=Me,Y=OH
FEAU X=Et,Y=OH
FIAC X=I,Y=NH2

Scheme 8.7.

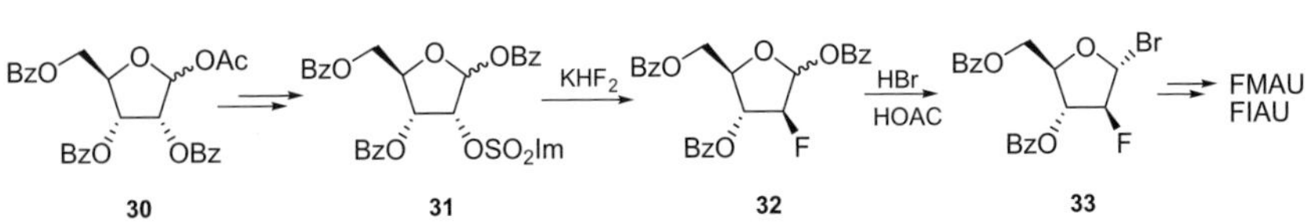

Scheme 8.8.

2′,3′-Dideoxy nucleosides (ddNs), such as 2′,3′-dideoxyinosine (ddI) and 2′,3′-dideoxyadenosine (ddA), exhibited potent anti-HIV activity. However, the toxicity and instability of these compounds limited their application. Thus, the fluorine atom was introduced into this class of nucleosides at the position C2′ to improve their acidic and enzymatic stability. Both 2′,3′-dideoxy-2′-β-fluoro-inosine (F-ara-ddI) and -adenosine (F-ara-ddA, **37**) were found to be as potent as their parent drugs (Chu *et al.*, 1989; Marquez *et al.*, 1990) with indefinite stability in acidic media that would decompose ddI and ddA within minutes. A convenient route to F-ara-ddA was developed by Marquez and coworkers (Scheme 8.9) (Siddiqui *et al.*, 1998). Their synthesis started with the facile introduction of fluorine at C2′ from the α-side of the protected 9-(β-D-arabinofuranosyl)adenine to afford 2′-α-fluoroadenosine **35**. Inversion of the stereochemistry at C2′ was accomplished via a stable olefinic intermediate **36**, which underwent stereoselective reduction of the double bond to give the desired F-ara-ddA with the opposite β-fluoro stereochemistry.

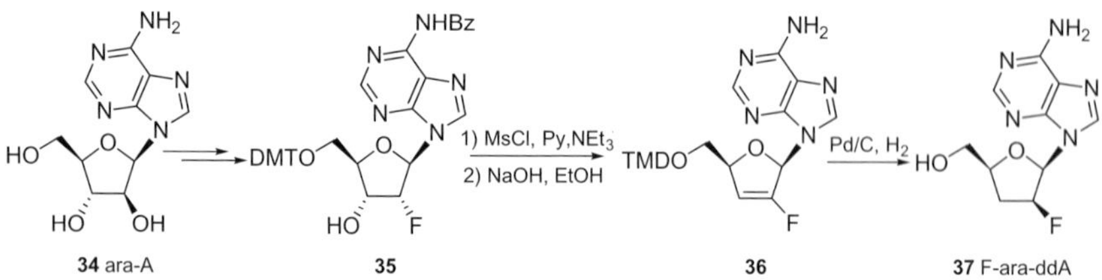

Scheme 8.9.

Starting from the commercially available 1-α-glycosyl bromide **33**, a number of 2′-deoxy-2′-β-fluoroarabinonucleosides **38** were synthesized. They were further converted to the corresponding 2′,3′-dideoxy-2′-β-fluoroarabinonucleosides **41** via Barton deoxygenation and then additionally to some interesting 2′-fluoro-2′,3′-unsaturated- and 2′,3′-dideoxy-2′-β-fluoro-3′-substituted nucleosides (Scheme 8.10) (Marquez *et al.*, 1990; Martin *et al.*, 1990; Sterzycki *et al.*, 1990; Watanabe *et al.*, 1990; Barchi *et al.*, 1991; Siddiqui *et al.*, 1992; Ford *et al.*, 1994; Bhattacharya *et al.*, 1995; Jeong *et al.*, 1995a; Shortnacy-Fowler *et al.*, 2001; Izawa *et al.*, 2003; Moukha-Chafiq *et al.*, 2005).

4′-Thionucleosides have been recognized as a novel and important class of antiviral agents ever since Secrist *et al.* (1991) and Walker and

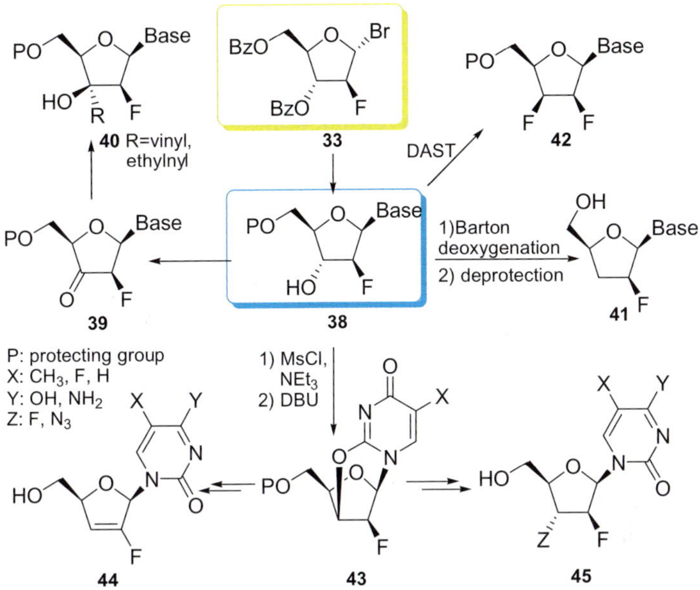

Scheme 8.10.

coworkers (Dyson *et al.*, 1991) independently reported the syntheses and antiviral activities of pyrimidine 2′-deoxy-4′-thionucleosides. Extensive studies on this class of nucleosides demonstrated that 4′-thionucleosides not only retain their potential biological activity, but also exhibited improved metabolic stability (Yokoyama, 2000). Thus a number of fluorinated thionucleosides were synthesized and evaluated. Several 2′-deoxy-2′-β-fluoro-4′-thionucleosides were found to exhibit potent antitumour and antiviral activities (Machida *et al.*, 1998; Zajchowski *et al.*, 2005). 2′-Deoxy-2′-fluoro-4′-thio-β-D-arabinofuranosyl-cytosine (4′-thioFAC) (Miura *et al.*, 1998, 2002; Zajchowski *et al.*, 2005) was highly active against various human solid tumour cell lines *in vitro*, especially the colon carcinoma SW48 (IC_{50} 0.018 mg/mL) and the gastric carcinoma MKN-45 (IC_{50} 0.057 mg/mL) and showed strong antitumour acitivity *in vivo*, even by oral administration. A series of pyrimidine and purine analogues of 4′-thioFAC were synthesized and tested as antiviral and antitumour agents by Yoshimura *et al.* (1997, 1999, 2000) (Scheme 8.11). The bicyclic compound **46** was prepared from D-glucose in eight steps, then transformed

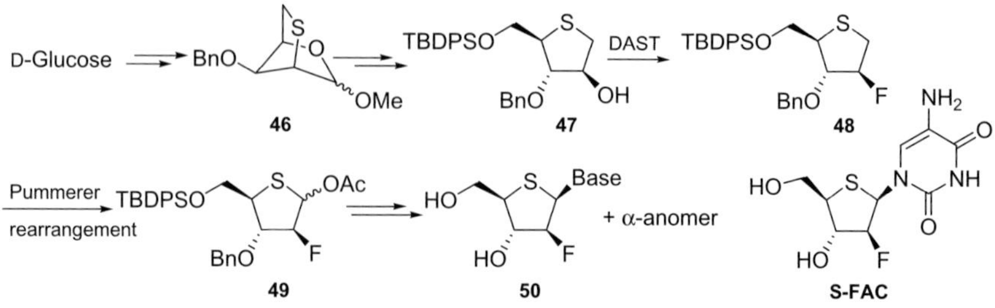

Scheme 8.11.

to the thiosugar **47**. Treatment of **47** with DAST introduced a fluorine atom with retention of the 2′-stereochemistry, yielding **48**. Acetate **49** was obtained via Pummerer reaction, which, after glycosylation with various nucleobases and deprotection, gave the target 2′-deoxy-2′-β-fluoro-4′-thionucleoside **50** together with its α-anomer. All nucleosides were tested for their antiviral and antitumour activities. Whereas the α-anomers were inactive, the β-anomers showed potent and selective anti-HSV-1 and HSV-2 activities *in vitro*.

8.2.3 *2′, 2′-Difluoronucleosides*

The introduction of a *gem*-difluoromethylene group into the sugar moiety of a nucleoside is of interest, as the fluorine could mimic both a hydrogen atom and hydroxyl group to some extent. This suggests that a *gem*-difluoromethylene-containing nucleoside, besides its own special qualities, might combine the characteristics of β-fluoro and α-fluoro nucleosides. Gemcitabine (2′-deoxy-2′,2′-difluorocytidine), a clinically effective anticancer agent for the treatment of pancreatic cancer, was prepared from the key 2-deoxy-2,2-difluoro-ribonic acid **53**, which is available by reaction of Reformatsky reagent $Zn/BrCF_2CO_2Et$ with 2,3-*O*-isopropylidene glyceraldehyde **51** (Hertel *et al.*, 1990; Noble *et al.*, 1997) (Scheme 8.12). Removal of the isopropylidene group resulted in spontaneous cyclization. The resultant lactone **54** was silylated to afford compound **55a**. After conversion of compound **55a** to mesylate **56a**, condensation with the persilylated cytosine provided gemcitabine **57** and its α anomer **58**. In

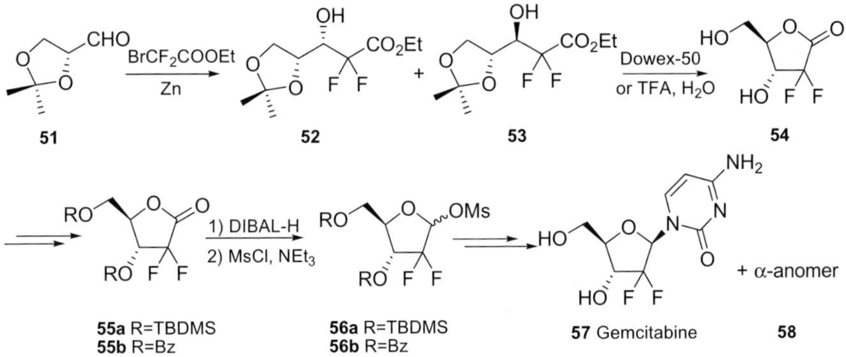

Scheme 8.12.

view of the fact that the above synthetic route needed separation of the isomers **52**/**53** and **57**/**58** (1:1) by HPLC, the synthetic method was improved by selecting Bz over TBDMS as the protecting group. The ratio of β-anomer **57** against its α-anomer increased to 4:1, and crystallization of the desired ribonolactone **55b** and gemcitabine **57** from their respective diastereomeric mixtures was also realized.

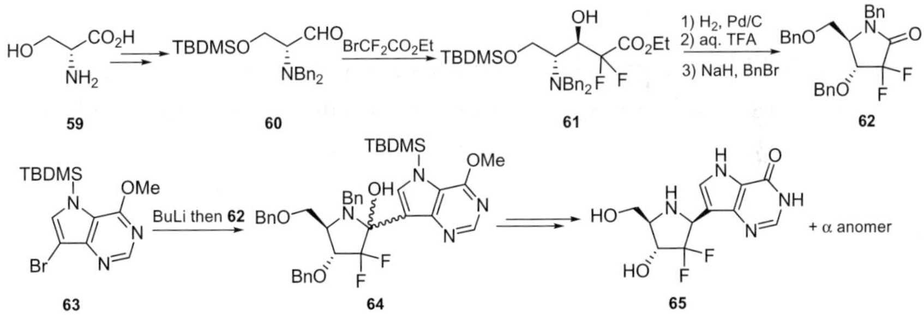

Scheme 8.13.

In 2003, Tyler and coworkers first reported the synthesis of 2′-deoxy-2′,2′-difluoro-aza-C-nucleoside (Evans *et al.*, 2003), an analogue of Immucillin-H (Kicska *et al.*, 2001) that inhibits the proliferation of human T lymphocytes. Starting with D-serine **59** and $BrCF_2CO_2Et$ as *gem*-difluoromethylene source, the lactam **62** was obtained in a straightforward fashion (Scheme 8.13). Lithiation of 9-bromo-9-deazahypoxanthine derivative **63**, by bromine–lithium exchange and subsequent addition of the lactam **62** to the reaction mixture provided the alcohol **64**. Reduction of compound **64** with $NaBH_3CN$ followed by hydrogenolysis and acidic hydrolysis afforded the nucleoside **65** and its α-anomer. Compared to the parent Immucillin-H, compound **65** and its α-anomer were poorer inhibitors.

8.2.4 *2′-Fluoro-2′,3′-didehydro-2′,3′-dideoxynucleosides*

The fluorinated 2′,3′-didehydro-2′,3′-dideoxynucleosides (Fd4Ns) have attracted considerable attention since Chu's group (Lee *et al.*, 1999, 2002; Choi *et al.*, 2002; Chong *et al.*, 2002; Choo *et al.*, 2003; Wang *et al.*, 2005) demonstrated that a series of 2′-fluoro-2′,3′-unsaturated nucleosides had interesting biological properties; L-2′F-d4C, L-2′F-d4FC are among the most potent HBV agents, without significant cytotoxicity. The L-2′F-d4Ns **70** was synthesized starting from (*S*)-glyceraldehyde acetonide **66** (Lee *et al.*, 1999) (Scheme 8.14). Aldehyde **66** was subjected to Horner–Emmons reaction with $(EtO)_2P(O)CHFCO_2Et$ to give 2-fluoro-α,β-unsaturated carboxylate **67** ($E/Z = 9{:}1$). Under acidic conditions for the removal of the isopropylidene group, ester **67** simultaneously cyclized to the corresponding 2-fluorobutenolide, which was transformed to the key intermediate, acetate **69**. Condensation of the fluorosugar **69** with various silylated nucleobases was performed to furnish L-2′F-d4Ns. Similarly, Chu and coworkers also accomplished the synthesis of D-2′F-d4Ns from compound **71** (Lee *et al.*, 2002).

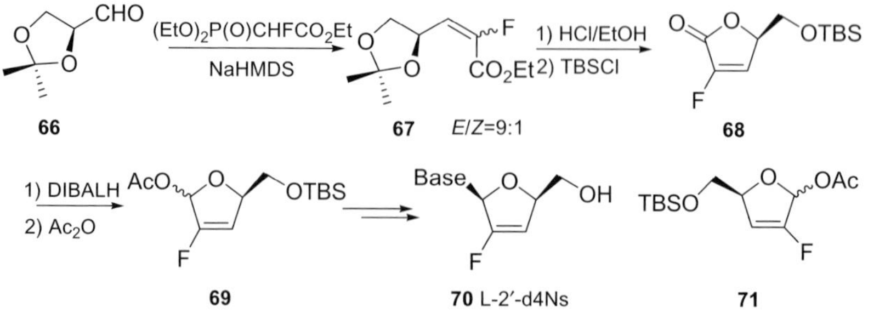

Scheme 8.14.

Chu's group extended their work to the synthesis of a series of D- and L-2′,3′-didehydro-2′,3′-dideoxy-2′-fluoro-4′-thionucleosides (D- and L-S-2′F-d4Ns) (Chong *et al.*, 2002; Choo *et al.*, 2003). The (*S*)-2-fluorobutenolide **72** was hydrogenated to give 2-fluoro-γ-butyrolactone **73** as a single stereoisomer, which was transformed to iodo esters **74** as an epimeric mixture in a ratio of 6:1 (Scheme 8.15). The iodo esters **74** were then converted to their corresponding thiolactone with **76** being formed as the major epimer. Then, after several manipulations, the key acetyl phenylselenyl 4-thio-β-D-ribofuranoside **78** was obtained. Glycosylation of the thiosugar **78** with various pyrimidine or purine bases followed by *m*CPBA-mediated elimination and deprotection affored the target thionucleosides **79**.

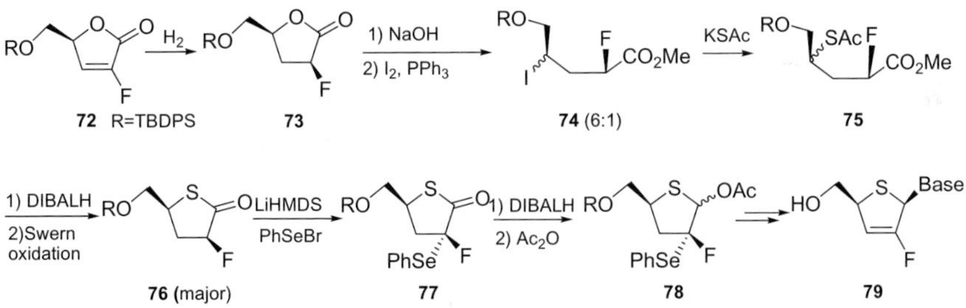

Scheme 8.15.

Chu's group also developed practical synthetic access routes to D- and L-2′,3′-didehydro-2′,3′-dideoxy-2′-fluorocarbocyclic nucleosides (D- and L-2′F-C-d4Ns) (Wang *et al.*, 2005). The ketone **80**, prepared from D-ribose in 17 steps, was fluorinated with DAST to give a difluorinated carbasugar, which underwent elimination to afford the allylic ether **81** (Scheme 8.16). After changing the protecting group and debenzylation, the key intermediate **82** was obtained and coupled with purines and pyrimidines under Mitsunobu conditions.

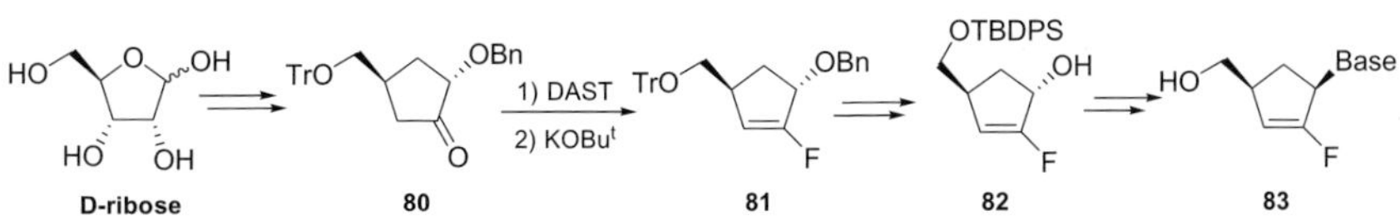

Scheme 8.16.

8.3 Nucleosides Fluorinated at C3′

8.3.1 *3′-α-Fluoro nucleosides*

A number of 3′-α-fluoro nucleosides display potent antiviral activities. D-β-2′,3′-dideoxy-3′-α-fluoro-thymidine (FLT) was found to be very active against human immunodeficiency virus (HIV) (Koshida *et al.*, 1989). The corresponding deoxyuridine, deoxycytidine, deoxyadenosine and deoxyguanosine (FLG) derivatives were prepared and were found to be less active than FLT (Herdewijn *et al.*, 1987a,b; Balzarini *et al.*, 1988). FLG has also been shown to inhibit HBV. All these compounds are potential inhibitors of the viral reverse transcriptase (RT) and chain terminators. Unfortunately, they were found to be highly cytotoxic. However, D-β-2′,3′-dideoxy-3′-α-fluoro-4-chlorouridine (FddClU) exhibited significant anti-HIV activity with low cytotoxity in human leukaemic cells, as well as in bone marrow progenitor cells. The L-counterparts of the above D-3′-α-fluoro nucleosides were also synthesized and evaluated (Sugimura *et al.*, 1991; von Janta-Lipinski *et al.*, 1998; Chun *et al.*, 2000); the 5′-triphosphates of several L-β-3′-α-fluoronucleosides (L-β-FNTPs) emerged as effective inhibitors of HBV/DHBV DNA polymerases. Nevertheless, all L-β-FdNTPs were inactive against HIV-RT, a result which contrasted sharply with the high efficiency of the D-β-FdNTPs against this polymerase.

De Clercq and coworkers pioneered the synthesis of 3′-deoxy-3′-α-fluoro-D-ribofuranosides **88** in 1989. Starting from the methyl glycoside **84**, epoxide **85** was provided in a straightforward manner over four steps (Scheme 8.17). Treatment of the epoxide **85** with KHF_2/NaF gave fluoro-sugar **86**, which was converted to **87** after several operations protecting groups including the inversion of configuration of C2 position via a nucleophilic substitution. The condensation between **87** and various silylated bases followed by deprotection afforded the desired nucleosides **88**.

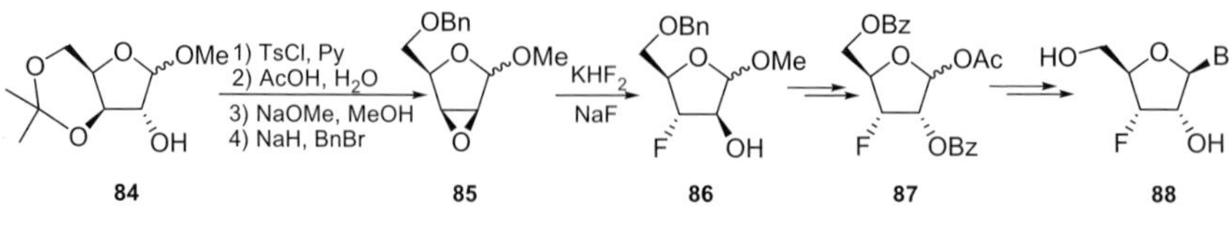

Scheme 8.17.

Komatsu and Araki (2003), described the first application of a chemoenzymatic strategy to synthesize 2′,3′-dideoxy-3′-α-fluoro-β-D-guanosine **91** (Scheme 8.18). The key intermediate, phosphate potassium salt **90**, was stereoselectively prepared from the methyl furanoside **89** in four steps. Then, in the presence of bacterial purine nucleoside phosphorylase (PNPase) and guanine, nucleoside **91** was obtained in 71% yield with the β-anomer as the only product.

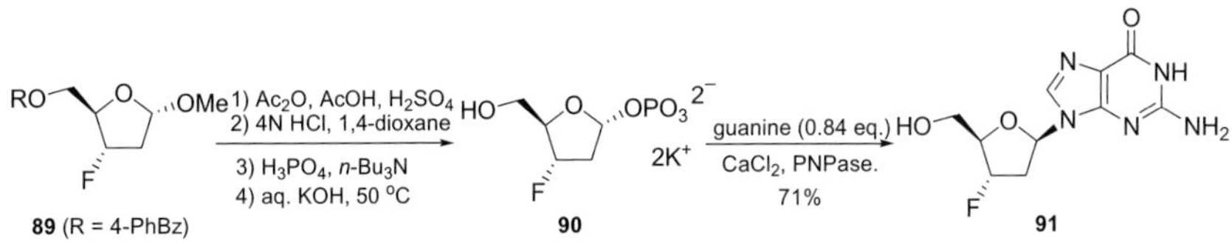

Scheme 8.18.

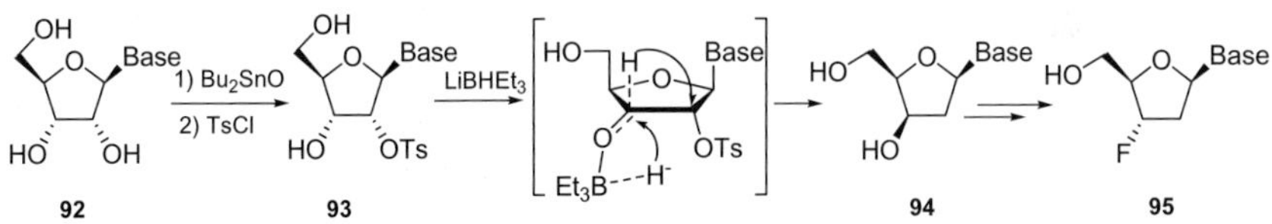

Scheme 8.19.

Robins and coworkers (Timoshchuk *et al.*, 2004; Robins *et al.*, 2007) developed a facile sequence including a deoxygenative [1,2]-H shift rearrangement as the key step to prepare 2′-deoxy-xylosyl-nucleosides **94** (Scheme 8.19), which could be easily transformed to 2′,3′-dideoxy-3′-α-fluoro-nucleosides **95** (3′-α-F-ddNs) (Liu *et al.*, 2008). Conversion of the ribonucleoside **93** to 2′-deoxy-xylonucleoside **94** consists of a [1,2]-hydride shift with the departure of a leaving group from the opposite face. Transient formation of a C=O group is followed by a rapid transfer of a hydride-equivalent from the triethylborohydride intermediate, which leads to an inversion of the stereochemistry at the C-3 position and triggers the [1,2]-hydride shift from C-3 to C-2 and the departure of the leaving group.

Takamatsu and coworkers (Takamatsu *et al.*, 2003; Torii *et al.*, 2005; Katayama *et al.*, 2006) reported an efficient method for the synthesis of 2′,3′-dideoxy-3′-α-fluoro-adenosine **99** (Scheme 8.20). Fluorination of both 3′-β-bromo-3′-deoxyadenosine **96** and 2′-β-bromo-2′-deoxyadenosine

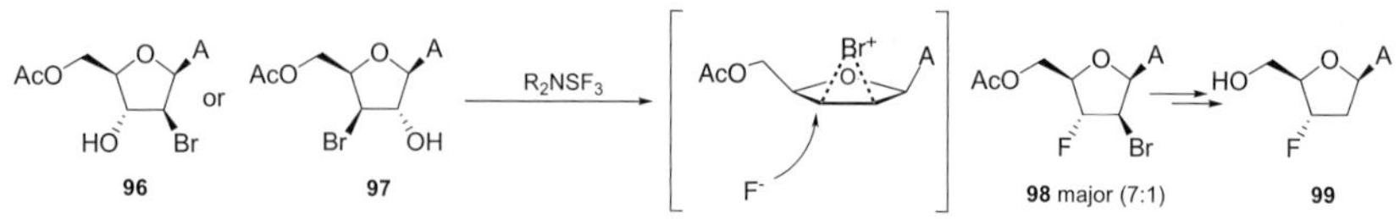

Scheme 8.20.

97 with R_2 NSF_3 fluorinating agent, gave compound **98** as a major product. Radical reduction of the 2′-bromide followed by deprotection afforded the desired 3′-fluoronucleoside **99** in good yield.

8.3.2 *3′-β-Fluoro nucleosides*

Just like the synthesis of 3′-α-fluoro nucleosides, 3′-β-fluoro nucleosides were prepared using standard methods. Recently, an efficient synthesis of 3′-fluoro-5′-thioxyl-furanosyl-nucleosides **105** was described by Komiotis and coworkers (Tsoukala *et al.*, 2007). 3′-β-Fluorine was introduced by treatment of the tosylate **100** with KF/acetamide (Scheme 8.21). The resulting fluoride **101** was subjected to periodate oxidation, borohydride reduction and sulfonylation to afford intermediate **102**. After thioacylation of the compound **102** and acetolysis, the resulting acetate **104** was condensed with silylated pyrimidine bases to give the target nucleosides **105**. These nucleosides were good candidates for the development of potential antiviral agents. In contrast to 3′-azido-3′-deoxythymidine (AZT), significantly lower concentrations of these agents were required to neutralise rotavirus infectivity. The most promising antitumour activity of these compounds was observed in the case of colon carcinoma treatment, where growth inhibition and cytotoxic effect were achieved at low concentration in comparison to

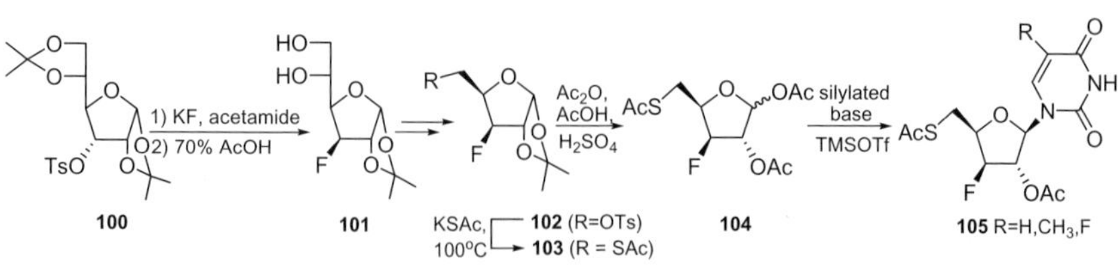

Scheme 8.21.

5-fluorouracil. However, the antitumour activity was found to be cell-type dependent.

In 2002, Fuentes *et al.* demonstrated that the nucleophilic opening of nucleoside-derived cyclic sulfates could be used as a regio- and stereoselective method for the preparation 3′-β-fluoro nucleoside derivatives. The nucleoside derivative **106** was treated with SO_2Cl_2/Et_3N to provide the cyclic sulfate **107** as the sole product (Scheme 8.22). Ring opening of compound **107** with tetraethylammonium fluoride dihydrate followed by acidic hydrolysis gave the two regioisomers **108** and **109** in a ratio of 11:1.

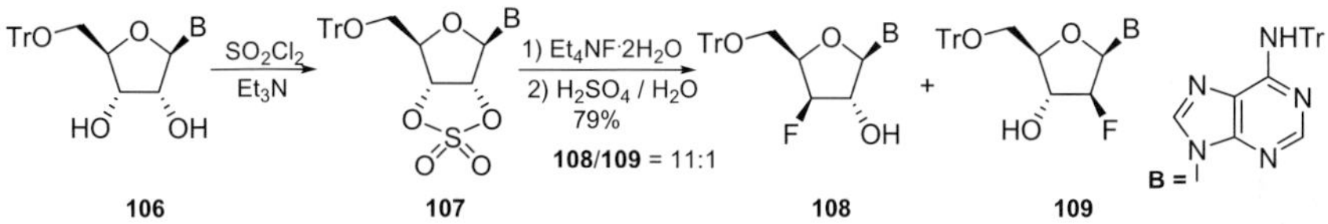

Scheme 8.22.

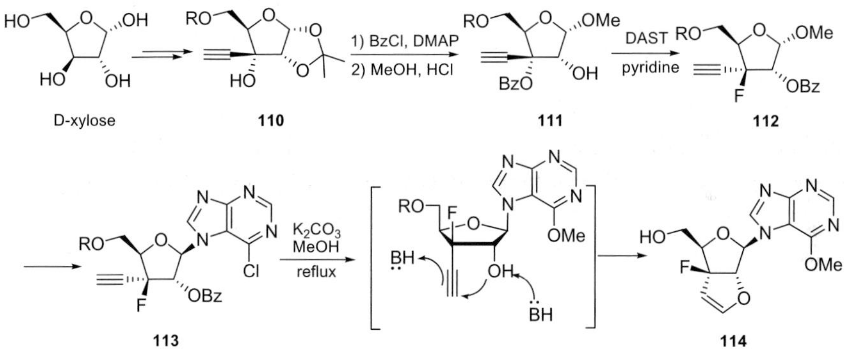

Scheme 8.23.

Zhao *et al.* (2008) accomplished the synthesis of 3′-deoxy-3′-fluoro-2′-O,3′-C-vinylidene-linked bicyclic purine nucleoside **114** (Scheme 8.23). Their synthesis commenced with D-xylose, which was converted to 3-β-C-ethynyl sugar **110** in five steps. Benzoylation of **110** and subsequent methanolysis with concentrated HCl gave the intermediate **111**. Fluorination of **111** with DAST produced the 3′-fluoro-3′-deoxy-3′-α-*C*-ethynyl sugar **112** by neighbouring-group participation. Acetolysis of

112 followed by condensation with persilylated 6-chloropurine furnished the purine derivative **113**. Treatment of compound **113** with K_2CO_3 in MeOH at 65°C finalized the synthesis of fluorinated bicyclic nucleoside **114** in 55% yield. They proposed that the formation of the bicyclic ring involved an intramolecular cycloaddition, facilitated by the highly electronegative fluorine.

8.3.3 *3′,3′-Difluoro nucleosides*

There are two general methods to introduce a CF_2 group into nucleosides: (i) direct difluorination of a carbonyl group (C=O) in the position with less steric hindrance; (ii) construction of appropriate sugar moieties with suitable CF_2 containing building blocks. Using the former strategy, Chu and coworkers (Gumina *et al.*, 2001; Chong *et al.*, 2003; Zhou *et al.*, 2004b) developed a general synthetic route to both D-and L-2′,3′-dideoxy-3′,3′-difluoro nucleosides and their analogues (Scheme 8.24). Difluorination of the readily available ketones, **115** and **121**, gave the key CF_2-containing sugar intermediates **117** and **122**, which were subsequently coupled with bases to afford the corresponding 3′,3′-difluoronucleosides **118** and **123**, respectively. Similarly, Chu and coworkers accomplished the synthesis of 4′–thiofuranosyl (Zhu *et al.*, 2004) as well as carbocyclic (Wang *et al.*, 2007) analogues of 2′,3′-dideoxy-3′,3′-difluoronucleosides.

For the preparation of 3′,3′-difluoronucleosides bearing a vicinal 2′-OH, Qing and coworkers (Zhang *et al.*, 2003; Xu *et al.*, 2006) developed a

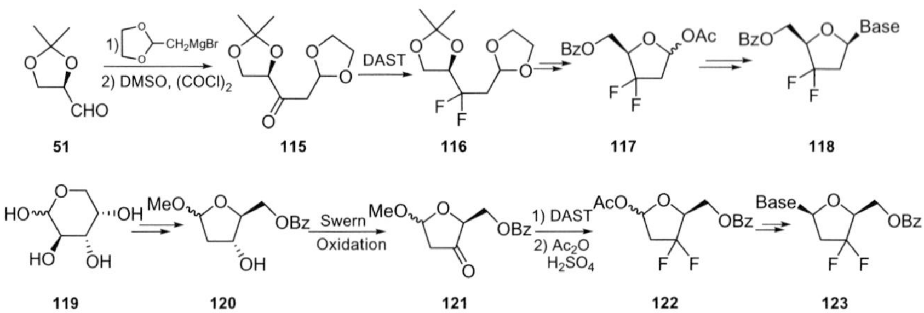

Scheme 8.24.

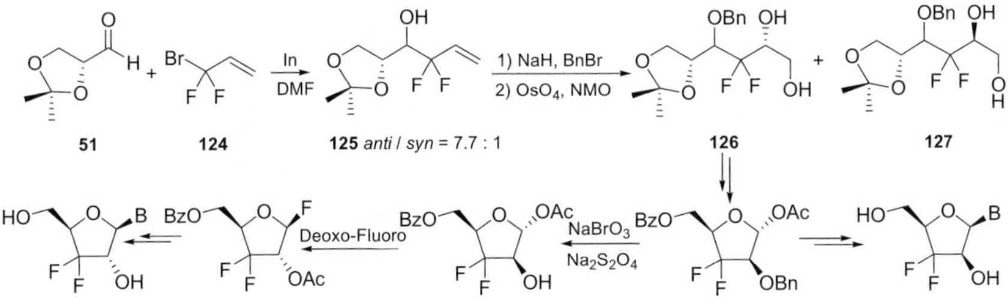

51 124 125 *anti* / *syn* = 7.7 : 1 126 127

131 130 129 128 132

Scheme 8.25.

stereoselective procedure using a versatile CF_2-containing building block (Scheme 8.25). Their synthesis featured the indium-mediated coupling reaction of 1′-(*R*)-glyceraldehyde acetonide **51** and 3-bromo-3,3-difluoropropene **124** to afford the key difluorohomoallyl alcohol **125** in 90% yield with 77% *de*. Protection of the hydroxyl group in **125** followed by osmium-catalysed dihydroxylation gave the separable diols **126** and **127**. Diol **126** was then converted to furanose **128**, which was coupled with silylated bases to give the 3′,3′-difluoroarabinosyl nucleosides **132**. Alternatively, starting from **128**, the 3′-deoxy-3′,3′-difluoro-ribosyl nucleosides **131** was also synthesized via sequential debenzylation, deoxo-fluoro-mediated reversion of the configuration of 2′-hydroxyl and condensation with silylated bases. The enantiomer of **131** was prepared efficiently from diol **127**.

Starting from the building block **126**, Qing's group (Zheng *et al.*, 2006) accomplished the synthesis of L-β-3′-deoxy-3′,3′-difluoro-4′-thionucleosides **136** (Scheme 8.26). This synthesis highlighted the installation of the thioacetyl group in high efficiency via CsF/DMF-mediated nucleophilic inversion of the secondary triflate.

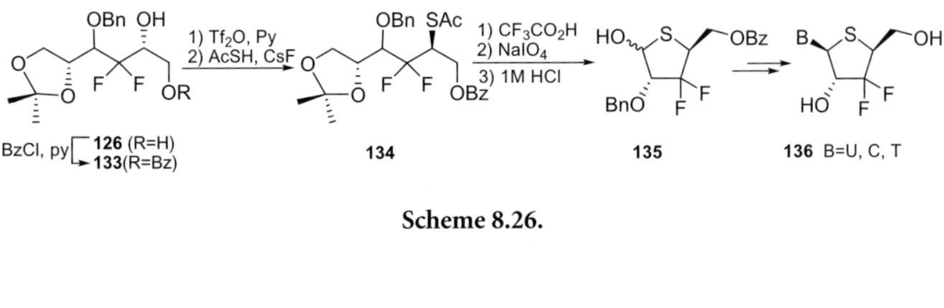

126 (R=H) 133(R=Bz) 134 135 136 B=U, C, T

Scheme 8.26.

8.3.4 *3′-Fluoro-2′, 3′-didehydro-2′, 3′-dideoxy nucleosides*

A number of 3′-fluoro-2′,3′-unsaturated nucleosides (3′F-d4Ns), their 4′-thiofuranosyl and carbocyclic analogues were synthesized in Chu′s group, (Gumina *et al.*, 2001; Chong *et al.*, 2003; Zhou *et al.*, 2004b; Zhu *et al.*, 2004; Wang *et al.*, 2007) via an elimination reaction of 3′, 3′-difluoronucleosides in the presence of ^{t}BuOK, like in the synthesis of 2′F-d4Ns (Scheme 8.27).

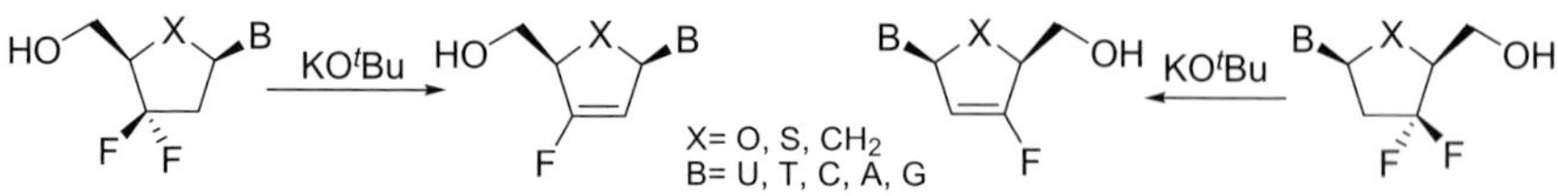

Scheme 8.27.

The comprehensive structure–activity relationship study of 2′F-d4Ns and 3′F-d4Ns showed that several compounds possess potent antiviral activity (Fig. 8.2) (Liu *et al.*, 2008). L-2′F-d4C and L-2′F-d4FC are among the most potent anti-HBV agents. L-3′F-d4C and L-3′F-d4FC their 4′-thio analogues L-S-3′F-d4C and L-S-3′F-d4FC exhibit potent anti-HIV activity without significant toxicity. Carbocyclic D-C-3′F-d4G also exhibits potent anti-HIV activity. In addition, D-C-3′F-d4G and L-3′F-d4C show significant activity against lamivudine-resistant viruses in peripheral blood mononuclear (PBM) cells. The additional hydrophobicity derived from the 2′,3′-double bond and the fluoro group may contribute to the enhanced antiviral activity of these nucleosides.

Ohrui *et al.* (2000) demonstrated that some 4′-ethynyl-2′-deoxy nucleosides are highly active against HIV by blocking the replication of a wide spectrum of laboratory and clinical HIV-1 strains. Inspired by these findings, Chu and coworkers (Chen *et al.*, 2004) synthesized

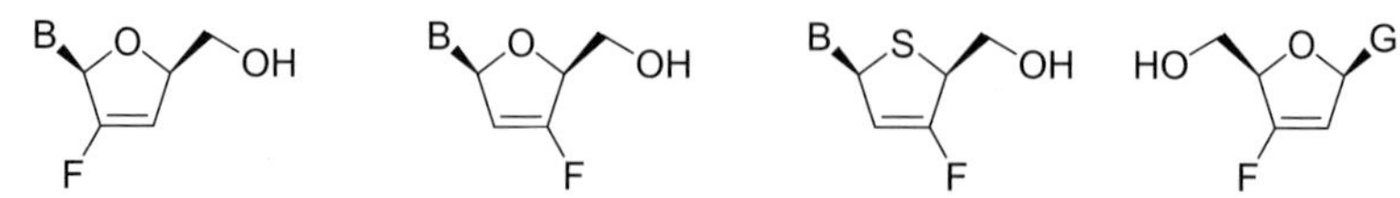

Figure 8.2. 2′F-d4Ns and 3′F-d4Ns possessing potent antiviral activity.

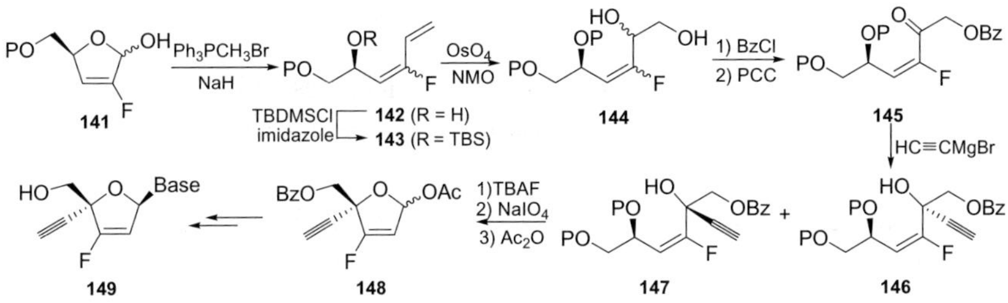

Scheme 8.28.

D- and L-4′-ethynyl-3′-fluoro 2′,3′-unsaturated nucleosides (Scheme 8.28). Fluorolactol **141** was transformed to the α,β-unsaturated ketone **145** in several steps. Ketone **145** was then subjected to the Grignard reaction with ethynylmagnesium bromide to afford the key intermediates **146** and **147**. After deprotection, oxidation and acetylation, compound **147** was converted to 2′, 3′-unsaturated 4′-ethynylfuranose **148**. Condensation of **148** with various silylated bases followed by deprotection gave the desired nucleoside **149**. The L-4′-ethynyl-3′-fluoro-2′,3′-unsaturated nucleosides were prepared from compounds **146**, using the same procedure. All the synthesized 4′-substituted nucleosides were tested for their activities against HIV. The D-adenine derivative showed moderate anti-HIV activity without significant cytotoxicity.

8.4 Nucleosides Fluorinated at C4′

Nucleosides with substituents at the 4′-position have been considered as good candidates for antiviral agents. For example, 4′-azidothymidine demonstrated very potent anti-HIV activity, but its high toxicity rendered it ineffective as an antiviral (Maag *et al.*, 1992). Other 4′-substitutions, such as fluoro (Guillerm *et al.*, 1995), cyano (O-Yang *et al.*, 1992) and ethynyl (Ohrui *et al.*, 2000) also afforded nucleosides with strong activity, including anti-HIV activity. Their biological activities may be due to the rigid North conformation of their sugar rings due to the 4′-substituents.

The natural nucleocidin **153**, an antitrypanosomal antibiotic, was first isolated in 1957. Almost 20 years later Jenkins *et al.* (1976) first accomplished the synthesis of nucleocidin (Scheme 8.29). The key step of their

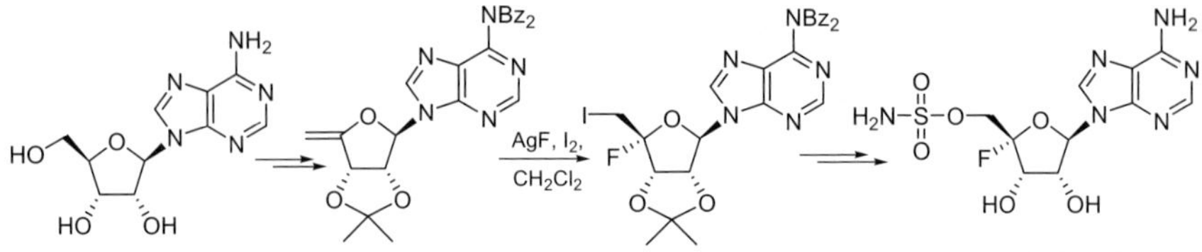

Scheme 8.29.

synthesis was the unselective electrophilic addition of the iodine fluoride, generated *in situ* from AgF/I_2, to the 4′,5′-olefin of the intermediate **151**. The desired 4′-fluoro-5′-iodo nucleoside **152** was then converted to the desired nucleocidin **153** in several steps.

Chu and coworkers published an interesting work on an asymmetric fluorination of tertiary carbon for the synthesis of 4′-fluorinated apionucleosides **157** (Hong *et al.*, 1998) and carbocyclic 2′,3′-dideoxy nucleosides **160** (Gumina *et al.*, 2000). They used a [3,3]-sigmatropic Claisen rearrangement reaction as the key step to introduce the required quaternary fluorinated carbon (Scheme 8.30).

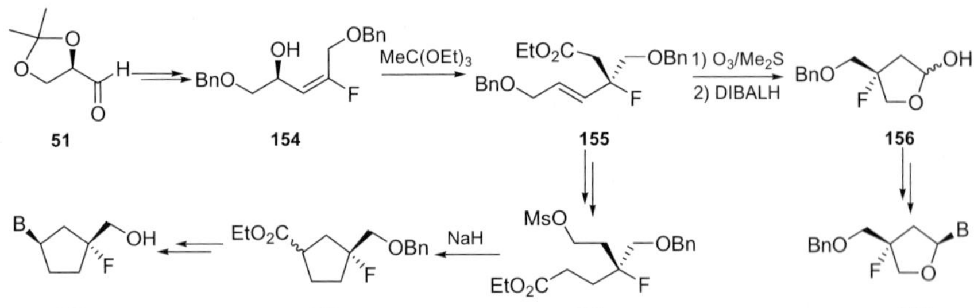

Scheme 8.30.

The Verdine group (Lee *et al.*, 2007) described a concise synthesis of 4′-fluoro nucleosides **164** in two to three steps by sequential bromination and fluorination of ribofuranoses and nucleosides (Scheme 8.31). Treatment of 1′-*O*-acetyl-2′,3′,5′-tri-*O*-benzoyl-β-D-ribose **161** with NBS under a sun lamp followed by direct fluorination of the resulting crude reaction mixture 4′-bromosugars using silver tetrafluoroborate, generated *in situ*

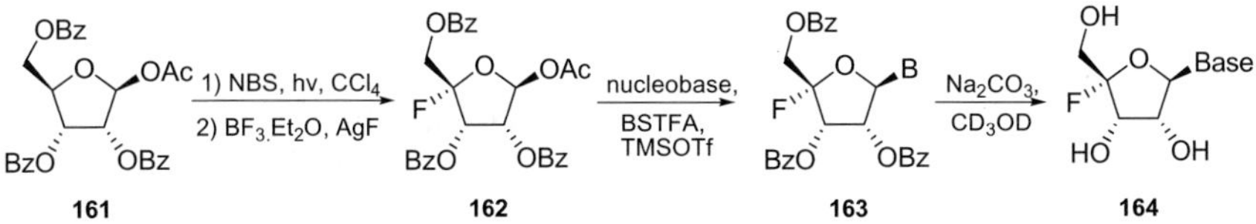

Scheme 8.31.

from $BF_3 \cdot Et_2O$ and AgF, afforded the desired 4-fluoro-β-D-ribofuranose **162**, together with an almost equal amount of 4-fluoro-α-L-lyxofuranose.

8.5 Nucleosides Fluorinated at C6′

Carbocyclic nucleosides have emerged as a particularly interesting class of antiviral agents and several derivatives with potent antiviral activity have been discovered (Marquez *et al.*, 1986). Using the nucleoside scaffold for medicinal agents and biochemical mechanistic investigations, the C6′ site of the carbocyclic nucleoside framework offers a centre for modification, not available in the normal ribofuranosyl-based nucleoside. Because of the promising properties displayed by the fluorinated carbocyclic nucleosides, several methods were developed to synthesize C6′ fluorinated nucleosides.

In 1987, Borthwick and coworkers (Biggadike *et al.*, 1987) described the preparation and anti-herpes activity of two carbocyclic 2′-deoxy-6′-fluoro-uridines **170** and **172** (Scheme 8.32). The epoxide **166** was opened

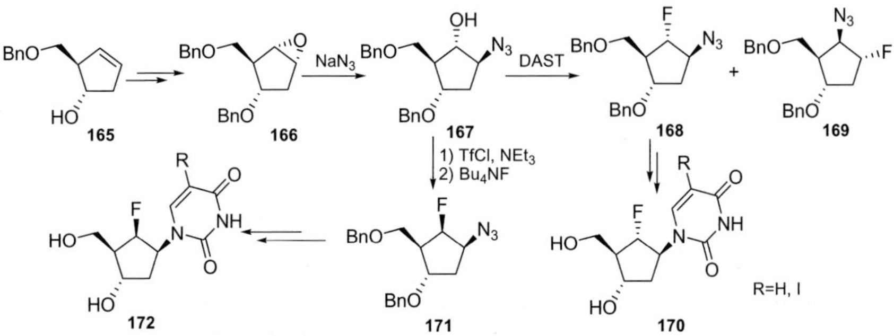

Scheme 8.32.

regioselectively using azide ions to give the alcohol **167**, which was subjected to DAST fluorination to give compound **168**, with retention of configuration, and its regioisomer **169**. Treatment of the corresponding triflate of compound **167** with Bu_4NF afforded the epimeric key intermediate **171**. Interestingly, the 6′-α-fluoro-compound **170** was highly active against herpes simplex virus type 1 (HSV-1)-infected cells in the microtitre assay, whereas the 6′-β-fluorocompound **172** was at least two orders of magnitude less active.

Roberts and coworkers reported a novel synthesis of 2′-deoxy-6′-fluoro-carbocyclic nucleosides **179** starting from the bicyclic ketone **173** (Payne *et al.*, 1992). Their synthesis featured the introduction of the fluorine atom via ring-opening of the strained tricyclic ketone **175** using triethylamine trihydrofluoride (NEt_3·3HF) (Scheme 8.33).

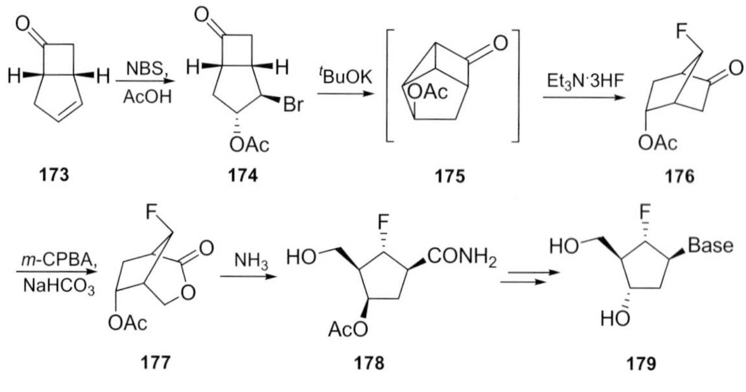

Scheme 8.33.

Samuelsson and coworkers (Wachtmeister *et al.*, 1997) described an asymmetric synthesis of 6′-α- and 6′-β-monofluorinated carbanucleosides **185** and **186** using an electrophilic Selectfluor-mediated α-fluorination of the cyclopentanone derivative **181** as the key step (Scheme 8.34). Thus, silylenol ether, **181** which had been prepared from cyclopentanone **180**, was treated with Selectfluor in DMF to give a 1:1 inseparable diastereomeric mixture of the fluoroketone **182**. Selective reduction of the ketone gave two diastereomeric alcohols, the key intermediates **183** and **184**.

The naturally occurring neplanocin A **187** is a potent inhibitor of S-adenosyl-L-homocysteine hydrolase (SAH) and vaccinia virus

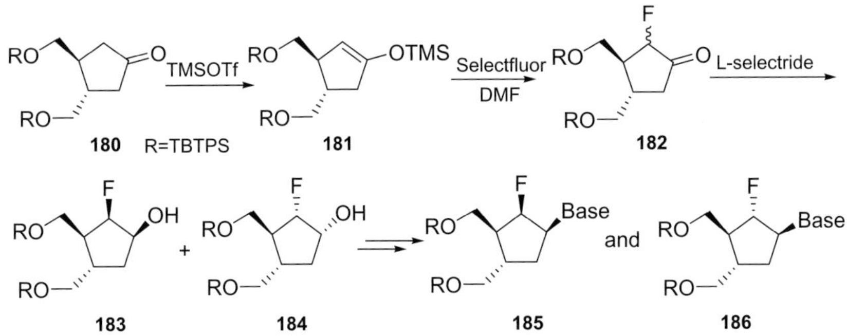

Scheme 8.34.

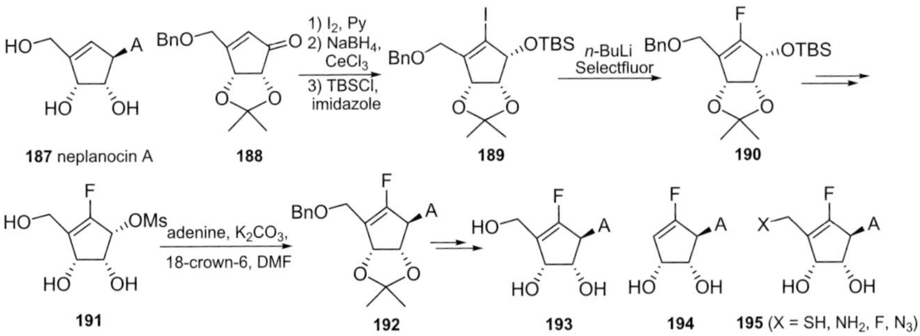

Scheme 8.35.

multiplication (Borchardt *et al.*, 1984). Thus fluoroneplanocin A **193** was designed as a novel mechanism-based inhibitor of SAH and efficiently synthesized via an electrophilic vinyl fluorination reaction (Jeong *et al.*, 2003). Treatment of the iodide **189**, obtained from cyclopentenone derivative **188** by iodination and subsequent Luche reduction, with Selectfluor/*n*-BuLi afforded the key intermediate **190** (Scheme 8.35). Fluoroneplanocin A **193** exhibited a twofold more potent SAH-inhibitory activity than its parent neplanocin A. Unlike neplanocin A showing reversible inhibition of SAH, fluoroneplanocin A exhibited a new type of irreversible inhibition. Furthermore, using fluoroneplanocin A as a template for the design of a new antiviral agent operating via SAH inhibition, fluoro-DHCeA **194** (Kim *et al.*, 2004) and 5′-substituted fluoroneplanocin A analogues **195** (Moon *et al.*,

2004) were also prepared and biologically evaluated in this group. Fluoro-DHCeA **194** was found to be as potent as DHCeA, but exhibited irreversible enzyme inhibition. The inhibitory activity of the fluoroneplanocin A series (**193** and **195**) against SAH was in the following order: OH > NH_2 > SH > F, N_3, indicating that a hydrogen bond donor such as OH, NH_2 was essential for the inhibitory activity.

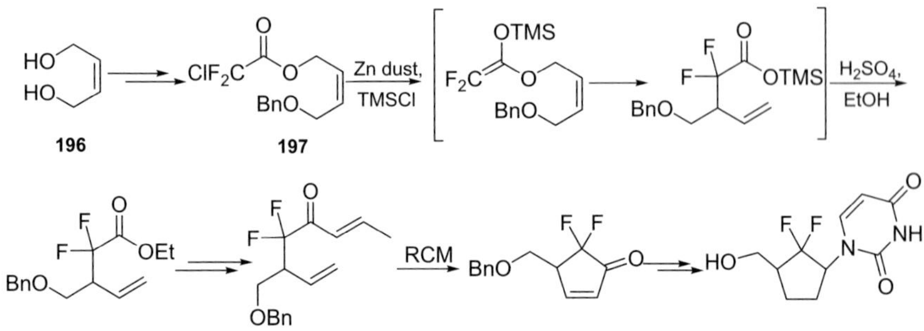

Scheme 8.36.

Gem-difluoromethylene (CF_2) as a replacement for the endocyclic oxygen is a neutral change as an electronegative group replaces an electronegative atom with only little difference in shape. Such a replacement has been proved successful in difluorophosphonates that mimic phosphate both geometrically and electronically (Blackburn *et al.*, 1981). Qing and coworkers (Yang *et al.*, 2004) introduced the CF_2 group into the carbocyclic ring to simulate the ring oxygen of furanose sugars based on this bioisosteric rationale and accomplished the synthesis of 2′,3′-dideoxy-6′,6′-difluorocarbocyclic nucleosides **201** (Scheme 8.36). Highlighted features of their synthesis are the construction of the carbocyclic ring via ring-closing metathesis (RCM) and the introduction of CF_2 group by means of silicon-induced Reformatsky–Claisen reaction of chlorodifluoroacetic ester **197**.

After the successful synthesis of the racemic 2′,3′-dideoxy-6′,6′-difluorocarbocyclic nucleosides, Yang *et al.* (2007) prepared the chiral difluorocarbocyclic nucleosides **209** (Scheme 8.37). The key fluorinated

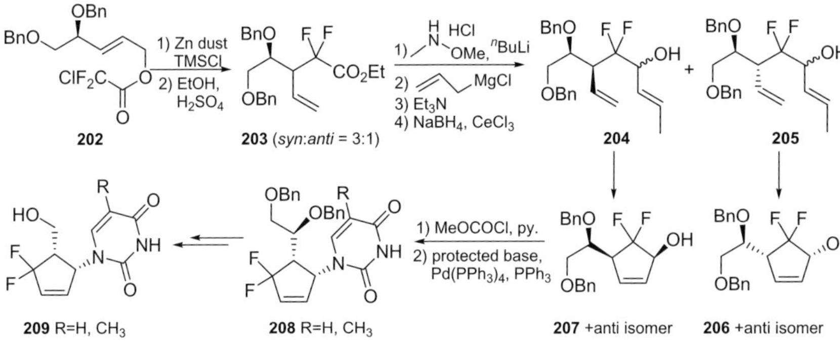

Scheme 8.37.

building block was obtained via silicon-induced Reformastsky–Claisen reaction of the chiral ester **202**. The rearrangement product **203** was then converted to the separable RCM precursors **204** and **205**, which were subjected to RCM reaction conditions to give the carbocyclic alcohol **206** and **207** together with their anti isomers. Exposure of intermediate **207** to MeOCOCl/pyridine produced the corresponding allylic carbonate which reacted with 3′-benzoyl thymine(uracil) in the presence of $Pd(PPh_3)_4$ to yield γ-substituted compound product **208** exclusively. Interestingly, the outcome of the regioselectivity was totally different to that of non-fluorinated substrates, which is probably due to the strong electron-withdrawing inductive effect of the CF_2 group.

The success of (−)-2′,3′-deoxy-3′-thiacytidine (3TC) (Doong *et al.*, 1991; Chang *et al.*, 1992) and (+)-2′,3′-deoxy-3′-oxacytidine (L-OddC) (Grove *et al.*, 1995) stimulated Qing and coworkers to synthesize a series of 2′,3′-dideoxy-6′,6′-difluoro 3′-thia- (Wu *et al.*, 2004), 3′-aza- (Yue *et al.*, 2008) and 2′-thia (Zheng *et al.*, 2009) nucleosides based on the bioisosteric rationale. Several versatile difluorinated building blocks were efficiently constructed for this purpose. During the synthesis of compounds **217** (Scheme 8.38) and **222** (Scheme 8.39), the basic azide group (N_3) was introduced into the neighbouring position of the CF_2 group without any dehydrofluorination, which was further reduced to amine to construct the nucleobase or to form the azasugar moiety. For the synthesis of 2′-thianucleosides, an improved Reformatsky–Claisen rearragement of secondary allyl chlorodifluoroacetate **223** was developed, which gave the key building block **224** stereoselectively (Scheme 8.40). From the same intermediate **224**, 2′-thianucleosides **226** and their enantiomer **229**, were prepared via similar routes.

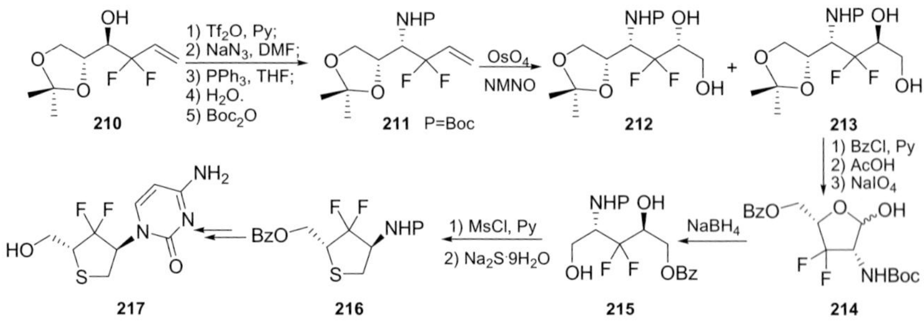

Scheme 8.38.

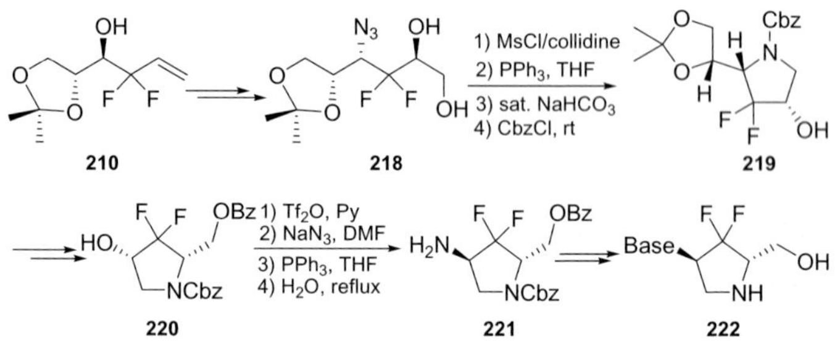

Scheme 8.39.

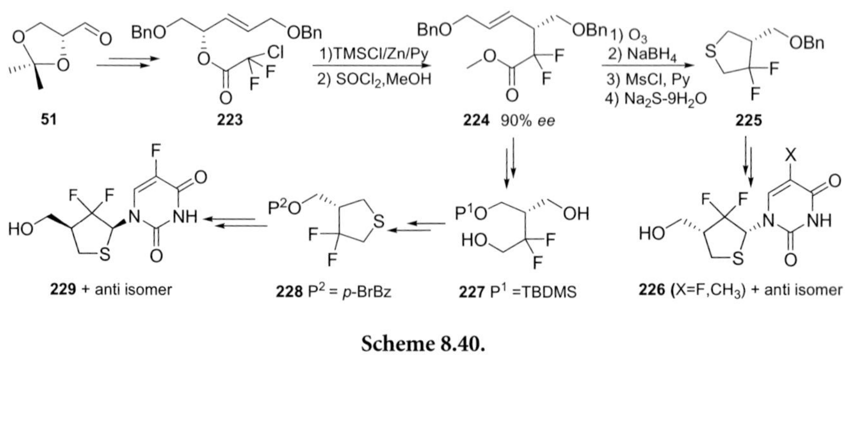

Scheme 8.40.

8.6 5′-Fluorinated and Phosphonodifluoromethylenated Nucleosides

Nucleosides bearing fluorine(s) at C5′ were designed and synthesized mainly for eliminating the possibility of phosphorylation of these nucleosides to the corresponding mono-, di- and triphosphates in cells. Next these 5′-deoxy-5′-fluoro compounds were investigated to explore if any would show activity that would not be dependent on their conversion into the corresponding nucleotides. Usually, these compounds were prepared either by condensation of suitable 5′-fluoro sugar moiety with an appropriate nucleobase (Sharma *et al.*, 1995) or by direct fluorination of nucleosides at C5′. A variety of methods were applied, such as nucleophilic displacement of mesylates (tosylates) with KF or tetrabutylammonium fluoride, as well as direct displacement with DAST (Herdewijn *et al.*, 1989).

Based on the enzymatic pathway for the conversion of SAH to adenosine, McCarthy *et al.* (1989) designed, synthesized and biologically evaluated a novel class of mechanism-based inhibitors of SAH hydrolase, 4′,5′-unsaturated-5′-fluoroadenosine nucleosides **234** and **235** (Scheme 8.41). The reaction of sulfoxide **231** with DAST provided the corresponding α-fluoro thioether via Pummerer rearrangement, which was further oxidized to the α-fluoro sulfoxide **232**. The intermediate **232** was transformed to the terminal vinyl fluoride **233** by thermolysis. Both **234** and **235** were effective inhibitors of the replication of the Moloney leukaemia virus *in vitro*.

More challenging was a replacement of the oxygen of the 5′-hydroxyl function with an isopolar and isosteric CF_2 group in order to synthesize difluoromethylene phosphonate nucleotides, e.g. to make the $-CH_2-CF_2-P-$ linkage a good mimic of the $-CH_2-O-P-$ moiety of natural nucleotides. Groups such as CHF and CF_2 have previously been incorporated in place

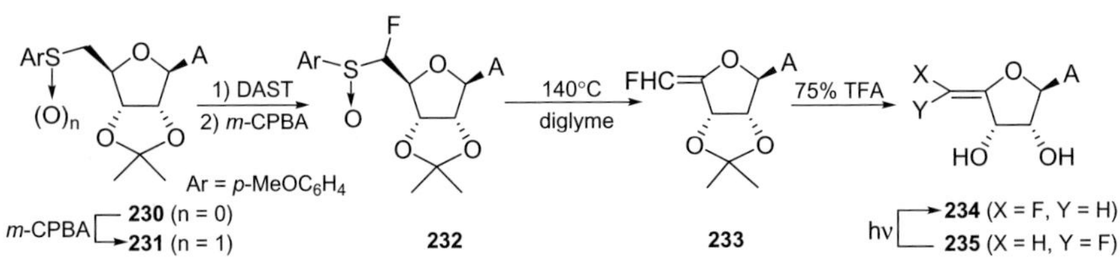

Scheme 8.41.

of 3′- or 5′-oxygens of nucleosides or as replacement for bridging oxygens in the corresponding di- and triphosphates. Fluorinated phosphonates play an important role as antiviral agents, biomedical agents, potential enzyme inhibitors and are useful probes for elucidation of biochemical processes (Romanenko *et al.*, 2006).

A general method for the synthesis of 5′-difluoromethylenephosphonates was developed by Matulic-Adamicetal *et al.* (1995). Treatment of the 5′-aldehyde **237** with $LiCF_2P(O)(OEt)_2$ followed by radical deoxygenation or conversion of the 5′-aldehyde **237** to the corresponding triflate derivative followed by reaction with $LiCF_2P(O)(OEt)_2$ afforded the key 5′-difluoromethylenated intermediate **238** (Scheme 8.42).

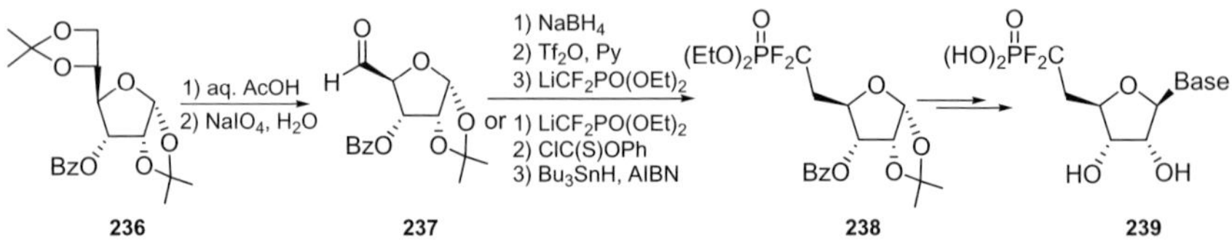

Scheme 8.42.

In 1998, Zard's group (Boivin *et al.*, 1998) utilized an expedient radical based approach to realize the synthesis of a 5′-difluorophosphonate analogue of thionucleosides. In their synthesis, the key radical addition took place upon the heating of olefin **240** and xanthate **241** with lauroyl peroxide as the initiator, and the desired product **242** was afforded in 60% yield (Scheme 8.43).

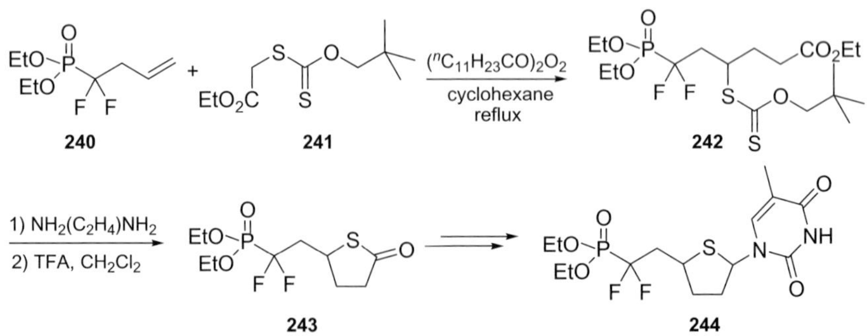

Scheme 8.43.

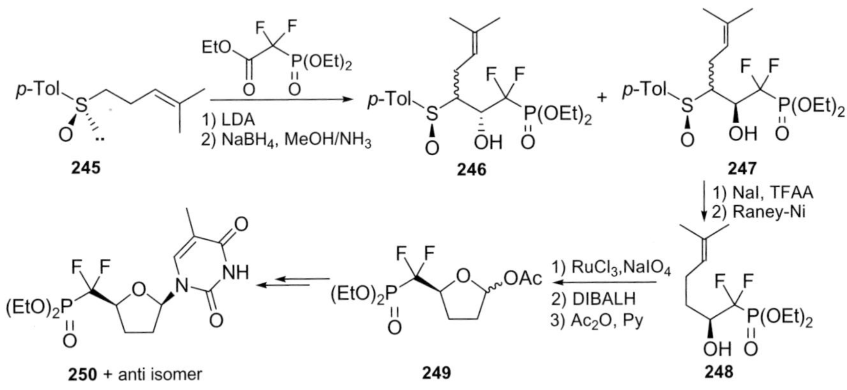

Scheme 8.44.

Bravo and coworkers (Arnone *et al.*, 1999) reported the synthesis of the enantiomerically pure 2′,3′,5′-trideoxy-5′-phosphono-5′,5′-difluoro thymidines via a building block approach (Scheme 8.44). The key steps of the synthetic sequence was the condensation of 2′-methyl-5′-(4′-methyl-phenylsulfinyl)- pent-2-ene **245** and ethyl 2′-(diethoxyphosphoryl)-2′,2′-difluoroacetate. Reduction of the resulting ketone to alcohols **246** and **247**, reductive removal of the sulfur moiety to hydroxyl phosphonate **248**, and oxidative cyclization gave furanose derivatives **249**.

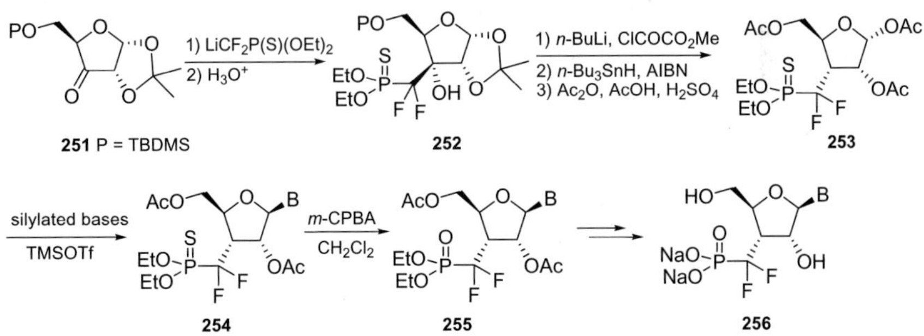

Scheme 8.45.

The first synthesis of 3′-difluorophosphonate analogues of nucleosides was described by Piettre and coworkers (Lopin *et al.*, 2002). The key steps of their synthesis involved the stereoselective addition of the lithium salt of difluoromethylphosphonothioate to the readily available ketone **251** and the conversion of the P=S bond of the phosphonothioate **254** to a P=O bond through oxidation with *m*-CPBA (Scheme 8.45). It is noteworthy that sulfur played a crucial role in the introduction of the phosphorus-substituted difluoromethylene unit onto the furanose ring.

8.7 Nucleosides Bearing Exocyclic Fluorocarbon Substituents at C2′, C3′ and C4′

Introduction of trifluoromethyl (CF_3), difluoromethyl (CF_2H), difluoromethylene (CF_2) or fluoromethylene (CHF) groups into the sugar moiety of nucleosides is another plausible strategy to discover new biologically active nucleoside analogues. These groups are favourite structural features used to induce both lipophilicity and metabolic stability.

8.7.1 *Nucleosides containing a trifluoromethyl group*

Nucleophilic addition of a CF_3 carbanion, generated *in situ* from $TMSCF_3$/TBAF, to a suitable keto sugar intermediate is a general method for introducing a CF_3 group into nucleosides. For example, Piccirilli′s group (Li *et al.*, 2001) reported the first synthesis of 2′-C-trifluoromethyl pyrimidine ribonucleosides **260** (Scheme 8.46). The key intermediate **258** was prepared from ribofuranose **257** in three steps including the trifluoromethylation using $TMSCF_3$/TBAF. Because of the *cis* relationship between the heterocycle and the large electron-withdrawing CF_3, Hilbert–Johnson glycosylation reaction of **258** with pyrimidines failed. Thus compound **258** was converted to ribofuranosyl bromide **259**, which was coupled with silylated bases in the presence of HgO/$HgBr_2$ to afford exclusively the β-anomers **260**.

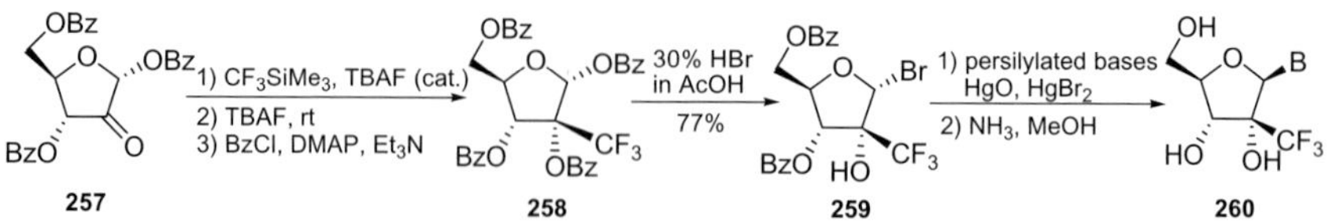

Scheme 8.46.

Interestingly, the synthesis of the 2′-trifluoromethyl-2′,3′-dideoxyuridines **263** and **264** was accidentally developed by Serafinowski and Brown (2000). They found that treatment of 2′-deoxy-2′-difluoromethylidene-uridine **261** in the presence of TBAF in THF gave the 2′,3′-unsaturated-2′-trifluoromethyl-uridine **262** in moderate yield (Scheme 8.47).

Qing's group presented another efficient synthetic route to D-2′,3′-dideoxy-2′-trifluoromethyl nucleosides (Zhang *et al.*, 2000) as well as L-2′,3′-dideoxy-2′-trifluoromethyl-4′-thiocytidines (Zhang *et al.*, 2002). The key building block, α-trifluoromethyl-α,β-unsaturated ester **266**, was obtained by the treatment of the α-bromo-α,β-unsaturated esters **265** with $FSO_2CF_2CO_2Me$/CuI in DMF/HMPA (Scheme 8.48).

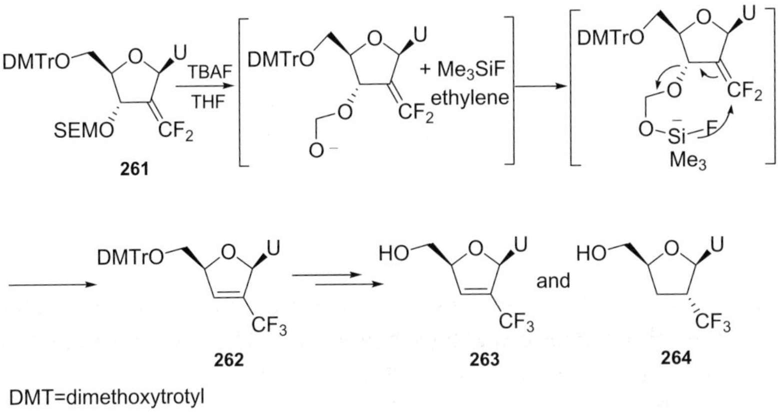

Scheme 8.47.

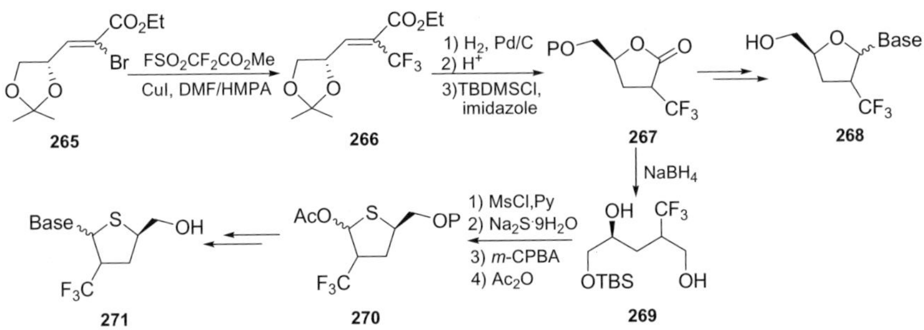

Scheme 8.48.

8.7.2 *Nucleosides containing a difluoromethylene, fluoromethylene or difluoromethyl group*

Introduction of a difluoromethylene or difluoromethyl group into the sugar moiety of nucleosides was also accomplished via a nucleophilic addition of a fluorine-containing reagent to a carbonyl of a suitable sugar derivative. Using this strategy, McCarthy and coworkers accomplished the synthesis of 2′-deoxy-2′-difluoromethylidene-cytidine **275** (Sabol *et al.*, 1992). They effectively built up the difluoromethylene group at C2′ via nucleophilic addition

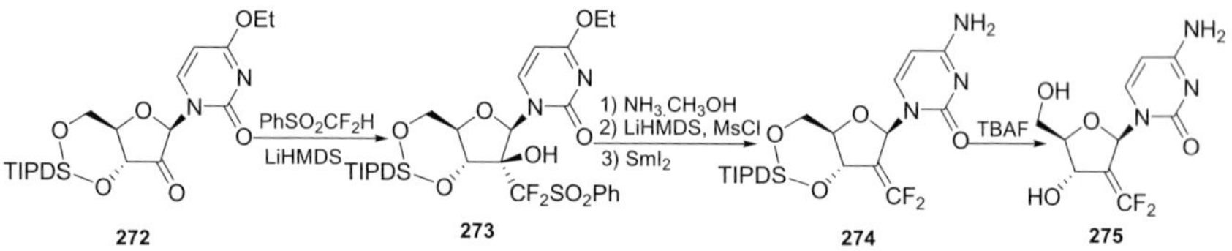

Scheme 8.49.

of difluoromethylphenylsulfone to 2′-ketocytidine **272** and subsequent reductive elimination using SmI_2 (Scheme 8.49). By choosing a different reducing agent, Na/Hg/H_2, Piccirilli's group (Ye *et al.*, 2005) converted the difluoromethylphenylsulfone to a difluoromethyl group and thus developed an efficient synthetic route to 2′-C-β-difluoromethyl ribonucleosides **279** (Scheme 8.50).

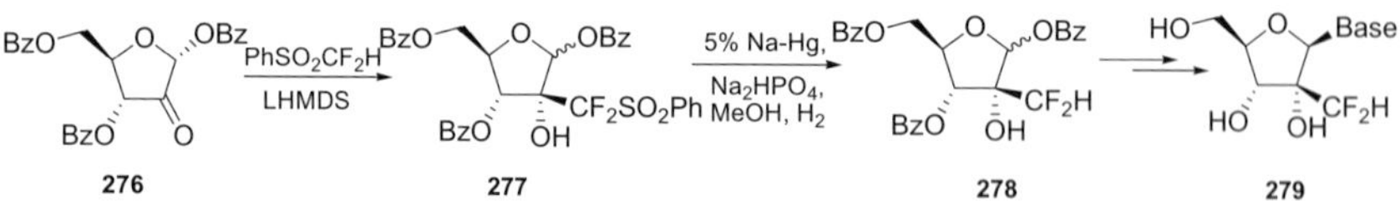

Scheme 8.50.

Besides the above-mentioned nucleophilic addition method, the Wittig-type olefination is another common method to introduce a fluoromethylene group into a sugar that bears a carbonyl group. McCarthy *et al.* (1991) reported a stereospecific method to *E* and *Z* terminal fluoro olefins and its application to the synthesis of 2′-deoxy-2′-fluoromethylidene nucleosides (Scheme 8.51). Using the Horner–Wittig

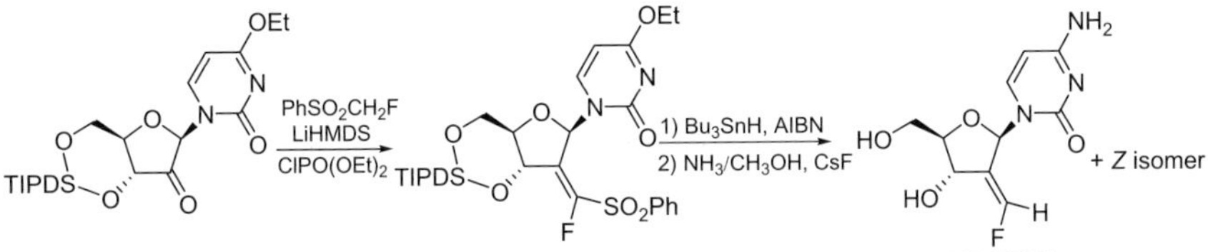

280

281 E:Z = 9.5 : 1

282 E-FMC

Scheme 8.51

reaction, 2′-ketonucleoside **280** was converted to a mixture of readily separable *E* and *Z* fluorovinyl sulfones **281**. Both of the isomers were transformed to (fluorovinyl)stannanes with retention of configuration using tributyltin hydride. The desired product **282** and its *Z* isomer were obtained directly from the (fluorovinyl)stannanes. Fluoroolefin nucleoside **282** was shown to be a potent cytotoxic agent against leukaemia and solid tumours, whereas its *Z* geometric isomer was substantially less active.

Using a similar Wittig-type olefination approach, Qiu and Qing (2004, 2005) carried out the synthesis of 2′,3′,4′-trideoxy-2′-difluoromethyl-4′-aza-nucleosides **287** and 3′,4′-dideoxy-3′-difluoromethyl-4′-aza-nucleosides **290** and **291**, starting with the same natural amino acid **283** (Scheme 8.52). Using different sequences, ketones **284** and **288**, were prepared and subjected to Wittig-type olefination conditions with CF_2Br_2/HMPT/Zn to afford the key fluorinated intermediates **285** and **289**, respectively.

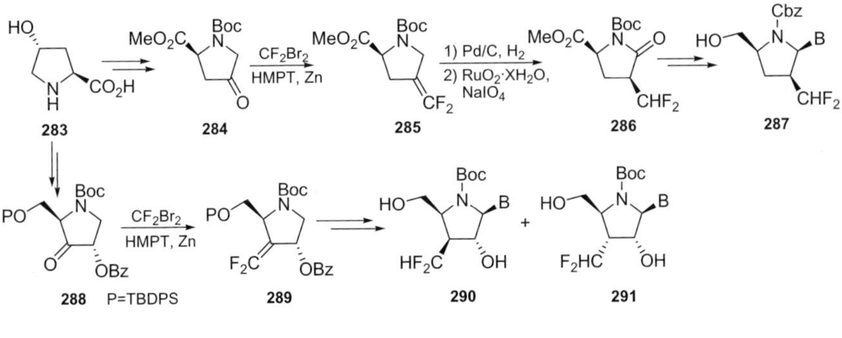

283 284 285 286 287

288 P=TBDPS 289 290 291

Scheme 8.52.

8.8 Other Fluorinated Nucleosides

Since the discovery of naturally occurring oxetanocin A, **292** (Hoshino *et al.*, 1987; Nishiyama *et al.*, 1988; Sakuma *et al.*, 1991) (Fig. 8.3), which exhibits high antiviral activity, nucleoside analogues with rigid and small sugar rings have received a great deal of attention. The carbocyclic analogue of oxetanocin A, cyclobut-G (BHCG, **293**) (Nishiyama *et al.*, 1989; Norbeck *et al.*, 1990; Bisacchi *et al.*, 1991) was demonstrated to be a highly potent inhibitor of a broad spectrum of the herpes viruses including the herpes simplex virus types 1 and 2 (HSV1, HSV2), VZV and human cytomegalovirus (HCMV). In addition, the cyclopropane analogue **294** (Sekiyama *et al.*, 1998) also exhibited extremely potent antiviral activity against HSV1 with good selectivity. Thus a number of fluorinated cyclopropyl, cyclobutyl and oxetanyl nucleoside analogues have been synthesized and structure–activity relationship studies of this class of analogues are under active investigation.

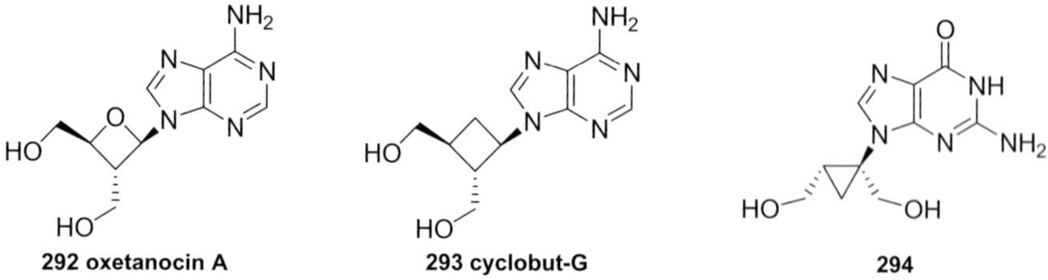

Figure 8.3. Some potent antiviral agents with rigid and small sugar ring analogues.

8.8.1 *Fluorinated cyclopropyl nucleosides*

In 2001, Kim and coworkers (Lee *et al.*, 2001) reported the synthesis and evaluation of fluorocyclopropanoid nucleosides (±)-**298** as potential antiviral agents. Introduction of a fluorine and the double bond for the installation of the cyclopropyl group was achieved by HWE reaction of the aldehyde **295** with triethyl 2-fluoro-2-phosphonoacetate (Scheme 8.53).

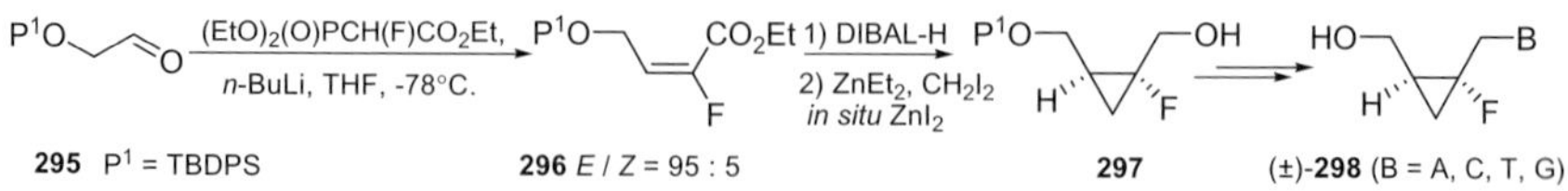

Scheme 8.53.

The formed α-fluoro-α,β-unsaturated ester **296** was reduced with DIBAL-H and the resulting allylic alcohol was subjected to ZnI_2-catalytic cyclopropanation to provide the key cyclopropane derivative **297**.

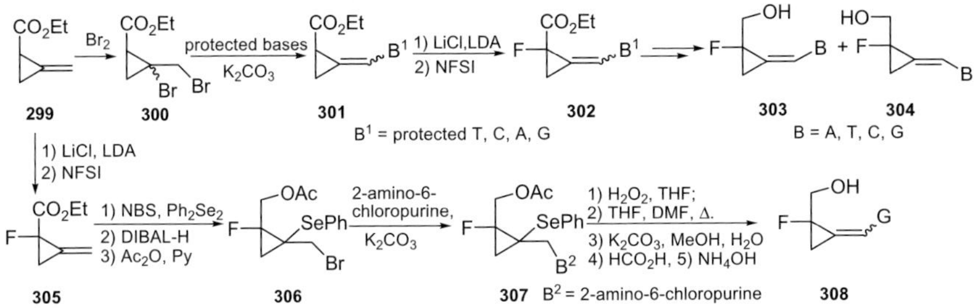

Scheme 8.54.

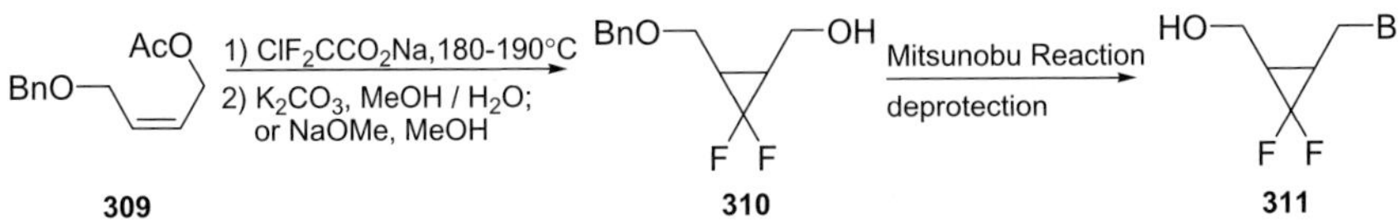

Scheme 8.55.

In 2004, Zemlicka's group (Zhou *et al.*, 2004a) described a structurally novel class of nucleoside analogues based on a methylenecyclopropane template. The purine (*Z,E*)-methylenecyclopropane carboxylates **301** was selectively fluorinated using lithium diisopropylamide, LiCl and NFSI to give (*Z,E*)-fluoroesters **302**. Reduction of **302** with $LiBH_4$ or DIBAL-H followed by a base transformation gave the *Z*-isomers **303** and *E*-isomers **304** (Scheme 8.54). A new method for the synthesis of fluoromethylenecyclopropanenucleosides, such as **308**, by an alkylation-elimination procedure, was also developed by this group (Zhou *et al.*, 2005). Most of these nucleosides showed potent antiviral activity *in vitro*.

In 1998, Csuk *et al.* reported the first synthesis of the difluorocyclopropyl homonucleoside analogue **311** via a very efficient way (Scheme 8.55). Addition of the difluorocarbene generated from sodium chlorodifluoroacetate to substrate **309** afforded the key intermediate **310**.

8.8.2 *Fluorinated cyclobutyl and oxetanosyl nucleosides*

DAST-mediated fluorination is also a common way to introduce fluorine atoms into a cyclobutyl or oxetanyl derivatives. Fluorinated oxetanocin analogue **312** (Wang *et al.*, 1991) and fluorinated cyclobut-A analogues **313** and **314** were synthesized using this approach. Unfortunately, neither of these compounds was found to be active against HIV (Fig. 8.4).

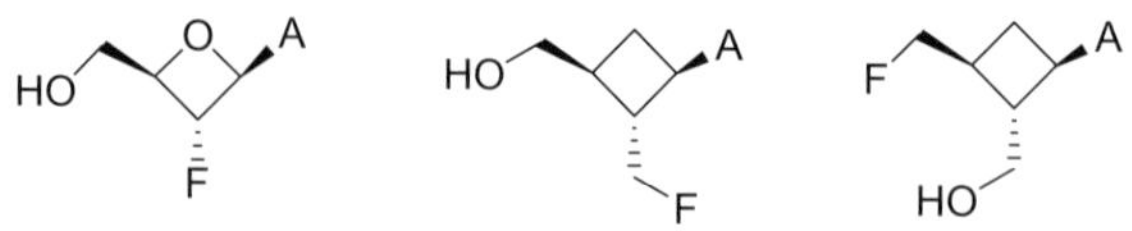

Figure 8.4. Fluorinated oxetanocin and cyclobut-A analogues.

In 2007, Liotta and coworkers (Li *et al.*, 2007) reported the synthesis of 2′-fluoro cyclobutyl nucleoside **318** using fluorination of the silyl enol ether **315** with Selectfluor as the key step (Scheme 8.56). Unfortunately, compound **318** showed no significant anti-HIV activity up to 100 μM in primary human lymphocytes infected with HIV-1.

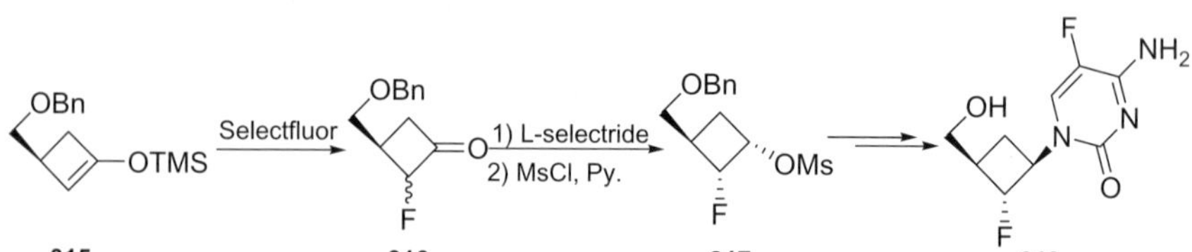

Scheme 8.56.

8.8.3 *Fluorinated pyranosyl nucleosides*

Nucleosides with a six-membered carbohydrate moiety have been evaluated for their potential antiviral and antibiotic properties, and as building blocks in nucleic acid synthesis. General methods for the introduction of fluorine into the pyranosyl sugar moiety involve either nucleophilic DAST-mediated or electrophilic N–F reagent based fluorination. For example, using the former approach, 4′-deoxy-4′-fluoropyranosylnucleoside **320** (Leclercq *et al.*, 1989) was prepared from nucleoside derivative **319**; the latter strategy was

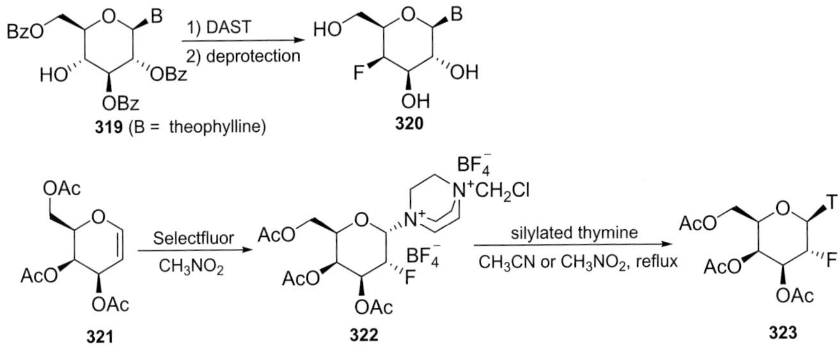

Scheme 8.57.

applied to the synthesis of 2′-deoxy-2′-fluoro-pyranosyl-thymine derivative **323** (Albert *et al.*, 1998), which highlighted the regioselective addition of Selectfluor to D-galactal derivative **321** (Scheme 8.57).

In view of the high bioactivities of gemcitabine against cancer cells, Fernández and Castillón (1999) synthesized 2′,3′-dideoxy-3′,3′-difluoro- and 2′,3′-dideoxy-2′,2′-difluoro-pyranosyl-nucleosides, analogues of gemcitabine (Scheme 8.58). D-Mannose and D-glucose were converted to the protected ketones **324** and **328**, respectively. After *gem*-difluorination of the ketones **324** and **328** with DAST, the resulting difluorinated derivatives **325** and **329** were subjected to glycosylation followed by deprotection to give the desired nucleosides **327** and **330** respectively.

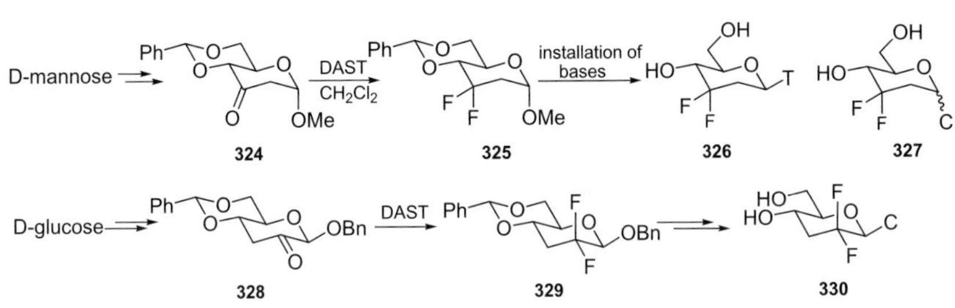

Scheme 8.58.

Recently, Qing and coworkers developed a novel synthetic route to 2′,3′,4′-trideoxy-4′,4′-difluoro-β-D-glucopyranosyl adenine **339** and

340 and their 2′,3′-unsatured analogues 337 and 338 (Xu *et al.*, 2009). The approach highlighted the highly regio- and stereoselective Pd-catalysed glycosylation of Boc-protected pyranose 334 prepared by the oxidation–cyclization of difluorinated diol 332 (Scheme 8.59).

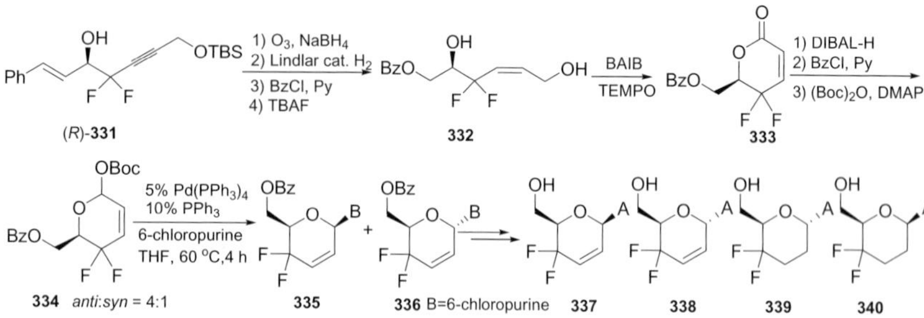

BAIB: bisacetoxyiodobenzene

Scheme 8.59.

8.9 Conformational Studies of Fluorinated Nucleosides

The potential biological activity of nucleosides is believed to depend strongly on the conformation of the sugar and nucleobase moieties. Even minor changes on sugar and nucleobase moieties can alter conformational preferences of the nucleosides and thereby profoundly impact their biological activity. The structural features (Barchi *et al.*, 2008) that may be altered by the presence of fluorine are those that define the nucleosides core structure: (i) the conformation of the furanose ring as defined by the concept of pseudorotation (Altona *et al.*, 1972) (Fig. 8.5, centre) which is described by the two pseudorotation parameters P (the phase angle of pseudorotation) and ν_{max}. (Figure 8.5(B), the maximum out-of-plane pucker of the ring corresponding to the radius of the circle); (ii) the glycosyl rotamer angle χ (Fig. 8.5(A), (C)) and (iii) the C4′–C5′ bond rotamer angle γ (Fig. 8.5(a), (D)). Altona and coworkers (Haasnoot *et al.*, 1980), Chattopadhyaya and coworkers (Thibaudeau *et al.*, 1998a), Marquez and coworkers (Sun *et al.*, 2004) and others have refined and added additional parameterization to the fundamental approach of Altona and Sundaralingham (1973) for the analysis of predominant conformer populations of the nucleosides. For the fluorinated sugar nucleosides, a

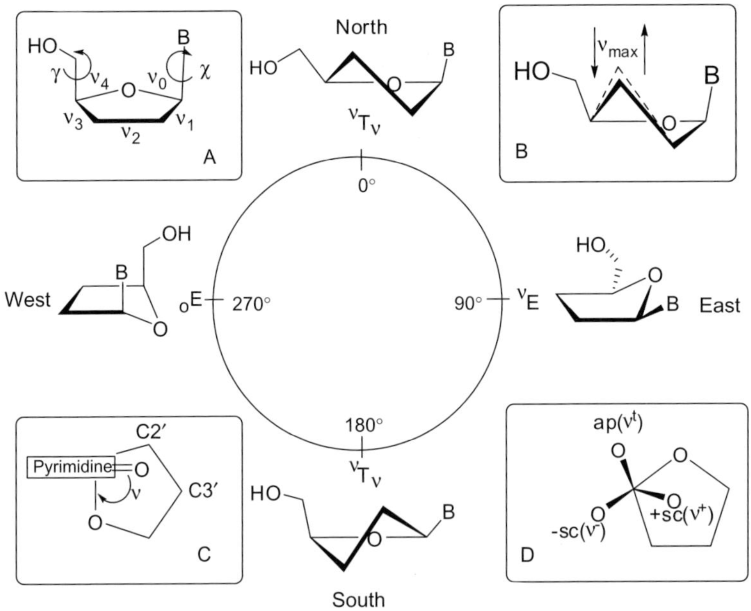

Figure 8.5. Structural parameters for nucleosides. The pseudo-rotational cycle (centre) describes the ring pucker of the furanose system where the endocyclic torsion angles (shown as ν_o–ν_4 in (A)) are varied systematically around a 'wheel' of twist and envelope conformations. The maximum out-of-plane pucker [ν_{max} in (B)] indicates the degree to which a specific atom in the five-membered ring will flex from the plane of the five atoms. Angles χ (C) and γ (D) describe the rotamer distribution about the anomeric and C4′–C5′ bonds, respectively (Barchi *et al.*, 2008).

new complicated seven-parameter Karplus relationship based on J_{HF} coupling constants was derived (Thibaudeau *et al.*, 1998b) and later modified by Mikhailopulo *et al.* (2003).

Many of the conformational adjustments imparted by substitution of electronegative atoms (or groups) into the carbohydrate ring have been rationalized based on arguments relating to the anomeric (Thibaudeau *et al.*, 1998b) (Fig. 8.6(a)) and the *gauche* effects (GE) (Amos *et al.*, 1992; Abraham *et al.*, 1994) (Fig. 8.6(c)). Recent reports studying difluorinated structures have invoked an alternative theory by Brunck and Weinhold (1979) which suggests that vicinal electronegative atoms prefer a *gauche* arrangement due to an antiperiplanar (*app*) σ to σ^* conjugative stabilization

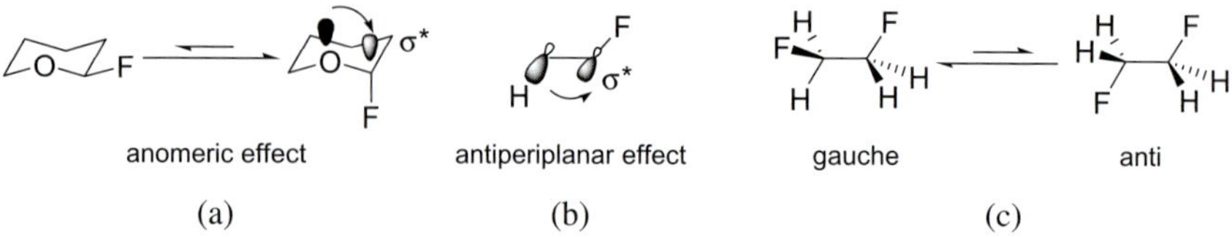

Figure 8.6. The anomeric effect involving n–$\sigma^*_{C\text{-}F}$ interaction (a), antiperiplanar effect involving $\sigma_{C\text{-}H}$–$\sigma^*_{C\text{-}F}$ interaction (b), and preferred *gauche* orientation of difluoroethane (c).

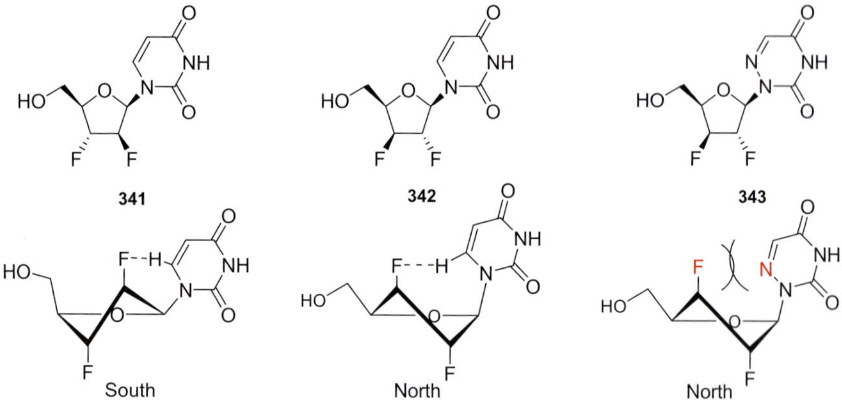

Figure 8.7. Structural studies of 2′,3′-difluoro uridine nucleosides. All the X-ray crystallography, NMR spectroscopy and *ab initio* calculations data demonstrated that the sugar pucker for all optimized structures had a tendency for the fluorine atoms to remain pseudoaxial with compound **341** in the 'Southern' hemisphere and **342** and **343** in the 'Northern' hemisphere. F–H interaction is shown with a dashed line.

(Fig. 8.6(b)) when the donating bond is the least polar one and the acceptor orbital is the most polarized bond.

Barchi *et al.* (2008) reported the comprehensive conformational studies of three 2′,3′-difluoro uridine nucleosides, **341**, **342** and **343** by X-ray crystallography, NMR spectroscopy and *ab initio* calculations (Fig. 8.7). They showed that the F–O4′ *gauche* effect along with two strong H′–O4′ *app* effects (H2′–O4′ and H3′–O4′) are able to afford sugar puckers that agree with what would be predicted with two *trans*-disposed fluorine atoms, namely the two electronegative fluorines are pseudoaxial. The weak interaction between the *endo* fluorine F2′ in compound **341** or F3′ in compound **342**, and H6 may contribute additional stabilization to the preferred structure and hold the χ angle in a high *anti*-rotamer population. The influence of the 6-azauracil base of compound **343** is evident from the calculated and experimental structures, and the F3′–N6 repulsion gives rise to the remarkable population of the 'South' pseudo-rotamer.

Recently, Qing′s group carried out the conformational analysis of D-2′-deoxy-2′,2′-difluoro-4′-dihydro-4′-thiouridine **344**. Their analysis suggested that crystal packing forces (hydrogen bond OH···O=C, dipole–dipole interaction C–F···C=O) allow the thiouridine to adopt the somewhat less stable 'Northern' conformation in the solid state (Fig. 8.8). In contrast, the antiperiplanar preference of C–H and C–C σ-bonds to σ^*C–F and σ^*C–O seemed to be responsible for the favoured 'Southern' conformation in solution, and the weak dipolar attractions between $^{\delta+}$C2–F$\beta^{\delta-}$ and $^{\delta+}$H6–C6$^{\delta-}$ may contribute further stabilization to the preferred structure and keep the base moiety in a high *anti*-rotamer population in solution (Fig. 8.9).

As deduced from X-ray crystallography, compound **344** exists in the solid state as a 'Northern'-type thiosugar pucker with *anti* conformation of the base, whereas from the NMR data and the high-level calculations it follows that the predominant conformer in methanol solution is of 'Southern' type with *anti* conformation of the base. The potential antiperiplanar

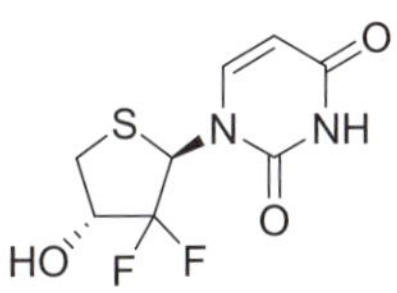

344

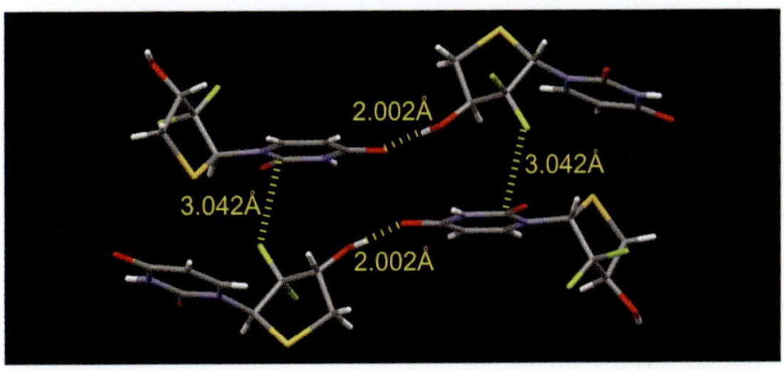

Figure 8.8. Packing patterns of **344**, which exhibits important intermolecular hydrogen bonding and F···C=O interaction.

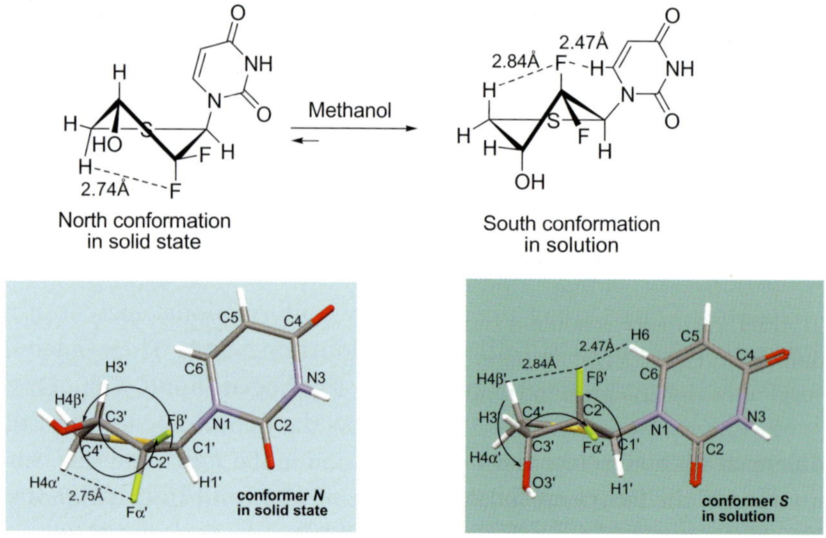

Figure 8.9. Conformational analysis of D-2′-deoxy-2′,2′-difluoro-4′-dihydro-4′-thionucleoside **344**.

effects and through-space F···H interaction are depicted by arrows and dashed lines respectively.

8.10 Conclusion

Fluorinated nucleosides make up an important part of the extensively studied field of nucleoside analogues, which are an important class of candidates for antiviral and antitumour agents. With the development of even more efficient fluorination methodologies and further enhancement of knowledge about structure–activity relationships, fluorinated nucleosides will continue to be explored, holding great promises for novel antiviral and antitumour agents.

References

Abraham, R.J., Chambers, E.J. & Thomas, W.A. (1994) Conformational analysis. Part 22. An NMR and theoretical investigation of the *gauche* effect in fluoroethanols. *J Chem Soc Perkin Trans 2*, 949–955.

Agrawal, S. & Zhao, Q. (1998) Antisense therapeutics. *Curr Opin Chem Biol* 2, 519–528.

Albert, M., Dax, K. & Ortner, J. (1998) A novel direct route to 2-deoxy-2-fluoro-aldoses and their corresponding derivatives. *Tetrahedron* 54, 4839–4848.

Altona, C. & Sundaralingam, M. (1972) Conformational analysis of the sugar ring in nucleosides and nucleotides: new description using the concept of pseudorotation. *J Am Chem Soc* 94, 8205–8212.

Altona, C. & Sundaralingam, M. (1973) Conformational analysis of the sugar ring in nucleosides and nucleotides: improved method for the interpretation of proton magnetic resonance coupling constants. *J Am Chem Soc* 95, 2333–2344.

Amos, R.D., Handy, N.C., Jones, P.G. *et al.* (1992) Bond length and reactivity: the *gauche* effect. A combined crystallographic and theoretical investigation of the effects of β-substituents on C–OX bond length. *J Chem Soc Perkin Trans 2*, 549–558.

Arnone, A., Bravo, P., Frigerio, M. *et al.* (1999) Synthesis of enantiomerically pure 2′,3′,5′-trideoxy-4′-[(diethoxyphosphoryl)difluoromethyl]thymidine analogues. *Eur J Org Chem* 1999, 2149–2157.

Baldwin, S.A., Mackey, J.R., Cass, C.E. *et al.* (1999) Nucleoside transporters: molecular biology and implications for therapeutic development. *Mol Med Today* 5, 216–224.

Balzarini, J., Baba, M., Pauwels, R. *et al.* (1988) Anti-retrovirus activity of 3′-fluoro- and 3′-azido-substituted pyrimidine 2′,3′-dideoxynucleoside analogues. *Biochem Pharmacol* 37, 2847–2856.

Barchi, J.J., Marquez, V.E., Driscoll, J.S. *et al.* (1991) Potential anti-AIDS drugs: lipophilic, adenosine deaminase-activated prodrugs. *J Med Chem* 34, 1647–1655.

Barchi, J.J., Karki, R.G., Nicklaus, M.C. *et al.* (2008) Comprehensive structural studies of 2′,3′-difluorinated nucleosides: comparison of theory, solution, and solid state. *J Am Chem Soc* 130, 9048–9057.

Bhattacharya, B.K., Ojwang, J.O., Rando, R.F. *et al.* (1995) Synthesis and anti-DNA viral activities *in vitro* of certain 2,4-disubstituted-7-(2-deoxy-2-fluoro-β-D-arabinofuranosyl)pyrrolo[2,3-D]pyrimidine nucleosides. *J Med Chem* 38, 3957–3966.

Biffinger, J.C., Kim, H.W. & DiMagno, S.G. (2004) The polar hydrophobicity of fluorinated compounds. *ChemBioChem* 5, 622–627.

Biggadike, K., Borthwick, A.D., Exall, A.M. *et al.* (1987) Synthesis of fluorinated carbocyclic nucleosides: preparation of carbocyclic 1-(2′-deoxy-6′-fluororibofuranosyl)-5-iodouracils. *J Chem Soc Chem Commun* 4, 255–256.

Bisacchi, G.S., Braitman, A., Cianci, C.W. *et al.* (1991) Synthesis and antiviral activity of enantiomeric forms of cyclobutyl nucleoside analogs. *J Med Chem* 34, 1415–1421.

Blackburn, G.M., England, D.A. & Kolkmann, F. (1981) Monofluoro- and difluoromethylenebisphosphonic acids: isopolar analogues of pyrophosphoric acid. *J Chem Soc Chem Commun* 17, 930–932.

Boivin, J., Ramos, L. & Zard, S.Z. (1998) An expedient radical based approach to difluorophosphonate analogues of thionucleosides. *Tetrahedron Lett* 39, 6877–6880.

Borchardt, R.T., Keller, B.T. & Patel-Thombre, U. (1984) Neplanocin A: a potent inhibitor of *S*-adenosylhomocysteine hydrolase and of vaccinia virus multiplication in mouse L929 cells. *J Biol Chem* 259, 4353–4358.

Brunck, T.K. & Weinhold, F. (1979) Quantum-mechanical studies on the origin of barriers to internal rotation about single bonds. *J Am Chem Soc* 101, 1700–1709.

Chang, C.N., Doong, S.L., Zhou, J.H. *et al.* (1992) Deoxycytidine deaminase-resistant stereoisomer is the active form of (+/–)-2′,3′-dideoxy-3′-thiacytidine in the inhibition of hepatitis B virus replication. *J Biol Chem* 267, 13938–13942.

Chen, X., Zhou, W., Schinazi, R.F. *et al.* (2004) Synthesis of 3′-Fluoro-2′,3′-dideoxy-2′,3′-didehydro-4′-ethynyl-D- and -L-furanosyl nucleosides. *J Org Chem* 69, 6034–6041.

Cheng, D.M., Kan, L.-S., Ts'o, P.O.P. *et al.* (1983) Multinuclear magnetic resonance studies of monomers and diemers containing 2′-fluoro-2′-deoxyadenosine. *Biopolymers* 22, 1427–1444.

Choi, Y., Choo, H., Chong, Y. *et al.* (2002) Synthesis and potent anti-HIV activity of L-2′,3′-didehydro-2′,3′-dideoxy-2′-fluoro-4′-thiocytidine. *Org Lett* 4, 305–307.

Chong, Y., Choo, H., Choi, Y. *et al.* (2002) Stereoselective synthesis and antiviral activity of D-2′,3′-didehydro-2′,3′-dideoxy-2′-fluoro-4′-thionucleosides. *J Med Chem* 45, 4888–4898.

Chong, Y., Gumina, G., Mathew, J.S. *et al.* (2003) L-2′,3′-Didehydro-2′,3′-dideoxy-3′-fluoronucleosides: synthesis, anti-HIV activity, chemical and enzymatic stability, and mechanism of resistance. *J Med Chem* 46, 3245–3256.

Choo, H., Chong, Y., Choi, Y. *et al.* (2003) Synthesis, anti-HIV activity, and molecular mechanism of drug resistance of L-2′,3′-didehydro-2′,3′-dideoxy-2′-fluoro-4′-thionucleosides. *J Med Chem* 46, 389–398.

Chu, C.K., Matulic-Adamic, J., Huang, J.-T. *et al.* (1989) Nucleosides. CXXXV. Synthesis of some 9-(2-deoxy-2-fluoro-β-D-arabinofuranosyl)-9*H*-purines and their biological activities. *Chem Pharm Bull* 37, 336–339.

Chu, C.K., Ma, T., Shanmuganathan, K. *et al.* (1995) Use of 2′-fluoro-5-methyl-beta-L-arabinofuranosyluracil as a novel antiviral agent for hepatitis B virus and Epstein–Barr virus. *Antimicrob Agents Chemother* 39, 979–981.

Chun, B.K., Schinazi, R.F., Cheng, Y.-C. *et al.* (2000) Synthesis of 2′,3′-dideoxy-3′-fluoro-ribonucleosides as potential antiviral agents from D-sorbitol. *Carbohydr Res* 328, 49–59.

Clark, J.L., Hollecker, L., Mason, J.C. *et al.* (2005) Design, synthesis, and antiviral activity of 2′-deoxy-2′-fluoro-2′-C-methylcytidine, a potent inhibitor of hepatitis C virus replication. *J Med Chem* 48, 5504–5508.

Clark, J.L., Mason, J.C., Hollecker, L. *et al.* (2006) Synthesis and antiviral activity of 2′-deoxy-2′-fluoro-2′-C-methyl purine nucleosides as inhibitors of hepatitis C virus RNA replication. *Bioorg Med Chem Lett* 16, 1712–1715.

Codington, J.F., Doerr, I.L., Van Praag, D. *et al.* (1961) Nucleosides XIV: synthesis of 2′-deoxy-2′-fluorouridine1. *J Am Chem Soc* 83, 5030–5031.

Codington, J.F., Doerr, I.L. & Fox, J.J. (1964) Nucleosides XVIII: synthesis of 2′-fluorothymidine, 2′-fluorodeoxyuridine, and other 2′-halogeno-2′-deoxy nucleosides[1,2]. *J Org Chem* 29, 558–564.

Csuk, R. & Eversmann, L. (1998) Synthesis of difluorocyclopropyl carbocyclic homo-nucleosides. *Tetrahedron* 54, 6445–6456.

De Clercq, E. (2004) Antiviral drugs in current clinical use. *J Clin Virol* 30, 115–133.

DiMagno, S.G. & Sun, H. (2006) The strength of weak interactions: aromatic fluorine in drug design. *Curr Top Med Chem* 6, 1473–1482.

Doerr, I.L. & Fox, J.J. (1967) Nucleosides XXXIX. 2′-deoxy-2′-fluorocytidine, 1-β-D-arabinofuranosyl-2-amino-1,4(2*H*)-4-iminopyrimidine, and related derivatives. *J Org Chem* 32, 1462–1471.

Doong, S.L., Tsai, C.H., Schinazi, R.F. *et al.* (1991) Inhibition of the replication of hepatitis B virus *in vitro* by 2′,3′-dideoxy-3′-thiacytidine and related analogues. *Proc Natl Acad Sci USA* 88, 8495–8499.

Du, J., Choi, Y., Lee, K. *et al.* (1999) A practical synthesis of L-FMAU from L-arabinose. *Nucleos Nucleot Nucl* 18, 187–195.

Dyson, M.R., Coe, P.L. & Walker, R.T. (1991) The synthesis and antiviral activity of some 4′-thio-2′-deoxy nucleoside analogs. *J Med Chem* 34, 2782–2786.

Evans, G.B., Furneaux, R.H., Lewandowicz, A. *et al.* (2003) Exploring structure–activity relationships of transition state analogues of human purine nucleoside phosphorylase. *J Med Chem* 46, 3412–3423.

Fernández, R. & Castillón, S. (1999) Synthesis of 2′,3′-dideoxy-3′,3′-difluoro and 2′,3′-dideoxy-2′,2′-difluoro-pyranosyl nucleosides analogues of gemcitabine. *Tetrahedron* 55, 8497–8508.

Ford, H., Driscoll, J.S., Siddiqui, M. *et al.* (1994) Chemistry and anti-HIV activity of 2′-β-fluoro-2′,3′-dideoxyguanosine. *Nucleos Nucleot* 13, 213–234.

Fuentes, J., Angulo, M. & Pradera, M.A. (2002) Fluoronucleosides, isothiocyanato *C*-nucleosides, and thioureylene di-*C*-nucleosides via cyclic sulfates. *J Org Chem* 67, 2577–2587.

Grove, K.L., Guo, X., Liu, S.-H. *et al.* (1995) Anticancer Activity of β-L-dioxolane-cytidine, a novel nucleoside analogue with the unnatural L configuration. *Cancer Res* 55, 3008–3011.

Guillerm, D., Muzard, M., Allart, B. *et al.* (1995) Synthesis of 4′-fluoroadenosine as an inhibitor of *S*-adenosyl-L-homocysteine hydrolase. *Bioorg Med Chem Lett* 5, 1455–1460.

Gumina, G., Chong, Y., Choi, Y. *et al.* (2000) Stereoselective synthesis of carbocyclic L-4′-fluoro-2′,3′-dideoxyadenosine. *Org Lett* 2, 1229–1231.

Gumina, G., Schinazi, R.F. & Chu, C.K. (2001) Synthesis and potent anti-HIV activity of L-3′-fluoro- 2′,3′-unsaturated cytidine. *Org Lett* 3, 4177–4180.

Guschlbauer, W. & Jankowski, K. (1980) Nucleoside conformation is determined by the electronegativity of the sugar substituent. *Nucl Acids Res* 8, 1421–1433.

Haasnoot, C.A.G., de Leeuw, F.A.A.M. & Altona, C. (1980) The relationship between proton–proton NMR coupling constants and substituent electronegativities I: an empirical generalization of the Karplus equation. *Tetrahedron* 36, 2783–2792.

Herdewijn, P., Balzarini, J., De Clercq, E. *et al.* (1987a) 3′-Substituted 2′,3′-dideoxynucleoside analogs as potential anti-HIV (HTLV-III/LAV) agents. *J Med Chem* 30, 1270–1278.

Herdewijn, P., Pauwels, R., Baba, M. *et al.* (1987b) Synthesis and anti-HIV activity of various 2′- and 3′-substituted 2′,3′-dideoxyadenosines: a structure–activity analysis. *J Med Chem* 30, 2131–2137.

Herdewijn, P., Van Aerschot, A. & Kerremans, L. (1989) Synthesis of nucleosides fluorinated in the sugar moiety — the application of diethylaminosulfur trifluoride to the synthesis of fluorinated nucleosides. *Nucleos Nucleot* 8, 65–96.

Hertel, L.W., Boder, G.B., Kroin, J.S. *et al.* (1990) Evaluation of the antitumor activity of gemcitabine. *Cancer Res* 50, 4417–4422.

Hong, J.H., Lee, K., Choi, Y. *et al.* (1998) Enantiomeric synthesis of 3′-fluoro-apionucleosides using Claisen rearrangement. *Tetrahedron Lett* 39, 3443–3446.

Hoshino, H., Shimizu, N., Shimada, N. *et al.* (1987) Inhibition of infectivity of human immunodeficiency virus by oxetanocin. *J Antibiotics* 40, 1077–1078.

Izawa, K., Takamatsu, S., Katayama, S. *et al.* (2003) An industrial process for synthesizing lodenosine (FddA). *Nucleos Nucleot Nucl* 22, 507–517.

Jenkins, I.D., Verheyden, J.P.H. & Moffatt, J.G. (1976) 4′-Substituted nucleosides. 2: synthesis of the nucleoside antibiotic nucleocidin. *J Am Chem Soc* 98, 3346–3357.

Jeong, L.S., Nicklaus, M.C., George, C. *et al.* (1994a) Facile fluorination of deoxy-4′-thiopyrimidine nucleosides with 'down' hydroxyl groups. Retention of configuration after fluoride opening of the quaternized N3-MEM anhydronucleosides. *Tetrahedron Lett* 35, 7573–7576.

Jeong, L.S., Nicklaus, M.C., George, C. *et al.* (1994b) Unanticipated retention of configuration in the DAST fluorination of deoxy-4′-thiopyrimidine nucleosides with 'up' hydroxyl groups. *Tetrahedron Lett* 35, 7569–7572.

Jeong, L.S. & Marquez, V.E. (1995a) First synthesis of a dideoxydifluoro nucleoside with a β-D-xlyo configuration. An unprecedented effect of the *cis*-fluorines on the reactivity of the aglycon. *J Org Chem* 60, 4276–4279.

Jeong, L.S. & Marquez, V.E. (1995b) The stereochemical outcome of the DAST fluorination of 4′-thiopyrimidine nucleosides with 'up' hydroxyl groups is controlled by the oxidation state of the sulfur atom. *Chem Lett* 24, 301–302.

Jeong, L.S., Yoo, S.J., Lee, K.M. *et al.* (2003) Design, synthesis, and biological evaluation of fluoroneplanocin A as the novel mechanism-based inhibitor of *S*-adenosylhomocysteine hydrolase. *J Med Chem* 46, 201–203.

Katayama, S., Takamatsu, S., Naito, M. *et al.* (2006) A synthesis of 3′-α-fluoro-2′,3′-dideoxyadenosine via a bromine rearrangement during fluorination with MOST reagent. *J Fluorine Chem* 127, 524–528.

Kawasaki, A.M., Casper, M.D., Freier, S.M. *et al.* (1993) Uniformly modified 2′-deoxy-2′-fluoro-phosphorothioate oligonucleotides as nuclease-resistant antisense compounds with high affinity and specificity for RNA targets. *J Med Chem* 36, 831–841.

Kicska, G.A., Long, L., Horig, H. *et al.* (2001) Immucillin-H, a powerful transition-state inhibitor of purine nucleoside phosphorylase, selectively inhibits human T-lymphocytes. *Proc Natl Acad Sci USA* 98, 4593–4598.

Kim, H.O., Yoo, S.J., Ahn, H.S. *et al.* (2004) Synthesis of fluorinated cyclopentenyladenine as potent inhibitor of *S*-adenosylhomocysteine hydrolase. *Bioorg Med Chem Lett* 14, 2091–2093.

Komatsu, H. & Araki, T. (2003) Chemo-enzymatic synthesis of 2′,3′-dideoxy-3′-fluoro-β-D-guanosine via 2,3-dideoxy-3-fluoro-α-D-ribose 1-phosphate. *Tetrahedron Lett* 44, 2899–2901.

Koshida, R., Cox, S., Harmenberg, J. *et al.* (1989) Structure–activity relationships of fluorinated nucleoside analogs and their synergistic effect in combination with phosphonoformate against human immunodeficiency virus type 1. *Antimicrob Agents Chemother* 33, 2083–2088.

Leclercq, F. & Antonakis, K. (1989) Direct fluorination at positions 3′ 4′, and 6′ of β-D-glucopyranosyltheophylline. *Carbohydr Res* 193, 307–313.

Lee, K., Choi, Y., Gullen, E. *et al.* (1999) Synthesis and anti-HIV and anti-HBV activities of 2′-fluoro-2′,3′-unsaturated L-nucleosides. *J Med Chem* 42, 1320–1328.

Lee, Y.R., Park, J.-H., Jeon, R. *et al.* (2001) Design and synthesis of novel fluorocyclopropanoid nucleosides. *Nucleos Nucleot Nucl* 20, 677–679.

Lee, K., Choi, Y., Gumina, G. *et al.* (2002) Structure–activity relationships of 2′-fluoro-2′,3′-unsaturated D-nucleosides as anti-HIV-1 agents. *J Med Chem* 45, 1313–1320.

Lee, S., Uttamapinant, C. & Verdine, G.L. (2007) A concise synthesis of 4′-fluoro nucleosides. *Org Lett* 9, 5007–5009.

Li, N.-S., Tang, X.-Q. & Piccirilli, J.A. (2001) 2′-*C*-branched ribonucleosides 2. Synthesis of 2′-*C*-β-trifluoromethyl pyrimidine ribonucleosides. *Org Lett* 3, 1025–1028.

Li, Y., Mao, S., Hager, M.W. *et al.* (2007) Synthesis and evaluation of 2′-substituted cyclobutyl nucleosides and nucleotides as potential anti-HIV agents. *Bioorg Med Chem Lett* 17, 3398–3401.

Li, F., Maag, H. & Alfredson, T. (2008) Prodrugs of nucleoside analogues for improved oral absorption and tissue targeting. *J Pharm Sci* 97, 1109–1134.

Liu, P., Sharon, A. & Chu, C.K. (2008) Fluorinated nucleosides: synthesis and biological implication. *J Fluorine Chem* 129, 743–766.

Lopin, C., Gautier, A., Gouhier, G. *et al.* (2002) First and efficient synthesis of phosphonodifluoromethylene analogues of nucleoside 3′-phosphates: crucial role played by sulfur in construction of the target molecules. *J Am Chem Soc* 124, 14668–14675.

Ma, T., Pai, S.B., Zhu, Y.L. *et al.* (1996) Structure–activity relationships of 1-(2-deoxy-2-fluoro-2-L-arabino-furanosyl)pyrimidine nucleosides as anti-hepatitis B virus agents. *J Med Chem* 39, 2835–2843.

Maag, H., Rydzewski, R.M., McRoberts, M.J. *et al.* (1992) Synthesis and anti-HIV activity of 4′-azido- and 4′-methoxynucleosides. *J Med Chem* 35, 1440–1451.

Machida, H., Ashida, N., Miura, S. *et al.* (1998) Anti-herpesvirus activity profile of 4′-thioarabinofuranosyl purine and uracil nucleosides and activity of 1-β-2′-fluoro-4′-thioarabinofuranosyl guanine and 2,6-diaminopurine against clinical isolates of human cytomegalovirus. *Antiviral Res* 39, 129–137.

Marquez, V.E. & Lim, M.I. (1986) Carbocyclic nucleosides. *Med Res Rev* 6, 1–40.

Marquez, V.E., Tseng, C.K.H., Mitsuya, H. *et al.* (1990) Acid-stable 2′-fluoro purine dideoxynucleosides as active agents against HIV. *J Med Chem* 33, 978–985.

Martin, J.A., Bushnell, D.J., Duncan, I.B. *et al.* (1990) Synthesis and antiviral activity of monofluoro and difluoro analogs of pyrimidine deoxyribonucleosides against human immunodeficiency virus (HIV-1). *J Med Chem* 33, 2137–2145.

Matulic-Adamic, J., Haeberli, P. & Usman, N. (1995) Synthesis of 5′-deoxy-5′-difluoromethyl phosphonate nucleotide analogs. *J Org Chem* 60, 2563–2569.

McAtee, J.J., Schinazi, R.F. & Liotta, D.C. (1998) A completely diastereoselective electrophilic fluorination of a chiral, noncarbohydrate sugar ring precursor: application to the synthesis of several novel 2′-fluoronucleosides. *J Org Chem* 63, 2161–2167.

McCarthy, J.R., Jarvi, E.T., Matthews, D.P. *et al.* (1989) 4′,5′-Unsaturated-5′-fluoro adenosine nucleosides: potent mechanism-based inhibitors of S-adenosyl-L-homocysteine hydrolase. *J Am Chem Soc* 111, 1127–1128.

McCarthy, J.R., Matthews, D.P., Stemerick, D.M. *et al.* (1991) Stereospecific method to (*E*) and (*Z*) terminal fluoroolefins and its application to the synthesis of 2′-deoxy-2′-fluoromethylenenucleosides as potential inhibitors of ribonucleoside diphosphate reductase. *J Am Chem Soc* 113, 7439–7440.

Meng, W.-D. & Qing, F.-L. (2006) Fluorinated nucleosides as antiviral and antitumor agents. *Curr Top Med Chem* 6, 1499–1528.

Mikhailopulo, I. A., Pricota, T.I., Sivets, G.G. *et al.* (2003) 2′-Chloro-2′,3′-dideoxy-3′-fluoro-D-ribonucleosides: synthesis, stereospecificity, some chemical transformations, and conformational analysis. *J Org Chem* 68, 5897–5908.

Miura, S., Yoshimura, Y., Endo, M. *et al.* (1998) Antitumor activity of a novel orally effective nucleoside,1-(2-deoxy-2-fluoro-4-thio-β-D-arabino-furanosyl) cytosine. *Cancer Lett* 129, 103–110.

Miura, S., Endo, Y., Yoshimura, Y. *et al.* (2002) Potent antitumor effect of 1-(2-deoxy-2-fluoro-4-thio-β-D-arabinofuranosyl)cytosine on peritoneal dissemination models of gastrointestinal cancers. *Oncol Rep* 9, 1319–1322.

Moon, H.R., Lee, H.J., Kim, K.R. *et al.* (2004) Synthesis of 5′-substituted fluoroneplanocin A analogues: importance of a hydrogen bonding donor at 5′-position for the inhibitory activity of *S*-adenosylhomocysteine hydrolase. *Bioorg Med Chem Lett* 14, 5641–5644.

Moukha-Chafiq, O., Tiwari, K.N. & Secrist, J.A. (2005) Synthesis of some 2′-fluoro-2′-deoxy-3′-C-ethynyl and 3′-C-vinyl-β-D-lyxofuranosyl nucleosides. *Nucleos Nucleot Nucl* 24, 713–715.

Murakami, E., Bao, H., Ramesh, M. *et al.* (2007) Mechanism of activation of β-D-2′-deoxy-2′-fluoro-2′-C-methylcytidine and inhibition of hepatitis C virus NS5B RNA polymerase. *Antimicrob Agents Chemother* 51, 503–509.

Nishiyama, Y., Yamamoto, N., Takahashi, K. *et al.* (1988) Selective inhibition of human cytomegalovirus replication by a novel nucleoside, oxetanocin-G. *Antimicrob Agents Chemother* 32, 1053–1056.

Nishiyama, Y., Yamamoto, N., Yamada, Y. *et al.* (1989) Anti-herpesvirus activity of carbocyclic oxetanocin-G *in vitro*. *J Antibiotics* 42, 1854–1859.

Noble, S. & Goa, K.L. (1997) Gemcitabine — a review of its pharmacology and clinical potential in non-small cell lung cancer and pancreatic cancer. *Drugs* 54, 447–472.

Norbeck, D.W., Kern, E., Hayashi, S. *et al.* (1990) Cyclobut-A and cyclobut-G: broad-spectrum antiviral agents with potential utility for the therapy of AIDS. *J Med Chem* 33, 1281–1285.

O-Yang, C., Wu, H.Y., Fraser-Smith, E.B. *et al.* (1992) Synthesis of 4′-cyanothymidine and analogs as potent inhibitors of HIV. *Tetrahedron Lett* 33, 37–40.

Ohrui, H., Kohgo, S., Kitano, K. *et al.* (2000) Syntheses of 4′-*C*-ethynyl-β-d-arabino- and 4′-*C*-ethynyl-2′-deoxy-β-D-ribo-pentofuranosylpyrimidines and -purines and evaluation of their anti-HIV. *J Med Chem* 43, 4516–4525.

Pankiewicz, K.W. (2000) Fluorinated nucleosides. *Carbohydr Res* 327, 87–105.

Payne, A.N. & Roberts, S.M. (1992) Synthesis of two 6′a-fluorocarbocyclic nucleosides and incorporation into short segments of DNA. *J Chem Soc Perkin Trans*, 2633–2641.

Qiu, X.-L. & Qing, F.-L. (2004) Synthesis of 2′,3′-dideoxy-2′-difluoromethyl azanucleosides. *Synthesis* 35, 334–340.

Qiu, X.-L. & Qing, F.-L. (2005) Synthesis of 3′-deoxy-3′-difluoromethyl azanucleosides from *trans*-4-hydroxy-L-proline. *J Org Chem* 70, 3826–3837.

Ranganathan, R. (1977) Modification of the 2′-position of purine nucleosides: syntheses of 2′-α-substituted-2′-deoxyadenosine analogs. *Tetrahedron Lett* 18, 1291–1294.

Reichman, U., Watanabe, K.A. & Fox, J.J. (1975) A practical synthesis of 2-deoxy-2-fluoro-arabinofuranose derivatives. *Carbohydr Res* 42, 233–240.

Robins, M.J., Nowak, I., Wnuk, S.F. *et al.* (2007) Deoxygenative [1,2]-hydride shift rearrangements in nucleoside and sugar chemistry: analogy with the [1,2]-electron shift in the deoxygenation of ribonucleotides by ribonucleotide reductases. *J Org Chem* 72, 8216–8221.

Rodriguez, J.B. & Comin, M.J. (2003) New progress in the enantioselective synthesis and biological properties of carbocyclic nucleosides. *Mini-Rev Med Chem* 3, 95–114.

Romanenko, V.D. & Kukhar, V.P. (2006) Fluorinated phosphonates: synthesis and biomedical application. *Chem Rev* 106, 3868–3935.

Sabol, J.S. & McCarthy, J.R. (1992) A new route to 1,1-difluoro olefins: application to the synthesis of 2′-deoxy-2′-difluoromethylene nucleosides. *Tetrahedron Lett* 33, 3101–3104.

Sakuma, T., Saijo, M., Suzutani, T. *et al.* (1991) Antiviral activity of oxetanocins against varicella-zoster virus. *Antimicrob Agents Chemother* 35, 1512–1514.

Schachter, J.B., Yasuda, R.P. & Wolfe, B.B. (1995) Adenosine receptor activation potentiates phosphoinositide hydrolysis and arachidonic acid release in DDT1-MF2 cells: putative interrelations. *Cell Signal* 7, 659–668.

Schneller, S.W. (2002) Carbocyclic nucleosides (carbanucleosides) as therapeutic leads. *Curr Top Med Chem* 2, 1087–1092.

Secrist III, J.A., Tiwari, K.N., Riordan, J.M. *et al.* (1991) Synthesis and biological activity of 2′-deoxy-4′-thio pyrimidine nucleosides. *J Med Chem* 34, 2361–2366.

Sekiyama, T., Hatsuya, S., Tanaka, Y. *et al.* (1998) Synthesis and antiviral activity of novel acyclic nucleosides: discovery of a cyclopropyl nucleoside with potent inhibitory activity against herpesviruses. *J Med Chem* 41, 1284–1298.

Serafinowski, P.J. & Brown, C.A. (2000) New method for the preparation of some 2′- and 3′-trifluoromethyl-2′,3′-dideoxyuridine derivatives. *Tetrahedron* 56, 333–339.

Sharma, M., Li, Y.X., Ledvina, M. *et al.* (1995) Synthesis of 5′-fluoro-5′-deoxy-and 5′-amino-5′-deoxytoyocamycin and sangivamycin and some related derivatives. *Nucleos Nucleot* 14, 1831–1852.

Shi, J.X., Du, J.F., Ma, T.W. *et al.* (2005) Synthesis and anti-viral activity of a series of D- and L-2′-deoxy-2′-fluororibonucleosides in the subgenomic HCV replicon system. *Bioorg Med Chem* 13, 1641–1652.

Shortnacy-Fowler, A.T., Tiwari, K.N., Montgomery, J.A. *et al.* (2001) Synthesis and biological activity of 4′-C-hydroxymethyl-2′-fluoro-D-arabinofuranosylpurine nucleosides. *Nucleos Nucleot Nucl* 20, 747–750.

Shryock, J.C. & Belardinelli, L. (1997) Adenosine and adenosine receptors in the cardiovascular system: biochemistry, physiology, and pharmacology. *Am J Cardiol* 79, 2–10.

Siddiqui, M.A., Driscoll, J.S., Marquez, V.E. *et al.* (1992) Chemistry and anti-HIV properties of 2′-fluoro-2′,3′-dideoxyarabinofuranosylpyrimidines. *J Med Chem* 35, 2195–2201.

Siddiqui, M.A., Driscoll, J.S. & Marquez, V.E. (1998) A new synthetic approach to the clinically useful, anti-HIV-active nucleoside, 9-(2,3-dideoxy-2-fluoro-β-D-threo-pentofuranosyl)adenine (β-FddA): introduction of a 2′-β-fluoro substituent via inversion of a readily obtainable 2′-α-fluoro isomer. *Tetrahedron Lett* 39, 1657–1660.

Sterzycki, R.Z., Ghazzouli, I., Brankovan, V. *et al.* (1990) Synthesis and anti-HIV activity of several 2′-fluoro-containing pyrimidine nucleosides. *J Med Chem* 33, 2150–2157.

Stuyver, L.J., McBrayer, T.R., Whitaker, T. *et al.* (2004) Inhibition of the subgenomic hepatitis C virus replicon in huh-7 cells by 2′-deoxy-2′-fluorocytidine. *Antimicrob Agents Chemother* 48, 651–654.

Sugimura, H., Osumi, K., Yamazaki, T. *et al.* (1991) Coupling reaction of 1-thiopentofuranosides with silylated pyrimidine bases by activation with *N*-bromosuccinimide: synthesis of 3′-azido-3′-deoxythymidine and its related nucleoside analogs. *Tetrahedron Lett* 32, 1813–1816.

Sun, G., Voigt, J.H., Filippov, I.V. *et al.* (2004) PROSIT: pseudo-rotational online service and interactive tool, applied to a conformational survey of nucleosides and nucleotides. *J Chem Inf Comput Sci* 44, 1752–1762.

Takamatsu, S., Katayama, S., Naito, M. *et al.* (2003) A facile synthetic method for 3′-fluoro-2′,3′-dideoxyadenosine. *Nucleos Nucleot Nucl* 22, 711–713.

Tann, C.H., Brodfuehrer, P.R., Brundidge, S.P. *et al.* (1985) Fluorocarbohydrates in synthesis. An efficient synthesis of 1-(2-deoxy-2-fluoro-β-D-arabinofuranosyl)-5-iodouracil (β-FIAU) and 1-(2-deoxy-2-fluoro-β-D-arabinofuranosyl) thymine (β-FMAU). *J Org Chem* 50, 3644–3647.

Thibaudeau, C., Kumar, A., Bekiroglu, S. *et al.* (1998a) NMR Conformation of (−)-β-D-aristeromycin and its 2′-deoxy and 3′-deoxy counterparts in aqueous solution. *J Org Chem* 63, 5447–5462.

Thibaudeau, C., Plavec, J. & Chattopadhyaya, J. (1998b) A new generalized Karplus-type equation relating vicinal proton-fluorine coupling constants to H-C-C-F torsion angles. *J Org Chem* 63, 4967–4984.

Timoshchuk, V.A., Hogrefe, R.I. & Vaghefi, M.M. (2004) Improved and reliable synthesis of 3′-azido-2′,3′-dideoxyguanosine derivatives. *Nucleos Nucleot Nucl* 23, 171–181.

Torii, T., Onishi, T., Tanji, S. *et al.* (2005) Synthetic study of 3′-α-fluoro-2′,3′-dideoxyguanosine. *Nucleos Nucleot Nucl* 24, 1051–1054.

Tsoukala, E., Agelis, G., Dolinsek, J. *et al.* (2007) An efficient synthesis of 3-fluoro-5-thio-xylofuranosyl nucleosides of thymine, uracil, and 5-fluorouracil as potential antitumor or/and antiviral agents. *Bioorg Med Chem* 15, 3241–3247.

Tsuchiya, T. (1990) Chemistry and developments of fluorinated carbohydrates. *Adv Carbohydr Chem Biochem* 48, 91–277.

Uesugi, S., Kaneyasu, T., Imura, J. *et al.* (1983) ^{1}H-NMR studies on the dinucleoside monophosphates containing 2′-halogeno-2′-deoxypurinenucleosides: effects of 2′-substitutes on conformation. *Biopolymers* 22, 1189–1202.

Van der Boogaart, J.E., Kalinichenko, E.N., Podkopaeva, T.L. *et al.* (1994) Conformation analysis of 3′-fluorinated A(2′-5′)A(2′-5′)A fragments: relation between conformation and biological activity. *Eur J Biochem* 221, 759–768.

von Janta-Lipinski, M., Costisella, B., Ochs, H. *et al.* (1998) Newly synthesized L-enantiomers of 3′-fluoro-modified β-2′-deoxyribonucleoside 5′-triphosphates inhibit hepatitis B DNA polymerases but not the five cellular DNA polymerases α, β, γ, δ, and ε nor HIV-1 reverse transcriptase. *J Med Chem* 41, 2040–2046.

Wachtmeister, J., Classon, B., Samuelsson, B. *et al.* (1997) Synthesis of carbocyclic 2′3′-dideoxy-2′-fluoro-3′-C-hydroxymethyl nucleoside analogues as potential inhibitors of HIV and HSV. *Tetrahedron* 53, 1861–1872.

Wang, Y., Fleet, G.W.J., Wilson, F.X. *et al.* (1991) Oxetane nucleosides with fluorine and azide substitutents: nucleophilic displacements on an oxetane ring. *Tetrahedron Lett* 32, 1675–1678.

Wang, J., Jin, Y., Rapp, K.L. *et al.* (2005) Synthesis, antiviral activity, and mechanism of drug resistance of D- and L-2′,3′-didehydro-2′,3′-dideoxy-2′-fluorocarbocyclic nucleosides. *J Med Chem* 48, 3736–3748.

Wang, J., Jin, Y., Rapp, K.L. *et al.* (2007) D- and L-2′,3′-didehydro-2′,3′-dideoxy-3′-fluoro-carbocyclic nucleosides: synthesis, anti-HIV activity and mechanism of resistance. *J Med Chem* 50, 1828–1839.

Watanabe, K.A., Reichman, U., Hirota, K. *et al.* (1979) Nucleosides 110. Synthesis and antiherpes virus activity of some 2′-fluoro-2′-deoxyarabinofuranosylpyrimidine nucleosides. *J Med Chem* 22, 21–24.

Watanabe, K.A., Harada, K., Zeidler, J. *et al.* (1990) Synthesis and anti-HIV-1 activity of 2′-‘up’-fluoro analogs of active anti-AIDS nucleosides 3′-azido-3′-deoxythymidine (AZT) and 2′,3′-dideoxycytidine (DDC). *J Med Chem* 33, 2145–2150.

Wright, J.A., Taylor, N.F. & Fox, J.J. (1969) Nucleosides LX. Fluorocarbohydrates XXII. Synthesis of 2-deoxy-2-fluoro-D-arabinose and 9-(2-deoxy-2-fluoro-α and β-D-arabinofuranosyl)adenines. *J Org Chem* 34, 2632–2636.

Wright, J.A., Wilson, D.P. & Fox, J.J. (1970) Nucleosides. LXIV. Fluoro sugar analogs of arabinosyl- and xylosylcytosines. *J Med Chem* 13, 269–272.

Wu, Y.-Y., Zhang, X., Meng, W.-D. *et al.* (2004) Synthesis of new 2′,3′-dideoxy-6′,6′-difluoro-3′-thionucleoside from *gem*-difluorohomoallyl alcohol. *Org Lett* 6, 3941–3944.

Xu, J., Zhang, X., Qiu, X.-L. *et al.* (2009) Synthesis of 2,3,4-trideoxy-4,4-difluoro-D-ribo-hexopyranose adenosines. *Synthesis* 2009, 602–608.

Xu, X.-H., Qiu, X.-L., Zhang, X. *et al.* (2006) Synthesis of L- and D-2′-3′-deoxy-3′,3′-difluoronucleosides. *J Org Chem* 71, 2820–2824.

Yamashita, M., Kato, Y., Suzuki, K. *et al.* (1998) A novel preparation and X-ray crystallographic analysis of phospha sugar nucleoside derivatives. *Heterocycl Commun* 4, 411–414.

Yang, Y.-Y., Meng, W.-D. & Qing, F.-L. (2004) Synthesis of 2′,3′-dideoxy-6′,6′-difluorocarbocyclic nucleosides. *Org Lett* 6, 4257–4259.

Yang, Y.-Y., Xu, J., You, Z.-W. *et al.* (2007) Synthesis of 3′,3′-difluoro-2′- hydroxymethyl-4′,5′-unsaturated carbocyclic nucleosides. *Org Lett* 9, 5437–5440.

Ye, J.-D., Liao, X. & Piccirilli, J.A. (2005) Synthesis of 2′-*C*-difluoromethylribonucleosides and their enzymatic incorporation into oligonucleotides. *J Org Chem* 70, 7902–7910.

Yokoyama, M. & Momotake, A. (1999) Synthesis and biological activity of azanucleosides. *Synthesis* 1999, 1541–1554.

Yokoyama, M. (2000) Synthesis and biological activity of thionucleosides. *Synthesis* 2000, 1637–1655.

Yoshimura, Y., Kitano, K., Yamada, K. *et al.* (1997) A novel synthesis of 2′-modified 2′-deoxy-4′-thiocytidines from D-glucose. *J Org Chem* 62, 3140–3152.

Yoshimura, Y., Endo, M., Miura, S. *et al.* (1999) An alternative synthesis of the antineoplastic nucleoside 4′-thioFAC and its application to the synthesis of 4′-thioFAG and 4′-thiocytarazid. *J Org Chem* 64, 7912–7920.

Yoshimura, Y., Kitano, K., Yamada, K. *et al.* (2000) Synthesis and biological activities of 2′-deoxy-2′-fluoro-4′-thioarabinofuranosylpyrimidine and -purine nucleosides. *Bioorg Med Chem* 8, 1545–1558.

Yue, X., Qiu, X.-L. & Qing, F.-L. (2008) Synthesis of 2′,3′-dideoxy-6′,6′-difluoro-3′-azanucleosides. *J Fluorine Chem* 129, 866–874.

Zajchowski, D.A., Biroc, S.L., Liu, H.-L. *et al.* (2005) Anti-tumor efficacy of the nucleoside analog 1-(2-deoxy-2-fluoro-4-thio-β-D-arabinofuranosyl)cytosine (4′-thio-FAC) in human pancreatic and ovarian tumor xenograft models. *Int J Cancer* 114, 1002–1009.

Zhang, X., Qing, F.-L. & Yu, Y. (2000) Synthesis of 2′,3′-dideoxy-2′-trifluoromethylnucleosides from α-trifluoromethyl-α,β-unsaturated ester. *J Org Chem* 65, 7075–7082.

Zhang, X. & Qing, F.-L. (2002) Synthesis of novel L-2′,3′-dideoxy-2′-trifluoromethyl-4′-thiocytidines from α-trifluoromethyl-α,β-unsaturated ester. *J Org Chem* 67, 1016–1019.

Zhang, X., Xia, H., Dong, X. *et al.* (2003) 3-Deoxy-3,3-difluoro-D-arabinofuranose: first stereoselective synthesis and application in preparation of *gem*-difluorinated sugar nucleosides. *J Org Chem* 68, 9026–9033.

Zhou, S., Kern, E.R., Gullen, E. *et al.* (2004a) (*Z*)- and (*E*)-[2-Fluoro-2-(hydroxymethyl)cyclopropylidene]methylpurines and -pyrimidines, a new class of methylenecyclopropane analogues of nucleosides: synthesis and antiviral activity. *J Med Chem* 47, 6964–6972.

Zhou, W., Gumina, G., Chong, Y. *et al.* (2004b) Synthesis, structure–activity relationships, and drug resistance of β-D-3′-fluoro-2′,3′-unsaturated nucleosides as anti-HIV agents. *J Med Chem* 47, 3399–3408.

Zhou, S. & Zemlicka, J. (2005) A new alkylation–elimination method for synthesis of antiviral fluoromethylenecyclopropane analogues of nucleosides. *Tetrahedron* 61, 7112–7116.

Zhao, B., Chang, J., Wang, Q. *et al.* (2008) Design and synthesis of 3′-deoxy-3′-fluoro-2′-O,3′-C-vinylene-linked bicyclic purine nucleoside. *Synlett* 2008, 2993–2996.

Zheng, F., Zhang, X.-H., Qiu, X.-L. *et al.* (2006) Synthesis of L-β-3′-deoxy-3′,3′-difluoro-4′-thionucleosides. *Org Lett* 8, 6083–6086.

Zheng, F., Zhang, X. & Qing, F.-L. (2009) Stereoselective Reformatsky–Claisen rearrangement: synthesis of 2′,3′-dideoxy-6′,6′-difluoro-2′-thionucleosides. *Chem Commun*, 1505–1507.

Zhu, W., Chong, Y., Choo, H. *et al.* (2004) Synthesis, structure–activity relationships, and mechanism of drug resistance of D- and L-β-3′-fluoro-2′,3′-unsaturated-4′-thionucleosides as anti-HIV agents. *J Med Chem* 47, 1631–1640.

9

Synthesis of Fluorinated Neurotransmitter Analogues

Margit Winkler and David O'Hagan†*

9.1 Introduction

Fluorinated analogues of natural and synthetic organic compounds have been a focus of interest for drug development (Kirk, 2006) and diagnostics for many years now. Organic fluorine is not generally found in metabolism and only a handful of natural fluorinated compounds have been identified. Nevertheless, a high proportion of active pharmaceutical ingredients developed by the pharmaceutical industry contain at least one fluorine atom (Isanbor and O'Hagan, 2006; Le Bars, 2006). It is a frequently applied strategy to exchange hydrogen atoms or hydroxyl groups of drugs by fluorine, largely maintaining the biodistribution behaviour of the parent compound. As diagnostic tools, fluorinated compounds can be exploited as non-invasive NMR reporters (^{19}F) (Passe *et al.*, 1995; Yu *et al.*, 2005) (see Chapters 4 and 12) or positron emission tomography (PET) tracers (^{18}F) (e.g. Couturier *et al.*, 2004; Le Bars, 2006) (see Chapters 10 and 11). With this review we aim at giving an overview about the synthesis and use of fluorinated non-peptidic small-molecule neurotransmitters.

* ACIB GmbH c/o Institution of Molecular Biotechnology, Graz University of Technology, Petersgasse 14, Austria; fax: +43 316 873 933. E-mail: margit.winkler@acib.at.

† Centre for Biomolecular Sciences and School of Chemistry, University of St. Andrews, North Haugh, St. Andrews KY16 9ST, UK.

9.2 Adenosine Receptors

Adenosine is an endogenous nucleoside which is present in all body fluids and exhibits a cytoprotective function within the body. To date, four different adenosine receptors have been classified: A_1, A_{2a}, A_{2b} and A_3, depending on their agonist and antagonist selectivity. Agonists and antagonists of adenosine receptors could prove useful for the treatment of a range of conditions such as neurodegenerative or cardiovascular diseases. There are several positions on adenosine which can be modified by fluorine and many of them have been prepared, but only very few have actually been tested as agonists/antagonists on adenosine receptors. The synthetic efforts towards fluorinated adenosine analogues are briefly summarised here.

The most recent of the few strategies for the preparation of 2-fluoroadenosine published so far involved the replacement of a nitro group on perbenzoylated adenosine using TBAF (Scheme 9.1). Starting from adenosine, this four-step sequence afforded the product in 57% overall yield (Braendvang and Gundersen, 2006).

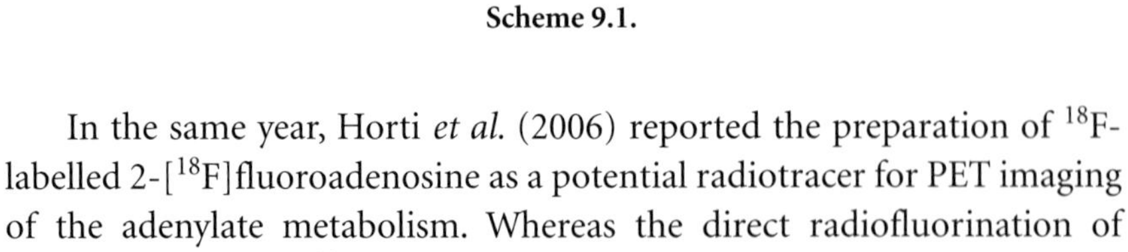

Scheme 9.1.

In the same year, Horti *et al.* (2006) reported the preparation of ^{18}F-labelled 2-[^{18}F]fluoroadenosine as a potential radiotracer for PET imaging of the adenylate metabolism. Whereas the direct radiofluorination of adenosine failed, 2-[^{18}F]fluoroadenosine was obtained in 1% radiochemical yield (RCY) in a no-carrier-added protocol and in 5% RCY under carrier-added conditions, starting from 2-iodo and 2-fluoroadenosine, respectively.

8-Fluoroadenosine was synthesised for the first time in the late 1990s. All earlier approaches were unsuccessful due to the lability of the C8–F bond, which caused decomposition during standard deprotection steps. Finally, Barrio *et al.* (1998) prepared the compound by a Balz–Schiemann

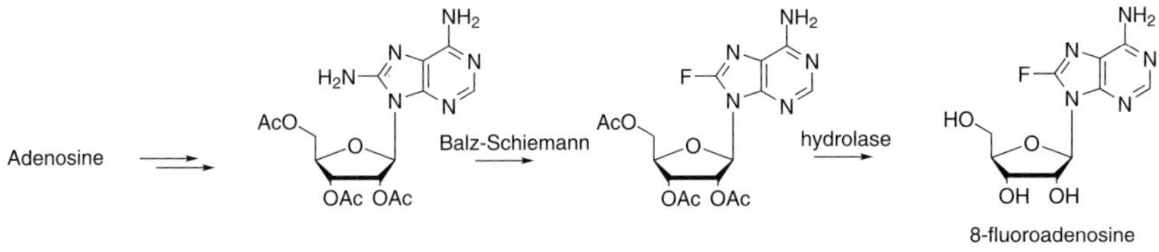

Scheme 9.2.

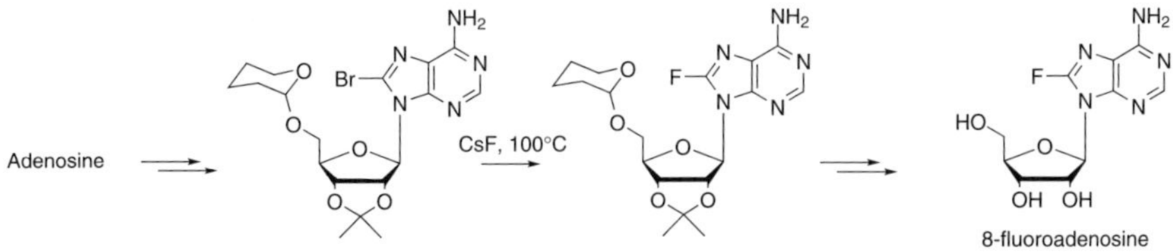

Scheme 9.3.

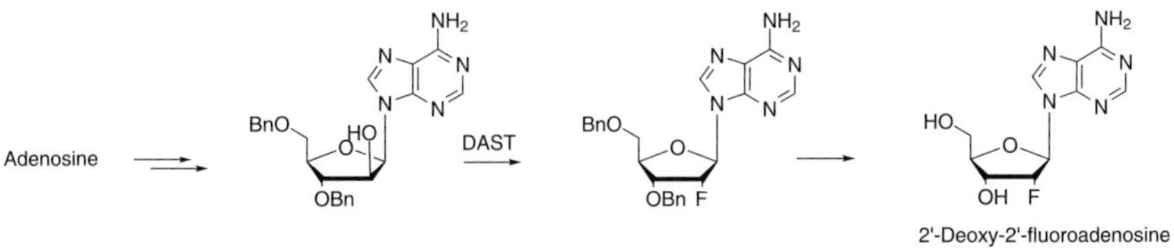

Scheme 9.4.

fluorination of 2′,3′,5′-triacetyladenosine and subsequent enzymatic deacetylation (see Scheme 9.2).

Just recently, an alternative approach overcame several synthetic obstacles arising from the highly functionalised nucleosides and a synthesis of 8-fluoroadenosine from 8-bromoadenosine in five steps was reported (Butora *et al.*, 2007). The final important step was the perchloric acid-catalysed removal of an acetonide protecting group in presence of the labile C8–F bond (see Scheme 9.3).

Three strategies have been followed for the preparation of 2′-deoxy-2′-fluoroadenosine. The key step in the first described synthesis was the introduction of the fluorine with diethylaminosulfur trifuoride (DAST) at RT in 82% yield (see Scheme 9.4) (Pankiewicz *et al.*, 1992). In general, nucleoside syntheses require a number of protection/deprotection steps for the oxygen as well as the nitrogen functionalities. To prevent protecting group chemistry, Tuttle *et al.* (1993) developed a protocol for enzymatic transglycosylation to give access to several 2′-deoxy-2′-fluoronucleosides, among them the adenosine derivative. A thymidine phosphorylase (TP) was applied to convert 2′-deoxy-2′-fluorouridine to 2-deoxy-2-fluoro-α-D-ribosephosphate as a reactive intermediate for the subsequent enzymatic reaction with purine nucleoside phosphorylase (PNP) in the presence of adenine. The reaction required 17 days at 50°C to yield 82% of the desired product (see Scheme 9.5). The anti-influenza activity as well as the cytotoxicity of the compound was evaluated in the same study (Tuttle *et al.*, 1993).

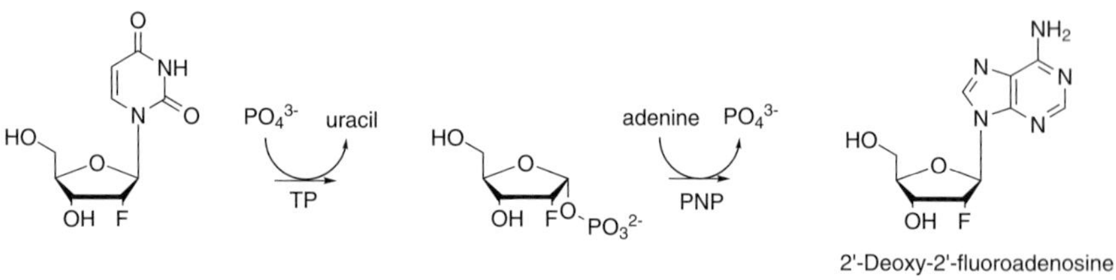

Scheme 9.5.

An ^{18}F-labelled stereomeric analogue of 2′-deoxy-2′-fluoroadenosine was prepared by Alauddin *et al.* (2003). Notably, the configuration at the sugar in this case changed from *ribo* to *arabino* to give 2′-deoxy-2′-[^{18}F]fluoro-1-β-D-arabinofuranosyl-adenine ([^{18}F]-FAA). Similarly, the only reported analogue of 3′-deoxy-3′-fluoroadenosine is 3′-deoxy-3′-[^{18}F]fluoro-1-β-D-xylofuranosyl-adenine ([^{18}F]-FXA) (Alauddin *et al.*, 2003). Both of the latter compounds were evaluated as imaging agents. [^{18}F]-FAA showed promising behaviour for tumour imaging, whereas potential as a heart imaging agent was ascribed to [^{18}F]-FXA (Alauddin *et al.*, 2007).

3′-Deoxy-3′-fluoroadenosine was prepared in a sequence of ten overall steps in a stereoselective manner starting from 3,5-*O*-isopropylidene-β-D-xylofuranoside. The fluorine was introduced to the 3′-position by replacement of the trifluoromethanesulfonate by TBAF. The coupling to adenine was accomplished as the last step before deprotection (see Scheme 9.6) (Morizawa *et al.*, 1989).

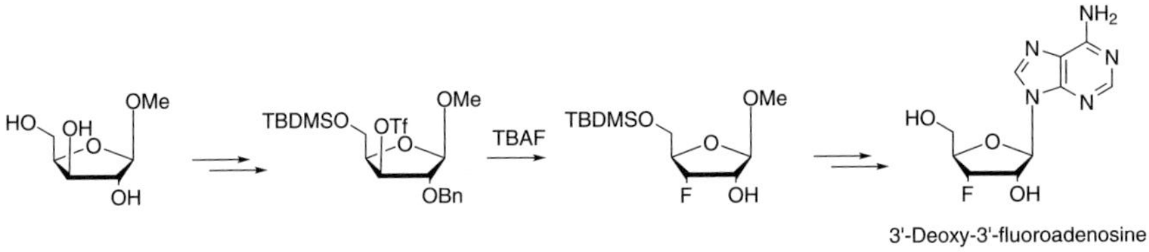

Scheme 9.6.

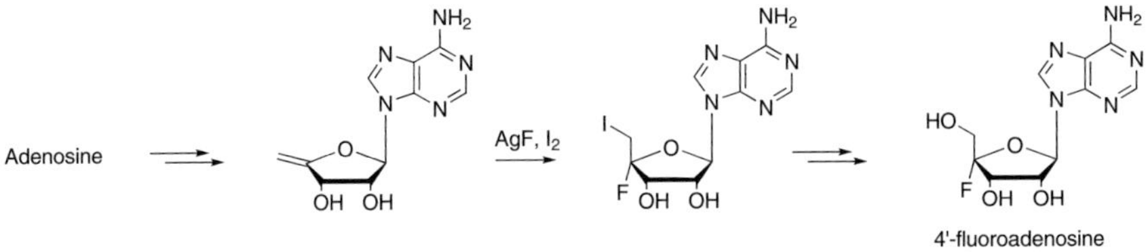

Scheme 9.7.

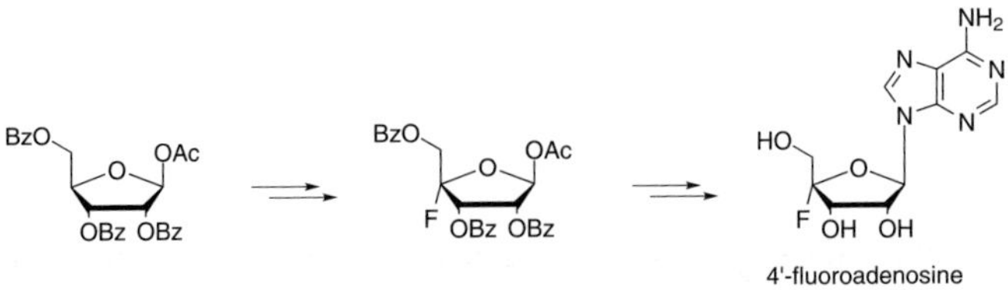

Scheme 9.8.

Although 4′-fluoroadenosine has appeared in the literature before, the unprotected compound was first reported in 1995. The eight-step synthetic route started from adenosine and the fluorine was introduced by addition of 'IF' to the enol ether (see Scheme 9.7) (Guillerm *et al.*, 1995). Recently, Lee *et al.* (2007) published an alternative four-step sequence starting from 1-*O*-acetyl-2,3,5-tri-*O*-benzoyl-β,D-ribose. The authors photobrominated the 4′-position and exchanged bromine for fluorine by using $BF_3·OEt_3$/AgF. Subsequent coupling to nucleobases and global deprotection afforded not only 4′-fluoroadenosine, but also other 4′-fluoronucleosides (see Scheme 9.8).

5′-Deoxy-5′-fluoroadenosine was first prepared biocatalytically from *S*-adenosyl-L-methionine using the cell-free extract from a *Streptomyces cattleya* culture (O'Hagan *et al.*, 2002). The compound is a key intermediate on the metabolic pathway to fluoroacetate and 4-fluorothreonine. The introduction of fluorine from inorganic fluoride ion is accomplished by the enzyme fluorinase (Scheme 9.9). In the same year, the first synthetic route to the compound was also published (see Scheme 9.10) (Schaffrath *et al.*, 2002). The efficacy of the synthesis, however, was greatly improved a few years later by masking the N6 position by chlorine (see Scheme 9.11) (Ashton and Scammells, 2005).

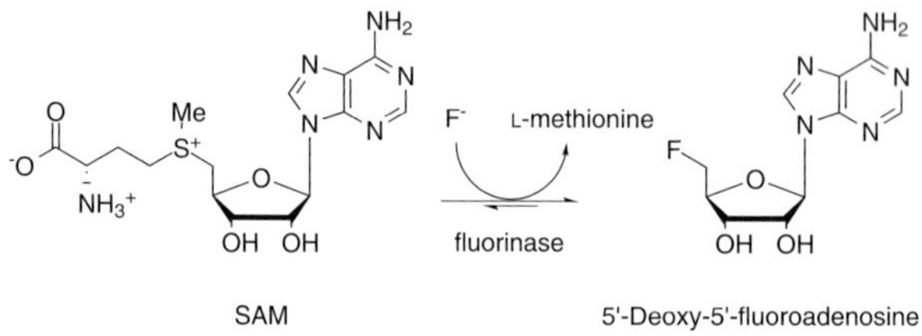

Scheme 9.9.

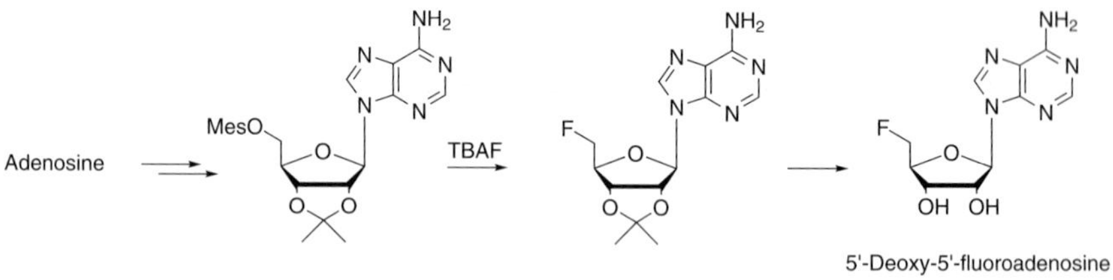

Scheme 9.10.

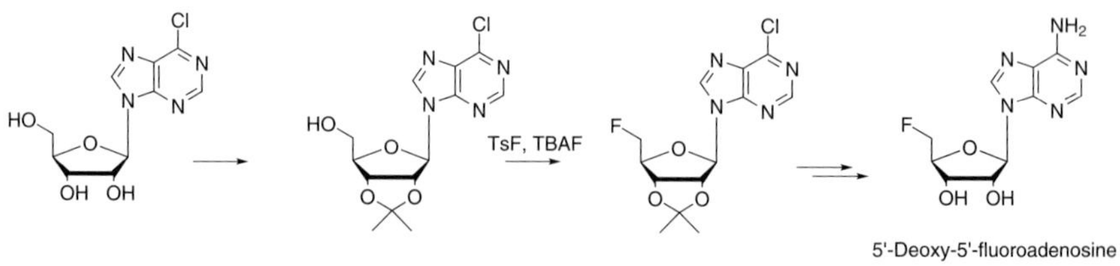

Scheme 9.11.

The first successful synthesis of an ^{18}F-labelled precursor of 5′-deoxy-5′-fluoroadenosine involved the nucleophilic exchange of sulfonylates with [^{18}F]fluoride from [^{18}F]KF/Kryptofix® 222 in low radiochemical yield (Lehel *et al.*, 2002). In the following year, the fluorinase enzyme was exploited to catalyse the one-step synthesis of 5′-deoxy-5′-[^{18}F]fluoroadenosine, originally in low radiochemical yields (Martarello *et al.*, 2003). The RCYs, however, were improved to >86% (corrected for decay) by the use of recombinant fluorinase in high concentrations in a coupled assay with L-amino acid oxidase (Deng *et al.*, 2006).

9.3 Adrenoreceptors

Adrenoreceptors belong, like adenosine receptors, to the superfamily of G-protein-coupled receptors and are classified into two main groups, α and β, which are, in turn, subdivided further. The effects caused by receptor activation include relaxation of gastrointestinal smooth muscle, platelet aggregation, increased cardiac rate and lipolysis, to mention only a few examples stemming from different subtypes of adrenoreceptors. Accordingly, abnormal adrenergic activities play a role in the pathogenesis of several diseases including, e.g. congestive heart failure.

9.3.1 *Epinephrine (adrenaline)*

Epinephrine — also known as adrenaline (Fig. 9.1) — causes vasoconstriction in certain blood vessels and vasodilation in others. Like

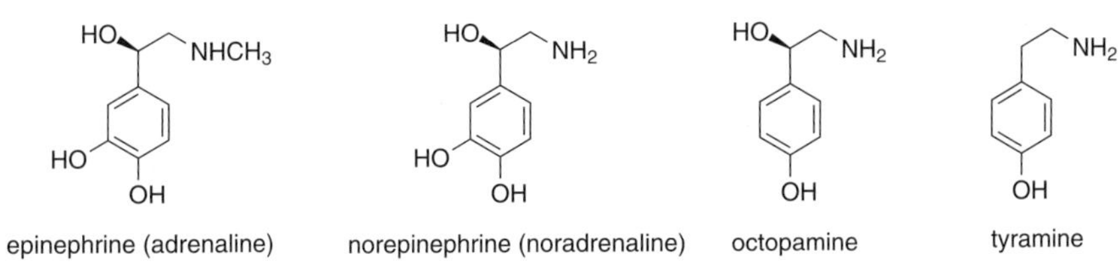

Figure 9.1.

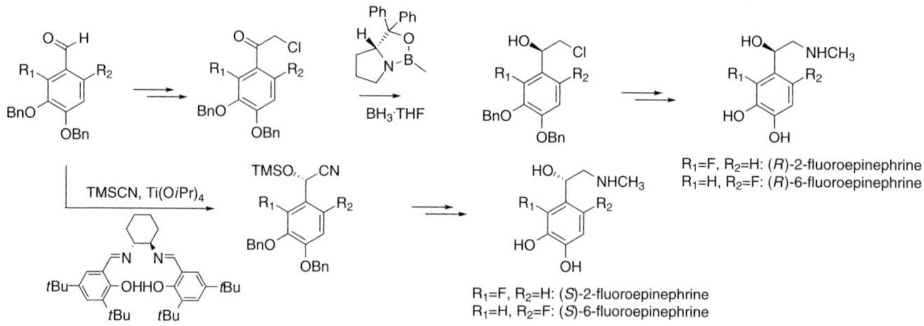

Scheme 9.12.

norepinephrine, epinephrine belongs to the family of catecholamines. Racemic 2- and 6-fluoroepinephrine were prepared and tested for adrenergic activity for the first time about twenty years ago (Adejare *et al.*, 1988). For the preparation of enantiomerically pure fluorinated epinephrines, the use of the chiral (*R*)-Me-CBS-oxazolidine developed by Corey *et al.* for the generation of the respective (*R*)-enantiomers in good yield and in an excellent enantiomeric excess was reported. Similarly, the same authors used a chiral Lewis acid (salen)Ti^{IV} complex as a catalyst for asymmetric cyanohydrin formation for the respective (*S*)-enantiomers in moderate yields and excellent *ee*'s (see Scheme 9.12) (Lu *et al.*, 2000). The position of the fluorine substitution on the aromatic ring revealed some interesting trends with respect to affinity towards those analogues α_1, α_2, β_1 and β_2-adrenergic receptors. Hence, (*R*)-2-F-epinephrine showed reduced activity at both α-receptors but enhanced activity at β-receptors, whereas fluorination of the 6 position in (*R*)-6-F-epinephrine reduced the activity at β-receptors. The behaviour of the respective (*S*)-enantiomers could not be summarised as easily (Lu *et al.*, 2000).

9.3.2 *Norepinephrine (noradrenaline)*

Norepinephrine — also known as noradrenaline (Fig. 9.1) — is a neurotransmitter of the sympathetic nervous system which causes the contraction of blood vessels by activating specific adrenoreceptors similarly to epinephrine. The racemic mixtures of 2-, 5- and 6-fluoronorepinephrine

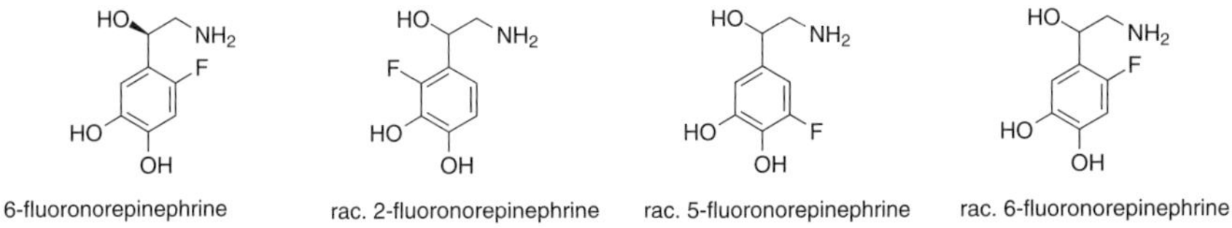

Figure 9.2.

were prepared as early as 1979 (Fig. 9.2). The respective fluorines were introduced by diazotation of the respective aminobenzaldehydes and subsequent photochemical decomposition of the diazonium fluoroborates (Kirk *et al.*, 1979). Recently, the single enantiomers of 2-fluoronorepinephrine and 6-fluoronorepinephrine have been prepared by asymmetric catalysis, similarly to the reactions for epinephrine enantiomers outlined above. In fact, they were intermediates of the syntheses described in Scheme 9.12 (Lu *et al.*, 2000). The biological evaluation of the affinity of the single enantiomers towards α_1, α_2, β_1 and β_2-adrenergic receptors revealed exactly the same trends as for the epinephrine enantiomers. Notably, substitution of the hydrogen at the 5-position of the aromatic ring by fluorine had little effect on adrenergic activity (Kirk *et al.*, 1979).

6-[^{18}F]Fluoronorepinephrine was prepared as the first no-carrier-added ^{18}F-labelled catecholamine in a four-step synthesis from a protected catecholic aldehyde, a nucleophilic aromatic substitution reaction with [^{18}F]fluoride being the first step of the sequence. The introduction of the side chain was achieved by a cyanohydrin reaction with subsequent reduction to a racemic mixture, which was directly used for an *in vivo* PET study of the myocardium of baboons (Ding *et al.*, 1991). A few years later, 6-[^{18}F]fluoronorepinephrine was prepared enzymatically from 6-[^{18}F]fluorodopamine in a stereoselective manner by the use of dopamine β-hydroxylase (DβH) (see Scheme 9.13) (Lui *et al.*, 1998).

9.3.3 *Octopamine*

Octopamine is a norepinephrine analogue (Fig. 9.1) which acts as neurotransmitter in invertebrates and some plants. Only the *meta*-fluorinated

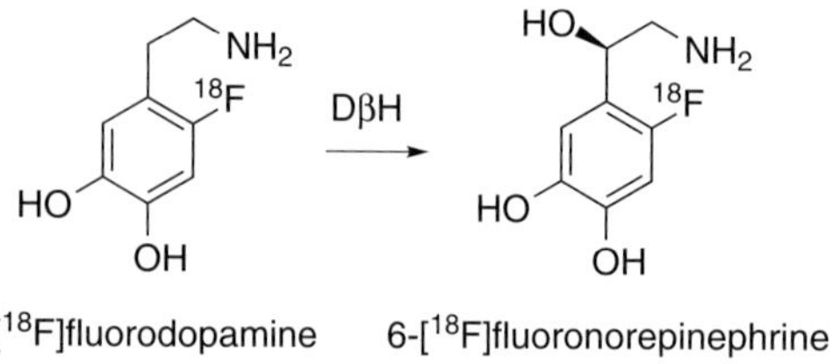

Scheme 9.13.

analogue of octopamine — which is identical to the 3-dehydroxy-6-fluoro analogue of norepinephrine — has been patented recently (Box *et al.*, 2005), the synthetic strategy being largely identical to that for the (*R*)-enantiomers outlined in Scheme 9.12.

9.3.4 *Tyramine*

The neuroactive tyramine is structurally related to epinephrine (Fig. 9.1). Similarly to octopamine, the only published fluorinated analogues are 3-fluorotyramine and 3,5-difluorotyramine, both obtained by photochemical decomposition of the respective aryldiazonium fluoroborates (Kirk, 1976b). It is not surprising that fluorination at the benzylic position has never been reported, as this position is known to be highly reactive due to a labile C–F bond as a consequence of the para-phenolic alcohol.

9.4 Cannabinoid Receptors

Cannabinoid receptors also belong to the family of G-protein-coupled receptors and they are subdivided in type 1 (CB1) and type 2 (CB2). The pharmacologic properties of both plant-derived cannabinoids, such as tetrahydrocannabinols (THC), and endogenous cannabinoids (endocannabinoids), are similar. Both ligand types can regulate a variety of physiological and pathological events such as bone formation, the cardiovascular system, appetite control and energy metabolism.

N-Arachidonoyl ethanol amide (AEA) was considered as the main endogenous ligand for cannabinoid receptors in the 1990s. Later, 2-arachidonoyl glycerol (2-AG) was recognised as even more potent agonist of the cannabinergic character of traditional plant derived cannabinoids such as THC. To study CB1 receptor activity, the metabolic stability of 2-AG was improved by the introduction of a methyl group α to the ester moiety as well as the introduction of one or two fluorines into the glycerol moiety of the molecule. Therefore, the arachidonic acid precursor was coupled to dehydroxy-fluoroglycerol (Scheme 9.14). A noteworthy CB1 receptor activity was only detected for the (*R*)-enantiomer of mono-fluorinated 2-AG (Parkkari *et al.*, 2006).

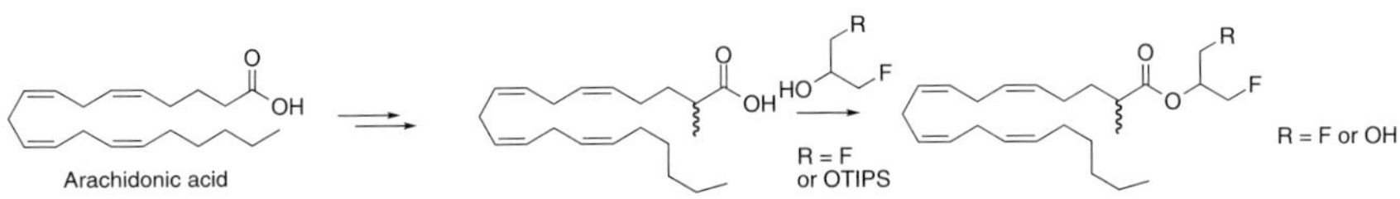

Scheme 9.14.

Several fluorinated THCs were studied as CB1 agonists in the early 1990s. The synthetic routes usually included a classical fluorination step by nucleophilic substitution of a leaving group by fluorine (Charalambous *et al.*, 1992; Usami *et al.*, 1998; Crocker *et al.*, 1999). Δ^8-THC analogues with fluorines at position 5′ or 11 (Fig. 9.3) were more potent than the parent Δ^8-THC in a range of pharmacological tests such as the mouse-tailflick assay (Charalambous *et al.*, 1992). In a recent study, 1-fluoro-1-deoxy-Δ^8-THC analogues were examined. 1-Fluoro-3,5-dimethoxybenzene served as the starting material for their synthesis. *In vitro* binding assays, however, revealed that an exchange of the

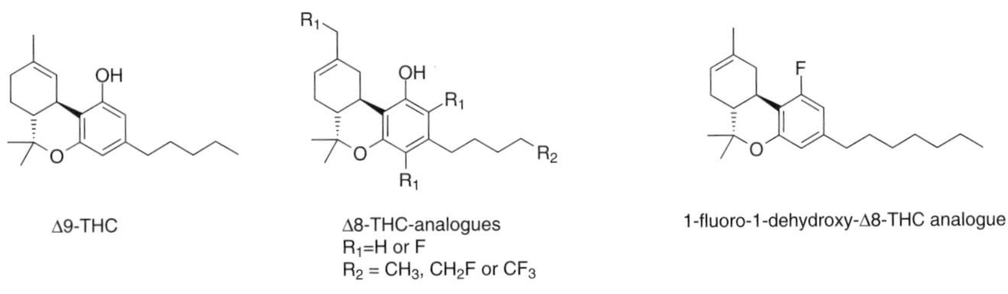

Figure 9.3.

hydroxyl group at position 1 has a detrimental effect on CB1 binding (Crocker *et al.*, 2007).

PET imaging of the brain cannabinoid receptor CB1 has been reviewed by Horti *et al.* (2008).

9.5 Dopamine Receptors

Originally, two types of dopamine receptors were distinguished, based on either activation or inhibition of adenylate cyclase. Recently, additional subtypes have been identified, which resulted in the classification of five dopamine receptors, D_1–D_5 to date. They belong to the family of G-protein-coupled transmembrane receptors and exhibit similar signal transduction mechanisms as other members of this family.

9.5.1 *Dopamine*

Dopamine is the major catecholamine neurotransmitter in the mammalian brain and is responsible for various functions such as locomotion, cognition and also emotion. Additionally, dopamine is also responsible for peripheric regulation, e.g., of renal function or hormone secretion. Dysregulation of the dopamine household is responsible for a number of conditions such as Parkinson's disease or schizophrenia. The biosynthesis of dopamine is outlined in Scheme 9.15.

The regioisomeric 2-, 5- and 6-fluorodopamines were obtained by photochemically induced decomposition of the respective diazonium ions in aqueous fluoroboric acid. 5-Fluorodopamine emerged as a by-product of the preparation of 3,5-difluorotyramine. 6-Fluorodopamine was prepared in the same study by selective nitration of *N*-trifluoroacetyl-3,4-dihydroxyphenethylamine at the 6 position,

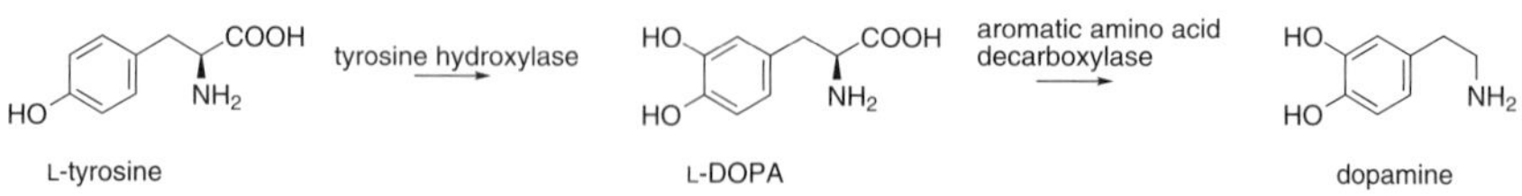

Scheme 9.15. Biosynthesis of dopamine.

followed by a reduction, fluorination and deprotection sequence. In a similar manner, the 2-fluoro isomer was also obtained (Kirk, 1976b). The binding affinities of all ring-fluorinated dopamine regioisomers including 2,6- and 5,6-difluorodopamine have been investigated towards D_1, D_2, D_3 and D_4 receptors. Briefly, 6-fluorodopamine showed reduced binding affinity whereas the 2- and 5-fluoroisomers showed negligible effects (Nie *et al.*, 1996). Side-chain fluorinated analogues have not been reported to our knowledge, probably for the same reasons as outlined for tyramine.

The first and most extensively reported ^{18}F-labelled fluorodopamine derivative is 6-[^{18}F]fluorodopamine. A number of strategies have been reported for the radiolabelling of fluorodopamine and the compound is now frequently used for medicinal studies (e.g. Timmers *et al.*, 2007), the exhaustive listing of which is clearly beyond the scope of this chapter. The reader should, however, be aware that in the literature 6-[^{18}F]fluorodopamine is abbreviated to ^{18}F-FDA as is also [^{18}F]fluoroadenosine.

9.5.2 *L-DOPA*

L-DOPA has a particular relevance in neurology because it is the direct precursor of the neurotransmitter dopamine (Scheme 9.15). Several fluorinated DOPA analogues were prepared throughout the past century, whereby the hydrogens of the aromatic moiety were more or less sequentially substituted by fluorine (Kaiser and Burger, 1957; Creveling and Kirk, 1985). A recent example comprises a route starting from commercial precursors with the fluorines already in place, followed by an asymmetric side-chain manipulation using Oppolzer's chiral sultams for the enantioselective synthesis of fluorinated L-DOPA derivatives (see Scheme 9.16) (Deng *et al.*, 2002).

A fluorine-tagged derivative of L-DOPA has been designed and proved useful as an indicator of the *in vivo* turnover rate of L-DOPA to dopamine (Scheme 9.15). The compound crosses the blood–brain barrier, gets taken up into dopaminergic neurons and is readily converted to the corresponding fluorinated dopamine analogue after amino acid decarboxylation *in vivo* (see Fig. 9.4) (Dingman *et al.*, 2004).

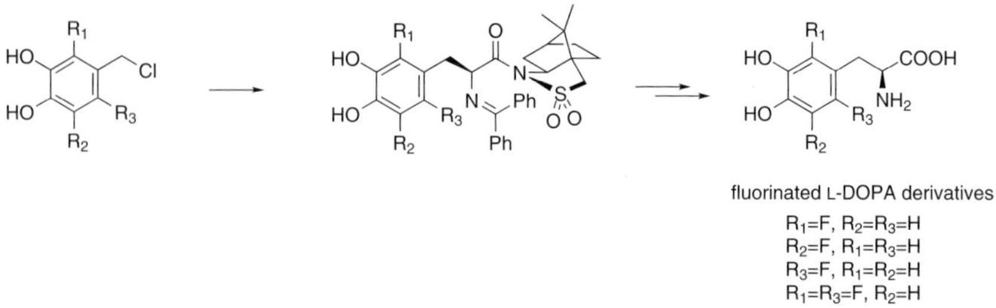

Scheme 9.16.

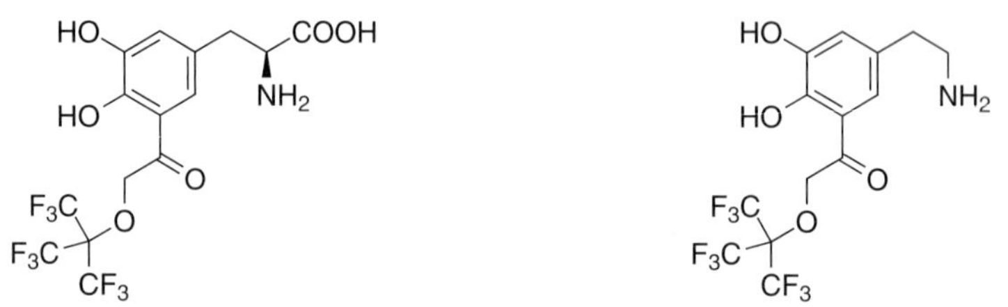

L-5-(2-(perfluoro-*t*-butoxy)acetyl)DOPA L-5-(2-(perfluoro-*t*-butoxy)acetyl)dopamine

Figure 9.4.

6-[^{18}F]FluoroDOPA appears like 6-[^{18}F]fluorodopamine extensively as an imaging tracer in the medical literature. Amongst several other significant ^{18}F-labelled neurotransmitters, it has recently been reviewed (Cai *et al.*, 2008) and the interested reader is referred to this review and more specific literature, e.g. Eshuis *et al.* (2009).

9.6 GABA Receptors

After glycine, γ-aminobutyric acid (GABA, Fig. 9.5) is the next major inhibitory neurotransmitter in the mammalian central nervous system (Krogsgaard-Larsen, 1988; Andersen *et al.*, 2001). GABA receptors are currently subdivided into $GABA_A$, $GABA_B$ and $GABA_C$ subtypes. Whereas $GABA_A$ is a ligand-gated channel, $GABA_B$ belongs to the G-protein-coupled receptors. $GABA_C$ receptors were defined as receptors insensitive to both bicuculline and baclofen. The dysfunction of GABA regulation is associated with a variety of neurological disorders, including anxiety and epilepsy.

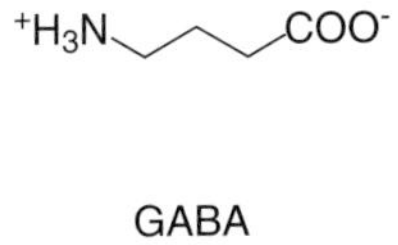

Figure 9.5.

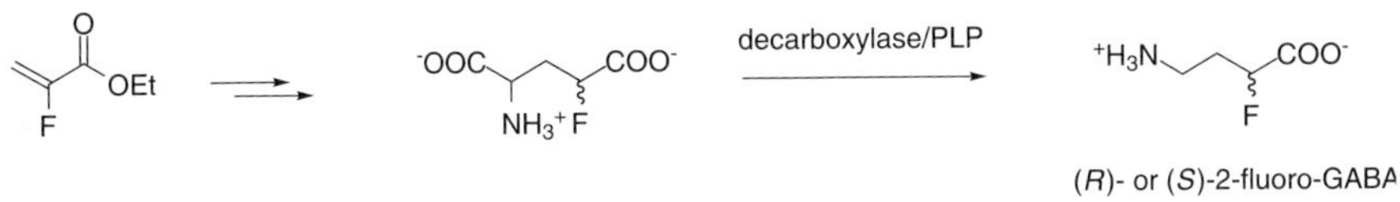

Scheme 9.17.

There is extensive literature on GABA analogues for the investigation of the binding properties to their respective GABA receptors, and in this regard there is a strong focus on conformationally restricted GABA analogues, a significant number of which contain fluorine (Qiu *et al.*, 2000; Wang *et al.*, 2006). Additionally, a large number of GABA analogues have been evaluated as inhibitors for γ-aminobutyric acid aminotransferase (GABA-AT) or as GABA uptake inhibitors. Fluorinated GABA analogues, where fluorine replaces hydrogen on the GABA backbone at the 2- and recently also the 3-position, have been reported. The precursor for the preparation of 2-fluoro-GABA — 4-fluoroglutamic acid — was synthesised by a Michael addition of diethyl acetamidomalonate to 2-fluoroacrylate and subsequent separation of the diastereoisomers (Tolman, 1993). After deprotection, the action of glutamate decarboxylase gave rise to the enantiomers of 2-fluoro-GABA (Scheme 9.17) but there is no report of biological studies so far (Tolman *et al.*, 2000).

The racemate of 3-fluoro-GABA has been prepared by replacement of the alcohol of DL-4-amino-3-hydroxy-butyric acid using SF_4/HF (Kollonitsch *et al.*, 1979). A recent synthetic sequence towards the enantiomers of 3-fluoro-GABA set out from D- or L-phenylalanine, respectively. The fluorination was accomplished using Deoxo-Fluor in a reaction that proceeds via an aziridinium ion. Subsequent oxidative cleavage of the phenyl ring gave the carboxylic acid functionality (Scheme 9.18). The fluorinated analogues displayed agonist activity towards cloned (human)

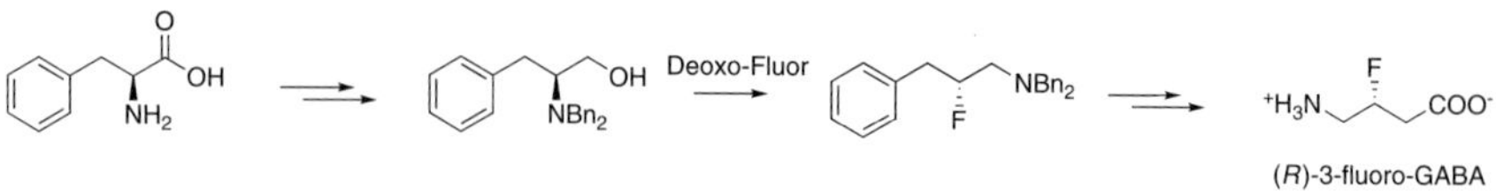

Scheme 9.18. Synthetic route to enantiomerically pure 3-fluoro-GABA, exemplified for the (*R*)-enantiomer.

GABA$_A$ receptors, albeit with significantly reduced activity when compared to native GABA (Deniau *et al.*, 2007). Notably, the (*R*)-enantiomer undergoes GABA-AT-mediated elimination of HF whereas the (*S*)-enantiomer does not (Clift *et al.*, 2007).

(*S*)-4-Amino-5-fluoropentanoic acid has been prepared from L-glutamic acid via the respective γ-lactam using AgF to accomplish a nucleophilic substitution of bromine for fluorine (see Scheme 9.19) (Silverman and Levy, 1980). (*S*)-4-Amino-5-fluoropentanoic acid acts as an inactivator of GABA-AT by irreversible binding to the pyridoxal-phosphate cofactor of the aminotransferase (Silverman and Invergo, 1986).

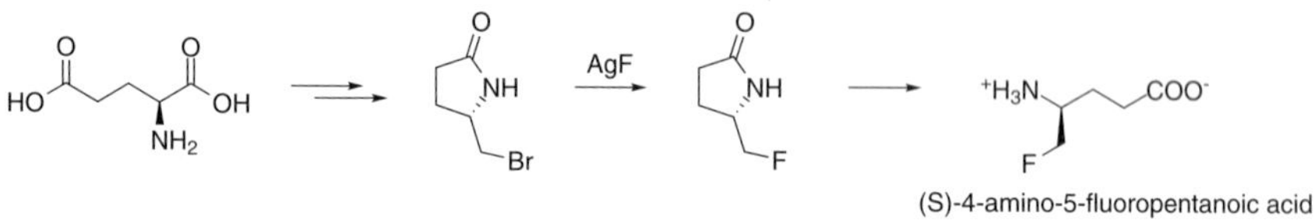

Scheme 9.19.

9.7 Glutamate Receptors

Glutamate receptors have been studied extensively with selective agonists and antagonists. Currently there are four major subtypes, namely NMDA, AMPA, kainate and metabotropic receptors. Glutamate is the principal and ubiquitous excitatory transmitter in the central nervous system. Aspartate has a similar role in certain brain regions and will therefore also be treated in this section. Another important compound in this series is *N*-methyl-D-aspartate (NMDA), after which one subtype of glutamate receptors has been named, although the compound is synthetic and not an endogenous neurotransmitter. The parent compounds are depicted in Fig. 9.6.

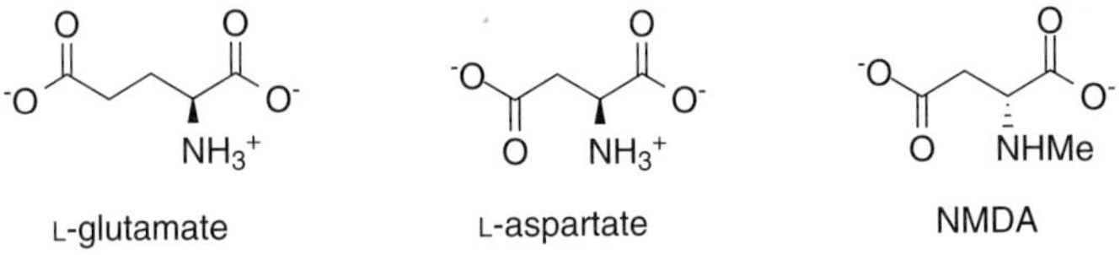

Figure 9.6.

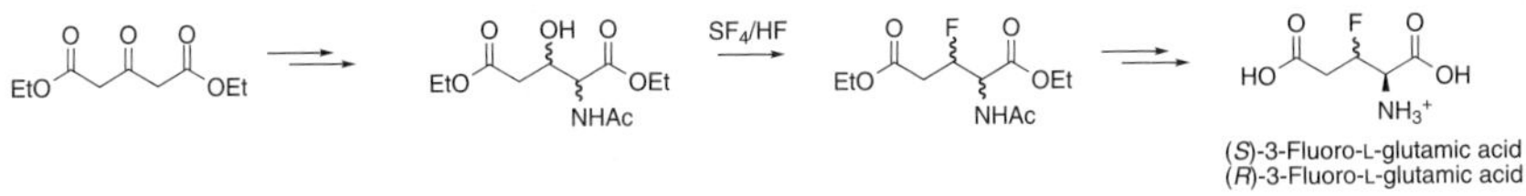

Scheme 9.20.

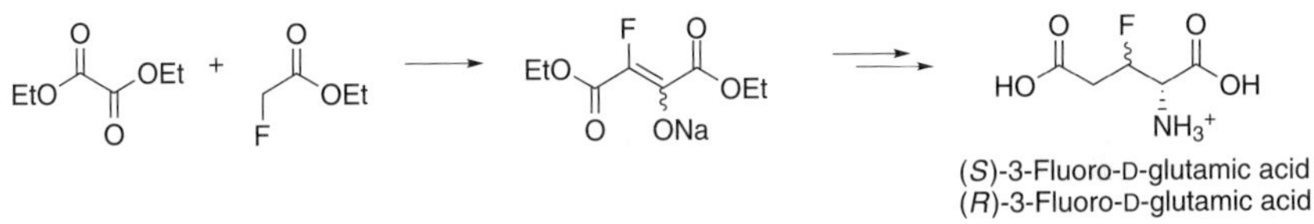

Scheme 9.21.

3-Fluoroglutamic acid was first reported in 1985. The authors started their nine-step synthesis from diethyl acetonedicarboxylate, which was converted to the isomers of 3-hydroxyglutamic acid. The alcohol moiety was replaced by fluorine in a reaction with SF_4 in liquid HF at –78°C. After diastereoisomer separation, the final step in the reaction sequence involved an (*S*)-selective enzymatic *N*-deacetylation (see Scheme 9.20) (Vidal-Cros *et al.*, 1985). The same research group developed a method for the 3-fluorinated enantiomers of D-glutamic acid which began with a Claisen condensation between diethyl oxalate and ethyl fluoroacetate, followed by three subsequent steps. The final step was an enzymatic reductive amination (see Scheme 9.21) (Vidal-Cros *et al.*, 1989). The 3,3-difluorinated derivative was also prepared from a masked 3-oxoproline where the fluorines were introduced by treatment with DAST (see Scheme 9.22) (Hart and Coward, 1993).

Racemic 4-fluoroglutamate is a key intermediate on the synthetic route to 4-fluoro-GABA which was outlined above (see Scheme 9.18) (Tolman, 1993), but other strategies have been used to obtain the racemate

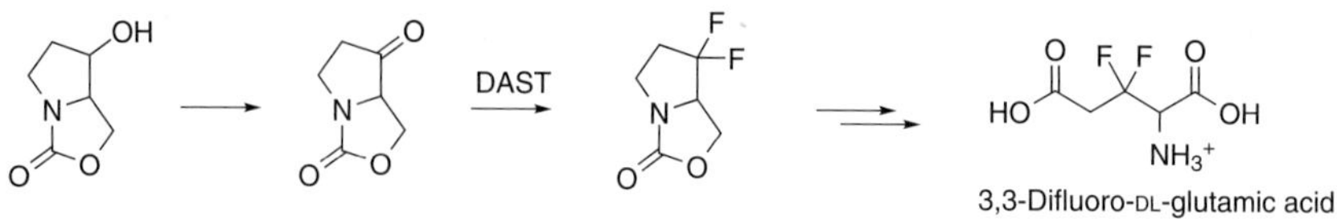

Scheme 9.22.

(Tsushima *et al.*, 1988) as well as single diastereoisomers (Konas *et al.*, 2001). Different multistep strategies lead to the 4,4-difluoro analogue of glutamic acid, two of which are outlined in Scheme 9.23. The key step in the first case is the addition of a difluorinated alkene to a chiral aldehyde precursor (Ding *et al.*, 2001), whereas the fluorines are introduced by electrophilic addition to a chiral enolate in the other strategy (Konas and Coward, 2001).

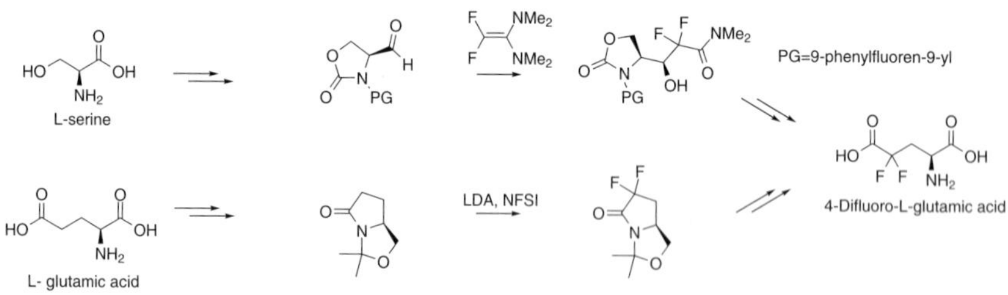

Scheme 9.23.

The racemate of 3-fluoroaspartic acid was first synthesised many years ago in a manner similar to that described for racemic 3-fluoro-GABA. This involved a direct fluorodehydroxylation using SF_4/HF (Kollonitsch *et al.*, 1979). Several other routes to racemic 3-fluoroaspartic acid were published in the 1980s. Both diastereoisomers were prepared in a stereocontrolled manner; however, two different strategies had to be applied in each case. For (*S*)-3-fluoro-D-aspartic acid, diethyl epoxysuccinate served as the starting material and the fluorine was introduced with DAST. This reaction proceeds formally with retention of configuration due to an intermediate aziridinium ion, similar to that proposed for the 3-fluoro-GABA synthesis. For the other fluoroaspartic acid diastereoisomer, a key intermediate involved a cyclic sulfate derived from D-tartrate which was ring-opened with Et_4NF (see Scheme 9.24) (Charvillon *et al.*, 1996).

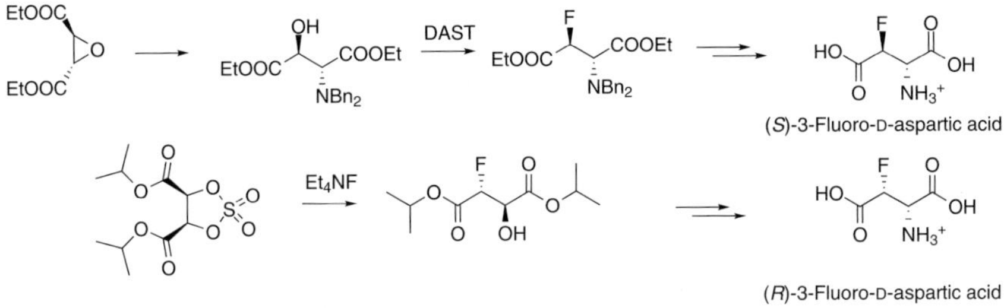

Scheme 9.24.

To date, no difluorinated variant of aspartic acid has been published and data on biological activity with respect to glutamate receptors of these mono-fluorinated compounds are not available.

9.8 Histamine Receptors

Based on their pharmacology, histamine receptors have been classified into four subtypes: H_1, H_2, H_3 and H_4. These receptors are found in smooth muscle, the central nervous system and the spleen. The preparation of side-chain fluorinated β-fluorohistamine (Kollonitsch *et al.*, 1979) and β,β-difluorohistamine (see Fig. 9.7) has been reviewed recently (Dolensky *et al.*, 2004). Briefly, the introduction of fluorine was *inter alia* accomplished by the addition of 'BrF' — which is formed *in situ* from NBS and e.g. $Et_3N.3HF$ — to a double bond.

A series of fluorine-containing 2-substituted histamines was prepared by the cyclisation of adequately fluorinated imidates or amidates in liquid ammonia with 2-oxo-4-phthalimido-1-butyl acetate or 2-bromo-4-phthalimido-1-butanal as substrates, and subsequent dephthaloylation

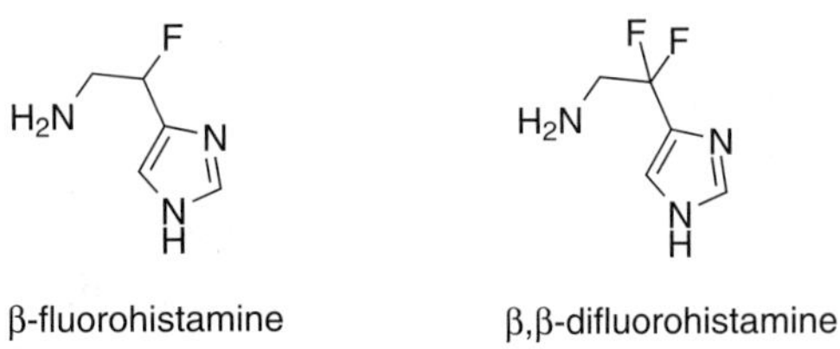

Figure 9.7.

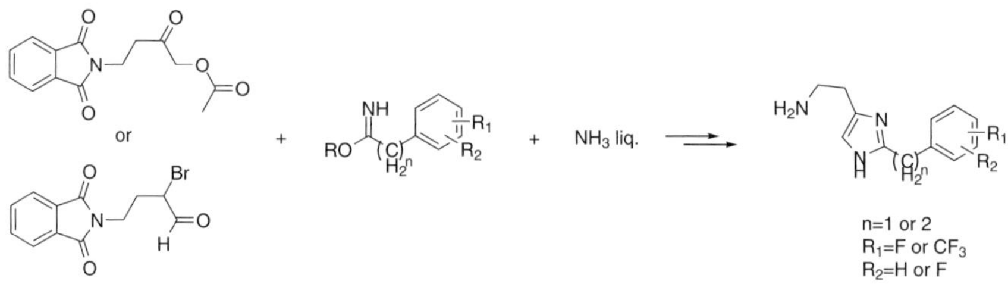

Scheme 9.25.

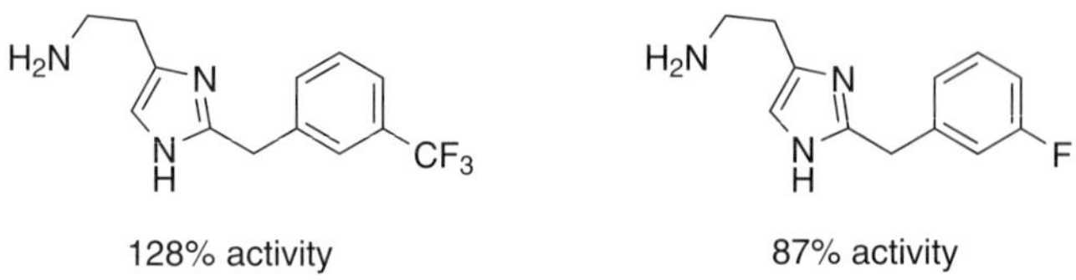

Figure 9.8.

(see Scheme 9.25) (Zingel *et al.*, 1990; Leschke *et al.*, 1995). The maleates of these novel compounds were tested for contractile effects on isolated guinea pig ileal segments and compared to the action of histamine itself. The 2-[3-(trifluoromethyl)benzyl] derivative (Fig. 9.8) was found to be a 128% more potent agonist than histamine towards the investigated H_1-receptor, while the 3-fluorophenyl derivative showed 87% activity compared to histamine. Only *meta* fluoro substitution in the aromatic ring resulted in diminished agonist activity and the authors deduced that this positioning is distinguished in terms of biological activity.

9.9 Muscarinic Receptors

Although acetylcholine is an important neurotransmitter which interacts with muscarinic receptors, the chemical structure itself limits the possibilities for fluorination (Fig. 9.9). No directly fluorinated acetylcholines have been reported to our knowledge, not even the trifluoroacetyl derivative. Muscarine is a fungal toxin which acts on the respective receptors. The fluorinated analogue 3-fluoro-muscarine has been described in the literature (Brown *et al.*, 1994) in addition to all four isomers of the 4-deoxy-4-fluoro

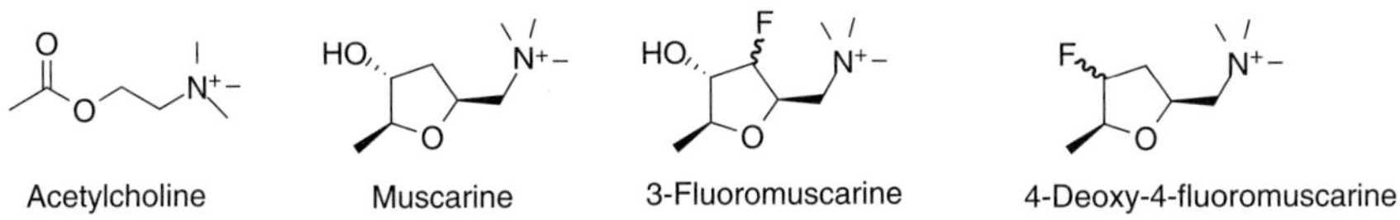

Figure 9.9.

derivative see Fig. 9.9 (Bravo *et al.*, 1992; Farina *et al.*, 1996). The synthetic pathway is outlined in Scheme 9.26 for the isomer with the best potency for muscarinic receptors. This isomer has the same absolute configuration as the hydroxyl group of natural muscarine and exhibits similar behaviour towards subtype 2 and 3 muscarinic receptors with only subtle differences (Bravo *et al.*, 1992).

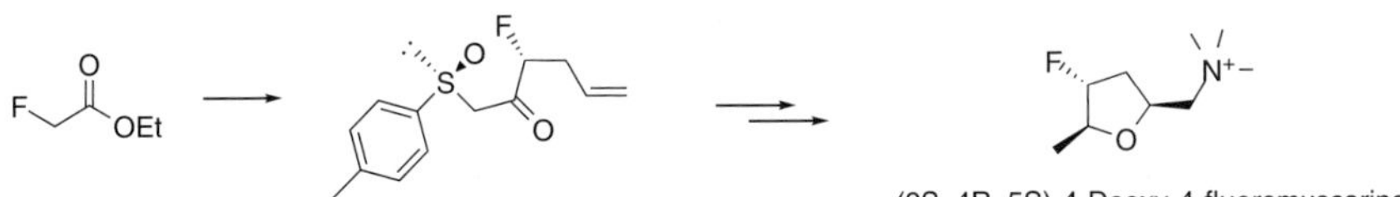

Scheme 9.26. Chiral sulfoxide-aided fluoromuscarine synthesis exemplified for the (4*R*)-isomer.

9.10 Nicotinic Acetylcholine Receptors

Three different classes of nicotinic receptors have been defined, namely the muscle-type, the ganglion-type and the CNS-type. They are all ligand-gated ion channels which are activated by nicotine and/or acetylcholine. The lack of fluorinated acetylcholine derivatives has already been mentioned above. However, fluorinated nicotinic derivatives have been widely studied. (*S*)-5-Fluoronicotine has been tested as an agonist for nicotine receptors in the neocortex of rats and found to increase the cortical release of acetylcholine and dopamine similarly to nicotine itself (Summers *et al.*, 1995). Racemic 6-fluoronicotine was prepared in a one-step Balz–Schiemann reaction by diazotisation of racemic 6-aminonicotine in the presence of HBF_4. The compound was evaluated amongst other nicotine analogues, with respect to binding affinity and in functional studies including a tail-flick

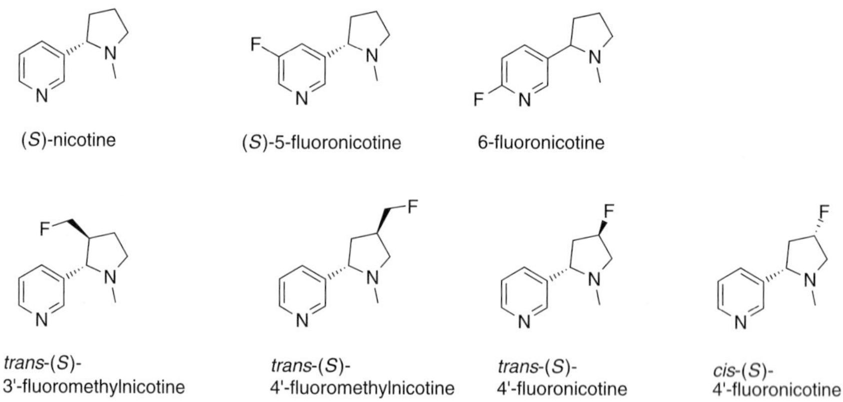

Figure 9.10.

assay, and was found to behave very similarly to *rac*-nicotine (see Fig. 9.10) (Dukat *et al.*, 1996). From these data it could be concluded that fluorine substitution on the pyridine core of nicotine has little effect on its biological activity. Such derivatives may therefore be considered as potential mimics of the parent compound, a feature that has been exploited in PET studies.

Alternatively, the hydrogens on the pyrrolidine ring have been substituted by fluorine, and the respective derivatives subjected to binding affinity studies on neuronal nicotinic acetylcholine receptors. However, details of their synthesis have not been described. The fluorinated analogues generally bind more weakly than (*S*)-nicotine. For example *trans*-(*S*)-4′-fluoromethylnicotine binds tenfold less and *trans*-(*S*)-3′-fluoromethylnicotine and *cis*-(*S*)-4′-fluoronicotine about 100-fold less. *Trans*-(*S*)- 4′-fluoronicotine is almost 400-fold weaker than nicotine in binding to nicotinic acetylcholine receptors (see Fig. 9.10) (Kim *et al.*, 1996). Unfortunately, the K_i values from this study cannot be directly compared to those from 6-fluoronicotine because the latter were determined on racemates.

The preparation of nicotine derivatives ^{18}F-labelled in position 2 and 6 was described in the 1980s in 23% and 8% decay-corrected radiochemical yield, respectively (Ballinger *et al.* 1984). However, no further imaging studies have been reported.

9.11 Serotonin Receptors

The classification of serotonin receptors is somewhat more complicated than for other receptor types, and seven groups based on pharmacological profiles, primary sequences and signal transduction mechanisms are currently accepted in the field. Needless to say, these groups are further distinguished by a number of subgroups. The neurotransmitter serotonin (5-hydroxytryptamine, 5-HT, Fig. 9.11) is involved in multiple physiological functions and is synthesised *in vivo* from L-tryptophan in serotonergic neurons and enterochromaffin cells of the gastrointestinal tract. Kirk and coworkers have reported synthetic strategies for several fluorinated analogues of serotonin over the past three decades (e.g. Dolensky *et al.*, 2004). Early work focused on fluorination of the indole core, which was mainly achieved by utilisation of the Abramovitch adaptation of the Fischer indole synthesis, with the fluorine substituents already on the desired positions of the precursors (Kirk, 1976a; Chen *et al.*, 1998). Recently, 7-fluoro-serotonin was prepared in this traditional manner in low overall yield. However, an alternative approach, based on the use of 4-aminobutyraldehyde diethyl acetal for the formation of the indole core, significantly improved the overall yield. This same approach was also chosen for the preparation of 6,7-difluoro-serotonin (see Scheme 9.27) (Heredia-Moya *et al.*, 2006). For

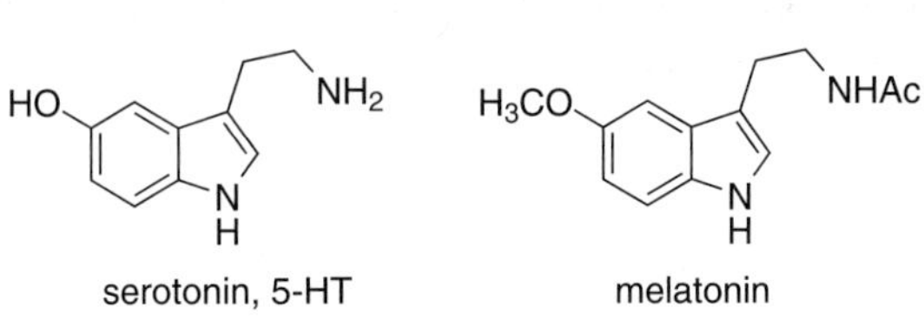

Figure 9.11.

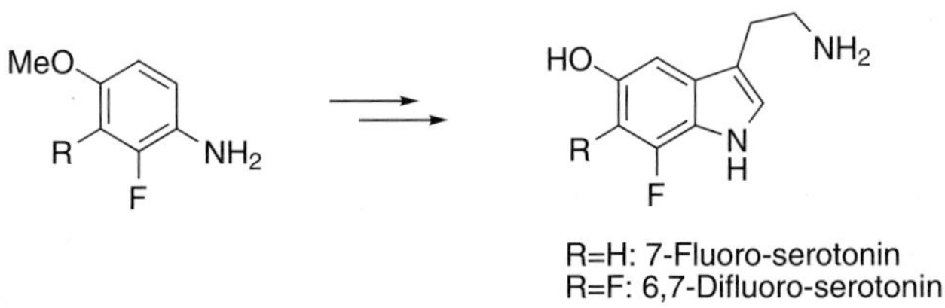

Scheme 9.27.

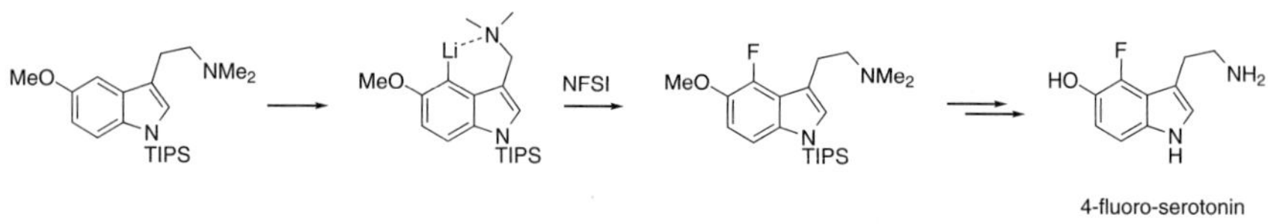

Scheme 9.28.

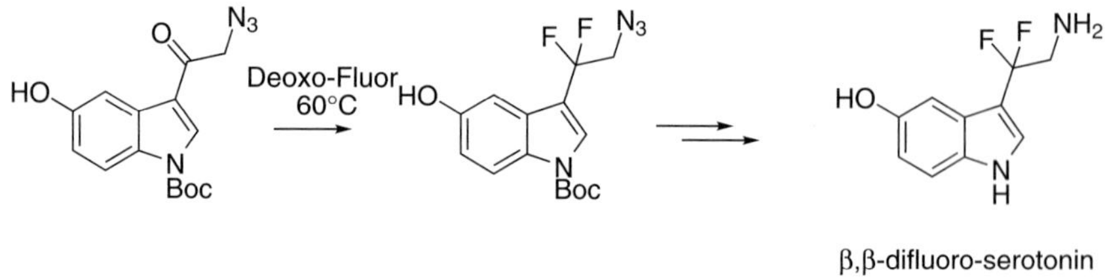

Scheme 9.29.

6-fluoro-DL-tryptophan 6-fluoro-serotonin

Scheme 9.30.

the selective preparation of 4-fluoro-serotonin, an elegant strategy involving a regioselective lithiation with subsequent electrophilic fluorination was chosen (see Scheme 9.28) (Hayakawa *et al.*, 1999).

Side-chain fluorinations have also been investigated in this arena (Deng *et al.*, 2003; Dolensky *et al.*, 2004). The approach adopted for β,β-difluorination was different from the strategy used for the analogous reaction on histamines. In this case it was accomplished by reaction of the respective ketone with Deoxo-Fluor at 60°C (Scheme 9.29).

6-Fluoro-serotonin (6F-5HT) was identified as the main metabolite of 6-fluoro-DL-tryptophan in rat brain after specific hydroxylation by the enzyme tryptophan hydroxylase and decarboxylation (see Scheme 9.30) (Peters, 1971). 6F-5-HT inhibits serotonin synthesis *in vivo*, but it is not clear whether the compound is a true substitute of 5-HT or a false transmitter (Chanut *et al.*, 1994).

9.12 Melatonin Receptors

Melatonin receptors, like serotonin receptors, belong to varying receptor types which reflects their changing classification. The molecular structure of melatonin is closely related to that of serotonin (Fig. 9.11). Thus, very similar synthetic strategies to fluorinated melatonin derivatives have been used (Hayakawa *et al.*, 1999; Deng *et al.*, 2003; Heredia-Moya *et al.*, 2006).

Although the synthesis of radiolabelled ^{18}F-melatonin was accomplished as early as 1988 by the use of $[^{18}F]F_2$ (Chirakal *et al.*, 1988), no further biological or imaging studies have been published.

9.13 Vanilloid Receptors

The vanilloid receptor is a typical ligand-gated cation channel. It is sensitive to structures derived from vanillic acid, hence the name, but also to heat and acidic conditions (Conway, 2008).

9.14 Capsaicin

Capsaicin is the major component responsible for the pungency of chilli peppers. It triggers the opening of Na^+ or Ca^{2+} channels and therefore a sensation of pain. The two enantiomers of 2-fluorocapsaicin have recently been prepared by stereoselective electrophilic fluorination using a chiral catalyst (Scheme 9.31). Their biological activity towards dorsal root

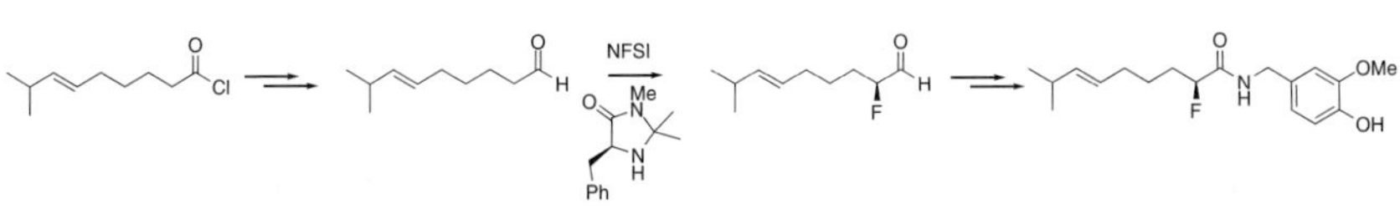

Scheme 9.31. Synthetic route exemplified for (*S*)-2-fluorocapsaicin.

ganglia neurons was found to be exactly the same as the unfluorinated parent compound which renders those derivatives good mimics of capsaicin for further studies (Winkler *et al.*, 2009). The only other report regarding fluorinated capsaicin derivatives carried the fluorine on the aromatic ring (Pooput *et al.*, 2008).

9.15 Anandamide

Arachidonylethanolamide (anandamide, Fig. 9.12) is a putative endogenous ligand for cannabinoid receptors, but also acts as an agonist for vanilloid receptors (Ross, 2003). A fluorinated derivative of anandamide has been reported recently albeit without description of its synthesis (Fig. 9.12). The compound was then tested as an agonist for cannabinoid receptors and was found to be 10 times more potent than the parent compound. Similar effects were found for α-methylated anandamide compared to the fluorinated counterpart (Wiley *et al.*, 1997).

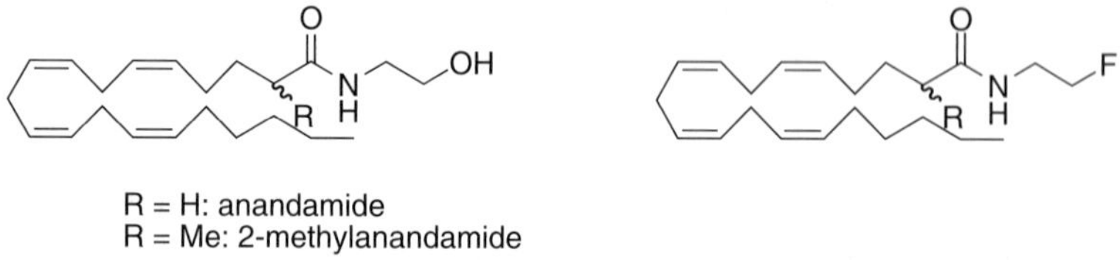

Figure 9.12.

9.16 Conclusion

It is evident from this survey that an impressive array of selectively fluorinated neurotransmitters have been synthesised over the past two or three decades. Many of them show interesting modifications of agonistic or antagonistic properties; others show essentially equal bioactivities compared to their non-fluorinated parents and are thus of particular interest as non-invasive NMR reports (^{19}F) or as PET tracers (^{18}F). Important aspects of potential conformational changes or modulation of physicochemical properties, in particular basicity, lipophilicity or membrane permeation, as a consequence of the exchange of hydrogen by fluorine, are not discussed in this overview, mostly because such effects have not been reported systematically. Specifically fluorinated neurotransmitter analogues will be designed and synthesised in the future; their further exploration in terms of biological activity as well as conformational and physicochemical properties will greatly advance our knowledge about fluorine-containing bioactive molecules.

References

Adejare, A., Gusovsky, F., Padgett, W. *et al.* (1988) Syntheses and adrenergic activities of ring-fluorinated epinephrines. *J Med Chem* 31, 1972–1977.

Alauddin, M.M., Fissekis, J.D. & Conti, P.S. (2003) Synthesis of [^{18}F]-labeled adenosine analogues as potential PET imaging agents. *J Labelled Cpd Radiopharm* 46, 805–814.

Alauddin, M.M., Shahinian, A., Park, R. *et al.* (2007) Biodistribution and PET imaging of [^{18}F]-fluoroadenosine derivatives. *Nucl Med Biol* 34, 267–272.

Andersen, K.E., Sorensen, J.L., Lau, J. *et al.* (2001) Synthesis of novel γ-aminobutyric acid (GABA) uptake inhibitors. 5.1 Preparation and structure–activity studies of tricyclic analogues of known GABA uptake inhibitors. *J Med Chem* 44, 2152–2163 and references therein.

Ashton, T.D. & Scammells, P.J. (2005) An improved synthesis of 5′-fluoro-5′-deoxyadenosines. *Bioorg Med Chem Lett* 15, 3361–3363.

Ballinger, J.R., Bowen, B.M., Firnau, G. *et al.* (1984) Radiofluorination with reactor-produced cesium [^{18}F]fluoride: no-carrier-added [^{18}F]2-fluoronicotine and [^{18}F]6-fluoronicotine. *Int J Appl Radiat Isot* 35, 1125–1128.

Barrio, J.R., Namavari, M., Keen, R.E. *et al.* (1998) The elusive 8-fluoroadenosine. *Tetrahedron Lett* 39, 7231–7234.

Box, P.C., Coe, D.M. & Hobbs, H. (2005) Preparation of substituted phenoxy aryl amides as β2-adrenoceptor agonists for the treatment of COPD. WO patent 2005040103, 56.

Braendvang, M. & Gundersen, L.-L. (2006) A novel method for the introduction of fluorine into the purine 2-position: synthesis of 2-fluoroadenosine and a formal synthesis of the antileukemic drug fludarabine. *Synthesis* 18, 2993–2995.

Bravo, P., Resnati, G., Angeli, P. *et al.* (1992) Synthesis and pharmacological evaluation of enantiomerically pure 4-deoxy-4-fluoromuscarines. *J Med Chem* 35, 3102–3110.

Brown, D., Liston, D., Mantell, S.J. *et al.* (1994) Synthesis and structure activity relationship of 3-substituted muscarines. *Carbohydrate Lett* 1, 31–36.

Butora, G., Schmitt, C., Levorse, D.A. *et al.* (2007) The elusive 8-fluoroadenosine: a simple non-enzymatic synthesis and characterization. *Tetrahedron* 63, 3782–3789.

Cai, L., Lu, S. & Pike, V.W. (2008) Chemistry with [^{18}F]fluoride ion. *Eur J Org Chem* 17, 2853–2873.

Chanut, E., Bonnet, J.J., Trouvin, J.H. *et al.* (1994) 6-Fluoro-serotonin as a substrate for the neuronal serotonin transporter. *J Neural Transm* 96, 105–112.

Charalambous, A., Lin, S., Marciniak, G. *et al.* (1991) Pharmacological evaluation of halogenated Δ^8-THC analogs. *Pharm Biochem Behav* 40, 509–512.

Charvillon, F.B. & Amouroux, R. (1996) Synthesis of nonracemic 3-fluoro-aspartic acids. *Tetrahedron Lett* 37, 5103–5106.

Chen, H.-J.C., Applewhite, T., Jayachandran, B. *et al.* (1998) Synthesis of 4,6-difluoro-5-hydroxy-(α-methyl)tryptamine and 4,6-difluoro-5-hydroxy-(β-methyl)tryptamine as potential selective monoamine oxidase B inhibitors. *J Fluorine Chem* 92, 41–44.

Chirakal, R., Sayer, B.G., Firnau, G. *et al.* (1988) Synthesis of F-18 labeled fluoromelatonins and 5-hydroxyfluorotryptophans. *J Labelled Compd* 25, 63–71.

Clift, M.D., Ji, H., Deniau, G.P. *et al.* (2007) Enantiomers of 4-amino-3-fluorobutanoic acid as substrates for γ-aminobutyric acid aminotransferase: conformational probes for GABA binding. *Biochemistry* 46, 13819–13828.

Conway, S. J. (2008) TRPing the switch on pain: an introduction to the chemistry and biology of capsaicin and TRPV1. *Chem Soc Rev* 37, 1530–1545.

Couturier, O., Luxen, A., Chatal J.-F. *et al.* (2004) Fluorinated tracers for imaging cancer with positron emission tomography. *Eur J Nucl Med Mol Imaging* 31, 1182–1206.

Creveling, C.R. & Kirk, K.L. (1985) The effect of ring-fluorination on the rate of *O*-methylation of dihydroxyphenylalanine (DOPA) by catechol-*O*-methyltransferase: significance in the development of ^{18}F-PET scanning agents. *Biochem Biophys Res Commun* 130, 1123–1131.

Crocker, P.J., Saha, B., Ryan, W.J. *et al.* (1999) Development of agonists, partial agonists and antagonists in the Δ^8-tetrahydrocannabinol series. *Tetrahedron* 55, 13907–13926.

Crocker, P.J., Mahadevan, A., Wiley, J.L. *et al.* (2007) The role of fluorine substitution in the structure–activity relationships (SAR) of classical cannabinoids. *Bioorg Med Chem Lett* 17, 1504–1507.

Deng, W.-P., Wong, K.A. & Kirk, K.L. (2002) Convenient syntheses of 2-, 5- and 6-fluoro- and 2,6-difluoro- L-DOPA. *Tetrahedron: Asymmetry* 13, 1135–1140.

Deng, W.-P., Nam, G., Fan, J. *et al.* (2003) Syntheses of β,β-difluorotryptamines. *J Org Chem* 68, 2798–2802.

Deng, H., Cobb, S.L., Gee, A.D. *et al.* (2006) Fluorinase mediated C–^{18}F bond formation, an enzymatic tool for PET labelling. *Chem Commun*, 652–654.

Deniau, G., Slawin, A.M.Z., Lebl, T. *et al.* (2007) Synthesis, conformation and biological evaluation of the enantiomers of 3-fluoro-γ-aminobutyric acid ((*R*)- and (*S*)-3F-GABA): an analogue of the neurotransmitter GABA. *ChemBioChem* 8, 2265–2274.

Ding, Y.-S., Fowler, J.S., Gatley, S.J. *et al.* (1991) Synthesis of high specific activity (+)- and (−)-6-[^{18}F]fluoronorepinephrine via the nucleophilic aromatic substitution reaction. *J Med Chem* 34, 767–771.

Ding, Y., Wang, J., Abboud, K.A. *et al.* (2001) Synthesis of L-4,4-difluoroglutamic acid via nucleophilic addition to a chiral aldehyde. *J Org Chem* 66, 6381–6388.

Dingman, S., Mack, D., Branch, S. *et al.* (2004) The fate of perfluoro-tagged metabolites of L-DOPA in mice brains. *J Immunoassay Immunochem* 25, 359–370.

Dolensky, B., Nam, G., Deng, W.-P. *et al.* (2004) Syntheses of side-chain fluorinated biologically important imidazoles and indoles. *J Fluorine Chem* 125, 501–508.

Dukat, M., Fiedler, W., Dumas, D. *et al.* (1996) Pyrrolidine-modified and 6-substituted analogs of nicotine: a structure–affinity investigation. *Eur J Med Chem* 31, 875–888.

Eshuis, S.A., Jager, P.L., Maguire, R.P. *et al.* (2009) Direct comparison of FP-CIT SPECT and F-DOPA PET in patients with Parkinson's disease and healthy controls. *Eur J Nucl Med Mol Imag* 36, 454–462.

Farina, A., Meille, S.V. & Resnati, G. (1996) The crystalline conformation and packing of fluoro-muscarine and *allo*-fluoro-muscarine iodides: the effect of hydroxyl group substitution by fluorine. *J Fluorine Chem* 80, 47–52.

Guillerm, D., Muzard, M., Allart, B. *et al.* (1995) Synthesis of 4′-fluoroadenosine as an inhibitor of *S*-adenosyl-L-homocysteine hydrolase. *Bioorg Med Chem Lett* 5, 1455–1460.

Hart, B.P. & Coward, J.K. (1993) The synthesis of DL-3,3-difluoroglutamic acid from a 3-oxoprolinol derivative. *Tetrahedron Lett* 34, 4917–4921.

Hayakawa, Y., Singh M., Shibata, N. *et al.* (1999) Regioselective electrophilic fluorination of indoles: syntheses of 4-fluoroserotonin and 4-fluoromelatonin. *J Fluorine Chem* 97, 161–164.

Heredia-Moya, J., Hayakawa, Y. & Kirk, K.L. (2006) Syntheses of 7-fluoro-and 6,7-difluoroserotonin and 7-fluoro- and 6,7-difluoromelatonin. *J Fluorine Chem* 127, 1256–1260.

Horti, A.G., Ravert, H.T., Mathews, W.B. *et al.* (2006) Synthesis of 2-[^{18}F]fluoroadenosine (2-[^{18}F]FAD) as potential radiotracer for studying malignancies by PET. *J Labelled Cpd Radiopharm* 49, 811–815.

Horti, A.G. & Van Laere, K. (2008) Development of radioligands for *in vivo* imaging of type 1 cannabinoid receptors (CB1) in human brain. *Curr Pharm Des* 14, 3363–3383.

Isanbor, C. & O'Hagan, D. (2006) Fluorine in medicinal chemistry. A review of anti-cancer agents. *J Fluorine Chem* 127, 303–319.

Kaiser, C. & Burger, A. (1957) 2-Substituted derivatives of 3,4-dihydroxyphenylalanine. *J Am Chem Soc* 79, 4365–4370.

Kim, K.H., Lin, N.-H. & Anderson, D.J. (1996) Quantitative structure–activity relationships of nicotine analogues as neuronal nicotinic acetylcholine receptor ligands. *Bioorg Med Chem* 4, 2211–2217.

Kirk, K.L. (1976a) Synthesis of ring-fluorinated serotonins and melatonins. *J Het Chem* 13, 1253–1256.

Kirk, K.L. (1976b) Photochemistry of diazonium salts, 4: synthesis of ring-fluorinated tyramines and dopamines. *J Org Chem* 41, 2373–2376.

Kirk, K.L., Cantacuzene, D., Nimitkitpaisan, Y. *et al.* (1979) Synthesis and biological properties of 2-, 5-, and 6-fluoronorepinephrines. *J Med Chem* 22, 1493–1497.

Kirk, K.L. (2006) Selective fluorination in drug design and development: an overview of biochemical rationales. *Curr Top Med Chem* 6, 1447–1456.

Kollonitsch, J., Marburg, S. & Perkins, L.M. (1979) Fluorodehydroxylation, a novel method for synthesis of fluoroamines and fluoroamino acids. *J Org Chem* 44, 771–777.

Konas, D.W. & Coward, J.K. (2001) Electrophilic fluorination of pyroglutamic acid derivatives: application of substrate-dependent reactivity and diastereoselectivity to the synthesis of optically active 4-fluoroglutamic acids. *J Org Chem* 66, 8831–8842.

Krogsgaard-Larsen, P. (1988) GABA synaptic mechanisms: stereochemical and conformational requirements. *Med Res Rev* 8, 27–56.

Le Bars, D. (2006) Fluorine-18 and medical imaging: radiopharmaceuticals for positron emission tomography. *J Fluorine Chem* 127, 1488–1493.

Lee, S., Uttamapinant, C. & Verdine, G.L. (2007) A concise synthesis of 4′-fluoro nucleosides. *Org Lett* 9, 5007–5009.

Lehel, S., Horvath, G., Boros, I. *et al.* (2002) The nucleophilic substitution reaction for [^{18}F]fluoride-ion on the series of N^6-benzoyl-2′,3′-isopropylideneadenosine-5′-sulfonates. *J Radioanal Nucl Chem* 251, 413–416.

Leschke, C., Elz, S., Garbarg, M. *et al.* (1995) Synthesis and histamine H_1 receptor agonist activity of a series of 2-phenylhistamines, 2-heteroarylhistamines, and analogs. *J Med Chem* 38, 1287–1294.

Lu, S., Herbert, B., Haufe, G. *et al.* (2000) Syntheses of (*R*)- and (*S*)-2- and 6-fluoronorepinephrine and (*R*)- and (*S*)-2- and 6-fluoroepinephrine: effect of stereochemistry on fluorine-induced adrenergic selectivities. *J Med Chem* 43, 1611–1619.

Lui, E., Chirakal, R. & Firnau, G. (1998) Enzymatic synthesis of (−)-6-[^{18}F]-fluoronorepinephrine from 6-[^{18}F]-fluorodopamine by dopamine β-hydroxylase. *J Labelled Cpd Radiopharm* 41, 503–521.

Martarello, L., Schaffrath, C., Deng, H. *et al.* (2003) The first enzymatic method for C–^{18}F bond formation: the synthesis of 5′- [^{18}F]-fluoro-5′-deoxyadenosine for imaging with PET. *J Labelled Cpd Radiopharm* 46, 1181–1189.

Morizawa, Y., Nakayama, T., Yasuda, A. *et al.* (1989) Stereoselective synthesis of 3′-deoxy-3′-fluoroadenosine. *Bull Chem Soc Jpn* 62, 2119–2120.

Nie, J.-Y., Shi, D., Daly, J.W. *et al.* (1996) Synthesis of fluorodopamines: effect of aryl fluoro substituents on affinities for adrenergic and dopaminergic receptors. *Med Chem Res* 6, 318–332.

O'Hagan, D., Schaffrath, C., Cobb, S.L. *et al.* (2002) Biosynthesis of an organofluorine molecule — A fluorinase enzyme has been discovered that catalyses carbon–fluorine bond formation. *Nature* 416, 279.

Pankiewicz, K.W., Krzeminski, J., Ciszewski, L.A. *et al.* (1992) A synthesis of 9-(2-deoxy-2-fluoro-β-D-arabinofuranosyl)adenine and hypoxanthine. An effect of C3′-endo to C2′-endo conformational shift on the reaction course of 2′-hydroxyl group with DAST. *J Org Chem* 57, 553–559.

Parkkari, T., Myllymäki, M., Savinainen, J.R. *et al.* (2006) α-methylated derivatives of 2-arachidonoyl glycerol: synthesis, CB1 receptor activity, and enzymatic stability. *Bioorg Med Chem Lett* 16, 2437–2440.

Passe, T.J., Charles, H.C., Rajagopalan, P. *et al.* (1995) Nuclear magnetic resonance spectroscopy: a review of neuropsychiatric applications. *Prog Neuro-Psychopharmacol Biol Psychiat* 19, 541–563.

Peters, D.A.V. (1971) Inhibition of serotonin biosynthesis by 6-halotryptophans in vivo. *Biochem Pharmacol* 20, 1413–1420.

Pooput, C., Pellett, P., Heredia-Moya, J. *et al.* (2008) 'Synthesis of ring fluorinated capsaicin analogs and their interactions with the vallinoid [*sic*] receptor'. Abstracts of papers, 235th ACS National Meeting, New Orleans, LA, USA, 6–10 April 2008.

Qiu, J. & Silverman, R.B. (2000) A new class of conformationally rigid analogues of 4-amino-5-halopentanoic acids, potent inactivators of γ-aminobutyric acid aminotransferase. *J Med Chem* 43, 706–720.

Ross, R.A. (2003) Anandamide and vanilloid TRPV1 receptors. *Br J Pharmacol* 140, 790–801.

Schaffrath, C., Cobb, S.L. & O'Hagan, D. (2002) Cell-free biosynthesis of fluoroacetate and 4-fluorothreonine in *Streptomyces cattleya. Angew Chem Int Edit* 41, 3913–3915.

Silverman, R.B. & Levy, M.A. (1980) Syntheses of (*S*)-5-substituted 4-aminopentanoic acids: a new class of γ-aminobutyric acid transaminase inactivators. *J Org Chem* 45, 815–818.

Silverman, R.B. & Invergo, B.J. (1986) Mechanism of inactivation of γ-aminobutyrate aminotransferase by 4-amino-5-fluoropentanoic acid: first example of an enamine mechanism for a γ-amino acid with a partition ratio of 0. *Biochemistry* 25, 6817–6820.

Summers, K.L., Lippiello, P., Verhulst, S. *et al.* (1995) 5-Fluoronicotine, noranhydroecgonine, and pyridyl-methylpyrrolidine release acetylcholine and biogenic amines in rat cortex *in vivo. Neurochem Res* 20, 1089–1094.

Timmers, H.J.L.M., Carrasquillo, J.A., Whatley, M. *et al.* (2007) Usefulness of standardized uptake values for distinguishing adrenal glands with pheochromocytoma from normal adrenal glands by use of 6-^{18}F-fluorodopamine PET. *J Nucl Med* 48, 1940–1944.

Tolman, V. (1993) Chemistry of 4-fluoroglutamic acid. Part 1. A critical survey of its syntheses: an attempt to optimize reaction conditions for large-scale preparation. *J Fluorine Chem* 60, 179–183.

Tolman, V. & Sedmera, P. (2000) Chemistry of 4-fluoroglutamic acid. Part 3. Preparation of the diastereomers of 4-fluoroglutamine and 4-fluoroisoglutamine: an enzymatic access to the antipodes of 4-amino-2-fluorobutyric acid. *J Fluorine Chem* 101, 5–10.

Tsushima, T., Kawada, K., Ishihara, S. *et al.* (1988) Fluorine-containing amino acids and their derivatives. 7. Synthesis and antitumor activity of α- and γ-substituted methotrexate analogs. *Tetrahedron* 44, 5375–5387.

Tuttle, J. V., Tisdale, M. & Krenitsky, T.A. (1993) Purine 2′-deoxy-2′-fluororibosides as antiinfluenza virus agents. *J Med Chem* 36, 119–125.

Usami, N., Kobana, K., Yoshida, H. *et al.* (1998) Synthesis and pharmacological activities in mice of halogenated Δ^9- tetrahydrocannabinol derivatives, *Chem Pharm Bull* 46, 1462–1467.

Vidal-Cros, A., Gaudry, M. & Marquet, A. (1985) L-*threo*- and L-*erythro*-3-fluoroglutamic acids. Synthesis by fluorodehydroxylation and enzymatic resolution. *J Org Chem* 50, 3163–3167.

Vidal-Cros, A., Gaudry, M. & Marquet, A. (1989) Facile synthesis of optically pure (2*R*,3*R*)- and (2*R*,3*S*)-3-fluoroglutamic acids using glutamate dehydrogenase. *J Org Chem* 54, 498–500.

Wang, Z. & Silverman, R.B. (2006) Syntheses and evaluation of fluorinated conformationally restricted analogues of GABA as potential inhibitors of GABA aminotransferase. *Bioorg Med Chem* 14, 2242–2252.

Wiley, J.L., Golden, K.M., Ryan, W.J. *et al.* (1997) Evaluation of cannabimimetic discriminative stimulus effects of anandamide and methylated fluoroanandamide in rhesus monkeys. *Pharm Biochem Behav* 58, 1139–1143.

Winkler, M., Moraux, T., Khairy, H.A. *et al.* (2009) Synthesis and vanilloid receptor (TRPV1) activity of the enantiomers of α-fluorinated capsaicin. *ChemBioChem* 10, 247–259.

Yu, J.-X., Kodibagkar, V.D., Cui, W. *et al.* (2005) ^{19}F: A versatile reporter for non-invasive physiology and pharmacology using magnetic resonance. *Curr Med Chem* 12, 819–848.

Zingel, V., Elz, S. & Schunack, W. (1990) Histamine analogues. 33 communication: 2-phenylhistamines with high histamine H_1-agonistic activity. *Eur J Med Chem* 25, 673–680.

Section 3

10

^{18}F-Radionuclide Chemistry

Romain Bejot and Véronique Gouverneur†*

10.1 Introduction

10.1.1 *Radioisotope ^{18}F*

^{18}F is a fluorine radioactive isotope that decays to ^{18}O by emission of a positron (β^+ 97%) and electron capture (EC 3%). The half-life of ^{18}F is 109.77 min. This radioisotope is used in physical chemistry and in the radiopharmaceutical industry because of its advantageous physical properties (Table 10.1) (Sonzogni, 2009).

10.1.2 *Nuclear reactions*

A nuclear reaction is a rapid process (less than a picosecond) in which a nucleus reacts with another nucleus, elementary particle or photon to produce other nuclei and/or particles (Friedlander *et al.*, 1981). In all reactions, with the exception of the annihilation process, the total number of nucleons is conserved, as well as the charge, energy, momentum, angular momentum, statistics and parity. This phenomenon was discovered by Rutherford in 1919 and the notation used for nuclear reactions is similar

* Singapore Bioimaging Consortium, Agency for Science, Technology and Research (A*STAR), 11 Biopolis Way, Singapore 138667.

† University of Oxford, Chemistry Research Laboratory, 12 Mansfield Road, Oxford OX1 3TA, UK.

Table 10.1. Physical characteristics of ^{18}F.

Half-life	Decay mode	Max. beta energy	Average beta energy	Max. positron range in tissue	Theoretical specific activity
109.77 min	96.73% β^+ 3.27% EC	634 keV	249.8 keV	2.35 mm	6.34×10^7 GBq/μmol

to that for chemical reactions (Equation 10.1). Alternatively a compact notation can be used (Equation 10.2).

$$^{18}O + p \rightarrow {}^{18}F + n \qquad (10.1)$$

or

$$^{18}O(p,n)^{18}F \qquad (10.2)$$

The symbols p, n, d, α represent proton, neutron, deuteron and alpha particle, respectively. In Equations 10.1 and 10.2, ^{18}O represents the target nucleus, p the accelerated particle of the cyclotron beam, n the side product of the nuclear reaction and ^{18}F the radionuclide produced in the process. The energy of the reaction corresponds to the energy of a single process, commonly expressed in electron volts (1 eV = 1.60×10^{-19} J).

Radioactive species always decay according to an exponential law independently of the decay mode (Equations 10.3 and 10.4):

$$N = N_0 e^{-\lambda t} \qquad (10.3)$$

or

$$A = A_0\, e^{-\lambda t} \quad \text{with } A = N\lambda \qquad (10.4)$$

In these equations, N and A represent the number of undecayed atoms and the measured activity at time t, respectively with N_0 and A_0 representing the initial numbers at $t = 0$. λ is the characteristic decay constant (probability of one atom decaying in one second) and is related to the half-life $t_{½}$ (the time necessary for half of a sample to decay) by the equation $\lambda = \ln 2/t_{½}$. Radioactivity units are the Becquerel (1 Bq corresponds to 1 disintegration per second) or the Curie (Ci); 1 Ci = 3.7×10^{10} Bq and is defined as the activity of 1 gram of ^{226}Ra.

10.1.3 *Production of ^{18}F*

^{18}F is produced with medical cyclotrons (10–18 MeV) (Strijckmans, 2001; Roberts *et al.*, 2005). The $^{18}O(p,n)^{18}F$ nuclear reaction is performed on either ^{18}O-enriched oxygen gas or water (Table 10.2). Target bodies are made from an inert metal such as silver, niobium or tantalum, designed to withstand the internal pressure and radiation, while maximising the quantity and radiochemical purity of the recovered ^{18}F (Berridge and Kjellström, 1999). ^{18}F can be subsequently recovered either directly as [^{18}F]fluoride ion in [^{18}O]water or as [^{18}F]F_2. For [^{18}F]F_2 production, after irradiation of [^{18}O]O_2, the target is emptied and filled with a trace amount of F_2 in an inert carrier gas (neon or krypton) and the target is irradiated again to retrieve the ^{18}F adsorbed on the target wall (O'Neil and VanBrocklin, 1999). Alternatively, the nuclear reaction $^{20}Ne(d,\alpha)^{18}F$ is carried out in a nickel, aluminium or stainless steel-body gas target filled with micromolar amounts of fluorine gas (F_2) in neon (Bishop *et al.*, 1996). The highly reactive [^{18}F]fluorine gas is removed from the target as a gas mixture of [^{18}F]F_2 and can be used directly in radiosyntheses or converted to a less reactive and more selective reagent. Various other nuclear reactions to produce

Table 10.2. Production routes for ^{18}F.[a]

Nuclear reaction	$^{18}O(p,n)^{18}F$	$^{16}O(^{3}He,p)^{18}F$	$^{20}Ne(d,\alpha)^{18}F$	$^{18}O(p,n)^{18}F$
Target	[^{18}O]H_2O	H_2O	Ne, F_2 (μmol)	[^{18}O]O_2, Kr (then F_2; μmol)
Particle energy (MeV)	$16 \to 3$	$36 \to 0$	$14 \to 0$	$16 \to 3$
Main product	[^{18}F]F^-	[^{18}F]F^-	[^{18}F]F_2	[^{18}F]F_2
Yield (GBq/μAh)	2.22	0.26	0.4	1.0

[a] Qaim, 2001; Qaim *et al.*, 1993.

^{18}F are known but scarcely used, such as $^{18}O(p,n)^{18}F$ [from $[^{18}O]O_2$)], $^{16}O(^3He,p)^{18}F$ or the combination of $^6Li(n,\alpha)^3H$–$^{16}O(^3H,n)^{18}F$ (Table 10.2).

10.1.4 *Positron emission tomography (PET)*

^{18}F is used extensively in PET, a non-invasive imaging technique benefiting from a very high sensitivity (picomolar range) and commonly used for both diagnosis and pharmaceutical drug evaluation. PET appears to be driving ^{18}F-radiochemistry since efficient radiolabelling techniques are required to advance nuclear medicine (Welch and Redvanly, 2005; Schubiger *et al.*, 2007; Ametamey *et al.*, 2008).

PET relies on the emission of a positron (the antiparticle of the electron) by the radionuclide, with subsequent annihilation of the positron with an electron and the detection of the two resulting gamma rays. The average distance travelled by a positron before its annihilation with an electron depends on its energy (maximum 634 keV for ^{18}F). The higher the energy, the further the positron travels before it combines with an electron, a process leading to the emission of two 511 keV gamma rays ($E = mc^2$) emitted in opposite directions (180°). ^{18}F is frequently used for PET as its low positron energy results in a shorter distance between the location of the radioisotope and the origin of gamma rays detected by the PET camera. This is an advantageous property allowing images of higher resolution to be acquired. The spatial resolution of PET is indeed limited by positron range but also by other factors such as detector size, non-collinearity and scattering effects, as well as patient motion (Sanchez-Crespo *et al.*, 2004). In 1999, it was proposed that the positron range has a 'cusplike' shaped distribution (Equation 10.5) and contributes a full width at half maximum (FWHM) of about 0.102 mm and a full width at tenth maximum (FWTM) of 1.03 mm in water (Levin and Hoffman, 1999).

$$P(x) = Ce^{-k_1 x} + (1 - C)e^{-k_2 x} \quad (10.5)$$

(with $C = 0.516$, $k_1 = 0.379\ mm^{-1}$, $k_2 = 0.031\ mm^{-1}$, for ^{18}F); $P(x)$ is the positron annihilation spread function.

In 2004, an effective FWHM of 0.54 mm was presented as a more realistic estimate, which takes into account broadening based on the entire range distribution (Lecomte, 2004). For the smaller detector and scanner diameter sizes, the ^{18}F positron range and detector size effects dominate and the resolution of a microPET system with ^{18}F is therefore typically of 1.5–2.0 mm FWHM and 3–4 mm FWTM (Levin and Zaidi, 2007; Visser *et al.*, 2009).

For PET, a suitable balance has to be found between administering sufficient activity to obtain good quality data and minimising the radiation dose to the patient, which depends on the radionuclide half-life, its emission characteristics and the biodistribution of the labelled compounds as well as daughter labelled metabolites. ^{18}F decays mainly by emission of positrons (97%), thereby minimising dosimetry problems compared to other positron-emitting radionuclides that also emit gamma rays, alpha or beta particles (Ekström and Firestone, 1999; Delacroix, 2002).

^{18}F is the most widely used PET radionuclide as its half-life of 109.77 min allows multi-step synthesis of the ^{18}F-labelled radiotracer, its commercial distribution and subsequent biodistribution studies. ^{18}F can be used to label any probe of interest as long as its half-life is compatible with the timescale of the biological event under investigation and its introduction does not affect the function of the probe itself. In some cases, as best exemplified with the most commonly used ^{18}F-radiotracer 2-deoxy-2-[^{18}F]fluoro-D-glucose ([^{18}F]FDG), the presence of the fluorine radionuclide is indeed essential for the probe to fulfil its role, in this case, the imaging of regions with high metabolic activity. Importantly, PET is an imaging modality that detects the presence of the radionuclide; therefore it is advantageous to understand the metabolic profile of the labelled molecule *in vivo*.

10.1.5 *Specific activity*

The specific activity measures the extent to which an ^{18}F-labelled compound is contaminated with the nonradioactive isotopic compound (Equation 10.6). It is usually expressed as the ratio of radioactivity relative to the mass or molar amount of the compound (e.g. Bq/g or Bq/mol).

Therefore it depends on the radioactive decay: the specific activity declines exponentially.

$$A_S = {}^{18}F[Bq]/({}^{18}F\ [mol] + {}^{19}F\ [mol]) \quad (10.6)$$

The maximum theoretical specific activity (activity/mole) is defined by Equation 10.7:

$$A_{S\ max} = \ln 2\ N_A/t_{1/2} \quad (10.7)$$

where N_A is the Avogadro constant.

For ^{18}F, the theoretical maximum specific activity is 6.34×10^4 GBq/μmol (or 1.7×10^3 Ci/μmol). In practice, the maximum specific activity is never attainable because of contamination with the stable isotope originating from radionuclide production, solvents, chemicals and impurities (Füchtner *et al.*, 2008; Berridge *et al.*, 2009). Direct nucleophilic ^{18}F-fluorination using no carrier added (n.c.a.) (which refers to a preparation of a radioactive isotope which is essentially free from stable isotopes of the element in question (Van Grieken and De Bruin, 1994)) [^{18}F]fluoride is the method of choice to obtain ^{18}F-labelled molecules with high specific activity (>100 GBq/μmol). In contrast, the production of the electrophilic fluorinating reagent [^{18}F]F_2 gas necessitates the use of the carrier-added reactant F_2. The amount of carrier F_2 gas depends on the internal volume and surface area of the cyclotron target and determines the specific activity of [^{18}F]F_2. Due to the equal likelihood of incorporating ^{18}F or ^{19}F, a maximum 50% radiochemical yield (RCY) is attainable. Therefore, electrophilic fluorination with [^{18}F]F_2 provides radiolabelled compounds consistently with lower specific activity <600 MBq/μmol (Hess *et al.*, 2000). Higher specific activity [^{18}F]F_2 is accessible upon $^{18}F/^{19}F$ exchange reaction involving [^{18}F]fluoromethane combined with the use of a low amount of carrier F_2, a process undertaken in an electrical discharge chamber. This 'post-target' synthesis of [^{18}F]F_2 (up to 55 GBq/μmol) was first reported by Bergman and Solin (1997).

The specific activity of a radiolabelled molecule is measured by determining the quantity of material for a known quantity of radioactivity, most often by HPLC using a calibration curve obtained by UV absorption or any other suitable detection method. The specific activity of [^{18}F]fluorine is typically measured by iodometric titration (Bishop *et al.*, 1996). A higher specific activity is preferable to minimise physiological perturbation induced by nonradioactive carriers. High specific activity is absolutely necessary to investigate the binding of the radiolabelled ligand to low concentration receptors without perturbation of the physiological equilibrium: low specific activity may lead to significant saturation of the binding sites, thereby decreasing the signal-to-noise ratio.

10.1.6 *Kinetics and radiochemical yield*

For radiochemistry involving the short half-life radionuclide ^{18}F, it is essential for the reactions used for ^{18}F-labelling to be as rapid as possible. Both reaction time and chemical yield are important parameters, since the radiochemical yield (the yield of a radiochemical transformation expressed as a fraction of the activity originally present) is also a function of radioactive decay (decay-corrected radiochemical yield) (Van Grieken and De Bruin, 1994). Hence the maximum radiochemical yield is always reached before the reaction has gone to completion (Welch and Redvanly, 2005). The stoichiometric ratio of the precursor to ^{18}F is typically in the order of 10^3. ^{18}F is hence consumed by pseudo first-order kinetics (Equation 10.8).

$$\text{Chemical yield} = 100 \times (1 - e^{-k' t}) \tag{10.8}$$

The effect of both reaction time and radioactive decay on radiochemical yield is illustrated in Fig. 10.1 for ^{18}F and an associated pseudo rate constant $k' = 0.001\ s^{-1}$.

The non-decay-corrected radiochemical yield is defined as the ratio of the radioactive product at the end of the synthesis (EOS) process to the amount of radioactivity typically obtained at the end of bombardment (EOB).

In this chapter, the quoted radiochemical yields are decay-corrected radiochemical yields. HPLC is usually the method of choice to measure

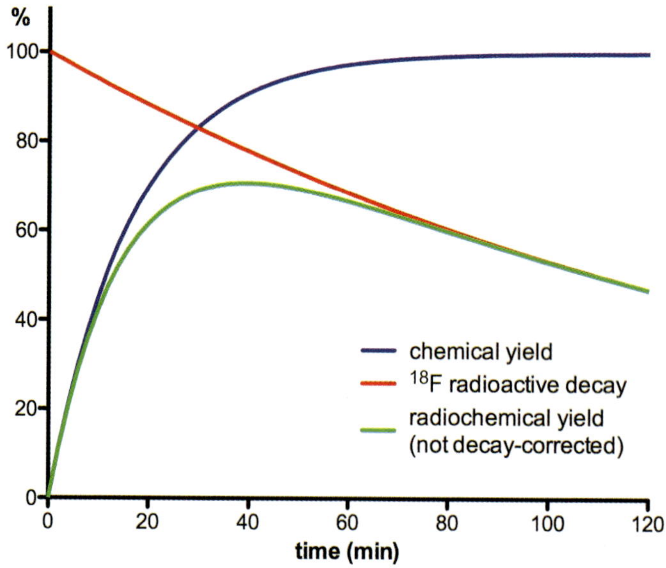

Figure 10.1. Radiochemical yield as a function of time: pseudo first-order reaction (k' = 0.001 s^{-1}).

conversion for a particular radiochemical transformation and to confirm the identity of the labelled product. Thin-layer chromatography methods (TLC) may be advantageous for fluorination processes using [^{18}F]fluoride because of the high tendency of fluoride to adhere on reverse stationary phases.

10.2 Carrier-Added ^{18}F-Labelled Probes

^{18}F Radioactive probes have found applications beyond the context of positron emission tomography.

10.2.1 *Carrier-added [^{18}F]fluoride*

^{18}F-Labelled molecules may be used as highly sensitive probes to study surface interactions by means of gamma radiation detection (1.58×10^{-20} mole ^{18}F per Bq). Winfield and coworkers have extensively investigated various surface-related phenomena with radioactive ^{18}F probes, especially

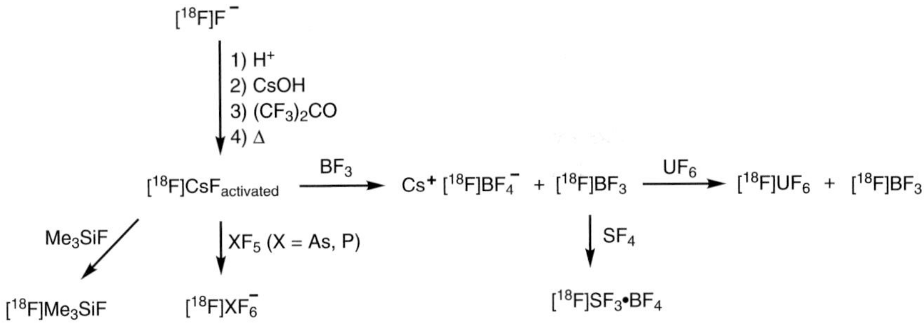

Scheme 10.1. Reactions leading to ^{18}F-labelled binary fluorides starting from solid [^{18}F]CsF.

on solid Lewis acids. They studied the reactivity of fluoride supported on γ-alumina and amorphous chromia using ^{18}F-radiolabelled alkali-metal fluoride, boron trifluoride, arsenic tetrafluoride and sulfur tetrafluoride (Nickkho-Amiry and Winfield, 2007). Activated [^{18}F]CsF, formed by the thermal decomposition of Cs alkoxides, was used as a tool to study isotopic exchange (Scheme 10.1). The protocol followed for these studies involves several steps. Initially, [^{18}F]fluoride is converted to [^{18}F]HF by distillation with 50% H_2SO_4. [^{18}F]HF is trapped and neutralised in aqueous CsOH at 0°C. Carrier HF is added and the solution evaporated to give finely divided carrier-added [^{18}F]CsF, which can be activated by the addition of hexafluoroacetone and thermal decomposition. Activated [^{18}F]CsF was then used to examine the lability of various element–fluorine bonds through isotopic exchange. Arsenic pentafluoride (AsF_5) and boron trifluoride (BF_3) can be labelled by isotopic exchange with activated [^{18}F]CsF, a process revealing that the exchange reactions between BF_3 (or AsF_5) and [^{18}F]CsF result in the formation of [^{18}F]BF_4^- (or of [^{18}F]AsF_6^-) as a by-product (Dixon and Winfield, 1989). ^{18}F-Radiolabelled sulfur tetrafluoride ([^{18}F]SF_4) was prepared from SF_4 and [^{18}F]BF_3 via the adduct [SF_3][BF_4]; [^{18}F]Me_3SiF was obtained by direct isotopic exchange with [^{18}F]CsF. ^{18}F was also used to demonstrate that $[MoF_7]^-$ and $[WF_7]^-$ anions coexist with their hexafluorides in solution in acetonitrile and under heterogeneous conditions with caesium fluoride at room temperature (Ghorab and Winfield, 1993). Winfield and coworkers reported that [^{18}F]fluoride exchange between ^{18}F-labelled fluorotrimethylsilane or boron trifluoride and substituted tungsten(VI), molybdenum(VI), uranium(VI and V) or iodine(V) fluorides occurs readily at room temperature. In contrast, tellurium(VI) does not seem responsive to isotopic exchange (Frazer *et al.*, 1975; Poole and Winfield, 1976; Sanyal and Winfield, 1984). [^{18}F]Fluoride exchange reactions between the hexafluorides of molybdenum, tungsten and uranium and various fluoro-anions (BF_4^-, PF_6^-, AsF_6^- and SbF_6^-) showed that SbF_6^- is inert and that, to some extent, fluorine in AsF_6^- is less labile than fluorine in BF_4^- and PF_6^- (Ghorab and Winfield, 1990).

10.2.2 *Surface interactions with ^{18}F-labelled probes*

Various catalytic processes involving [^{18}F]HF have been investigated, for example the interactions between heavily fluorinated chromia catalysts and [^{18}F]HF. Treatment of amorphous chromia with carrier-added [^{18}F]HF leads to the displacement of oxygen by fluorine in Cr–O bonds. Zinc or nickel(II)-doped catalysts follow similar behaviour toward [^{18}F]HF, but unlike chromia, are readily converted to their fluorides (Bonniface *et al.*, 1999). Fluorination of calcinated γ-alumina can be performed using [^{18}F]SF_4 leading to a fluorinated surface with both Brønsted and Lewis acid character: the uptake level depends on the temperature and conditions of [^{18}F]SF_4 addition (Bendada *et al.*, 1996b; Bozorgzadeh *et al.*, 2001). With BF_3, fluorination is less efficient (Klapotke *et al.*, 2006). Fluorinated γ-alumina demonstrated its ability to exchange ^{18}F with [^{18}F]HF, [^{18}F]BF_3 and [^{18}F]SF_4 (Bendada *et al.*, 1996a). It is thus an acidic surface that acts as an active catalyst for room temperature fluorine–chlorine exchange reactions on hydrochlorocarbons (Thomson *et al.*, 1993). Additionally, isotopic exchange and surface complexation was demonstrated between [^{18}F]SF_4 and aluminium(III), chromium(III), niobium(V), tantalum(V) and uranium(V) fluorides, at room temperature, rationalised on the basis of $[SF_3]^+$ formation (Dixon *et al.*, 1987). Finally, investigation of polycrystalline diamond surface interactions with HF, revealed that either HF displacement or an ^{18}F/^{19}F isotopic exchange occurs between surface-adsorbed [^{18}F]HF and free HF, whereas no isotopic exchange could be observed between

HF and covalently bound [^{18}F]fluorine on the diamond surface (Kealey *et al.*, 2001).

10.2.3 *Catalytic fluorination*

The catalytic fluorination of CF_3CH_2Cl to CF_3CH_2F with a zinc or nickel(II)-doped catalyst was proved to be feasible (Bonniface *et al.*, 1999) as well as the multi-step catalytic nucleophilic fluorination of trichloroethene yielding [^{18}F]1,1,1,2-tetrafluoroethane in the presence of [^{18}F]HF (Baker *et al.*, 2000) (Scheme 10.2).

Scheme 10.2. Synthesis of $CF_3CH_2[^{18}F]F$.

10.3 Nucleophilic ^{18}F-Radiolabelling

^{18}F-Radiolabelling methods with either nucleophilic [^{18}F]fluoride or electrophilic sources of [^{18}F]fluorine have been extensively reviewed (Schlyer, 2004; Cai *et al.*, 2008; Miller *et al.*, 2008). In this section, the emphasis of the discussion is therefore on defining the scope and limitation of currently available ^{18}F-radiochemistry instead of cataloguing all ^{18}F-labelling processes known to date. Substitution of a hydrogen or a hydroxyl group for a fluorine in biologically active molecules is well tolerated as the van der Waals radius of fluorine r_v (1.47 Å) lies between that of oxygen (1.52 Å) and hydrogen (1.20 Å). Fluorine substitution therefore exerts only minor steric demand at receptor sites. Fluorine, however, is highly electronegative ($\chi_P = 4.0$), and its introduction can alter conformation and physicochemical properties. Fluorination can therefore modulate significantly the biological properties of a molecule, in either a beneficial or detrimental way (Purser *et al.*, 2008). Although the carbon–fluorine bond is strong (485 kJ/mol), aliphatic fluorinated molecules can be subject to *in vivo* defluorination, a transformation which can cause complications for PET imaging due to the high propensity of fluoride to accumulate within the bones (Zhang and Suzuki, 2007).

10.3.1 *Reactive [^{18}F]fluoride*

10.3.1.1 *Water removal*

Nucleophilic [^{18}F]fluoride is commonly produced as a solution in [^{18}O]water from the ^{18}O(*p*,*n*)^{18}F nuclear reaction. In the gas phase, the nonsolvated fluoride ion is a potent nucleophile (DePuy *et al.*, 1990; Vlasov, 1993). Aqueous fluoride is, however, inert as a nucleophile due to hydration. [^{18}F]Fluoride of sufficient nucleophilicity for ^{18}F-labelling can be obtained by removal of water and solubilisation in a suitable organic solvent. This is commonly achieved by adsorption onto an anion exchange resin (Schlyer *et al.*, 1990), followed by elution of the fluoride ion with a small volume of an aqueous weak base, and successive azeotropic evaporation with acetonitrile under a flow of inert gas. Commonly used resins are silica-based quaternary ammonium-modified strong anion exchangers (Waters QMA) or polystyrene–divinylbenzene copolymers-based strong anion exchangers (Dowex 1X8). In microfluidic devices, small anion exchange columns (down to 2 μL column volume) allow the trapping of several millilitres of aqueous [^{18}F]fluoride target solution and subsequent release in a reduced volume of just a few microlitres (e.g. 5 μL) (Bejot *et al.*, 2010; Elizarov *et al.*, 2010). Alternatively, the aqueous [^{18}F]fluoride solution from the target may be directly dried by azeotropic evaporation, but this is typically a longer procedure. The use of anion exchange resins also presents the advantage of purifying the [^{18}F]fluoride solution from contaminants arising from the irradiation of water and radiolysis (e.g. free radicals and metal ions) that may decrease the reactivity of the fluoride or compete as nucleophiles (Solin *et al.*, 1988). Solid-supported *n*-tetradecyltrimethylammonium bicarbonate provides an alternative method to trap and recover [^{18}F]fluoride in an organic solvent, a process not involving a drying step (Aerts *et al.*, 2010).

The efficiency of the [^{18}F]fluoride-drying protocol is not well known; it is assumed that some water is always present and therefore so-called dry 'naked' fluoride is not produced. This is not too much of a concern as [^{18}F]fluoride dried using these methods is sufficiently nucleophilic to participate in ^{18}F-radiolabelling. In some cases, the presence of a limited amount of water in [^{18}F]fluoride does not prevent its use as a nucleophile, at least for sufficiently reactive substrates, and is actually beneficial as some

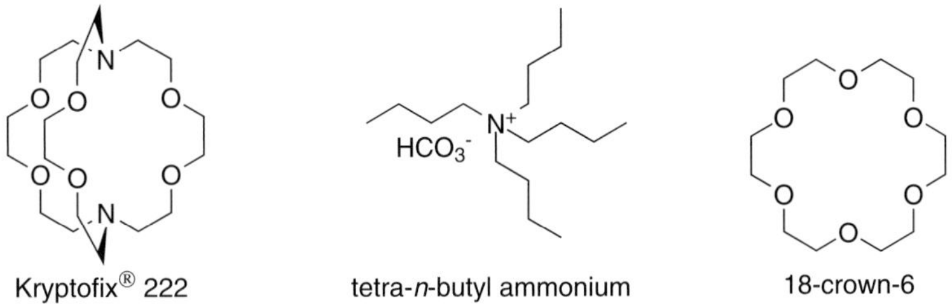

Figure 10.2. Phase transfer catalysts.

level of hydration prevents adsorption onto the reaction vessel (Briard and Pike, 2004). Alternative methods for removing water from [^{18}F]fluoride include electrochemical techniques: [^{18}F]fluoride in water can be adsorbed on a vitreous carbon electrode, on the wall of a glassy carbon vessel or in a microfluidic cell, then released in an organic solvents such as acetonitrile, DMF or DMSO with the assistance of a phase transfer catalyst, typically Kryptofix® 222 or tetrabutyl ammonium (Fig. 10.2) (Saiki *et al.*, 2010). $Et_3N{\cdot}3HF$ and $Et_3N{\cdot}HCl$ were also used to facilitate the desorption process, with the addition of $Et_3N{\cdot}3HF$ automatically leading to lower specific activity (Hamacher *et al.*, 2002; Reischl *et al.*, 2002). An electrochemical method avoiding the use of phase transfer catalysts employs a graphite electrode (Saito *et al.*, 2007). Beyond proof of concept, it has been shown by Coenen and coworkers that these electrochemical methods can provide [^{18}F]fluoride that is sufficiently reactive to allow the efficient production of radiotracers, such as [^{18}F]altanserin (Hamacher and Coenen, 2006).

10.3.1.2 *Solvents and phase-transfer catalysts*

The solubilisation of fluoride ion in polar aprotic solvents (acetonitrile, DMSO, DMF or THF) is typically facilitated by adding a cryptand such as Kryptofix® 222 (K_{222}), which chelates K^+ and therefore provides a more soluble and more reactive form of fluoride (Lehn and Sauvage, 1971). A common alternative to the potassium–K_{222} complex is the phase transfer catalyst tetrabutylammonium ion: [^{18}F]tetrabutylammonium fluoride can be obtained by eluting [^{18}F]fluoride from the anion exchange resin with an aqueous solution of tetrabutylammonium bicarbonate (Culbert *et al.*,

1995). Large cations like rubidium or caesium have also been considered (Jewett *et al.*, 1988; Pascali *et al.*, 1990; Karramkam *et al.*, 2003). When using cationic additives, carbonate is the most commonly used counter-anion as it does not compete with fluoride in nucleophilic substitution or cause side reactions (Smith and March, 2001). Recently, Chi and coworkers reported an efficient nucleophilic fluorination method using [^{18}F]caesium fluoride or [^{18}F]tetrabutylammonium fluoride in protic hindered alcohols such as *tert*-butyl alcohol and *iso*-amyl alcohol. Under these conditions, the ^{18}F-fluorinated radiolabelled molecules were obtained in higher yields and chemical purity as well as in shorter reaction times in comparison with conventional syntheses (Kim *et al.*, 2006, 2008).

In 2004, it was demonstrated that a complex between [^{18}F]hydrogen fluoride and 1,8-*bis*(dimethylamino)naphthalene, so-called proton sponge, can be used as an alternative to the potassium–K_{222} complex (Pascali *et al.*, 2004).

10.3.1.3 *Microwaves*

Microwave reactors have proved useful in facilitating ^{18}F-radiochemistry at several levels. The technology typically allows a decrease in the time required for the ^{18}F-drying process as well as for the ^{18}F-fluorination process itself and subsequent chemical transformations performed on the ^{18}F-labelled molecules. Microwave radiochemistry may also be beneficial to avoid decomposition or improve radiochemical purity. The technology is most often used for nucleophilic aromatic fluorination reactions, which typically require harsher reaction conditions, although applications have been reported for the ^{18}F-radiolabelling of aliphatic precursors (Stone-Elander and Elander, 2002).

Interestingly, microwaves could also be used to facilitate aliphatic fluorination with a nucleophile assisting leaving group without the assistance of a cryptand: the leaving group itself is used as a chelator of the [^{18}F] fluoride's counter-cation (Lu *et al.*, 2009).

10.3.2 *Nucleophilic carbon–fluorine bond formation*

The most widely used technique for ^{18}F-radiolabelling is the formation of a C–F bond by nucleophilic substitution with n.c.a. [^{18}F]fluoride.

10.3.2.1 *Nucleophilic aliphatic fluorination*

Direct nucleophilic aliphatic ^{18}F-fluorination is a suitable strategy to prepare ^{18}F-labelled molecules with high specific activity (>100 GBq/μmol). Radiolabelling upon stereospecific S_N2 displacement relies on the availability of a sufficiently reactive [^{18}F]fluoride complex and a substrate with a suitable leaving group, most often a sulfonate (triflate, tosylate, nosylate, mesylate) or a halide (iodo, bromo). As one might expect, the reactivity of the substrate for [^{18}F]fluoride substitution decreases in the order primary > secondary > tertiary alkyl sulfonates/halides. This strategy may require protection of nucleophilic functional groups that might compete with fluoride (e.g. acid, alcohol, amine group), a complication adding deprotection and purification steps and extending significantly synthesis times. [^{18}F]FDG, the most commonly prepared ^{18}F-labelled molecule, is synthesised within 30 min in >50% RCY by aliphatic nucleophilic fluorination of peracetylated mannose triflate followed by a deprotection step (Hamacher *et al.*, 1986) (Scheme 10.3).

Recently, it was reported that the addition of ionic liquids (1-butyl-3-methylimidazolium tetrafluoroborate [bmim][OTf] or 1-ethyl-3-methylimidazolium triflate [emim][BF_4]: 20–50 μL) improved significantly

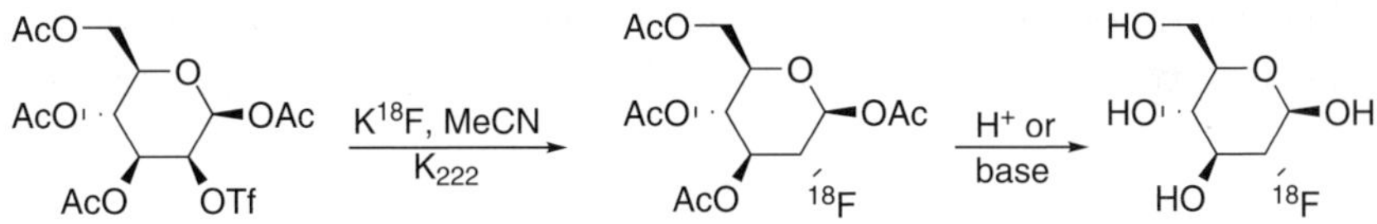

Scheme 10.3. Radiosynthesis of [^{18}F]FDG.

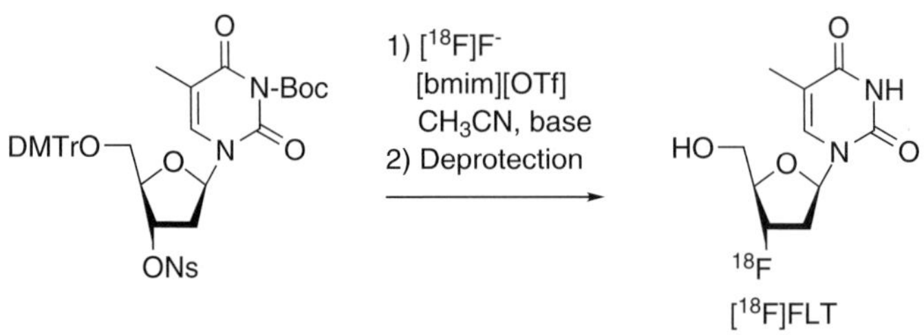

Scheme 10.4. Radiosynthesis of [^{18}F]FLT in ionic liquid [bmim][OTf].

the synthesis of 1-(2′-[^{18}F]-fluoroethoxy)-2,5-bis(4′-methoxystyryl)benzene ([^{18}F]FESB), the use of tetrafluoroborate potentially affecting the specific activity (Kumar *et al.*, 2005). More interestingly, 1-butyl-3-methylimidazolium triflate ([bmim][OTf]) in the presence of carbonate provides an efficient medium for direct radiolabelling with aqueous [^{18}F]fluoride (Kim *et al.*, 2003). The presence of water does not seem to prevent fluorination since the method proved successful for the synthesis of [^{18}F]FDG and [^{18}F]FLT (75% and 61% RCY, respectively) without the need to dry the [^{18}F]fluoride (Kim *et al.*, 2004; Moon *et al.*, 2006) (Scheme 10.4). Palladium-catalysed allylic fluorination has been developed and its applicability to ^{18}F-radiolabelling demonstrated. The use of *p*-nitrobenzoate as the leaving group is significant to the success of this catalytic organometallic fluorination process. This reaction represents the first example of Pd-mediated ^{18}F–C bond forming process (Hollingworth *et al.*, 2011).

10.3.2.2 *Nucleophilic aromatic fluorination*

Carbon–fluorine bond formation by nucleophilic aromatic substitution is a very important transformation for the preparation of ^{18}F-labelled aromatic compounds. This fluorinated structural sub-motif benefits from *in vivo* stability and this radiochemical transformation allows the radiotracer to be made available in high specific activity. Direct nucleophilic aromatic ^{18}F-fluorination proceeds via an S_NAr mechanism and therefore requires aromatic precursors activated by electron-withdrawing substituents (nitro, cyano or carbonyl: 3-NO_2 < 4-Ac < 4-CHO < 4-CN ≈ 4-CF_3 < 4-NO_2) on the *ortho* or *para* positions with respect to the leaving group. These ^{18}F-fluorinations are typically carried out by the displacement of a halide, nitrite, trimethylamine or iodoaryl (I < Br < Cl < F < NO_2 ≈ N^+Me_3 < I^+Ar). The reaction conditions are much harsher than for nucleophilic aliphatic substitution (e.g. 120–180°C in DMSO) and phase transfer catalysts are required. The dimethylsulfonium moiety has also been reported as a leaving group in aromatic fluorination using [^{18}F]TBAF (Maeda *et al.*, 1987).

Trimethylammonium salt precursors are often considered for aromatic nucleophilic ^{18}F-fluorination as they can be readily separated from the ^{18}F-labelled product, a significant advantage with respect to overall radiosynthesis time. The use of trimethylammonium as a leaving group (chloride, perchlorate or triflate salt) in DMSO or dimethylacetamide has proven beneficial in some cases for slightly less activated aromatics, for example aryl ketones or haloarenes (Hansch *et al.*, 1991; Ermert *et al.*, 2004). A commonly observed side reaction when using less activated trimethylammonium salt precursors (aromatic substituents with poor electron withdrawing ability, i.e. with lower Hammett σ constants) (Hansch *et al.*, 1991) is an aliphatic nucleophilic substitution on the NMe_3 group itself leading to the formation of [^{18}F]fluoromethane (Ermert *et al.*, 2004; Sun and DiMagno, 2007). The nucleophilic aromatic fluorination by displacement of trimethylammonium can be performed under mild conditions on highly activated substrates (cyano and carbonyl groups in the *ortho* and *para* positions) allowing the direct ^{18}F-labelling of peptides at 50–90°C in up to 90% RCY (Becaud *et al.*, 2009).

Iodonium salt precursors are excellent candidates for the nucleophilic ^{18}F-fluorination of aromatics (Pike and Aigbirhio, 1995; Chun *et al.*, 2010). A range of simple [^{18}F]fluoroaryls were prepared in good RCYs from weakly unactivated and even electron-rich precursors. The reaction can be carried out under relatively mild conditions, i.e. with lower boiling point solvents, and the presence of water does not prevent fluorination (Wadsworth and Devenish, 2005). The fluorination of unsymmetrical diaryl iodonium salts could lead to two possible ^{18}F-labelled arenes; however, the ^{18}F-labelled product distribution depends on the electronic and steric properties of the individual aromatic ring with the less electron-rich ring being labelled preferentially (Pike and Aigbirhio, 1995). The desired ^{18}F-labelled arene can be obtained as the major or sole labelled product when using either *p*-methoxyphenyl or 2-thienyl derivatives to direct the ^{18}F nucleophilic substitution onto the less electron-rich aryl ring (Ross *et al.*, 2007). Aromatic rings with *ortho* substituents direct the ^{18}F fluorination toward that ring with increased RCYs (known as the *ortho* effect) (Ross *et al.*, 2007). Radical scavengers, such as TEMPO, may improve the RCYs of ^{18}F fluorination of diaryliodonium salts, by preventing precursor decomposition (Carroll *et al.*, 2007). To date, good RCYs have only been obtained with relatively simple substituted benzene derivatives, possibly because of the difficulties associated with the preparation of structurally complex or functionalised iodonium salt precursors.

A study examining the reactivity of various commonly used leaving groups for nucleophilic aromatic fluorination was conducted with the aim of accessing 1-bromo-4-[^{18}F]fluorobenzene (Scheme 10.5) (Ermert *et al.*,

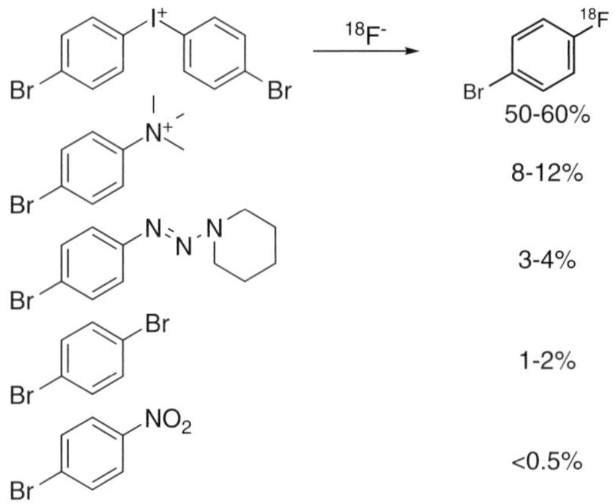

Scheme 10.5. Radiosynthesis of 1-bromo-4-[^{18}F]fluorobenzene.

2004). Notably, the symmetrical (4-bromophenyl)iodonium salt gave much better RCYs than the corresponding trimethylammonium salt, the triazene precursor (Pages and Langlois, 2001; Pages *et al.*, 2001; Huiban *et al.*, 2007), dibromobenzene or *para*-nitro bromobenzene. Interestingly, it was observed that the bromine is preferentially substituted by the [^{18}F]fluoride in *p*-nitro bromobenzene, despite the fact the nitro group is usually considered to be a better leaving group.

10.3.2.3 *Nucleophilic heteroaromatic fluorination*

Heteroaromatics can be efficiently ^{18}F-labelled, especially electron-deficient pyridines (Dollé, 2005). Very high RCYs are achieved using either a nitro or a trimethylammonium substituent as the leaving group. For example, 2-[^{18}F]fluoropyridine was obtained in 88% and 77% RCY after 5 min at 180°C in DMSO from its trimethylammonium and nitro precursor respectively (Scheme 10.6) (Dolci *et al.*, 1999). Direct ^{18}F-labelling at the 4-position is less efficient whilst very little fluorination is observed at the 3-position without the presence of additional electron-withdrawing substituents (Karramkam *et al.*, 2003). However, the presence of a cyano or an amido group is sufficient to activate the ring towards the nucleophilic displacement of chloride or bromide at the 3-position with [^{18}F]fluoride. (Abrahim *et al.*, 2006). Other heteroaromatics were successfully labelled by

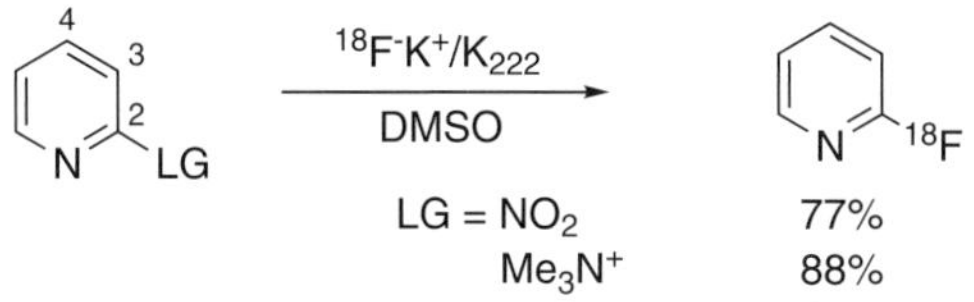

Scheme 10.6. Nucleophilic pyridine fluorination.

direct nucleophilic fluorination. For example, 2-[^{18}F]fluoro-1,3-thiazoles were obtained from the corresponding brominated precursors (Simeon and Pike, 2005).

10.3.2.4 *Enzymatic radiolabelling*

Fluorinase reactions

Fluorinated natural products are scarce, and the discovery of the first fluorinase enzyme by O'Hagan and coworkers was a spectacular advance in the field of enzymatic halogenation (O'Hagan and Harper, 1999; Onega *et al.*, 2009). This enzyme catalyses carbon–fluorine bond formation by nucleophilic fluorination (Zechel *et al.*, 2001; O'Hagan *et al.*, 2002). In 2003, it was demonstrated that the fluorinase enzyme isolated from the bacterium *Streptomyces cattleya* allows direct nucleophilic introduction of [^{18}F]fluoride into organic molecules (Martarello *et al.*, 2003). Fluorinase E.C. 2.5.1.63 was successfully used for the radiosynthesis of [^{18}F]-5′-fluoro-5′-deoxyadenine ([^{18}F]5′-FDA) upon nucleophilic displacement of L-methionine in (*S*)-adenosyl-L-methionine (SAM) with [^{18}F]fluoride ion. The enzymatic fluorination reaction is reversible and the formation of the ^{18}F-labelled compound was favoured by *in situ* conversion of the primary products of fluorination — [^{18}F]5′-FDA or L-methionine — into compounds that are not substrates of the fluorinase (Scheme 10.7) (Deng *et al.*, 2006). For example, *in situ* oxidation of L-methionine led to the synthesis of [^{18}F]5′-FDA in 95% RCY within two hours at 35°C, while coupled enzyme systems proved successful for the synthesis of the ^{18}F-labelled derivatives [^{18}F]-5′-fluoro-5′-deoxyinosine ([^{18}F]5′-FDI; RCY 75%, 4 h, adenylate deaminase), [^{18}F]-5-fluoro-5-deoxy-D-ribose ([^{18}F]5-FDR; RCY 80%, 2 h nucleoside hydroxylase) or various [^{18}F]-5′-deoxy-5′-fluorouridines

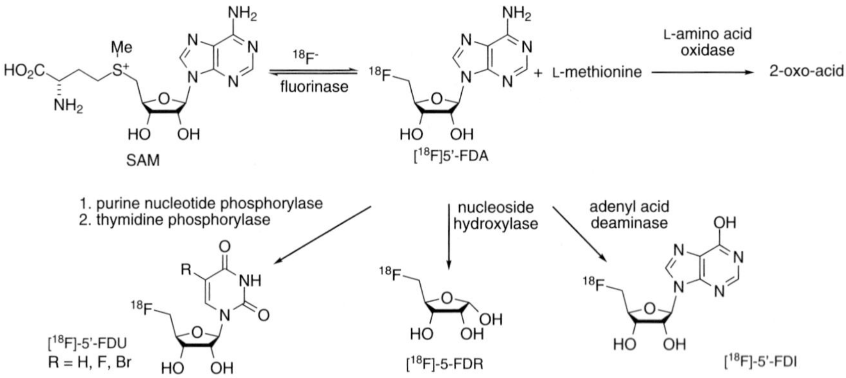

Scheme 10.7. Enzymatic ^{18}F-labelling of SAM.

([^{18}F]5′-FDU; RCY 20–35%, 4 h, thymidine phosphorylase) (Winkler *et al.*, 2008; Onega *et al.*, 2010). This enzymatic approach, an attractive method for radiolabelling with [^{18}F]fluoride under aqueous conditions and at room temperature, benefits from high chemospecificity, but suffers from relatively long reaction times (2–4 h) and a very limited scope due to the substrate-specificity of the enzyme. Mechanistically, it was noted that hydrogen-bonding involving the enzyme, the fluoride and the substrate is likely to be essential to assist C–F bond formation (Dong *et al.*, 2004), an observation reminiscent of the beneficial effect of protic *tert*-alcohols (Scheme 10.7) (Kim *et al.*, 2006).

Enzymatic transformations

In addition to catalysing ^{18}F–C bond formation, enzymes can be used to functionalise ^{18}F-labelled intermediates (Lin *et al.*, 2005). This is illustrated in Scheme 10.8 with the enzymatic radiosynthesis of 6-[^{18}F]fluoro-L-DOPA

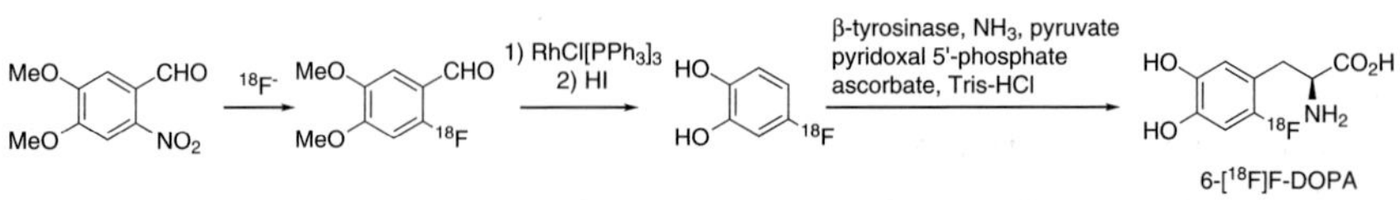

Scheme 10.8. Enzymatic synthesis of n.c.a. 6-[^{18}F]fluoro-L-DOPA.

([^{18}F]F-DOPA). An alternative procedure to the commonly used electrophilic fluorination protocol consists of a nucleophilic ^{18}F-labelling process leading firstly to 4-[^{18}F]fluorocatechol and its subsequent enzymatic conversion to [^{18}F]F-DOPA with β-tyrosinase (Kaneko *et al.*, 1999). Although the overall RCY is very low (2% from [^{18}F]fluoride after 150 min synthesis), the enzymatic transformation itself was efficient (60%) and [^{18}F]F-DOPA was obtained with a specific activity superior to 200 GBq/μmol, which is significantly higher than the specific activity typically obtained upon electrophilic fluorodestannylation (<100 MBq/μmol).

Another example of an enzyme-mediated transformation in the synthesis of ^{18}F-radiotracers is the stereospecific conversion of 6-[^{18}F]-fluorodopamine to (–)6-[^{18}F]-fluoronorepinephrine (>90% *ee*), a reaction successfully performed within 70 min in the presence of the dopamine β-hydroxylase (67% RCY) (Lui *et al.*, 1998). In another enzymatic process, uridine-5′-diphospho-2-deoxy-2-[^{18}F]fluoro-α-D-glucopyranose (UDP-[^{18}F]FDG), a glycosyl donor for the glycosylation of biomolecules, was prepared by the coupling of UTP to [^{18}F]FDG-1-phosphate with UDP glucose-pyrophosphorylase (UDP-Glc PP). UDP-[^{18}F]FDG was synthesised within 110 min in 20% RCY from [^{18}F]fluoride (Scheme 10.9) (Prante *et al.*, 2007).

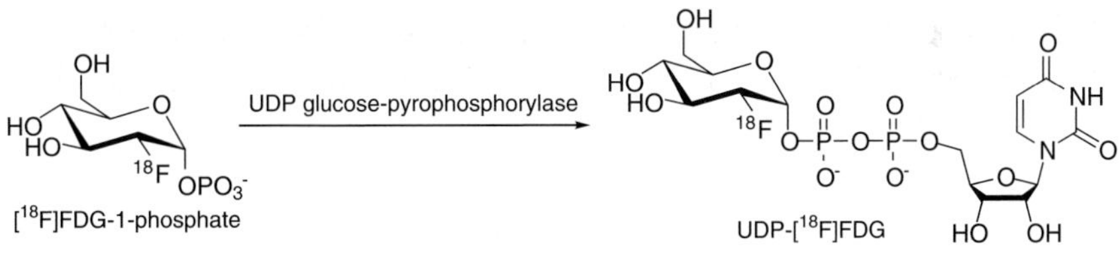

Scheme 10.9. Enzymatic synthesis of uridine-5′-diphospho-2-deoxy-2-[^{18}F]fluoro-α-D-glucopyranose.

10.3.2.5 *Nucleophilic addition*

Due to the fact that perfluorinated alkenes are prone to nucleophilic addition, [^{18}F]fluoride was used to convert trifluoroethylene into 1,1,1,2-tetrafluoroethane selectively labelled at position 1 (HFC-134a) (Aigbirhio *et al.*, 1995). This reaction is limited in scope and only a handful of examples have been reported. The conversion of hexafluoropropene into a mixture of

[^{18}F]2*H*-heptafluoropropane (HFC-227ea), [^{18}F]hexafluoropropene and [^{18}F]perfluoro-isohex-2-ene is another rare example of nucleophilic addition for ^{18}F-radiolabelling (Aigbirhio and Pike, 1995) (Scheme 10.10).

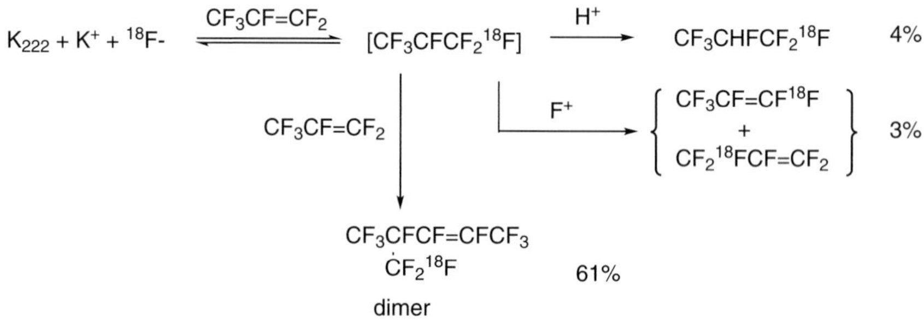

Scheme 10.10. Radiosynthesis of [1-^{18}F]HFC-227*ea*.

10.3.2.6 *Isotopic exchange*

^{18}F-Labelled molecules can be obtained from their stable ^{19}F-fluorinated analogues by aliphatic or aromatic exchange reaction with [^{18}F]fluoride. This method suffers from low specific activity as it is impossible to separate the precursor from the ^{18}F-labelled product, but can be of interest for the radiosynthesis of probes for which specific activity is not critical (Cacace *et al.*, 1981; Kilbourn and Subramanian, 1990; Satter *et al.*, 1994; Blom *et al.*, 2009). This chemistry was applied to the synthesis of [^{18}F]L-DOPA in 22% RCY (3.2 MBq/μmol) (Wagner *et al.*, 2009).

10.3.2.7 *Alternative nucleophilic ^{18}F-reagents*

Diethylaminosulfur trifluoride (DAST) is a common reagent for the nucleophilic fluorination of hydroxyl and carbonyl groups. The preparation of [^{18}F]DAST was reported by isotopic exchange at –78°C with [^{18}F]HF using sulfur tetrafluoride and trimethylsilyldiethylamine in a chlorofluorocarbon solvent ($CFCl_3$: freon-11) (Straatmann and Welch, 1977). This reagent is of low specific activity and has not been further exploited in the context of ^{18}F-radiolabelling. ^{18}F-Diazonium tetrafluoroborates or trichlorofluoroborates can be prepared by isotopic or halide exchange with [^{18}F]fluoride.

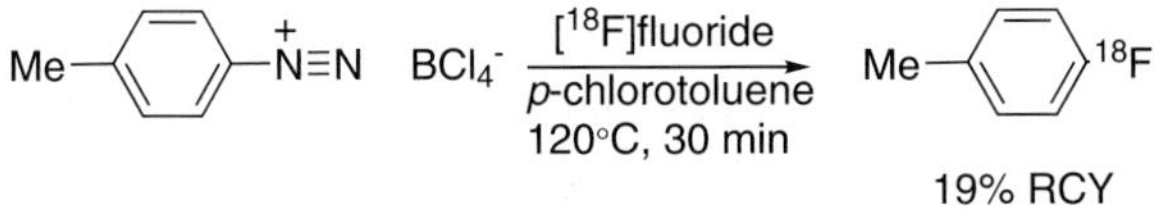

Scheme 10.11. Aromatic n.c.a. labelling with $^{18}F^-$ by modified Balz–Schiemann decomposition.

Balz–Schiemann fluorination using these diazonium salts afforded aryl [^{18}F]fluorides in low RCYs (Scheme 10.11) (Knöchel and Zwernemann, 1991; Argentini *et al.*, 1994). Due to the presence of more than one fluoride (or halide) substituents in these reagents, the RCYs are limited to 33% for [^{18}F]DAST and 25% for the diazonium salt of [^{18}F]BF_4^-.

10.3.3 *Silicon–fluorine bond formation*

Fluorosilanes have only recently been investigated for ^{18}F-radiolabelling. Due to the high affinity of silicon for fluorine, direct nucleophilic ^{18}F-fluorination of both small molecules and biomolecules can be achieved under milder conditions than typically required for carbon–fluorine bond formation (564 kJ/mol for Si–F *vs* 485 kJ/mol for C–F) (Anslyn and Dougherty, 2006). The stability of the fluorosilane depends on various parameters but mainly on steric hindrance at the silicon centre (Ting *et al.*, 2005; Schirrmacher *et al.*, 2006; Mu *et al.*, 2008); a theoretical model for assessing the hydrolytic stability of organofluorosilanes was recently developed based on DFT calculations and may serve as a predictive tool to estimate the hydrolytic half-life of ^{18}F-labelled silane derivatives (Höhne *et al.*, 2009). Aqueous [^{18}F]KF-K_{222} was used to efficiently access [^{18}F]*tert*-butyldiphenylfluorosilane from *tert*-butyldiphenylmethoxysilane within 5 min at room temperature (Choudhry *et al.*, 2007). This promising methodology allows the one-step radiolabelling of unprotected peptides with [^{18}F]KF-K_{222} by displacement of hydride, alkoxy or hydroxy groups in up to 53% RCY within 15 min at 90°C (Mu *et al.*, 2008). The direct isotopic exchange of a [^{19}F]fluorosilane-functionalised peptide using aqueous [^{18}F]fluoride is also possible and proceeds in high RCYs after 30 min at 95°C, but the method inherently suffers from low specific activity (Schirrmacher *et al.*, 2006). The silicon–fluorine

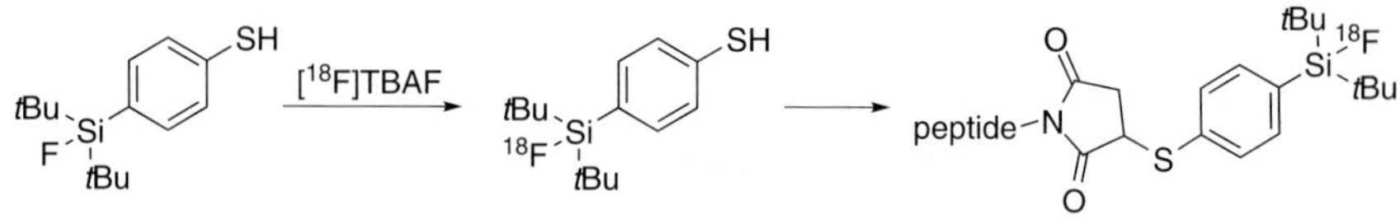

Scheme 10.12. Silicon-fuorine ^{18}F-fluorination for protein labelling.

radiolabeling strategy was also applied for the ^{18}F-fluorination of proteins. 4-(Di-*t*-butyl[^{18}F]fluorosilyl)benzenethiol (Si[^{18}F]FASH) was obtained in 40–60% RCY by isotopic exchange from anhydrous [^{18}F]TBAF (5 min, at room temperature) and subsequently coupled to a sulfomaleimide-functionalised protein (10 min, at room temperature) in an overall RCY of 12% within 20–30 min (Scheme 10.12) (Wängler *et al.*, 2009). This method requires the preparation of dry [^{18}F]TBAF.

10.3.4 *Boron–fluorine bond formation*

Biotin derivatives functionalised with a pending pinacol phenylboronate diester were directly labelled with carrier added aqueous [^{18}F]fluoride ([^{18}F]KHF_2) (Ting *et al.*, 2005). The nucleophilic displacement of the alkoxy groups results in the formation of fluoroborates with three fluorine substituents leading to low specific activity. In a similar vein, the labelling of the fluorescent BODIPY-functionalised aryl [^{18}F]trifluoroborate was also recently reported (Scheme 10.13) (Ting *et al.*, 2008).

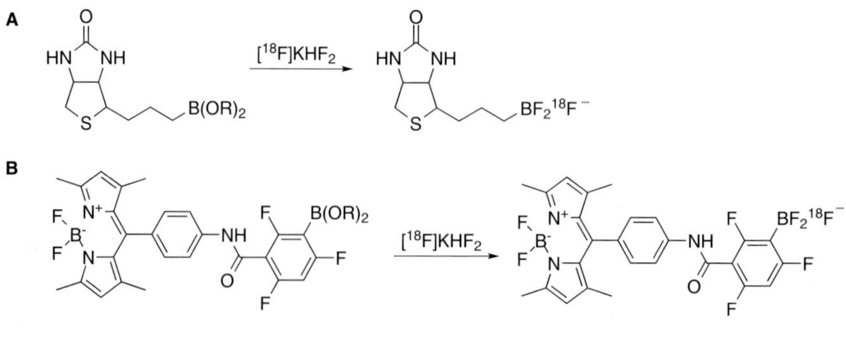

Scheme 10.13. Synthesis of carrier added [^{18}F]trifluoroborate-functionalised biotin (A) and BODIPY (B).

10.3.5 *Aluminium–fluorine bond formation*

Aluminium–fluorine bond formation was successfully used in the context of ^{18}F-radiolabelling (McBride and Goldenberg, 2008; McBride *et al.*, 2009). Aqueous [^{18}F]fluoride can directly bind to aluminium and attachment of the resulting ^{18}F-labelled aluminium complex to the molecule of interest allows the simple and rapid labelling of sensitive biomolecules. Various ligands were evaluated for peptide ^{18}F-labelling and these studies revealed that aluminium demonstrated promising reactivity with diethylenetriamine-pentaacetic acid (DTPA)- and 1,4,7-triazacyclononane-1,4,7-triacetic acid (NOTA)-functionalised peptides, providing up to 98% ^{18}F uptake as a metal–fluoride complex upon heating at 100°C. Although aluminium–fluoride complexes are reported to be stable *in vivo*, DTPA-Al-^{18}F proved unstable in serum and NOTA-Al-^{18}F was therefore preferred for imaging. Complementing the direct labelling method, isothiocyanate benzyl NOTA can be labelled with Al-^{18}F and subsequently conjugated to proteins or antibodies that are heat-sensitive (Fig. 10.3).

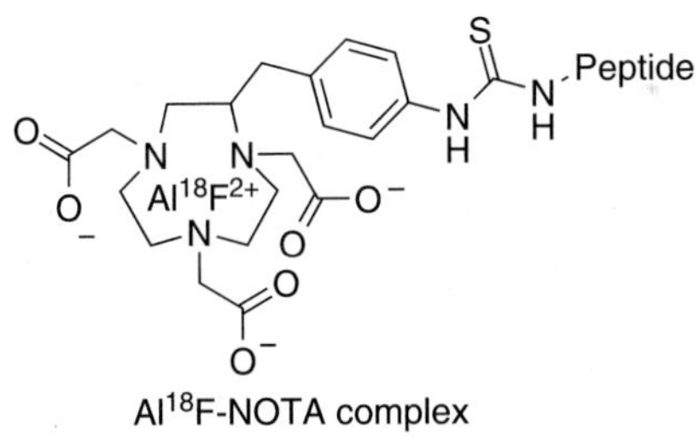

Figure 10.3. The Al^{18}F–NOTA complex.

10.3.6 *Phosphorus–fluorine bond formation*

A n.c.a. radiolabelling method by means of phosphorus–[^{18}F]fluorine bond formation was reported in 2005 (Studenov *et al.*, 2005). The cholinesterase inhibitor *N,N,N′,N′*-tetramethylphosphorodiamidic fluoride was ^{18}F-labelled by nucleophilic substitution in 96% RCY by letting the phosphorus chloride precursor react with anhydrous [^{18}F]TBAF (10 min, 75°C).

10.4 Electrophilic ^{18}F-Radiolabelling

10.4.1 *Electrophilic fluorination agents*

10.4.1.1 *Elemental fluorine [^{18}F]F_2*

[^{18}F]F_2 is generally produced from gas phase or liquid phase targets with the addition of carrier F_2 gas, a protocol leading to the production of ^{18}F-labelled radiotracers of low specific activity. [^{18}F]F_2 gas is the common reactant for electrophilic fluorination of organic molecules. When using this reagent, there is an equal likelihood of incorporating ^{18}F or ^{19}F, a limitation resulting in a maximum achievable RCY of 50%. This highly reactive fluorinating reagent is particularly suitable for the preparation of [^{18}F]fluoroaryl compounds, and other electron-rich systems which are difficult to access via direct nucleophilic fluorination. The overwhelming reactivity of [^{18}F]F_2 may of course lead to unselective reactions resulting in the formation of side products, which may cause problems during purification. Therefore, less reactive and more selective secondary radiolabelling reagents have been prepared from [^{18}F]F_2 to investigate their suitability as electrophilic sources of ^{18}F for labelling.

10.4.1.2 *Post-target production of [^{18}F]F_2*

The chemical oxidation of fluoride into F_2 would be an ideal solution to the generic problem of producing ^{18}F-labelled tracers in high specific activity when prepared by electrophilic fluorination, but this has not been possible due to the need to overcome the high redox potential ($E^o = 2.87$ V). In an attempt to improve the specific activity of [^{18}F]F_2 gas, a post-target production of [^{18}F]F_2 was developed providing labelled material of significantly higher specific activity (e.g. 6-[^{18}F]fluoro-L-DOPA was synthesised with a specific activity of 3.7 GBq/µmol) (Forsback *et al.*, 2008). [^{18}F]F_2 can be produced from n.c.a. methyl [^{18}F]fluoride and F_2 in a neon matrix carrier by atomisation in an electric discharge. After rearrangement and isotopic exchange, the desired radiolabelled gas is reproducibly obtained in up to 55 GBq/µmol (Scheme 10.14) (Bergman and Solin, 1997).

More recently, Langström and Ulin (2009) reported the production of high specific activity [^{18}F]F_2 from electrolytically dried [^{18}F]fluoride,

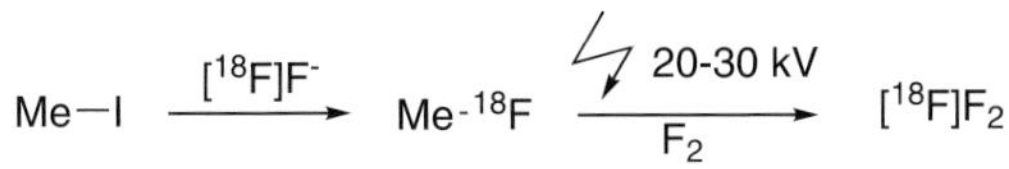

Scheme 10.14. Post-target production of $[^{18}F]F_2$.

followed by addition of a carrier (F_2 gas or metal fluoride) and plasma-induced scrambling. Although no specific activity was reported for this procedure, it is expected to be in the GBq/μmol range.

10.4.1.3 *$[^{18}F]$Hypofluorite reagents*

Trifluoromethyl $[^{18}F]$hypofluorite is produced by the direct reaction between caesium fluoride, F_2 and carbonyl fluoride with target-bound ^{18}F in the target chamber for 35 min at 100°C (Neirinckx *et al.*, 1978). Acetyl $[^{18}F]$hypofluorite is a widely used secondary electrophilic ^{18}F-labelling agent as it is of significantly milder reactivity than $[^{18}F]F_2$, thereby allowing the selective introduction of ^{18}F into a wider range of compounds. The current method to prepare acetyl $[^{18}F]$hypofluorite involves a gas–solid-phase reaction. $[^{18}F]F_2$ gas is passed through a stationary phase with bound complexes of acetic acid and alkali metal acetate. $[^{18}F]$AcOF is subsequently washed off the stationary phase (Jewett *et al.*, 1984). The reaction is quantitative but, as expected, only half of ^{18}F is available as the electrophilic agent (Scheme 10.15).

Scheme 10.15. Radiosynthesis of acetyl $[^{18}F]$hypofluorite.

10.4.1.4 *Perchloryl $[^{18}F]$fluoride*

Perchloryl fluoride owes its reactivity to the fact that fluorine is bound to a chlorine atom in its highest oxidation state. Gaseous $[^{18}F]$perchloryl fluoride ($[^{18}F]FClO_3$) is produced by passing $[^{18}F]F_2$ gas through a stationary phase with $KClO_3$ at 90°C (Ehrenkaufer and MacGregor, 1983). The conversion is total but only 50% of the radioactivity is available because the electrophilic $[^{18}F]FClO_3$ is produced with equimolar amounts of $[^{18}F]KF$ as a by-product. Perchloryl fluoride poses a constant threat of explosion when used in organic solvents. As a result its chemistry has not been widely explored. Interestingly, Hiller *et al.* (2008) attempted to develop a procedure to synthesise n.c.a. $[^{18}F]FClO_3$ starting from $[^{18}F]$fluoride in superacidic media and in the presence of $HClO_4$ (Scheme 10.16). The preparation of $[^{18}F]FClO_3$ was validated but the low RCYs (1–6%) and poor reproducibility were identified as significant drawbacks.

$$[^{18}F]F^-_{aq} \xrightarrow{\text{fuming sulfuric acid}} [^{18}F]HSO_3F \xrightarrow{KClO_4} [^{18}F]FClO_3$$

Scheme 10.16. Preparation of n.c.a. $[^{18}F]FClO_3$ from $[^{18}F]$fluoride.

10.4.1.5 *Xenon [^{18}F]difluoride*

The reaction of $[^{18}F]F_2$ and xenon in a sealed nickel reactor at 390°C for 40 min provides access to the secondary fluorination agent $[^{18}F]XeF_2$ (Chirakal *et al.*, 1984). Alternatively, an isotopic exchange between XeF_2 and $[^{18}F]F^-$, XeF_2 and $[^{18}F]SiF_4$, or XeF_2 and $[^{18}F]AsF_5$ also led to the formation of this ^{18}F-labelled reagent (Schrobilgen *et al.*, 1981; Constantinou *et al.*, 2001).

10.4.1.6 *N-[^{18}F]Fluorinated reagents*

N-Fluorinated reagents are stable and easy to handle electrophilic fluorinating reagents. They have found a wide range of applications in recent years and are incredibly useful for selective fluorination (Lal *et al.*, 1996). However, their application in the context of ^{18}F-labelling remains limited, with only few examples of ^{18}F-labelled reagents known to date, e.g. *N*-$[^{18}F]$fluoropyridinium triflate, $[^{18}F]$-1-fluoro-2-pyridone and various *N*-$[^{18}F]$fluoro-*N*-alkylsulfonamides. The syntheses of *N*-$[^{18}F]$fluorobis-trifluoromethanesulfonimide and *N*-$[^{18}F]$fluoro-*O*-benzenesulfonimide reagents were reported, but these reagents were found unsuitable for subsequent ^{18}F-fluorinations (Oberdorfer and Dietzel, 2003). The radiolabelling agent *N*-$[^{18}F]$fluoropyridinium triflate was prepared in good (46%) RCY by direct fluorination of *N*-trimethylsilylpyridinium triflate with $[^{18}F]F_2$ at –40°C (Scheme 10.17A) (Oberdorfer *et al.*, 1988a). This reagent reacts

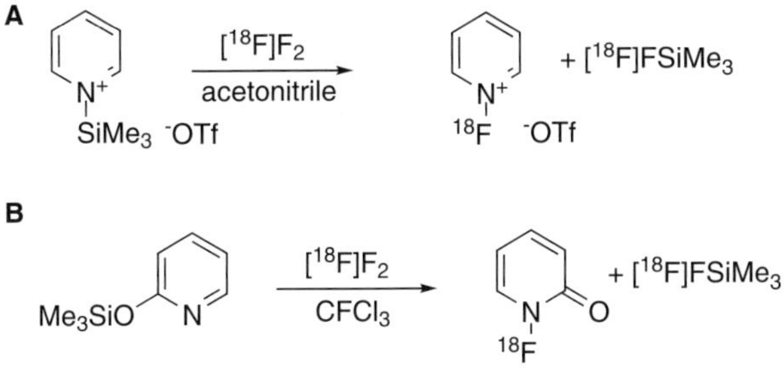

Scheme 10.17. Synthesis of *N*-[^{18}F]fluoropyridinium triflate (A) and 1-[^{18}F]fluoro-2-pyridone (B).

with aryl Grignard reagents or enolates. Similarly, [^{18}F]-1-fluoro-2-pyridone is prepared from 2-trimethylsilyloxypyridine in 48% RCY by bubbling [^{18}F]F_2 through a solution in chlorofluorocarbon at −78°C (Scheme 10.17B) (Oberdorfer *et al.*, 1988b). Even though this reagent was shown quantitatively to fluorinate alkyl lithium derivatives, further applications have not been reported.

The standard method for the production of *N*-[^{18}F]fluoro-*N*-alkylsulfonamides involves bubbling [^{18}F]F_2 through a solution of the corresponding parent sulfonamide in chlorofluorocarbon at −78°C (Scheme 10.18). The reaction is instantaneous with RCYs of up to 45%. The reactivity of these reagents is due to reduction of the electron density on the nitrogen atom by the electron-withdrawing groups (Satyamurthy *et al.*, 1990). More recently, *N*-[^{18}F]fluorobenzenesulfonimide ([^{18}F]NFSI) was prepared from sodium dibenzenesulfonimide in 50% RCY and this reagent was found suitable for the fluorination of silyl enol ethers and allylsilanes leading to ^{18}F-labelled fluorinated ketones and allylic fluorides, respectively. For allylsilanes, the formation of the ^{18}F-labelled product occurs with clean

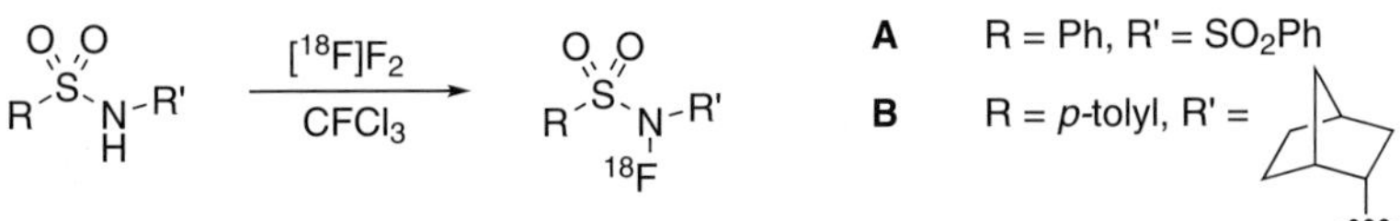

Scheme 10.18. Synthesis of [^{18}F]-*N*-fluoro-*N*-alkylsulfonamides: [^{18}F]NFSI (A) and [^{18}F]-*N*-fluoro-endo-norbornyl-*p*-tolylsulfonamide (B).

double bond transposition according to an S_E2' mechanism (Teare *et al.*, 2007). Selectfluor, one of the most reactive and commonly used electrophilic fluorinating N-F reagents, has also been radiolabelled with ^{18}F. The resulting new [^{18}F]-labelled N-F reagent is safe, non-toxic, and easy to handle. The combined use of [^{18}F]Selectfluor bis(triflate) and AgOTf allows for the preparation of electron-rich ^{18}F-aromatic compounds through a simple 'shake and mix' protocol at room temperature (Teare *et al.*, 2010).

All these secondary reagents are prepared from [^{18}F]F_2 and hence suffer from low specific activity. Therefore, carrier-added radiopharmaceuticals prepared by electrophic fluorination may not be suitable for PET applications if such compounds are highly toxic or if competitive saturation of the binding site by the nonradioactive carrier for receptor-mediated uptake processes occurs.

10.4.2 *Electrophilic carbon–fluorine bond formation*

Reagents for electrophilic ^{18}F-fluorination allow the labelling of electron-rich systems (e.g. alkenes, aromatic substrates, carbanions), which cannot be obtained by direct nucleophilic fluorination.

10.4.2.1 *Electrophilic addition*

The first synthesis of [^{18}F]FDG was based on the electrophilic fluorination of 3,4,6-tri-*O*-acetyl-D-glucal with molecular [^{18}F]F_2 or acetyl [^{18}F]hypofluorite (Ido *et al.*, 1978; Ehrenkaufer *et al.*, 1984). Nowadays, the electrophilic fluorination of enol ethers or alkenes is scarcely used. A radiopharmaceutical still produced using this chemistry is the hypoxia-selective biomarker 2-(2-nitro-1*H*-imidazol-1-yl)-*N*-(2,2,3,3,3-[^{18}F]pentafluoropropyl)-acetamide ([^{18}F]EF5) which is prepared by addition of [^{18}F]F_2 to a trifluorinated alkene (Bach and Henneike, 1970; Ziemer *et al.*, 2003). Recently, Dolbier and coworkers reported that the presence of small amounts of iodine, bromine or boron trifluoride increases the RCY of [^{18}F]F_2 addition to alkenes (Kachur *et al.*, 2010).

10.4.2.2 *Aliphatic fluorodemetallation*

Very few examples of electrophilic fluorodemetallation have been reported for the preparation of aliphatic fluorinated compounds as these compounds can usually be obtained by nucleophilic substitution. As proof of principle, the fluorination of various aliphatic compounds was described. For example, the reaction of 1-[^{18}F]fluoro-2-pyridone with methyl lithium yields [^{18}F]fluoromethane quantitatively (Oberdorfer *et al.*, 1988b). 1-[^{18}F]Fluorohexane and [^{18}F]fluorocyclohexane are obtained from their Grignard precursors in 78% RCY (with *N*-[^{18}F]fluoropyridinium triflate) and 29% RCY (with *N*-[^{18}F]fluoro-endo-norbornyl-*p*-tolylsulfonamide) respectively (Oberdorfer *et al.*, 1988a; Satyamurthy *et al.*, 1990).

10.4.2.3 *Fluorination of enols and derivatives*

The reactivity of *N*-[^{18}F]fluoropyridinium triflate with enolates was demonstrated with the synthesis of diethyl 2-[^{18}F]fluoro-2-phenylmalonate (58% RCY) and ethyl 3-amino-2-[^{18}F]fluoropropionate (23% RCY) (Oberdorfer *et al.*, 1988a). The electrophilic reaction of [^{18}F]F_2 with a keto–enol was exemplified with the ^{18}F-fluorination of a lactam to yield a diazepam derivative containing an α-fluoroketone in 25% RCY (Scheme 10.19A) (Luxen *et al.*,

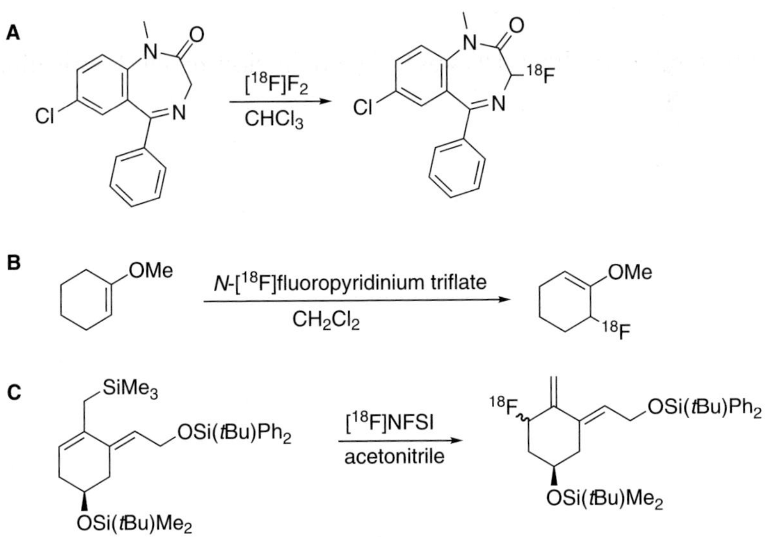

Scheme 10.19. Electrophilic fluorination of diazepam (enol fluorination: A), methyl enol ether (allylic fluorination: B) and A ring fragment of vitamin D_3 (fluorodesilylation: C).

1987). Alternatively, *N*-[^{18}F]fluoropyridinium triflate reacts with methyl enol ethers to yield allylic fluorides in good RCYs (Scheme 10.19B) (Oberdorfer *et al.*, 1988a). α-Fluoroketones can be prepared by fluorodesilylation of the corresponding silyl enol ethers with [^{18}F]NFSI. This reagent also allowed the ^{18}F-labelling of the fluorinated analogue of the A ring fragment of vitamin D_3 in 87% RCY (Scheme 10.19C) (Teare *et al.*, 2007).

10.4.2.4 *Electrophilic aromatic fluorination*

Electrophilic ^{18}F reagents are most useful for the synthesis of aryl [^{18}F]fluorides, which are not accessible upon nucleophilic aromatic substitution. Direct fluorination is carried out using elemental [^{18}F]fluorine, [^{18}F]hypofluorite reagents or xenon [^{18}F]difluoride, usually with poor selectivity. For example, the direct radiolabelling of L-DOPA precursor with [^{18}F]F_2 yields a mixture of the 2-, 5- and 6-[^{18}F]F-DOPA isomers (Firnau *et al.*, 1984). The reactivity and selectivity can be modulated using an acidic solvent (AcOH, HF, TFA), a less reactive fluorinating reagent and bulkier substrates. The reaction of the *O*-pivaloyl-protected F-DOPA precursor with acetyl [^{18}F]hypofluorite in acetic acid provided 6-[^{18}F]F-DOPA with improved RCY and purity (Ishiwata *et al.*, 1993; Azad *et al.*, 2007). Fluorination upon electrophilic fluorodemetallation is more attractive due to increased selectivity. It has been demonstrated that the fluorodemetallation of a protected 6-trimethylstannyl aryl precursor with [^{18}F]F_2 or [^{18}F]AcOF in $CFCl_3$ leads to the synthesis of [^{18}F]F-DOPA in up to 33% RCY after hydrolysis and purification (Namavari *et al.*, 1992; de Vries *et al.*, 1999) (Scheme 10.20). A variation of this method involving a fluorodesilylation reaction has been investigated but did not prove as efficient (8% RCY) (Diksic and Farrokhzad, 1985). The reaction of elemental fluorine has been investigated with aryl lithium, aryl Grignard derivatives, aryl

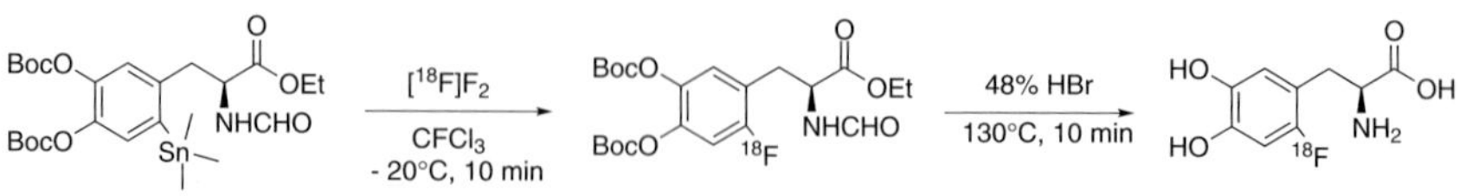

Scheme 10.20. Current radiosynthesis of [^{18}F]F-DOPA.

mercury and aryl metals from group IVb (tin, germanium, silicon, lead) (Adam *et al.*, 1983; Coenen and Moerlein, 1987). Amongst those various organometallic precursors, the fluorodestannylation of aryl tin derivatives with electrophilic ^{18}F-reagents (typically [^{18}F]F_2 or [^{18}F]AcOF) generally displays the best reactivity profile (Adam, 1986; Forsback *et al.*, 2009). The fluorodemetallation reaction is usually not influenced by aromatic substituents, due to the strong ionic character of the carbon–metal bond, but side products may be produced resulting from competitive direct fluorodeprotonation or benzylic substitution (Coenen and Moerlein, 1987). *N*-[^{18}F]Fluorinated reagents react with a variety of carbanions and organometallic compounds. *N*-[^{18}F]Fluoro-endo-norbornyl-*p*-tolylsulfonamide (see Scheme 10.18) was found to be the most reactive electrophilic reagent for the rapid and regioselective fluorination of Grignard and organolithium species (Satyamurthy *et al.*, 1984, 1990). Carrier added [^{18}F]$FClO_3$ proved useful for the fluorination of lithium salts derived from protected aniline, anisole and veratrole in moderate (21–34%) RCYs (Ehrenkaufer and MacGregor, 1983).

10.5 Prosthetic Groups

Direct ^{18}F-labelling of complex biomolecules such as peptides or proteins is highly challenging. Since nucleophilic carbon–fluorine bond construction typically requires harsh reaction conditions, this radiochemistry is not suitable for more complex and sensitive substrates. Alternative ^{18}F-labelling strategies have therefore been developed relying on the nucleophilic radiolabelling of aromatic or aliphatic prosthetic groups, their activation and subsequent coupling to a functional group present on the probe to be labelled. Various reviews have discussed in depth this so-called indirect ^{18}F-labelling strategy (Wester and Schottelius, 2007), so this section will cover the most commonly used ^{18}F-labelled prosthetic groups only. 4-Nitrophenyl 2-[^{18}F]fluoropropionate was successfully used for the labelling of peptides (Liu *et al.*, 2009). 4-[^{18}F]Fluorobenzoate is more lipophilic but remains one of the most versatile prosthetic groups because of its advantageous *in vivo* stability and coupling efficiency (Wüst *et al.*, 2003). Recently, a convenient one-pot, three-step synthesis was reported

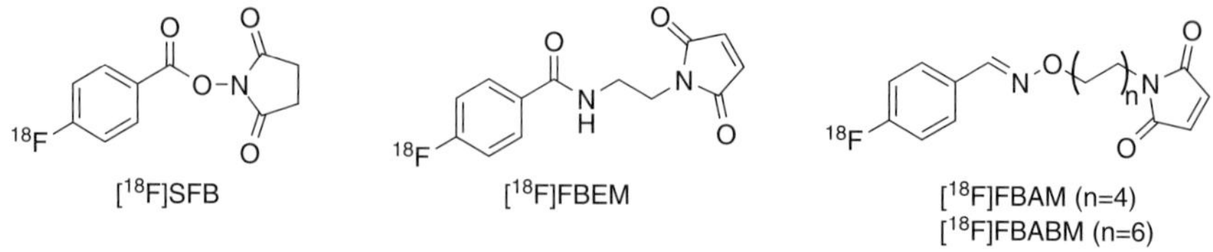

Figure 10.4. Representative ^{18}F-labelled prosthetic groups.

allowing the synthesis, in less than 1 h, of the labelling agent *N*-succinimidyl 4-[^{18}F]fluorobenzoate ([^{18}F]SFB) (Fig. 10.4). This labelled prosthetic group can be coupled with a free amino group of the molecule of interest (Tang *et al.*, 2008).

^{18}F-Labelled maleimides are thiol-specific labelling agents reacting chemoselectively and displaying a very advantageous reactivity profile. Various maleimide reagents have been developed and can be obtained in multi-step syntheses *inter alia* by oxime formation between 4-[^{18}F]fluorobenzaldehyde ([^{18}F]FBA) and aminooxyalkylmaleimide (to produce [^{18}F]FBAM and [^{18}F]FBABM), or amide bond formation between [^{18}F]SFB and aminoalkylmaleimide (to produce [^{18}F]FBEM) (Fig. 10.4) (Toyokuni *et al.*, 2003; Cai *et al.*, 2006; Berndt *et al.*, 2007; Li *et al.*, 2008;). [^{18}F]FBA can be directly used for hydrazone or oxime bond formation with hydrazine- or aminooxyl-functionalised probes.

An alternative approach to labelling small molecules, peptides or other sensitive probes involves the 1,3-dipolar Huisgen cycloaddition between alkynes and azides (click reaction). The reaction is very attractive because of its high chemoselectivity, its mild reaction conditions and its efficiency when catalysed by copper(I). This chemistry requires functionalisation of the molecule to be labelled with either an azide or alkyne group. The synthesis of various ^{18}F-labelled azides or alkynes has been reported in the literature and these prosthetic groups have been used successfully to selectively label functionalised peptides or proteins in high RCYs (Glaser and Årstad, 2007; Wangler *et al.*, 2010).

Prosthetic group labelling is not limited to two components reactions. So-called radio-multi-component reactions are possible leading to various ^{18}F-radiolabelled heterocycles and peptide units not accessible by direct fluorination with [^{18}F]fluoride (Lei Li *et al.*, 2011).

10.6 Purification

In radiosynthesis, the nonradioactive precursor is often used in large excess (μmol–mmol) relative to the amount of radiolabelling agent (pmol–nmol). Therefore, the radiolabelled molecule must be separated from the excess precursor before clinical use or further chemical transformations. The purification of ^{18}F-labelled compounds is conventionally performed by one or a combination of the following techniques. HPLC is a very powerful technique and probably the most commonly used, but can be time-consuming and lead to a significant loss of radioactivity. Distillation has proved useful for volatile radiolabelled compounds, but can also be time-consuming, difficult to implement and may suffer from poor reproducibility. In addition, heating of the reaction mixtures may result in decomposition (Tewson, 1997; Glaser and Årstad, 2007).

Solid-phase labelling has emerged as an attractive alternative technology. The nucleophilic fluorination reaction allows the radiotracer to be released from the insoluble solid-supported substrate, which itself is removed by simple filtration. The solid-phase synthesis of [^{18}F]FDG has been validated in high RCY and high chemical purity (Scheme 10.21) (Brown *et al.*, 2007, 2009). An electrophilic variant was reported for the synthesis of [^{18}F]F-DOPA from the solid-supported organotin precursor (Luthra *et al.*, 2006). Although conceptually elegant, the radiolabelling reaction itself typically requires extensive optimisation and could display an unfavourable kinetic profile because of the heterogeneity of the reaction mixture.

Based on the same principle as HPLC, chromatography by means of solid-phase extraction is much faster but fairly narrow in scope, as the

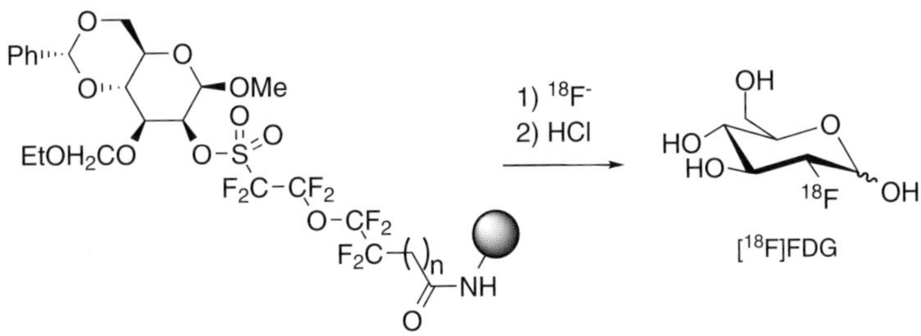

Scheme 10.21. Solid-phase synthesis of [^{18}F]FDG.

technique requires the precursor and the radiotracer to have significantly different affinities for the stationary phase. Fluorous solid-phase extraction (FSPE) has recently emerged as a new technique for the separation of the radiolabelled molecule from its precursor (Bejot *et al.*, 2009). Similar to the solid-phase methods, a fluorous-tagged precursor is detagged upon nucleophilic introduction of [^{18}F]fluoride. The ^{18}F-labelled molecule is subsequently separated from its precursor on FSPE, due to the affinity of the tag for the fluorous stationary phase. Noteworthy is the fact that the fluorous radiolabelling reaction can be performed in homogeneous phases, allowing more favourable reaction kinetics than solid-phase synthesis. The method however suffers from ^{19}F leakage and reduced specific activity.

Acknowledgements

The authors gratefully acknowledge the contribution of Dr Harriet Teare (University of Oxford), Dr Matthias Glaser (GE Healthcare Medical Diagnostics Discovery), Ida Stenhagen, Matthew Hopkinson, Lei Li, Dr Rodrigue Yeuna Loma and Dr Matthew Tredwell (University of Oxford).

References

Abrahim, A., Angelberger, P., Kletter, K. *et al.* (2006) Synthesis of fluorine-18-labelled 5- and 6-fluoro-2-pyridinamine. *J Labelled Compd Rad* 49, 345–356.

Adam, M.J., Berry, J.M., Hall, L.D. *et al.* (1983) The cleavage of aryl-metal bonds by elemental fluorine: synthesis of aryl-fluorides. *Can J Chem* 61, 658–660.

Adam, M.J. (1986) The demetallation reaction in radiohalogen labelling: synthesis of bromine and fluorine labelled compounds. *Int J Radiat Appl Instrum Part A* 37, 811–815.

Aerts, J., Voccia, S., Lemaire, C. *et al.* (2010) Fast production of highly concentrated reactive [^{18}F]fluoride for aliphatic and aromatic nucleophilic radiolabelling. *Tetrahedron Lett* 51, 64–66.

Aigbirhio, F.I. & Pike, V.W. (1995) Labelling of the CFC-alternative, 2H-heptafluoropropane (HFC 227ea), with fluorine-18. *J Fluorine Chem* 75, 67–73.

Aigbirhio, F.I., Pike, V.W., Waters, S.L. *et al.* (1995) Efficient regioselective labelling of the CFC alternative 1,1,1,2-tetrafluoroethane (HFC-134a) with fluorine-18. *J Fluorine Chem* 70, 279–287.

Ametamey, S.M., Honer, M. & Schubiger, P.A. (2008) Molecular imaging with PET. *Chem Rev* 108, 1501–1516.

Anslyn, E.V. & Dougherty, D.A. (2006) *Modern Physical Organic Chemistry*, University Science Books, Sausalito, CA.

Argentini, M., Wiese, C. & Weinreicht, R. (1994) Syntheses of 5-fluoro-D/L-dopa and [^{18}F]5-fluoro-L-dopa. *J Fluorine Chem* 68, 141–144.

Azad, B.B., Chirakal, R. & Schrobilgen, G.J. (2007) Trifluoromethanesulfonic acid, an alternative solvent medium for the direct electrophilic fluorination of DOPA: new syntheses of 6-[^{18}F]fluoro-L-DOPA and 6-[^{18}F]fluoro-D-DOPA. *J Labelled Compd Rad* 50, 1236–1242.

Bach, R.D. & Henneike, H.F. (1970) Molecular orbital approach to the mechanism of electrophilic additions to olefins. *J Am Chem Soc* 92, 5589–5602.

Baker, A.W., Bonniface, D., Klapötke, T.M. *et al.* (2000) Catalytic fluorination of trichloroethene by anhydrous hydrogen fluoride in the presence of fluorinated chromia under static conditions. Synthesis of [^{18}F]-labelled CF_3CH_2F and [^{36}Cl]-labelled CF_3CH_2Cl: catalytic dehydrofluorination of CF_3CH_2F and CF_3CH_2Cl. *J Fluorine Chem* 102, 279–284.

Becaud, J., Mu, L., Karramkam, M. *et al.* (2009) Direct one-step ^{18}F-labeling of peptides via nucleophilic aromatic substitution. *Bioconjugate Chem* 20, 2254–2261.

Bejot, R., Fowler, T., Carroll, L. *et al.* (2009) Fluorous synthesis of ^{18}F-radiotracers with the [^{18}F]fluoride ion: nucleophilic fluorination as the detagging process. *Angew Chem Int Edit* 48, 586–589.

Bejot, R., Elizarov, A.M., Ball, E. *et al.* (2011) Batch-mode microfluidic radiosynthesis of *N*-succinimidyl-4-[^{18}F]fluorobenzoate for protein labeling. *J Labelled Compd Rad* 54, 117–122.

Bendada, A., Bonniface, D.W., McMonagle, F. *et al.* (1996a) Role of chlorocarbon oligomers in the room-temperature heterogeneous catalytic fluorination of hydrochlorocarbons to hydrochlorofluorocarbons. *Chem Commun*, 1947–1948.

Bendada, A., Webb, G. & Winfield, J.M. (1996b) Fluorination of gamma-alumina by sulphur tetrafluoride, thionyl fluoride, carbonyl fluoride or anhydrous hydrogen fluoride: a radiotracer study. *Eur J Sol State Inorg* 33, 907–916.

Bergman, J. & Solin, O. (1997) Fluorine-18-labeled fluorine gas for synthesis of tracer molecules. *Nucl Med Biol* 24, 677–683.

Berndt, M., Pietzsch, J. & Wuest, F. (2007) Labeling of low-density lipoproteins using the ^{18}F-labeled thiol-reactive reagent *N*-[6-(4-[^{18}F]fluorobenzylidene)aminooxyhexyl]maleimide. *Nucl Med Biol* 34, 5–15.

Berridge, M.S. & Kjellström, R. (1999) Designs and use of silver [^{18}O]water targets for [^{18}F]fluoride production. *Appl Radiat Isotopes* 50, 699–705.

Berridge, M.S., Apana, S.M. & Hersh, J.M. (2009) Teflon radiolysis as the major source of carrier in fluorine-18. *J Labelled Compd Rad* 52, 543–548.

Bishop, A., Satyamurthy, N., Bida, G. *et al.* (1996) Identification and quantitation of gaseous compounds of fluorine generated in [^{18}F]F_2 target systems. *Nucl Med Biol* 23, 391–405.

Blom, E., Karimi, F. & Långström, B. (2009) [^{18}F]/^{19}F exchange in fluorine containing compounds for potential use in ^{18}F-labelling strategies. *J Labelled Compd Rad* 52, 504–511.

Bonniface, D.W., Scott, J.D., Watson, M.J. *et al.* (1999) Halogen exchange reactions for CFC alternatives: the behaviour of Fluorine-18 labelled hydrogen fluoride towards prefluorinated chromia containing nickel(II) or zinc(II). *Green Chem* 1, 9–11.

Bozorgzadeh, H., Kemnitz, E., Nickkho-Amiry, M. *et al.* (2001) Conversion of 1,1,2-trichlorotrifluoroethane to 1,1,1-trichlorotrifluoroethane and 1,1-dichlorotetrafluoroethane over aluminium-based catalysts. *J Fluorine Chem* 107, 45–52.

Briard, E. & Pike, V.W. (2004) Substitution–reduction: an alternative process for the [^{18}F]*N*-(2-fluoroethylation) of anilines. *J Labelled Compd Rad* 47, 217–232.

Brown, L.J., Bouvet, D.R., Champion, S. *et al.* (2007) A solid-phase route to ^{18}F-labeled tracers, exemplified by the synthesis of [^{18}F]2-fluoro-2-deoxy-D-glucose. *Angew Chem Int Edit* 46, 941–944.

Brown, L.J., Ma, N., Bouvet, D.R. *et al.* (2009) Synthesis of the positron-emitting radiotracer [^{18}F]-2-fluoro-2-deoxy-D-glucose from resin-bound perfluoroalkylsulfonates. *Org Biomol Chem* 7, 564–575.

Cacace, F., Speranza, M., Wolf, A.P. *et al.* (1981) Labelling of fluorinated aromatics by isotopic exchange with [^{18}F]fluoride. *J Labelled Compd Rad* 18, 1721–1730.

Cai, W., Zhang, X., Wu, Y. & Chen, X. (2006) A thiol-reactive ^{18}F-labeling agent, *N*-[2-(4-^{18}F-fluorobenzamido)ethyl]maleimide, and synthesis of RGD peptide-based tracer for PET imaging of $\alpha_V\beta_3$ integrin expression. *J Nucl Med* 47, 1172–1180.

Cai, L., Lu, S. & Pike, V.W. (2008) Chemistry with [^{18}F]fluoride ion. *Eur J Org Chem* 39, 2853–2873.

Carroll, M.A., Nairne, J., Smith, G. *et al.* (2007) Radical scavengers: a practical solution to the reproducibility issue in the fluoridation of diaryliodonium salts. *J Fluorine Chem* 128, 127–132.

Chirakal, R., Firnau, G., Schrobilgen, G.J. *et al.* (1984) The synthesis of [^{18}F]xenon difluoride from [^{18}F]fluoride gas. *Int J App Radiat Is* 35, 401–404.

Choudhry, U., Paul, R., Blower, P. *et al.* (2007) Rapid reaction of alkoxysilanes with [F-18]fluoride as a basis for biomolecule labelling for PET. *Eur J Nucl Med Mol I* 34, P337.

Chun, J.-H., Lu, S., Lee, Y.-S. *et al.* (2010) Fast and high-yield microreactor syntheses of ortho-substituted [^{18}F]fluoroarenes from reactions of [^{18}F]fluoride ion with diaryliodonium salts. *J Org Chem* 75, 3332–3338.

Coenen, H.H. & Moerlein, S.M. (1987) Regiospecific aromatic fluorodemetallation of group IVb metalloarenes using elemental fluorine or acetyl hypofluorite. *J Fluorine Chem* 36, 63–75.

Constantinou, M., Aigbirhio, F.I., Smith, R.G. *et al.* (2001) Xenon difluoride exchanges fluoride under mild conditions: a simple preparation of [^{18}F]xenon difluoride for PET and mechanistic studies. *J Am Chem Soc* 123, 1780–1781.

Culbert, P.A., Adam, M.J., Hurtado, E.T. *et al.* (1995) Automated synthesis of [^{18}F] FDG using tetrabutylammonium bicarbonate. *Appl Radiat Isotopes* 46, 887–891.

de Vries, E.F.J., Luurtsema, G., Brüssermann, M. *et al.* (1999) Fully automated synthesis module for the high yield one-pot preparation of 6-[^{18}F]Fluoro-L-DOPA. *Appl Radiat Isotopes* 51, 389–394.

Delacroix, D. (2002) *Radionuclide and Radiation Protection Data Handbook*, Nuclear Technology Publishing, Ashford.

Deng, H., Cobb, S.L., Gee, A.D. *et al.* (2006) Fluorinase mediated C–^{18}F bond formation, an enzymatic tool for PET labelling. *Chem Commun* 14, 652–654.

DePuy, C.H., Gronert, S., Mullin, A. *et al.* (1990) Gas-phase S_N2 and E2 reactions of alkyl halides. *J Am Chem Soc* 112, 8650–8655.

Diksic, M. & Farrokhzad, S. (1985) New synthesis of fluorine-18-labeled 6-fluoro-L-dopa by cleaving the carbon–silicon bond with fluorine. *J Nucl Med* 26, 1314–1318.

Dixon, K.W., Ghorab, M.F. & Winfield, J.M. (1987) Radiotracers in fluorine chemistry, part XI. The behaviour of [^{18}F]fluorine and [^{35}S]sulphur labelled sulphur tetrafluoride at Lewis acid fluoride surfaces. *J Fluorine Chem* 37, 357–370.

Dixon, K.W. & Winfield, J.M. (1989) Radiotracers in fluorine chemistry. Part 12. Reactions between cesium fluoride and arsenic pentafluoride, boron-trifluoride, sulfur-tetrafluoride, carbonyl fluoride, and carbon-dioxide under heterogeneous conditions at room-temperature: radiotracer studies using F-18, S-35, and C-14. *J Chem Soc, Dalton Trans,* 937–942.

Dolci, L., Dollé, F., Jubeau, S. *et al.* (1999) 2-[^{18}F]Fluoropyridines by no-carrier-added nucleophilic aromatic substitution with [^{18}F]FK-K_{222} — a comparative study. *J Labelled Compd Rad* 42, 975–985.

Dollé, F. (2005) Fluorine-18-labelled fluoropyridines: advances in radiopharmaceutical design. *Curr Pharm Design* 11, 3221–3235.

Dong, C., Huang, F., Deng, H. *et al.* (2004) Crystal structure and mechanism of a bacterial fluorinating enzyme. *Nature* 427, 561–565.

Ehrenkaufer, R.E. & MacGregor, R.R. (1983) Synthesis of [^{18}F]perchloryl fluoride and its reactions with functionalized aryl lithiums. *Int J Appl Rad I* 34, 613–615.

Ehrenkaufer, R.E., Potocki, J.F. & Jewett, D.M. (1984) Simple synthesis of F-18-labeled 2-fluoro-2-deoxy-D-glucose: concise communication. *J Nucl Med* 25, 333–337.

Ekström, L.P. & Firestone, R.B. (1999) Online database: Table of Radioactive Isotopes. http://ie.lbl.gov/toi/

Elizarov, A.M., van Dam, R.M., Shin, Y.S. *et al.* (2010) Design and optimization of coin-shaped microreactor chips for PET radiopharmaceutical synthesis. *J Nucl Med* 51, 282–287.

Ermert, J., Hocke, C., Ludwig, T. *et al.* (2004) Comparison of pathways to the versatile synthon of no-carrier-added 1-bromo-4-[^{18}F]fluorobenzene. *J Labelled Compd Rad* 47, 429–441.

Firnau, G., Chirakal, R. & Garnett, E.S. (1984) Aromatic radiofluorination with [^{18}F]fluorine gas: 6-[^{18}F]fluoro-L-dopa. *J Nucl Med* 25, 1228–1233.

Forsback, S., Eskola, O., Haaparanta, M. *et al.* (2008) Electrophilic synthesis of 6-[F-18]fluoro-L-DOPA using post-target produced [F-18]F_2. *Radiochim Acta* 96, 845–848.

Forsback, S., Eskola, O., Bergman, J. *et al.* (2009) Alternative solvents for electrophilic synthesis of 6-[^{18}F]fluoro-L-DOPA. *J Labelled Compd Rad* 52, 286–288.

Fouzi Ghorab, M.F. & Winfield, J.M. (1990) Radiotracers in fluorine chemistry. Part [1]. The lability of the fluoroanions BF_4^-, PF_6^-, AsF_6^-, SbF_6^-, NbF_6^- and TaF_6^- in acetonitrile solution: a fluorine-18 radiotracer study. *J Fluorine Chem* 49, 367–383.

Fouzi Ghorab, M.F. & Winfield, J.M. (1993) Radiotracers in fluorine chemistry. Part 16[1]. A fluorine-18 study of the formation and lability of heptafluorotungstate(VI) and heptafluoromolybdate(VI) anions in acetonitrile and under heterogeneous conditions. *J Fluorine Chem* 62, 101–110.

Frazer, C.J.W., Majid, A., Oates, G. *et al.* (1975) ^{18}F exchange between fluorotrimethylsilane and tungsten(VI) fluorides. *J Inorg Nucl Chem* 37, 1535–1537.

Friedlander, G., Macias, E.S., Kennedy, J.W. *et al.* (1981) *Nuclear and Radiochemistry*, John Wiley & Sons, Chichester.

Füchtner, F., Preusche, S., Mäding, P. *et al.* (2008) Factors affecting the specific activity of [^{18}F]fluoride from a [^{18}O]water target. *Nucl Med-Nukl* 47, 116–119.

Glaser, M. & Årstad, E. (2007) 'Click Labeling' with 2-[^{18}F]fluoroethylazide for positron emission tomography. *Bioconjugate Chem* 18, 989–993.

Hamacher, K., Coenen, H.H. & Stocklin G. (1986) Efficient stereospecific synthesis of no-carrier-added 2-[^{18}F]-fluoro-2-deoxy-D-glucose using aminopolyether supported nucleophilic substitution. *J Nucl Med* 27, 235–238.

Hamacher, K., Hirschfelder, T. & Coenen, H.H. (2002) Electrochemical cell for separation of [^{18}F]fluoride from irradiated ^{18}O-water and subsequent no carrier added nucleophilic fluorination. *Appl Radiat Isotopes* 56, 519–523.

Hamacher, K. & Coenen, H.H. (2006) No-carrier-added nucleophilic ^{18}F-labelling in an electrochemical cell exemplified by the routine production of [^{18}F]altanserin. *Appl Radiat Isotopes* 64, 989–994.

Hansch, C., Leo, A. & Taft, R.W. (1991) A survey of hammett substituent constants and resonance and field parameters. *Chem Rev* 91, 165–195.

Hess, E., Blessing, G., Coenen, H.H. *et al.* (2000) Improved target system for production of high purity [F-18]fluorine via the O-18(*p*,*n*)F-18 reaction. *App Radiat Isotopes* 52, 1431–1440.

Hiller, A., Fischer, C., Jordanova, A. *et al.* (2008) Investigations to the synthesis of n.c.a [^{18}F]$FClO_3$ as electrophilic fluorinating agent. *Appl Radiat Isotopes* 66, 152–157.

Höhne, A., Yu, L., Mu, L. *et al.* (2009) Organofluorosilanes as model compounds for ^{18}F-labeled silicon-based PET tracers and their hydrolytic stability: experimental data and theoretical calculations (PET = positron emission tomography). *Chem Eur J* 15, 3736–3743.

Hollingworth, C., Hazari, A., Hopkinson, M.N. *et al.* (2011) Palladium-catalyzed allylic fluorination. *Angew Chem Int Edit* 50, 2613–2617.

Huiban, M., Gee, A.D., Martarello, L. *et al.* (2007) Comparison of diaryliodonium salts and triazenes as precursors for radiofluorination. *J Labelled Compd Rad* 50, S130.

Ido, T., Wan, C.-N., Casella, V. *et al.* (1978) Labeled 2-deoxy-D-glucose analogs. ^{18}F-labeled 2-deoxy-2-fluoro-D-glucose, 2-deoxy-2-fluoro-D-mannose and ^{14}C-2-deoxy-2-fluoro-D-glucose. *J Labelled Compd Rad* 14, 175–183.

Ishiwata, K., Ishii, S.-I., Senda, M. *et al.* (1993) Electrophilic synthesis of 6-[^{18}F]fluoro-L-dopa: use of 4-*O*-pivaloyl-L-dopa as a suitable precursor for routine production. *Appl Radiat Isotopes* 44, 755–759.

Jewett, D.M., Potocki, J.F. & Ehrenkaufer, R.E. (1984) A gas–solid-phase microchemical method for the synthesis of acetyl hypofluorite. *J Fluorine Chem* 24, 477–484.

Jewett, D.M., Toorongian, S.A., Mulholland, G.K. *et al.* (1988) Multiphase extraction: rapid phase-transfer of [^{18}F]fluoride ion for nucleophilic radiolabeling reactions. *Int J Rad Appl Instrum A* 39, 1109–1111.

Kachur, A.V., Dolbier Jr., W.R., Xu, W. *et al.* (2010) Catalysis of fluorine addition to double bond: an improvement of method for synthesis of ^{18}F PET agents. *Appl Radiat Isotopes* 68, 293–296.

Kaneko, S., Ishiwata, K., Hatano, K. *et al.* (1999) Enzymatic synthesis of no-carrier-added 6-[^{18}F]fluoro-L-dopa with β-tyrosinase. *Appl Radiat Isotopes* 50, 1025–1032.

Karramkam, M., Hinnen, F., Vaufrey, F. *et al.* (2003) 2-, 3- and 4-[^{18}F]Fluoropyridine by no-carrier-added nucleophilic aromatic substitution with K[^{18}F]F-K222: a comparative study. *J Labelled Compd Rad* 46, 979–992.

Kealey, C.P., Klapötke, T.M., McComb, D.W. *et al.* (2001) Fluorination of polycrystalline diamond films and powders: an investigation using FTIR spectroscopy, SEM, energy-filtered TEM, XPS and fluorine-18 radiotracer methods. *J Mater Chem* 11, 879–886.

Kilbourn, M.R. & Subramanian, R. (1990) Synthesis of fluorine-18 labeled 1,1-difluoro-2,2-dichloroethyl aryl ethers by ^{18}F-for-^{19}F exchange. *J Labelled Compd Rad* 28, 1355–1361.

Kim, D.W., Choe, Y.S. & Chi, D.Y. (2003) A new nucleophilic fluorine-18 labeling method for aliphatic mesylates: reaction in ionic liquids shows tolerance for water. *Nucl Med Biol* 30, 345–350.

Kim, H.W., Jeong, J.M., Lee, Y.-S. *et al.* (2004) Rapid synthesis of [^{18}F]FDG without an evaporation step using an ionic liquid. *Appl Radiat Isotopes* 61, 1241–1246.

Kim, D.W., Ahn, D.-S., Oh, Y.-H. *et al.* (2006) A new class of S_N2 reactions catalyzed by protic solvents: facile fluorination for isotopic labeling of diagnostic molecules. *J Am Chem Soc* 128, 16394–16397.

Kim, D.W., Jeong, H.-J., Lim, S.T. *et al.* (2008) Facile nucleophilic fluorination reactions using tert-alcohols as a reaction medium: significantly enhanced reactivity of alkali metal fluorides and improved selectivity. *J Org Chem* 73, 957–962.

Klapötke, T.M., McMonagle, F., Spence, R.R. *et al.* (2006) γ-Alumina-supported boron trifluoride: catalysis, radiotracer studies and computations. *J Fluorine Chem* 127, 1446–1453.

Knöchel, A. & Zwernemann, O. (1991) Aromatic n.c.a labelling with ^{18}F- by modified Balz–Schiemann decomposition. *Appl Radiat Isotopes* 42, 1077–1080.

Kumar, P., Zheng, W., McQuarrie, S.A. *et al.* (2005) ^{18}F-FESB: synthesis and automated radiofluorination of a novel ^{18}F-labeled PET tracer for β-amyloid plaques. *J Labelled Compd Rad,* 48, 983–996.

Lal, G.S., Pez, G.P. & Syvret, R.G. (1996) Electrophilic NF fluorinating agents. *Chem Rev* 96, 1737–1756.

Langström, B. & Ulin, J. (2009) Production of [^{18}F]F_2 from [^{18}F]Fluoride Using a Plasma Induced Scrambling Procedure. US patent app 20090104087.

Lecomte, R. (2004) Technology challenges in small animal PET imaging. *Nucl Instrum Meth A* 527, 157–165.

Lehn, J.M. & Sauvage, J.P. (1971) Cation and cavity selectivities of alkali and alkaline-earth 'cryptates'. *J Chem Soc Chem Commun* 440–441.

Lei Li, Hopkinson, M.N., Leuma Yona, R. *et al.* (2011) Convergent ^{18}F radiosynthesis: a new dimension for radiolabelling. *Chem Sci* 2, 123–131.

Levin, C.S. & Hoffman, E.J. (1999) Calculation of positron range and its effect on the fundamental limit of positron emission tomography system spatial resolution. *Phys Med Biol* 44, 781–799.

Levin, C.S. & Zaidi, H. (2007) Current trends in preclinical PET system design. *PET Clin* 2, 125–160.

Li, X., Link, J.M., Stekhova, S. *et al.* (2008) Site-specific labeling of annexin V with F-18 for apoptosis imaging. *Bioconjugate Chem* 19, 1684–1688.

Lin, K.-S., Ding, Y-.S., Kim, S.-W. *et al.* (2005) Synthesis, enantiomeric resolution, F-18 labeling and biodistribution of reboxetine analogs: promising radioligands for imaging the norepinephrine transporter with positron emission tomography. *Nucl Med Biol* 32, 415–422.

Liu, Z., Liu, S., Wang, F. *et al.* (2009) Noninvasive imaging of tumor integrin expression using ^{18}F-labeled RGD dimer peptide with PEG4 linkers. *Eur J Nucl Med Mol I* 36, 1296–1307.

Lu, S., Lepore, S.D., Li, S.Y. *et al.* (2009) Nucleophile assisting leaving groups: a strategy for aliphatic ^{18}F-fluorination. *J Org Chem* 74, 5290–5296.

Lui, E., Chirakal, R. & Firnau, G. (1998) Enzymatic synthesis of (–)-6-[^{18}F]fluoronorepinephrine from 6-[^{18}F]fluorodopamine by dopamine β-hydroxylase. *J Labelled Compd Rad* 41, 503–521.

Luthra, S.K., Brady, F. & Wadsworth, H.J. (2006) Solid-phase Electrophilic Fluorination. US patent app 10482542.

Luxen, A., Barrio, J.R., Satyamurthy, N. *et al.* (1987) Electrophilic and nucleophilic approaches to the synthesis of 3-fluorodiazepam. *J Fluorine Chem* 36, 83–92.

Maeda, M., Fukumura, T. & Kojima, M. (1987) The dimethylsulfonium moiety as a leaving group in aromatic radiofluorination using tetra-*n*-butylammonium [^{18}F]fluoride. *Appl Radiat Isotopes* 38, 307–310.

Martarello, L., Schaffrath, C., Deng, H. *et al.* (2003) The first enzymatic method for C–^{18}F bond formation: the synthesis of 5′-[^{18}F]-fluoro-5′-deoxyadenosine for imaging with PET. *J Labelled Compd Rad* 46, 1181–1189.

McBride, W.J. & Goldenberg, D.M. (2008) Methods and Compositions for F-18 Labeling of Proteins, Peptides and Other Molecules. US patent app 20080253964.

McBride, W.J., Sharkey, R.M., Karacay, H. *et al.* (2009) A novel method of ^{18}F radiolabeling for PET. *J Nucl Med* 50, 991–998.

Miller, P.W., Long, N.J., Vilar, R. *et al.* (2008) Synthesis of ^{11}C, ^{18}F, ^{15}O, and ^{13}N radiolabels for positron emission tomography. *Angew Chem Int Edit* 47, 8998–9033.

Moon, B.S., Lee, K.C., An, G.I. *et al.* (2006) Preparation of 3′-deoxy-3′-[^{18}F]fluorothymidine ([^{18}F]FLT) in ionic liquid, [bmim][OTf]. *J Labelled Compd Rad* 49, 287–293.

Mu, L., Höhne, A., Schubiger, P.A. *et al.* (2008) Silicon-based building blocks for one-step ^{18}F-radiolabeling of peptides for PET Imaging. *Angew Chem Int Edit* 47, 4922–4925.

Namavari, M., Bishop, A., Satyamurthy, N. *et al.* (1992) Regioselective radiofluorodestannylation with [^{18}F]F_2 and [^{18}F]CH_3COOF: a high yield synthesis of 6-[^{18}F]fluoro-L-dopa. *Int J Rad Appl Instrum A* 43, 989–996.

Neirinckx, R.D., Lambrecht, R.M. & Wolf, A.P. (1978) Cyclotron isotopes and radiopharmaceuticals XXV. An anhydrous ^{18}F-fluorinating intermediate: trifluoromethyl hypofluorite. *Int J Appl Radiat Isotopes* 29, 323–327.

Nickkho-Amiry, M. & Winfield, J.M. (2007) Investigation of fluorinated surfaces by means of radio-labelled probe molecules. *J Fluorine Chem* 128, 344–352.

O'Hagan, D. & Harper, D.B. (1999) Fluorine-containing natural products. *J Fluorine Chem* 100, 127–133.

O'Hagan, D., Schaffrath, C., Cobb, S.L. *et al.* (2002) Biochemistry: biosynthesis of an organofluorine molecule. *Nature* 416, 279–279.

O'Neil, J.P. & VanBrocklin, H.F. (1999) Preparation of fluorine-18 gas from an 11MeV cyclotron: a target system for the CTI RDS 111 cyclotron. *Nucl Instrum Meth A* 438, 166–172.

Oberdorfer, F., Hofmann, E. & Maier-Borst, W. (1988a) Preparation of ^{18}F-labelled *N*-fluoropyridinium triflate. *J Labelled Compd Rad* 25, 999–1005.

Oberdorfer, F., Hofmann, E. & Maier-Borst, W. (1988b) Preparation of a new ^{18}F-labelled precursor: 1-[^{18}F]fluoro-2-pyridone. *Int J Rad Appl Instrum A* 39, 685–688.

Oberdorfer, F. & Dietzel, G. (2003) Methodical study of direct flourination using [^{18}F]F_2 in presence of amides and imides. *J Labelled Compd Rad* 46, S216.

Onega, M., Winkler, M. & O'Hagan, D. (2009) Fluorinase: a tool for the synthesis of ^{18}F-labeled sugars and nucleosides for PET. *Future Med Chem* 1, 865–873.

Onega, M., Domarkas, J., Deng, H. *et al.* (2010) An enzymatic route to 5-deoxy-5-[^{18}F]fluoro-D-ribose, a [^{18}F]fluorinated sugar for PET imaging. *Chem Commun* 46, 139–141.

Pages, T. & Langlois, B.R. (2001) Fluorination of aromatic compounds from 1-aryl-3,3-dimethyltriazenes and fluoride anions in acidic medium: a model for ^{18}F labelling. *J Fluorine Chem* 107, 321–327.

Pages, T., Langlois, B.R., Le Bars, D. *et al.* (2001) Fluorination of aromatic compounds from 1-aryl-3,3-dimethyltriazenes and fluoride anions in acidic medium: synthesis of (*S*)-[^{18}F]-3-fluoro-α-methylphenylalanine. *J Fluorine Chem* 107, 329–335.

Pascali, C., Luthra, S.K., Pike, V.W. *et al.* (1990) The radiosynthesis of [^{18}F]PK 14105 as an alternative radioligand for peripheral type benzodiazepine binding sites. *Int J Rad Appl Instrum A* 41, 477–482.

Pascali, G., Kiesewetter, D.O., Salvadori, P.A. *et al.* (2004) Use of 1,8-bis-(dimethylamino)-naphthalene/H^{18}F complex as new radiofluorinating agent. *J Labelled Compd Rad* 47, 373–383.

Pike, V.W. & Aigbirhio, F.I. (1995) Reactions of cyclotron-produced [^{18}F]fluoride with diaryliodonium salts: a novel single-step route to no-carrier-added fluoroarenes. *J Chem Soc Chem Commun* 2215–2216.

Poole, R.T. & Winfield, J.M. (1976) Radiotracers in fluorine chemistry. Part IV. F-18 exchange between labeled alkylfluorosilanes and fluorides, or fluoride methoxides, of tungsten(VI), molybdenum(VI), tellurium(VI), and iodine(V). *J Chem Soc Dalton* 1557–1560.

Prante, O., Hamacher, K. & Coenen, H.H. (2007) Chemoenzymatic NCA synthesis of the coenzyme UDP-2-deoxy-2-[^{18}F]fluoro-α-D-glucopyranose as substrate of glycosyltransferases. *J Labelled Compd Rad* 50, 55–63.

Purser, S., Moore, P.R., Swallow, S. *et al.* (2008) Fluorine in medicinal chemistry. *Chem Soc Rev* 37, 320–330.

Qaim, S.M., Clark, J.C., Crouzel, C. *et al.* (1993) 'PET radionuclide production', in Stöcklin, G. & Pike, V.W. (Eds), *Radiopharmaceuticals for Positron Emission Tomography — Methodological Aspects*, Kluwer Academic, Dordrecht.

Qaim, S.M. (2001) Nuclear data relevant to the production and application of diagnostic radionuclides. *Radiochim Acta* 89, 223–232.

Reischl, G., Ehrlichmann, W. & Machulla, H.-J. (2002) Electrochemical transfer of [^{18}F]fluoride from [^{18}O]water into organic solvents ready for labeling reactions. *J Radioanal Nucl Chem* 254, 29–31.

Roberts, A.D., Barnhart, T.E. & Nickles, R.J. (2005) 'Isotope production for medical applications', in Hellborg, R. (Ed.), *Electrostatic Accelerators*, Springer, Heidelberg.

Ross, T.L., Ermert, J., Hocke, C. *et al.* (2007) Nucleophilic ^{18}F-fluorination of heteroaromatic iodonium salts with no-carrier-added [^{18}F]fluoride. *J Am Chem Soc* 129, 8018–8025.

Saiki, H., Iwata, R., Nakanishi, H. *et al.* (2010) Electrochemical concentration of no-carrier-added [^{18}F]fluoride from [^{18}O]water in a disposable microfluidic cell for radiosynthesis of ^{18}F-labeled radiopharmaceuticals. *Appl Radiat Isotopes* 68, 1703–1708.

Saito, F., Nagashima, Y., Goto, A. *et al.* (2007) Electrochemical transfer of ^{18}F from ^{18}O water to aprotic polar solvent. *Appl Radiat Isotopes* 65, 524–527.

Sánchez-Crespo, A., Andreo, P. & Larsson, S.A. (2004) Positron flight in human tissues and its influence on PET image spatial resolution. *Eur J Nucl Med Mol I* 31, 44–51.

Sanyal, D.K. & Winfield, J.M. (1984) Radiotracers in fluorine chemistry. Part VIII. Fluorine-18 exchange reactions involving uranium(VI) or uranium(V) fluorides. Evidence for surface complexation and comparisons with ligand exchange reactions. *J Fluorine Chem* 24, 75–92.

Satter, M.R., Martin, C.C., Oakes, T.R. *et al.* (1994) Synthesis of the fluorine-18 labeled inhalation anesthetics. *Appl Radiat Isotopes* 45, 1093–1100.

Satyamurthy, N., Bida, G.T., Barrio, J.R. *et al.* (1984) Regioselective syntheses of [F-18]aryl fluorides via easily accessible [F-18] *N*-fluoro-*N*-alkylsulfonamides. *J Labelled Compd Rad* 21, 1196–1301.

Satyamurthy, N., Bida, G.T., Phelps, M.E. *et al.* (1990) *N*-[^{18}F]Fluoro-*N*-alkylsulfonamides: novel reagents for mild and regioselective radiofluorination. *Int J Rad Appl Instrum A* 41, 733–738.

Schirrmacher, R., Bradtmöller, G., Schirrmacher, E. *et al.* (2006) ^{18}F-Labeling of peptides by means of an organosilicon-based fluoride acceptor. *Angew Chem Int Edit* 45, 6047–6050.

Schlyer, D.J., Bastos, M.A.V., Alexoff, D. *et al.* (1990) Separation of [^{18}F]fluoride from [^{18}O]water using anion exchange resin. *Int J Rad Appl Instrum A* 41, 531–533.

Schlyer, D.J. (2004) PET tracers and radiochemistry. *Ann Acad Med* Singapore 33, 146–154.

Schrobilgen, G., Firnau, G., Chirakal, R. *et al.* (1981) Synthesis of [^{18}F]XeF_2, a novel agent for the preparation of ^{18}F-radiopharmaceuticals. *J Chem Soc Chem Commun* 198–199.

Schubiger, P.A., Lehmann, L. & Friebe, M. (2007) *PET Chemistry: The Driving Force in Molecular Imaging*, Springer, Berlin.

Simeon, F.G. & Pike, V.W. (2005) Radiosyntheses of 2-[^{18}F]fluoro-1,3-thiazoles. *J Labelled Compd Rad* 48, S158.

Smith, M. & March, J. (2001) *March's Advanced Organic Chemistry: Reactions, Mechanisms, and Structure*, John Wiley & Sons, New York.

Solin, O., Bergman, J., Haaparanta, M. *et al.* (1988) Production of ^{18}F from water targets: specific radioactivity and anionic contaminants. *Int J Rad Appl Instrum A* 39, 1065–1071.

Sonzogni, A.A. (2009) Online database: Interactive Chart of Nuclides. http://www.nndc.bnl.gov/chart/

Stone-Elander, S. & Elander, N. (2002) Microwave applications in radiolabelling with short-lived positron-emitting radionuclides. *J Labelled Compd Rad* 45, 715–746.

Straatmann, M.G. & Welch, M.J. (1977) Fluorine-18-labeled diethylaminosulfur trifluoride (DAST): an F-for-OH fluorinating agent. *J Nucl Med* 18, 151–158.

Strijckmans, K. (2001) The isochronous cyclotron: principles and recent developments. *Comput Med Imag Graph* 25, 69–78.

Studenov, A.R., Adam, M.J., Wilson, J.S. *et al.* (2005) New radiolabelling chemistry: synthesis of phosphorus-[^{18}F]fluorine compounds. *J Labelled Compd Rad* 48, 497–500.

Sun, H. & DiMagno, S.G. (2007) Competitive demethylation and substitution in *N,N,N*-trimethylanilinium fluorides. *J Fluorine Chem* 128, 806–812.

Tang, G., Zeng, W., Yu, M. *et al.* (2008) Facile synthesis of *N*-succinimidyl 4-[^{18}F]fluorobenzoate ([^{18}F]SFB) for protein labeling. *J Labelled Compd Rad* 51, 68–71.

Teare, H., Robins, E.G., Årstad, E. *et al.* (2007) Synthesis and reactivity of [^{18}F]-*N*-fluorobenzenesulfonimide. *Chem Commun* 2330–2332.

Teare, H., Robins, E.G., Kirjavainen, A. *et al.* (2010) Radiosynthesis and evaluation of [^{18}F]selectfluor bis(triflate). *Angew Chem Int Edit* 49, 6821–6824.

Tewson, T.J. (1997) Synthesis of [^{18}F]fluoroetanidazole: a potential new tracer for imaging hypoxia. *Nucl Med Biol* 24, 755–760.

Thomson, J., Webb, G., Winfield, J. *et al.* (1993) Ambient-temperature catalytic fluorination of C-1 chlorohydrocarbons to C-3 chlorohydrocarbons and related-compounds using oxide-supported organic layer catalysts. *Appl Catal A* 97, 67–76.

Ting, R., Adam, M.J., Ruth, T.J. *et al.* (2005) Arylfluoroborates and alkylfluorosilicates as potential PET imaging agents: high-yielding aqueous biomolecular ^{18}F-labeling. *J Am Chem Soc* 127, 13094–13095.

Ting, R., Lo, J., Adam, M.J. *et al.* (2008) Capturing aqueous [^{18}F]fluoride with an arylboronic ester for PET: synthesis and aqueous stability of a fluorescent [^{18}F]labeled aryltrifluoroborate. *J Fluorine Chem* 129, 349–358.

Toyokuni, T., Walsh, J.C., Dominguez, A. *et al.* (2003) Synthesis of a new heterobifunctional linker, *N*-[4-(aminooxy)butyl]maleimide, for facile access to a thiol-reactive ^{18}F-labeling agent. *Bioconjugate Chem* 14, 1253–1259.

Van Grieken, R. & De Bruin, M. (1994) Nomenclature for radioanalytical chemistry (IUPAC Recommendations 1994). *Pure Appl Chem* 66, 2513–2526.

Visser, E.P., Disselhorst, J.A., Brom, M. *et al.* (2009) Spatial resolution and sensitivity of the inveon small-animal PET scanner. *J Nucl Med* 50, 139–147.

Vlasov, V.M. (1993) Fluoride ion as a nucleophile and a leaving group in aromatic nucleophilic substitution reactions. *J Fluorine Chem* 61, 193–216.

Wadsworth, H.J. & Devenish, P.A. (2005) Fluoridation Method. US patent app 10559878.

Wagner, F.M., Ermert, J. & Coenen, H.H. (2009) Three-step, 'one-pot' radiosynthesis of 6-fluoro-3,4-dihydroxy-L-phenylalanine by isotopic exchange. *J Nucl Med* 50, 1724–1729.

Wängler, B., Quandt, G., Iovkova, L. *et al.* (2009) Kit-like ^{18}F-labeling of proteins: synthesis of 4-(di-*tert*-butyl[^{18}F]fluorosilyl)benzenethiol (Si[^{18}F]FA-SH) labeled rat serum albumin for blood pool imaging with PET. *Bioconjugate Chem* 20, 317–321.

Wängler, C., Schirrmacher, R., Bartenstein, P. *et al.* (2010) Click-chemistry reactions in radiopharmaceutical chemistry: fast easy introduction of radiolabels into biomolecules for in vivo imaging. *Curr Med Chem* 17, 1092–1116.

Welch, M.J. & Redvanly, C.S. (2005) *Handbook of Radiopharmaceuticals: Radiochemistry and Applications*, John Wiley & Sons, Chichester.

Wester, H.J. & Schottelius, M. (2007) 'Fluorine-18 labeling of peptides and proteins', in Schubiger, P.A., Lehmann, L. & Friebe, M. (Eds), *PET Chemistry: The Driving Force in Molecular Imaging*, Springer, Berlin.

Winkler, M., Domarkas, J., Schweiger, L.F. *et al.* (2008) Fluorinase-coupled base swaps: synthesis of [^{18}F]-5′-deoxy-5′-fluorouridines. *Angew Chem Int Edit* 47, 10141–10143.

Wüst, F., Hultsch, C., Bergmann, R. *et al.* (2003) Radiolabelling of isopeptide N^{ε}-(γ-glutamyl)-L-lysine by conjugation with *N*-succinimidyl-4-[^{18}F]fluorobenzoate. *Appl Radiat Isotopes* 59, 43–48.

Zechel, D.L., Reid, S.P., Nashiru, O. *et al.* (2001) Enzymatic synthesis of carbon — fluorine bonds. *J Am Chem Soc* 123, 4350–4351.

Zhang, M.R. & Suzuki, K. (2007) [^{18}F]Fluoroalkyl agents: synthesis, reactivity and application for development of PET ligands in molecular imaging. *Curr Top Med Chem* 7, 1817–1828.

Ziemer, L.S., Evans, S.M., Kachur, A.V. *et al.* (2003) Noninvasive imaging of tumor hypoxia in rats using the 2-nitroimidazole ^{18}F-EF5. *Eur J Nucl Med Mol I* 30, 259–266.

11

^{18}F-Labelled Tracers for PET Oncology and Neurology Applications

Sajinder K. Luthra and Edward G. Robins**

11.1 Introduction to Molecular Imaging

Over the past four decades various imaging modalities have emerged for the diagnosis and evaluation of human diseases. *In vivo* imaging can be divided into structural and functional techniques. Structural techniques produce excellent anatomical images (X-ray/CT, MRI) and functional techniques provide information about biochemical and physiological processes in living subjects by the use of specific radiopharmaceuticals, such as in single photon emission computed tomography (SPECT) and positron emission tomography (PET). The hybrid or fusion imaging of PET/MRI and PET/CT will further refine the application of molecular imaging probes as co-registration with a high-resolution CT will allow better localisation of the specific molecular signal from PET. Advances in technology and the availability of selective and specific imaging agents are increasing the clinical utility of PET imaging across a wide range of disease areas in oncology, neurology and cardiology (Jones, 1996; Phelps, 2000, 2004; Pither, 2003; Margolis *et al.*, 2007; Torigian *et al.*, 2007).

* GE Healthcare Medical Diagnostics Discovery, Cyclotron Building, Hammersmith Hospital, London W12 0NN, UK.

Molecular imaging allows the development of radiolabelled or optically labelled probes to observe, measure, and understand biology in cells, tissues and living organisms. The hallmark of disease is the disruption of these biological systems which are characterised by changes in cellular replication, receptor expression and internal and external enzyme systems. Leverage of the core disciplines of physics, mathematics, engineering, biology, chemistry and medicine to identify candidate molecules for development into tracers to investigate these systems *in vivo* is essential. The benefit of developing these targeted molecular imaging probes is to improve the clinical diagnosis of patients and to improve the fundamental understanding of disease processes (Jones, 1996; Phelps, 2000, 2004; Massoud and Gambir, 2003; Margolis *et al.*, 2007; Torigian, 2007).

11.2 Positron Emission Tomography (PET)

PET is a non-invasive imaging technique which uses radiopharmaceuticals labelled with cyclotron-produced short-lived positron emitting radioisotopes, such as ^{11}C, $t_{1/2} = 20$ min, and ^{18}F, $t_{1/2} = 110$ min. The sensitivity and resolution of PET enables the detection of picomolar concentrations of intravenously administered radiopharmaceuticals and the imaging of molecular processes occurring at low concentrations (Jones, 1996). Although a wide range of molecules have been labelled with ^{11}C for clinical research applications, more recently the effort has been directed towards the incorporation of ^{18}F into molecules of interest. The advantage of the longer half-life allows the production of ^{18}F-radiopharmaceuticals from a central manufacturing site and transportation of material to other medical centres with PET cameras but not cyclotron facilities. Furthermore, biological processes occurring over longer time periods, e.g. hours rather than minutes, can be imaged with ^{18}F-labelled compounds.

The rapid growth in the use of PET has been to a large degree based on the imaging characteristics of a single tracer, ^{18}F fluorodeoxyglucose (FDG), a marker of glucose metabolism (see below for more details). PET imaging has been used both as a diagnostic (primarily oncology) and a research tool, especially for the mapping of normal human brain (Jones, 1996; Ametamey and Honer, 2007; Ametamey *et al.*, 2008) and heart function (Phelps, 2004; Kopka *et al.*, 2008). Additionally, PET is increasingly being accepted as a crucial imaging modality for the early detection of disease, precise staging of disease progression and accurate assessment of the effects of therapy (Phelps, 2004; Margolis *et al.*, 2007). Increasingly, the trend has been to develop PET imaging agents which target specific systems and pathways in disease processes.

11.3 Biological Imaging Targets

As technology advances, the goal of molecular imaging is changing. For a long time, the major emphasis has been on the provision of diagnostic and prognostic information based on identification of the molecular events associated with a biochemical or pathological process. Recently, the shift has been towards the potential of guiding individually tailored pharmacological, cell-based or genetic therapeutic regimes. Molecular imaging probes (research and diagnostic) and drugs (therapeutics) share common concepts of design based on biochemical principles targeting enzymes, receptors, neurotransmitter systems, genetic material and pathological depositions. The basic concept behind several targets of functional imaging is shown in Fig. 11.1. For diseases with receptor overexpression, imaging agents that bind to either extracellular or intracellular targets offer insight into the expression levels of the receptor and possible downstream pathways, whilst enzyme-activated reporters indicate the function of the process. Targeting of these intracellular and extracellular systems will ultimately facilitate early disease detection, establishment of novel therapies, and selection of patients for treatment based on their individual disease biology (the paradigm of 'personalised medicine').

11.4 Tracer Development

The development of a novel PET tracer can be considered as following a similar pathway to that of the development of a therapeutic agent. It is a complex process and can be divided into discrete stages as shown in Fig. 11.2.

The development of a new PET tracer for clinical evaluation and as a potentially marketable product requires several key steps. In the early

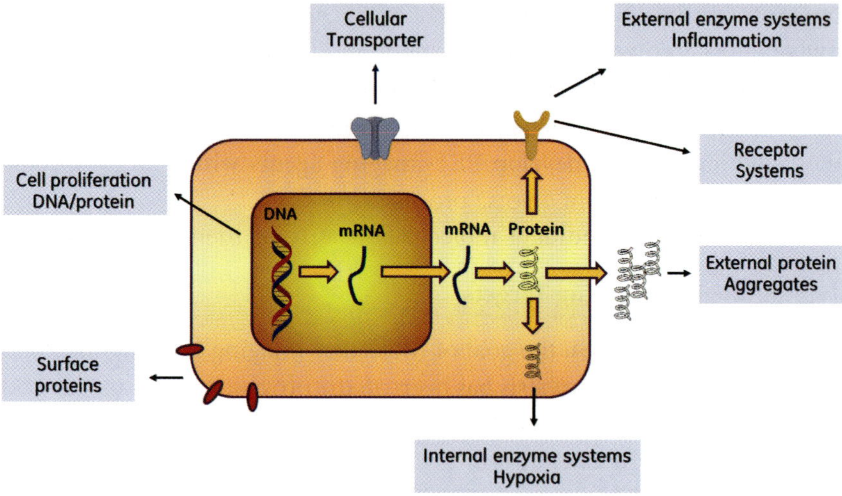

Figure 11.1. PET and SPECT nuclear imaging targets. Major classifications of *in vivo* imaging agents for evaluating function.

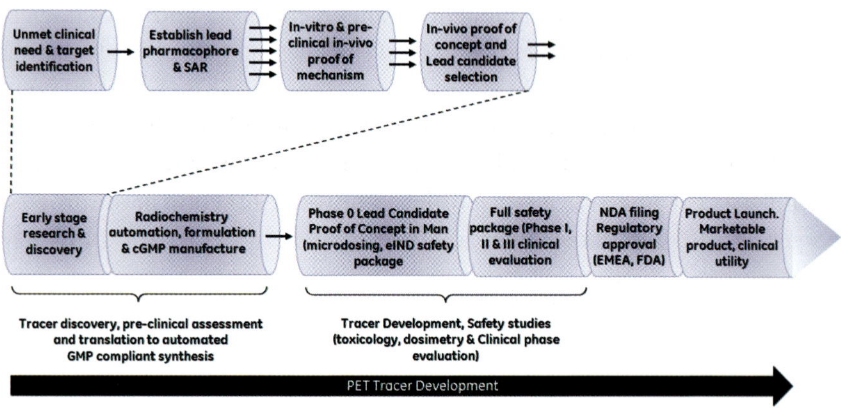

Figure 11.2. Tracer development pathway from early discovery through development to product launch.

stages of research and development, the first and possibly most crucial step is to define the unmet clinical need. This needs to take into account any limitations of existing tracers or other imaging modalities. Having established an understanding of the disease, suitable biological targets need to be identified. The next step is to define or select one or more suitable pharmacophores that will provide the basis for the generation of a focused library of candidate molecules and establish structure–activity relationships (SAR). The outcome of such SAR studies provides candidate tracers with suitable *in vitro* properties required for the chosen target (e.g. target affinity, lipophilicity).

Once suitable candidate molecules have been identified, a radiolabelling assessment can be undertaken. In general, selection of the appropriate molecular candidates is driven by an understanding of available radiolabelling techniques. In order to reduce the risk of developing candidate molecules that are not amenable to incorporation of the preferred radionuclide, radiolabelling methods are often developed alongside the molecular libraries using model compounds exhibiting similar structural and functional features as the chosen pharmacophore.

Once the candidate tracers have been successfully radiolabelled and isolated chemically and radiochemically pure, preliminary biological evaluation is then possible. Initially, many *in vivo* experiments focus on the biodistribution and biological fate of candidates in naïve animals, in particular, to demonstrate selective targeting of the region of interest by the candidate tracers. If successful, the next stage is the proof-of-concept phase where candidate tracers are assessed using animal disease models where appropriate. As a result, if one or more of the candidates appear promising and progress through a lead selection phase based on *in vitro* and *in vivo* performance, further work to automate and optimise the radiolabelling procedure can be initiated.

With the lead candidate selected and protocols for the GMP preparation defined and validated, the most comprehensive way to confirm the potential of any novel PET tracer is to examine its performance in a human disease population. Since only a very low quantity of the active chemical species is present in any PET tracer formulation, it has been possible to assess the first-in-man (FiM) efficacy of novel tracers by employing the PET-microdosing concept (Bergström *et al.*, 2003). As such, PET radiotracers assessed under a microdosing regime require a reduced safety and toxicity assessment with respect to therapeutic drugs before initial evaluation in the clinic. If these early clinical phase trials result in a successful outcome then the tracer enters a more traditional clinical assessment (requiring full toxicity and safety screening). In the final stages, potential commercial PET tracers enter a formal registration and regulatory approval phase similar to that for therapeutic drugs. The submission of a new drug application (NDA) to the appropriate regulatory body (such as the United States Food and Drug Administration (FDA) or the European Medicines Agency (EMA)) is the final hurdle before a PET tracer can be formally launched as a commercial product. Approval of a radiopharmaceutical typically involves submission of an NDA by a manufacturer or a company, clearly documenting two major aspects of the drug: (i) manufacturing of PET drug using current good manufacturing practices and (ii) the safety and effectiveness of a drug with specific indications. To date, only $[^{18}F]$FDG has received formal approval as a diagnostic imaging agent for reflecting levels of glucose metabolism *in vivo*. As the chemistry and biology of PET and clinical imaging becomes increasingly mainstream, it is likely that an increasing number of PET tracers will become commercially available in the near future.

11.5 Oncology Applications

Recently, an important article entitled 'The hallmarks of cancer' (Hanahan and Weinberg, 2000) described the phenotypic differences between healthy and cancerous cells. They suggest that six cellular alterations or 'hallmarks' are essential to malignant growth and these are believed to be common to most, if not all, human tumours. These 'hallmarks' of cancer cells can be described as (i) uncontrollable growth in the absence of growth stimulatory signals that normal cells require from the environment, (ii) evading death or apoptosis, (iii) becoming angiogenic, (iv) indefinite proliferation, (v) invasion of tissue and metastasis, and (vi) self-sufficiency in growth signals. Genome instability is an additional factor which may be necessary to explain the high incidence of cancer.

To date, a number of imaging agents have been described that aim at measuring fundamental biological processes known to be disregulated in tumours, including glucose utilisation, proliferation, apoptosis, hypoxia and angiogenesis (Vallabhajosula, 2007). While many of these imaging agents have been investigated fairly extensively in the preclinical setting and some to a lesser extent clinically, the vast majority of molecular imaging in clinical oncology consists of assessment of glucose utilisation using [^{18}F]fluorodeoxyglucose ([^{18}F]FDG). The new era of molecular imaging is supported by the widespread use and acceptance of [^{18}F]FDG as a tool for early cancer detection; it is hoped that new tracers will emerge that are approved by the regulatory authorities and become widely distributed. Therefore, in the next section, imaging agents that have progressed to clinical studies or have the potential to enter clinical evaluation will be described.

11.6 2-[^{18}F]Fluoro-2-Deoxy-D-Glucose ([^{18}F]FDG)

[^{18}F]FDG, the most widely used PET tracer in clinical practice, is a glucose analogue in which the hydroxyl group at the second carbon position is substituted by ^{18}F (Fig. 11.3). The radiosynthesis of [^{18}F]FDG was first described by Ido *et al* (1977, 1978) using [^{18}F]fluorine gas with 3,4,6-tri-*O*-acetyl-D-glucal which produced a 3:1 mixture of the ^{18}F-labelled isomers.

Typically, [^{18}F]FDG is now prepared by the nucleophilic substitution method (Hamacher *et al.*, 1986), where 1,3,4,6-tetra-*O*-acetyl-2-*O*-trifluoromethanesulfonyl-β-D-mannopyranose (mannose triflate) is reacted with [^{18}F]fluoride. Since 1986, there has been a focused effort to increase the yield of the manufacturing process by reduction of synthesis time and improving the efficiency of each of the synthetic steps. The refinement and optimisation of [^{18}F]FDG synthesis has also included the development of solid-phase ^{18}F-fluorination methods (Brown *et al.*, 2007), integrated microfluidic devices (Lee *et al.*, 2005), as well as advanced automated synthesis platforms. Scheme 11.1 describes

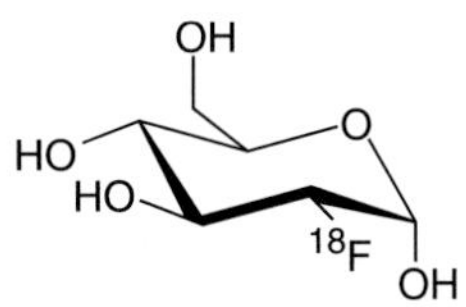

Figure 11.3. Structure of [^{18}F]fluorodeoxyglucose ([^{18}F]FDG).

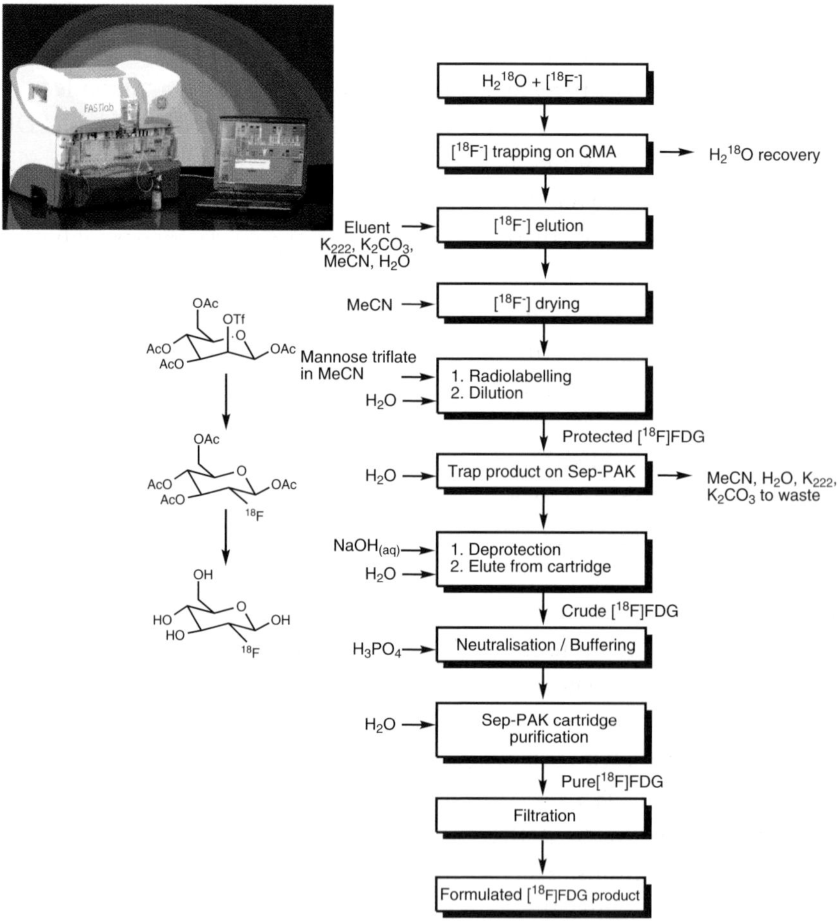

Scheme 11.1. GEHC FASTlab platform for [^{18}F]FDG synthesis.

the GE Healthcare platform FASTlab, a cassette-based module that typically delivers 70% non-decay-corrected yields of [^{18}F]FDG in a short synthesis time of about 23 min.

[^{18}F]FDG simply visualises a basic biochemical function of most living cells: their ability to utilise glucose as an energy substrate. Entry into cells occurs via the same GLUT family of membrane transporters as used by glucose, but unlike glucose, [^{18}F]FDG cannot be metabolised after it is phosphorylated by hexokinase. Thus [^{18}F]FDG remains trapped after it is taken up by the cell. The amount of intracellular [^{18}F]FDG reflects glucose uptake and is a useful proxy for cellular glucose metabolism. Disease processes can either up-regulate or down-regulate glucose metabolism.

[^{18}F]FDG PET has been widely used for the assessment of glucose metabolism in heart, lungs and the brain. As cancer cells have high metabolic activity, [^{18}F]FDG PET has been used for diagnosis, staging and monitoring of cancer treatments (Hicks *et al.*, 2001; Weber, 2005; de Geus-Oei *et al.*, 2007). Numerous reviews have been written on the potential benefits of using [^{18}F]FDG PET as a diagnostic/management tool in a wide range of cancers, e.g. lung, colorectal, lymphoma, melanoma, and head and neck (some reviews are cited here — Bingham, 2002; Kelloff *et al.*, 2005; Westerterp *et al.*, 2005). Also, [^{18}F]FDG PET reimbursement by Medicare has evolved substantially over the past few years and coverage for [^{18}F]FDG PET procedures for a variety of cancers has increased with time (details can be found in Bietendorf, 2004). Additionally, [^{18}F]FDG has been shown to be useful in detection of inflammatory processes, such as in lung (Zhuang *et al.*, 2005; Chen and Schuster, 2006) and musculoskeletal infection (Crymes *et al.*, 2004), as well as being able to show different patterns of hypometabolism in neurodegenerative disorders (Foster *et al.*, 2007) and being able to provide an objective and sensitive support to the clinical diagnosis of early dementia (Mosconi *et al.*, 2008).

Despite the excellent diagnostic performance of [^{18}F]FDG, it has recognised limitations. Firstly, the non-specificity of glucose-metabolic changes and secondly, the lack of contrast between physiological and pathological uptake limit sensitivity. For example, [^{18}F]FDG PET is unable to distinguish between proliferating tumour cells and inflammatory lesions. This has led to the development of other more specific ^{18}F-radiopharmaceuticals, some of which are described below.

11.7 3′-Deoxy-3′-[^{18}F]Fluoro-L-Thymidine ([^{18}F]FLT)

In oncology, the measurement of tumour growth and DNA synthesis are attractive targets for imaging. A variety of DNA precursors, primarily based on thymidine and its analogues, have been labelled with a range of isotopes (Shields, 2003; Bading and Shields, 2008); examples include

[^{11}C]thymidine, 5-[^{18}F]fluorouracil and its related nucleosides [^{18}F]fluorodeoxyuridine, 5-[^{123}I]iododeoxyuridine and 5-[^{76}Br]bromodeoxyuridine, most of which have suffered from rapid *in vivo* catabolism, limiting their use for routine imaging (Bergström *et al.*, 1998; Mangner *et al.*, 2003; Bading, 2008). The thymidine analogue, 3′-deoxy-3′-[^{18}F]fluoro-L-thymidine ([^{18}F]FLT), which resists *in vivo* degradation but undergoes glucuronidation, appears to be the most promising agent for imaging cellular growth. Shields *et al.* (1998) first reported the potential utility of [^{18}F]FLT for the detection of cellular proliferation by PET. Transport of [^{18}F]FLT across the cell membrane is facilitated by the nucleoside transport system (ENT1) (Fig. 11.4). [^{18}F]FLT is directly phosphorylated by the cytosolic enzyme thymidine kinase (TK1) (Seitz *et al.*, 2002; Grierson *et al.*, 2004a), which is exclusively expressed in the synthesis-phase (S-phase) of the cell cycle. It is further metabolised into [^{18}F]FLT triphosphate, which is not a substrate for DNA polymerase; hence very little becomes incorporated into DNA (Mercer, 2007). Thus, changes in [^{18}F]FLT uptake can be correlated directly with cellular proliferation.

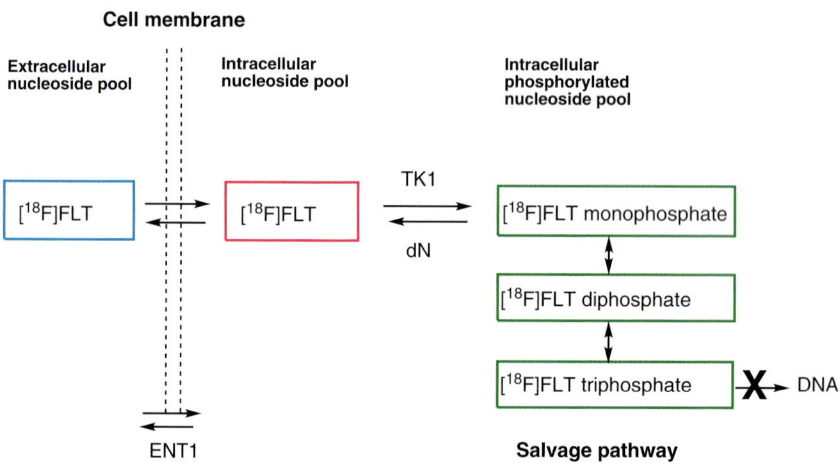

Figure 11.4. Transport of [^{18}F]FLT across the cell membrane is facilitated by ENT1 (equilibrative nucleoside transporter) and cellular phosphorylation through the salvage pathway (adapted from Mercer, 2007).

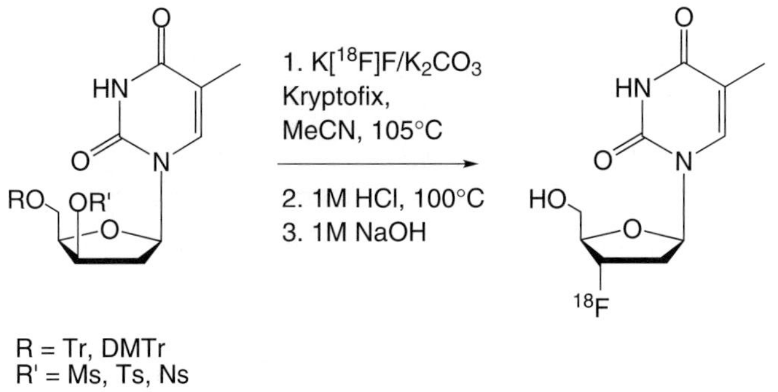

Scheme 11.2. Two-step radiosynthesis of [^{18}F]FLT starting from protected precursor.

The radiolabelling of [^{18}F]FLT has been investigated by several groups using precursors with different combinations of *N*-/*O*-protecting and leaving groups as shown in Scheme 11.2 (Martin *et al.*, 2002; Glaser *et al.*, 2003). A reliable radiosynthesis of [^{18}F]FLT has been developed based on [^{18}F]fluoride displacement of a protected nosylate precursor (Grierson and Shields, 2000). Recent approaches are amenable to automation and GMP production to provide clinical doses.

The majority of studies of [^{18}F]FLT PET have focused on validating the tracer as a means of quantifying cellular proliferation and testing its ability to accurately stage cancer (Vesselle *et al.*, 2002; Cobben *et al.*, 2003; Kenny *et al.*, 2007, 2009; Salskov *et al.*, 2007; Yamamoto *et al.*, 2008). [^{18}F]FLT is now being used at several centres around the world to image cellular proliferation in human cancers. Clinical studies have focused on assessment of tumour aggressiveness, prediction of outcome, therapy planning and monitoring response to treatment.

Recent clinical studies have reported that [^{18}F]FLT PET can accurately predict response very early after the initiation of chemotherapy. A recent study by Kenny *et al.* (2007) has demonstrated that [^{18}F]FLT PET can be used to measure response as early as one week after treatment with combination of 5-fluorouracil, epirubicin and cyclophosphamide (aFEC) chemotherapy in patients with stage II–IV breast cancer (Fig. 11.5).

[^{18}F]FLT PET is also finding an important role in the assessment of new cancer therapeutics in development for targeting specific cancer cell

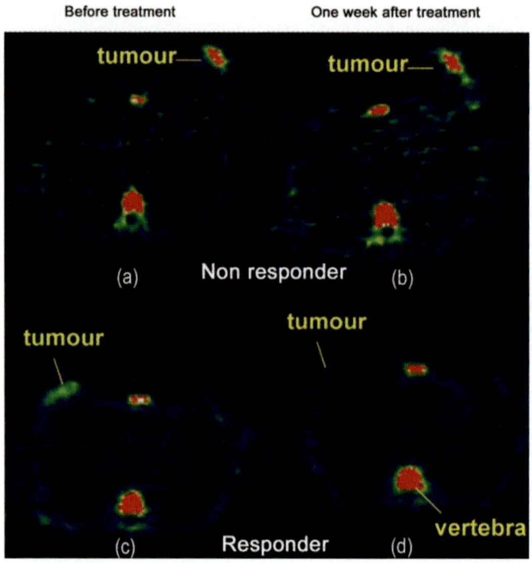

Figure 11.5. [^{18}F]FLT PET images in responding and non-responding patients. (a) Pre-treatment and (b) post-treatment images of a patient with grade II invasive ductal carcinoma who did not respond to treatment. (c) Pre-treatment and (d) post-treatment images of a patient with grade II lobular carcinoma who responded to treatment. Reproduced with permission from E. Aboagye (Kenny *et al.*, 2007).

enzymatic pathways, e.g. thymidylate synthase (TS) inhibitors. A recent PET study in breast cancer patients following treatment with a TS inhibitor, capecitabine has shown that [^{18}F]FLT could be used to measure the pharmacodynamics of TS inhibitors as well as for identifying patients who are unlikely to benefit from this type of treatment, as shown in Fig. 11.6 (Kenny *et al.*, 2009).

To further develop and understand the clinical value of [^{18}F]FLT as an imaging biomarker, the SNM Clinical Trials Group is establishing a multicentre investigational new drug (IND) with [^{18}F]FLT for cross-reference by developers of therapeutic drugs and biologics.

To address the limitations of [^{18}F]FLT as a marker of DNA synthesis, several other labelled tracers have been developed to measure DNA synthesis by modification of pyrimidine nucleoside as shown in Fig. 11.7.

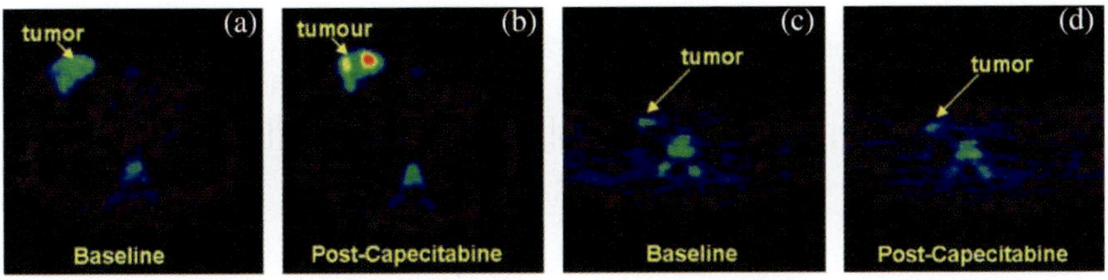

Figure 11.6. Representative baseline and postcapecitabine FLT PET images. Baseline (a) and post-treatment (b) images of a primary breast tumour which showed an increase in FLT uptake postcapecitabine. Baseline (c) and post-treatment (d) images in a patient with a cervical node which showed minimal change in FLT uptake. Reproduced with permission from E. Aboagye (Kenny *et al.*, 2009).

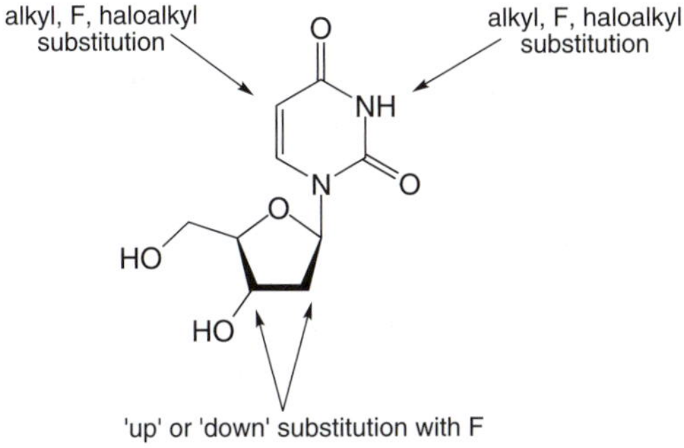

Figure 11.7. Sites of chemical modification for development of new ^{18}F labelled pyrimidine nucleoside derivatives.

FMAU (1-(2′-deoxy-2′-[^{18}F]fluoro-β-D-arabinofuranosyl)thymine) and FBAU (1-(2′-deoxy-2′-[^{18}F]fluoro-β-D-arabinofuranosyl)-5-bromouracil) have also been labelled with ^{18}F as shown in Fig. 11.8 (Mangner *et al.*, 2003; Bading and Shields, 2008). Unlike [^{18}F]FLT these compounds are incorporated into DNA (Lu *et al.*, 2002). Further studies will be needed to determine whether such tracers offer practical advantages over [^{18}F]FLT. Differences in metabolism may be more important in determining which tracer may best be used for different tumours or areas of the body. For example, the high retention of FMAU in the liver may impair imaging in the upper abdomen compared with [^{18}F]FLT, while FMAU is less readily cleared into the bladder than [^{18}F]FLT, leading to improved FMAU

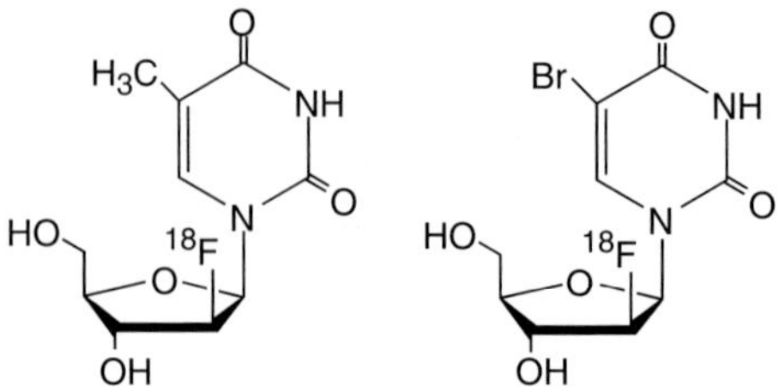

Figure 11.8. Chemical structures of [^{18}F]FMAU and [^{18}F]FBAU.

imaging of the pelvis. In summary, the development of PET will require new tracers that track different cellular pathways for use in oncology.

11.8 Imaging Tumour Angiogenesis

Angiogenesis is a vital component of both normal physiological processes and a number of disease states, particularly for the growth of primary malignant tumours and for the development of metastases. Tumour neovascularisation via angiogenesis is the proliferation of a network of blood vessels that penetrates into cancerous tissues, supplying nutrients and oxygen and removing waste products. One of the most important cell surface receptors in tumour development is the αvβ3 integrin receptor. Integrins, so called because they integrate the function of the cell with the extracellular matrix, are a family of membrane-spanning adhesion receptors composed of non-covalently linked α and β subunits (Humphries, 2000). Integrins αvβ3 and αvβ1 play a key role in angiogenesis by serving as receptors for a variety of extracellular matrix proteins containing an arginine-glycine-aspartic acid (RGD) sequence (Fig. 11.9).

These integrins mediate migration of endothelial cells into the basement membrane and regulate their growth, survival and differentiation. Integrin αvβ3 is expressed at low levels on epithelial cells and mature endothelial cells, but it is over-expressed on the activated endothelial cells of tumour neovasculature and some tumour cells (Brakebusch *et al.*, 2002). The upregulated expression of αvβ3 receptors on growing tumours and tumour cells of various origins provides a basis for the development of αvβ3-specific molecular imaging agents as well as the development of

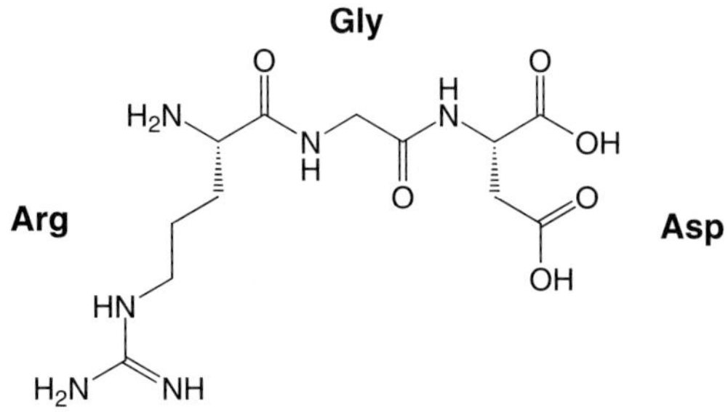

Figure 11.9. RGD (Arg-Gly-Asp) amino acid sequence.

antiangiogenic and antimetastatic therapeutic strategies. As αvβ3 integrins are an attractive target for cancer treatment, high-affinity ligands containing the RGD sequence have been developed. In the past decade, many labelled linear and cyclic RGD peptide antagonists have been evaluated as αvβ3 integrin-targeted radiotracers (Liu, 2006; Beer and Schwaiger, 2008; Dijkgraaf *et al.*, 2009).

A recent survey of ligands for mapping αvβ3-integrin expression *in vivo* (Schottelius *et al.*, 2009) describes numerous examples of labelled RGD peptide molecular imaging agents. Already, a wide range of radioisotopes including ^{99m}Tc, ^{68}Ga, ^{64}Cu and ^{111}In, have been conjugated to RGD peptides through the introduction of pendant chelator groups such as 1,4,7,10-tetracyclodecane-1,4,7,10-tetraacetic acid (DOTA). Radioiodinated cyclo(RGDyV) and cyclo(RGDfY) were the first labelled peptides examined *in vivo* (Haubner *et al.*, 1999). Following this, Haubner *et al.* (2001a) developed [^{123}I]- and [^{125}I]-labelled RGD peptides which displayed superior *in vivo* pharmacokinetics to previous derivatives and led to the development of the ^{18}F derivative [^{18}F]galacto-RGD as shown in Fig. 11.10 (Haubner *et al.*, 2001b, 2004). There are also numerous examples of other ^{18}F-labelled RGD peptides which have been highlighted in several in-depth reviews (see, for example, Liu, 2006; Niu and Chen, 2009; Schottelius *et al.*, 2009). In general, introduction of the ^{18}F radioisotope has been achieved through chemoselective labelling of RGD peptides with a variety of ^{18}F prosthetic groups. It is imperative that the ^{18}F-labelled prosthetic group is capable of forming a stable bond in a site-specific manner and having little or no impact on the biological activity of the parent peptide. In 2001, one of the first

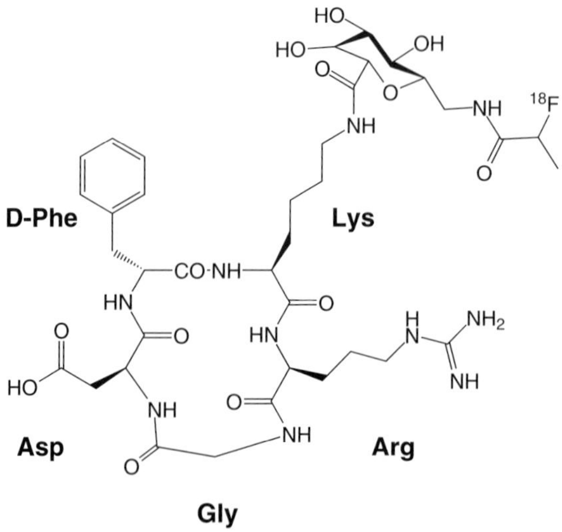

Figure 11.10. Structure of [^{18}F]fluoropropanamide c(RGDfK(SAA)). SAA = 7-amino-L-glycero-L-galacto-2,6-anhydro-7-deoxyheptanamide (Haubner *et al.*, 2001b).

^{18}F-labelled labelled RGD peptides, [^{18}F]galacto-RGD, was reported (Haubner *et al.*, 2001b), which was then also the first ^{18}F-labelled RGD peptide to be investigated in human subjects. Radiolabelling of [^{18}F]galacto-RGD was carried out by reacting a glycopeptide precursor with 4-nitrophenyl-2-[^{18}F]fluoropropionate (Haubner *et al.*, 2001b, 2004).

A variety of different prosthetic groups have been used to radiolabel the RGD peptide; the most commonly used are shown in Fig. 11.11. Chen and coworkers reported the [^{18}F]fluorobenzoylation of an RGD-peptide derivative using [^{18}F]*N*-succinimidyl-4-fluorobenzoate ([^{18}F]SFB), which reacts selectively with ε-amino functional groups (Chen *et al.*, 2004). In 2006, Cai and coworkers demonstrated the chemoselective labelling of monomeric and dimeric thio-functionalised cyclic-RGD derivatives using *N*-{2-(4-^{18}F-fluorobenzamido)ethyl}maleimide ([^{18}F]-FBEM). Reaction of [^{18}F]SFB with *N*-(2-aminoethyl)maleimide provided the radiolabelled maleimide which reacts site-specifically via Michael-addition with thiol-functionalised peptides (Cai *et al.*, 2006).

Oxime formation with [^{18}F]fluorobenzaldehyde has been extensively used as a method for ^{18}F-labelling of peptides containing the aminooxy

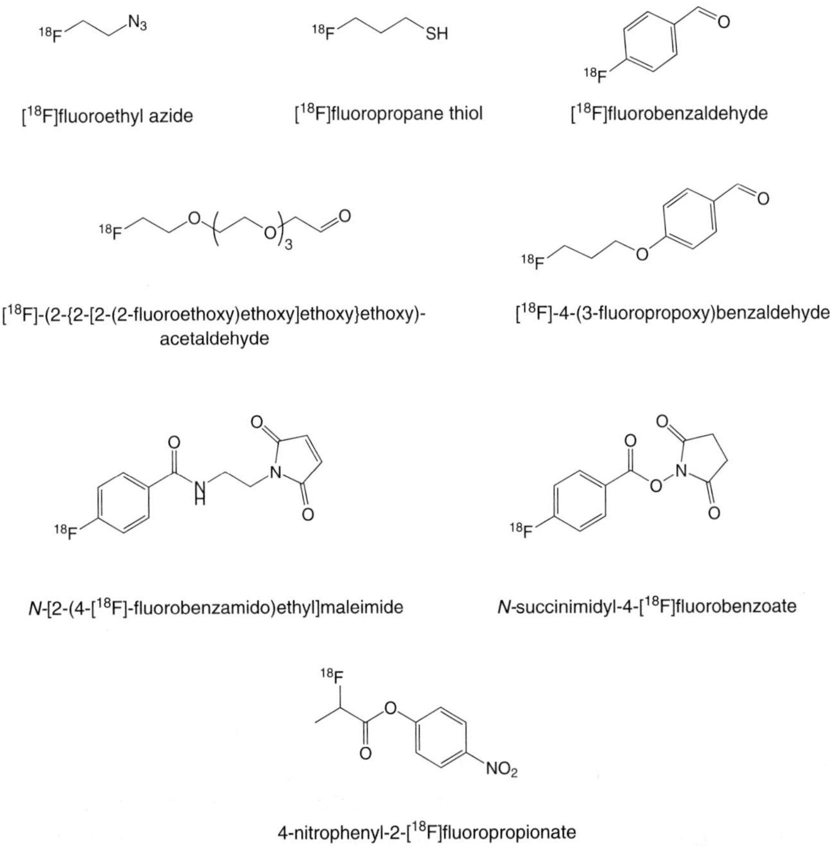

Figure 11.11. ^{18}F prosthetic groups used for radiolabelling RGD peptides.

functionality. Aminooxy groups rapidly condense with aldehydes to form stable oximes under relatively mild reaction conditions in aqueous media across a wide pH range. In 2008, Glaser and coworkers further investigated the oxime labelling strategy by comparing the conjugation of [^{18}F]fluorobenzaldehyde, [^{18}F]-(2-{2-[2-(2-fluoroethoxy)ethoxy]ethoxy}ethoxy)-acetaldehyde and [^{18}F]-4-(3-fluoropropoxy)benzaldehyde to the same cyclic-RGD peptide (Glaser *et al.* 2008). Whilst all ^{18}F-aldehydes readily condensed with the aminooxy-bearing cyclic-RGD peptide, the different prosthetic groups were shown to directly impact the biodistribution and tumour uptake of the labelled peptides in mice. Whilst each of the labelled conjugates localised to the tumour, the poly(ethylene glycol)-containing conjugate possessed superior tumour to blood, lung, liver and muscle ratios compared to the more hydrophobic derivatives.

Further to the methods already described, in 2009, Glaser *et al.* reported the copper(I)-catalysed Huisgen cycloaddition of [^{18}F]fluoroethyl azide with an alkyne functionalised RGD peptide (Glaser *et al.*, 2009). The authors reported that mild reaction conditions and short reaction times suggest the 'Huisgen ("click") reaction' provides a potentially attractive alternative to aminooxy aldehyde condensation.

Of the many RGD peptides evaluated in preclinical studies, still relatively few ^{18}F-labelled agents remain to be evaluated in a clinical setting. [^{18}F]Galacto-RGD was the first ^{18}F-labelled RGD peptide to successfully image αvβ3 expression in human tumours with good tumour to background ratios (Beer *et al.*, 2005). *In vivo*, [^{18}F]galacto-RGD was rapidly cleared from the blood pool and is primarily cleared through the kidneys. Biodistribution and dosimetry studies demonstrated that [^{18}F]galacto-RGD background activity in lung and muscle tissue is low and the calculated effective dose is comparable to that of an [^{18}F]FDG scan. PET imaging using [^{18}F]galacto-RGD was also found to correlate with the intensity of immunohistochemical staining of αvβ3 expression (Beer and Schwaiger, 2008). Tumour to background ratios with [^{18}F]galacto-RGD PET also have been measured in squamous cell carcinoma of the head and neck with a widely varying intensity of tracer uptake (Beer *et al.*, 2007). In comparative studies between [^{18}F]galacto-RGD PET and [^{18}F]FDG, no clear correlation between the uptake of the two tracers in patients with various tumours was observed. This result indicates that αvβ3 expression and glucose metabolism are not closely correlated in tumour lesions and that [^{18}F]galacto-RGD can provide different information to [^{18}F]FDG in cancer patients (Beer *et al.*, 2008).

More recently, two other novel RGD peptide derivatives labelled with ^{18}F have been evaluated in clinical trials, specifically [^{18}F]fluciclatide (previously referred to as AH111585 (Fig. 11.12)) and [^{18}F]RGD-K5. Developed by Siemens Molecular Imaging, [^{18}F]RGD-K5 is a [^{18}F]fluoroalkyl triazole-labelled RGD derivative prepared via the copper(I)-catalysed Huisgen cycloaddition of an RGD-K5-azide with [^{18}F]fluoropentyne (Kolb *et al.*, 2009a,b). Although only limited information is currently available in the literature, [^{18}F]RGD-K5 binds with high affinity (7.9 nM) to the αvβ3 integrin receptor (Kolb *et al.* 2009a,b). *In vivo*, the primary route for [^{18}F]RGD-K5 clearance is through the kidneys. The tracer is metabolically stable (98% intact tracer remains 1 h post injection) and microPET imaging reveals preferential tumour uptake in U87MG xenografts with a tumour to muscle ratio of more than 5:1 after 2 h. These results demonstrate that [^{18}F]RGD-K5 is a promising tracer for imaging αvβ3 integrin expression *in vivo*. Further evaluation of [^{18}F]RGD-K5 in patients with primary or metastatic breast cancer has recently been reported (Cho *et al.*, 2009). The authors concluded that [^{18}F]RGD-K5 was a useful marker for integrin expression in breast cancer, but it does not appear to be well correlated with angiogenesis with maximum microvessel density (hot-spot), since the RGD peptide binds not only the αvβ3 receptor but also to other integrin subtypes expressed in breast cancer cells.

[^{18}F]AH111585 (Fig. 11.12) was developed by GE Healthcare and is a cyclic-PEGylated-RGD peptide (Kenny *et al.*, 2008). The peptide structure has been optimised with the introduction of a disulfide and a sulfide bridge to stabilise the molecule and with the introduction of a polyethylene glycol spacer at the C-terminus to stabilise the peptide against carboxypeptidases and increase the *in vivo* circulation lifetime. Radiosynthesis of [^{18}F]fluciclatide was achieved by oxime formation by condensation of the aminooxy-functionalised peptide with [^{18}F]fluorobenzaldehyde (Kenny *et al.*, 2008; McParland *et al.*, 2008). In a Phase I trial, the biodistribution of [^{18}F]fluciclatide was assessed in 18 tumour lesions from seven patients with metastatic breast cancer (Kenny *et al.*, 2008).

The radiopharmaceutical and PET procedures were well tolerated in all patients and all 18 tumours detected by CT were visible on the PET images as distinct increases in uptake compared with the surrounding normal tissue (Fig. 11.13). Liver metastases were observed as regions of deficit uptake because of the high background activity in normal liver tissue. [^{18}F]Fluciclatide was either homogeneously distributed in the tumours or appeared within the tumour rim, consistent with the pattern of viable peripheral tumour and central necrosis often seen in association with angiogenesis (Kenny *et al.*, 2008).

Following the initial Phase I proof-of-concept study, a complete safety biodistribution and internal radiation dosimetry of [^{18}F]fluciclatide was reported (McParland *et al.*, 2008). Injection of [^{18}F]fluciclatide was

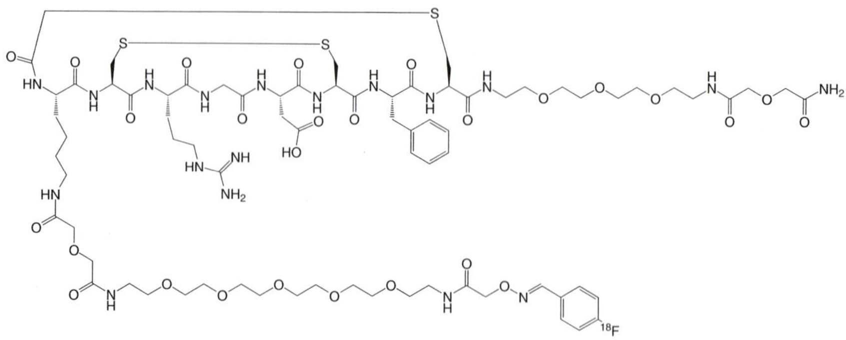

Figure 11.12. Structure of αvβ3 integrin-targeting RGD peptide imaging agent [^{18}F]Fluciclatide ([^{18}F]AH111585).

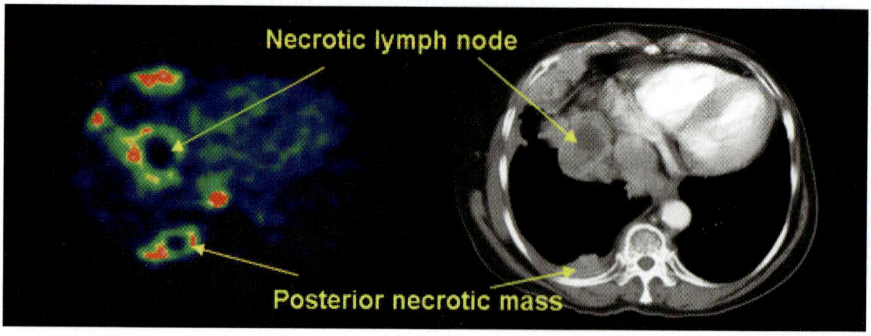

Figure 11.13. [^{18}F]Fluciclatide ([^{18}F]AH111585) PET image of metastatic lesions from breast cancer (LHS) and corresponding CT image (RHS). Reproduced with permission from E. Aboagye (Kenny *et al.*, 2008).

well tolerated in all subjects, with no serious or drug-related adverse events reported. The main route of ^{18}F excretion was renal (37%), and the three highest initial uptakes were by the liver, combined walls of the small, upper large and lower large intestines and kidneys. The three highest absorbed doses were received by the urinary bladder wall, kidneys and cardiac wall (Fig. 11.14).

Following on from these preliminary clinical studies, [^{18}F]fluciclatide has been assessed for its ability to non-invasively image tumour vasculature following antitumour therapy (Morrisson *et al.*, 2009). Small-animal

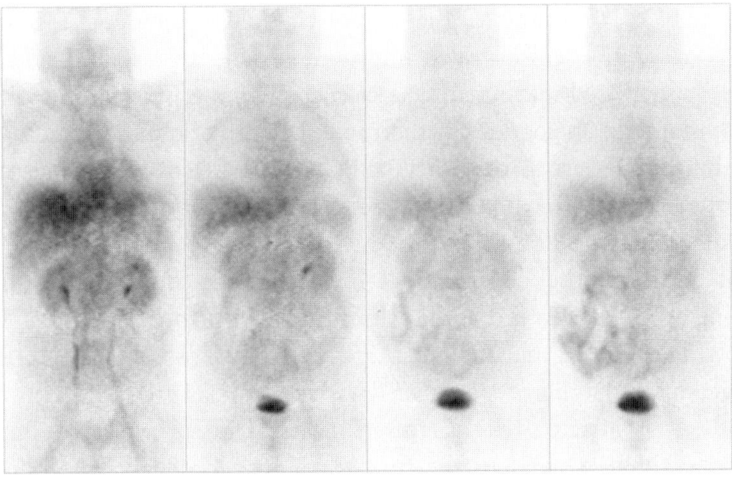

Figure 11.14. [^{18}F]Fluciclatide biodistribution and dosimetry time course in healthy volunteers.

PET imaging of Calu-6 tumours allowed visualisation of tumours above background tissue with mean baseline uptake of 2.2%ID/g. Paclitaxel therapy reduced the microvessel density in Lewis lung carcinoma (LLC) tumour-bearing mice and resulted in significantly reduced [^{18}F]fluciclatide tumour uptake. ZD4190 therapy resulted in a significant (31.8%) decrease in [^{18}F]fluciclatide uptake in Calu-6 tumours, compared with the vehicle control-treated Calu-6 tumours, which had a 26.9% increase in [^{18}F]fluciclatide uptake over the same period. [^{18}F]Fluciclatide is a promising ^{18}F-labelled RGD tracer that may reveal important information in the assessment of antitumour therapies, in particular those that predominantly target tumour blood vessels. [^{18}F]Fluciclatide is currently in Phase II clinical trials and will be tested both for its ability to be used diagnostically, for pre-selection of patients, and for monitoring of early therapy response after diagnosis and initial treatment.

Attempts to further improve upon the effectiveness of monomeric RGD derivatives have led to the development of a number of multimeric RGD peptides (see, for example, Wu *et al.*, 2007; Niu and Chen, 2009; Schottelius *et al.*, 2009). For example, the tetrameric RGD peptide [^{18}F]FPRGD4 (Wu *et al.*, 2007) demonstrated improved *in vitro* affinity and specificity as well as increased tumour uptake *in vivo* in comparison with its dimeric and monomeric analogues. Furthermore, heteromeric dimers such as the RGD-bombesin derivative, [^{18}F]FB-PEG$_3$-Glu-RGD-BBN, have also been developed to target $\alpha v\beta 3$ and gastrin-releasing peptide receptor positive tumours (Liu *et al.*, 2009).

In addition, the development of $\alpha v\beta 3$ integrin selective nuclear imaging agents has found applications outside oncology imaging. There is growing evidence that RGD peptides can play a role in the PET imaging of delayed-type hypersensitivity reaction (Lewis, 2005). More recently, [^{18}F]galacto-RGD was used to assess integrin expression, as a marker of angiogenesis in the process of myocardial repair following myocardial infarction (Makowski *et al.*, 2008a,b). The authors report that the [^{18}F]galacto-RGD PET signal localised in the infarcted region and that the observed signal is consistent with angiogenesis within the healing area.

Over the past decade, there has been considerable effort to develop molecular imaging agents for $\alpha v\beta 3$ integrin. The clinical evaluation of PET imaging agents such as [^{18}F]galacto-RGD, [^{18}F]RGD-K5 and [^{18}F]fluciclatide will be useful, not just in management of patients receiving currently approved antiangiogenic therapy, but also in advancing the development of novel antiangiogenic and antimetastatic drugs.

11.9 Choline Metabolism

Choline is an essential substrate for the synthesis of phosphatidylcholine (PC), which is a major component of mammalian cell membranes (Zeisel, 1981). Choline kinase (CK) is the first enzyme in the metabolic pathway responsible for the generation of PC, the major phospholipid of all eukaryotic membranes, and therefore an essential component in the machinery which controls cell proliferation. Whilst in cancer there is often an increase in the cellular transport and phosphorylation of choline (Zeisel, 1981, 1993), to date there are no other known inherited diseases in humans affecting this pathway (Kwee *et al.*, 2007). Increase in the expression levels of CK has been reported in several cancers, e.g. prostate, breast, lung, ovarian and colon cancers (de Molina *et al.*, 2002). Therefore, this pathway has become a target for developing imaging agents and anti-cancer drugs which are metabolised by CK.

The PET analogue of endogenous choline, [^{11}C-*methyl*]choline (Fig. 11.15), was first examined as a putative radiotracer for the imaging of brain and prostate cancer by Hara *et al.* (1997, 1998). [^{11}C]Choline has found the greatest utility in the imaging of bladder (Picchio *et al.*, 2006) and prostate cancers (de Jong *et al.*, 2002) for which [^{18}F]FDG imaging is unattractive (Hoh *et al.*, 1998). The need for longer-lived agents has led to the development of ^{18}F-labelled analogues, such as [^{18}F]fluoromethylcholine (FCH), [^{18}F]fluoroethylcholine (FECH) and [^{18}F]fluoropropylcholine (FPCH) as shown in Fig. 11.15 (DeGrado *et al.*, 2000; Hara *et al.*, 2002). Generally, these tracers have been labelled via alkylation of the amine precursor *N,N*-dimethylethanolamine using either [^{18}F]fluoroalkylhalide or [^{18}F]fluoroalkyltosylate synthons (DeGrado *et al.*, 2000, 2001; Hara *et al.* 2002).

In vitro studies have shown that fluorinated choline analogues are good substrates for choline kinase and are phosphorylated to varying degrees (DeGrado *et al.*, 2001; Bansal *et al.*, 2008). A relatively recent review by Kwee *et al.* (2007) gives a broad overview of the development

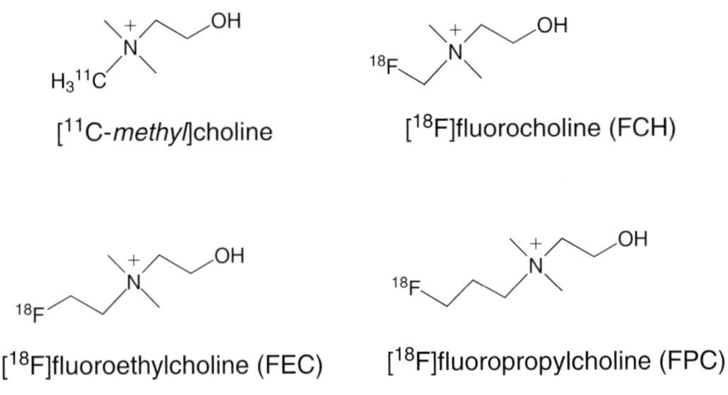

Figure 11.15. The major choline-based PET radiotracers.

of ^{18}F-labelled choline derivatives and their clinical use in imaging a variety of neoplasms including those of the breast, prostate, liver and brain.

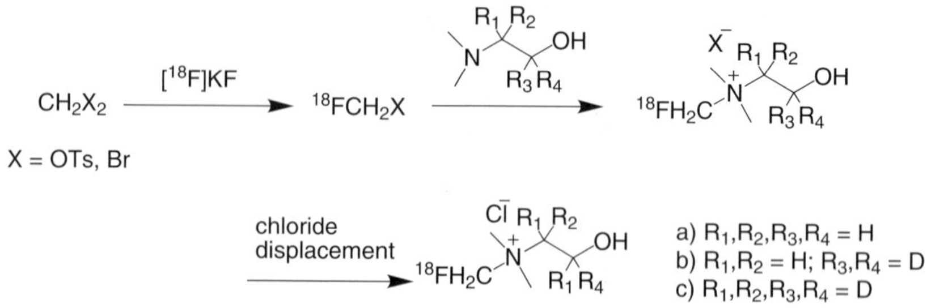

Scheme 11.3. Radiolabelling of [^{18}F]fluoro-[1,2-$^{2}H_4$]choline (D_4-FCH).

Although, clinical studies are underway, there are limitations with the use of choline and its fluoro analogues. These molecules are also substrates for the enzyme choline oxidase, and can undergo oxidation to the unwanted metabolite betaine or fluorinated derivatives of betaine to differing levels (Bansal *et al.*, 2008; Leyton *et al.*, 2009). To reduce the rate of *in vivo* oxidative metabolism of FCH, a tetra-deuterated analogue, [^{18}F]fluoromethyl-[1,2-$^{2}H_4$]-choline, has been recently developed by Leyton *et al.* (2009). The radiolabelling approach used is shown in Scheme 11.3. Briefly, [^{18}F]fluoro-[1,2-$^{2}H_4$]choline (D_4-FCH) is synthesised by alkylation of *N,N*-dimethylaminoethanol-d_4 precursor with either [^{18}F]fluorobromomethane or [^{18}F]fluoromethyl tosylate. The product is isolated using a cation exchange Sep-Pak cartridge in good radiochemical yield and purity.

Both FCH and [^{18}F]fluoro-[1,2-$^{2}H_4$]choline (D_4-FCH) have been compared in HCT116 tumour cells and in *in vivo* imaging studies in SKMEL-28 tumour xenograft mice (Fig 11.16).

Analysis of mouse plasma and tissue samples has shown that higher levels of phosphorylation and lower levels of oxidation are observed for D_4-FCH compared to FCH. Additionally, the uptake of D_4-FCH in the SKMEL-28 tumour was higher compared to FCH. This preliminary data indicates that D_4-FCH is a very promising metabolically relatively stable radiotracer for imaging choline metabolism in tumours.

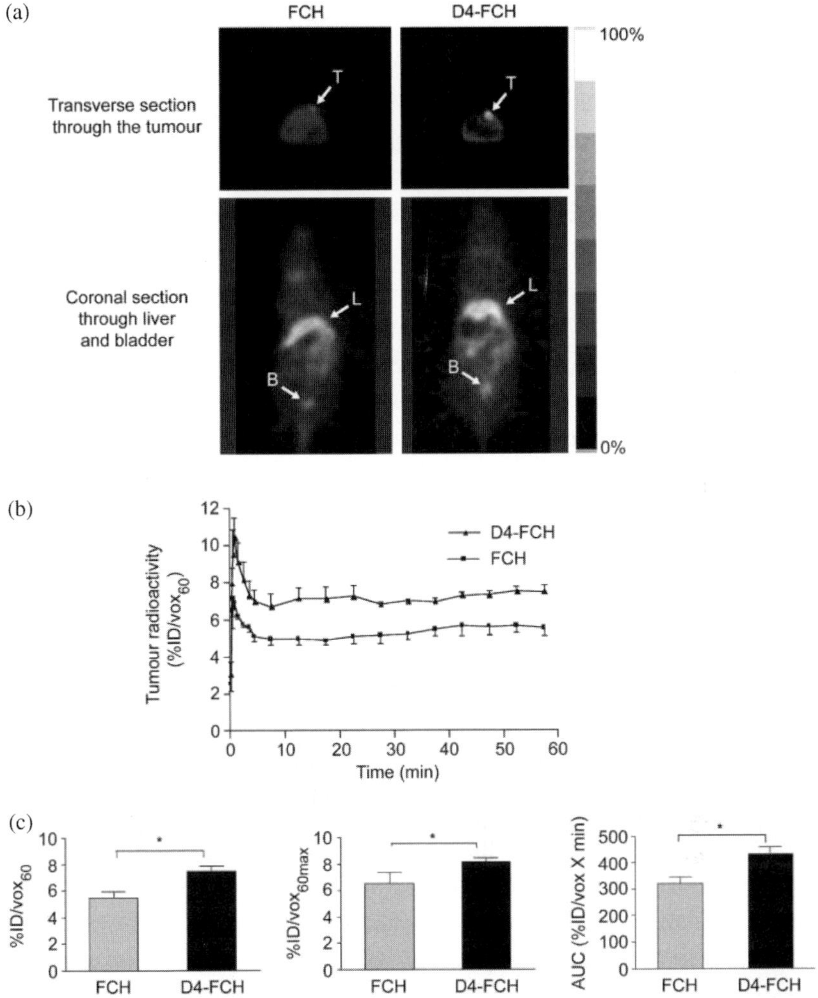

Figure 11.16. (a) FCH and D_4-FCH in SKMEL-28 tumour xenograft mice. Arrows: tumours (T), liver (L), bladder (B). (b) Comparison of time versus radioactivity curves for FCH & D_4-FCH in tumours. (c) The delivery and retention of D_4-FCH were quantitatively higher than of FCH. Reproduced with permission from E. Aboagye (Leyton *et al.*, 2009).

11.10 Apoptosis

Apoptosis is a process of programmed cell death in living organisms. It was originally defined by Kerr *et al.* (1972) and is a key process in controlling cell growth, cancer development and progression. Dysregulation of this process has been implicated in multiple diseases such as cancer, autoimmunity, neurodegeneration, ischaemia and transplant rejection (Lahorte *et al.*, 2004; Blank and Shiloh, 2007). Evasion of apoptosis is a hallmark of almost every human tumour (Hanahan and Weinberg, 2000) and pathways involved in apoptosis are targets for cancer treatment (Kerr *et al.*, 1994; Ashkenazi, 2002). Essentially, all treatments for cancer, whether radiotherapy, chemotherapy or immunotherapy, are intended to induce apoptosis in tumour cells. Monitoring the rate and extent to which apoptosis occurs could provide clinicians with important information on disease activity and direct assessment of the effectiveness of tumour treatment which may fundamentally alter the way cancer patients are managed (Cummings *et al.*, 2004; Call *et al.*, 2008).

Recently, several reviews have been published on the development of imaging agents for the *in vivo* assessment of apoptosis targeting biochemical changes in different pathways (Lahorte *et al.*, 2004; Wang, 2007; Smith *et al.*, 2009; Zhao, 2009). As such, only a few examples will be covered in this chapter.

To date, the most common apoptosis pathways for tracer development have been, (i) externalised phosphatidylserine (PS) residues on the cell surface (Lahorte *et al.*, 2004) and (ii) cysteine aspartate-specific proteases known as the 'caspases' (Smith *et al.*, 2009).

11.10.1 *PS targeting radiotracers*

Annexin V is a 36 kDa protein which binds to externalised PS residues on the cell surface in the presence of calcium ions (Walker *et al.*, 1992). It binds selectively to membrane-bound PS residues with nanomolar affinity ($K_d \approx 0.5$–7 nM). Annexin V has been labelled with different fluorescent tags and a wide range of radioisotopes, including ^{99m}Tc, $^{123/125}$I and ^{18}F (Lahorte *et al.*, 2004). 4-[^{18}F]Fluorobenzoylannexin V has been prepared by a multi-step synthesis from *N*-succinimidyl-4-[^{18}F]fluorobenzoate ([^{18}F]SFB) (Zijlstra *et al.*, 2003; Grierson *et al.*, 2004b). *In vitro* assessment in cell-based assays and *in vivo* studies in rodents have demonstrated that 4-[^{18}F]fluorobenzoylannexin V does bind to PS residues and a fourfold increase in tracer uptake was observed in a myocardial ischaemia-reperfusion model (Murakami *et al.*, 2003). However, human PET studies have not been reported to date.

Recently, a number of small molecular-weight molecules have been developed. Compounds containing a fluorescent dansyl core (Damianovich *et al.*, 2006; Reshef *et al.*, 2007) have demonstrated great promise. A dansylhydrazone derivative, DFNSH, which selectively binds to paclitaxel-induced apoptotic cancer cells has been ^{18}F-labelled using a two-step procedure (Zeng *et al.*, 2008). The radiosynthesis involved the preparation of [^{18}F]fluorobenzaldehyde followed by reaction with dansylhydrazine at 100°C for 30 min. [^{18}F]DFNSH was isolated in 50–60% radiochemical yield with a radiochemical purity of >99% (Scheme 11.4). The authors have reported that PET imaging studies of tumour apoptosis in rodent models are currently underway with [^{18}F]DFNSH.

Another small molecule probe from the *ApoSense* family, [^{18}F]ML-10 (5-[^{18}F]fluoropentyl-2-methyl-malonic acid) has been developed as a novel PET tracer for the imaging of apoptosis *in vivo* (Reshef *et al.*, 2008). This class of compounds responds to alterations in plasma membrane potential and phospholipid scrambling, which are hallmarks of apoptotic cells (Aloya *et al.*, 2006; Damianovich *et al.*, 2006; Reshef *et al.*, 2007). It has been shown that [^{18}F]ML-10 accumulates specifically in apoptotic cells and tissues, whereas it has significantly lower binding to non-target healthy tissues.

[^{18}F]ML-10 was synthesised from its respective precursor, ML-10-mesylate, by a nucleophilic substitution reaction using [^{18}F]fluoride (Scheme 11.5). This approach provided [^{18}F]ML-10 in good radiochemical

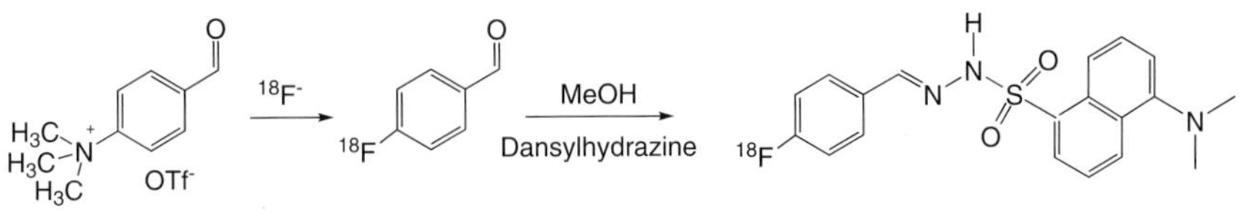

Scheme 11.4. Radiosynthesis of [^{18}F]DFNSH from [^{18}F]fluorobenzaldehyde.

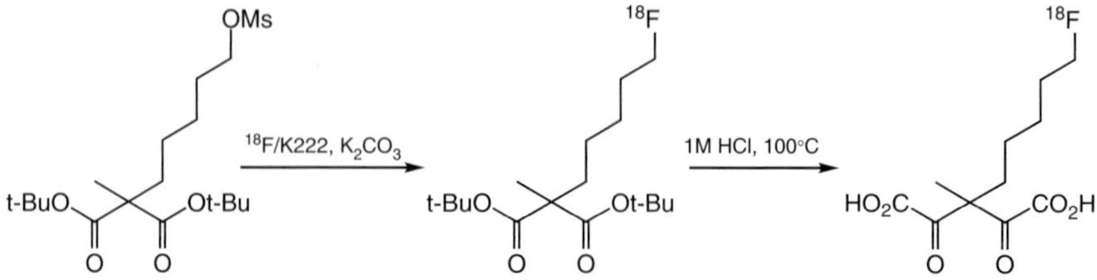

Scheme 11.5. Radiolabelling of ML-10 with ^{18}F-fluoride *via* mesylate precursor.

yield (30–40% at end of synthesis (EOS)) with high radiochemical purity (>99%). The specific radioactivity reported was >40 GBq/μmol.

Biodistribution studies in rats showed that [^{18}F]ML-10 cleared rapidly from blood; there was a lack of binding to healthy tissues and rapid elimination through the kidneys (Reshef *et al.*, 2008). Very little metabolism either in blood or in the brain was observed up to 90 min after tracer injection. Using small animal scanning, imaging of neurovascular cell death in an experimental ischaemic stroke mouse model was conducted. The images showed selective uptake of the tracer specifically in the infarct region. Uptake measurements *ex vivo* revealed twofold-higher uptake in the affected hemisphere and six- to tenfold higher uptake in the region of interest of the infarct. Additionally, there was good correlation between cerebral uptake and histological evidence in cell death. However, one of the limitations of this assessment is that ischaemic cerebral neurovascular damage is characterised by both apoptosis and blood–brain barrier (BBB) disruption. Therefore, this study could not adequately distinguish the specific uptake of [^{18}F]ML-10 in target apoptotic cells and the contribution of a defective BBB (Reshef *et al.*, 2008).

This promising data has led *ApoSense* to investigate [^{18}F]ML-10 further, both preclinically and clinically, as a suitable imaging agent of apoptosis or cell death. In a recent clinical PET study with [^{18}F]ML-10 in patients with metastatic brain tumours undergoing radiotherapy, it has been reported that clear images were obtained of both endogenous and irradiation-induced apoptosis and that [^{18}F]ML-10 might be able to monitor the effect of radiotherapy in individual patients (Shirvan *et al.*, 2009). Further clinical trials are underway to fully evaluate this tracer as a potential marker of apoptosis.

11.11 Caspase Targeting Radiotracers

The caspases involved in apoptosis can be divided into 'initiator' (caspases 8 and 9) and 'effector' (caspases 3, 6 and 7), and both small and large molecules (peptides) have been synthesised with differing specificity (Smith *et al.*, 2009). Recently, several groups have labelled a class of non-peptide small-molecule inhibitors based on an isatin lead candidate (Fig. 11.17) which is potent and binds selectively to the active site of caspase-3/7 (Lee *et al.*, 2001). Through structural optimisation, the groups of Mach (Chu *et al.*, 2005) and Kopka (2006) independently developed a range of highly potent isatin sulfonamide analogues which could be labelled with PET and SPECT radioisotopes.

^{18}F-labelled fluoroethyl phenyl ether, also known as [^{18}F]WC-II-89, (Fig. 11.17) was the first candidate to be labelled and evaluated *in vivo* as a putative tracer for imaging of apoptosis (Zhou *et al.*, 2006). [^{18}F]WC-II-89 was labelled via a nucleophilic substitution of the corresponding mesylate precursor with [^{18}F]fluoride in high yield and radiochemical purity. Biodistribution studies using [^{18}F]WC-II-89 in cycloheximide-treated rats, an animal model of apoptosis, revealed higher uptake in the liver and spleen relative to control animals. However, preclinical studies have also shown that the phenyl ethyl moiety is a primary site of metabolic instability and this has led to the development of next generation ^{18}F-labelled isatin sulfonamides, such as (*S*)-1-((1-(2-fluoroethyl)-1*H*-[1,2,3]-triazol-4-yl)methyl)-5-(2-(3,5-difluorophenoxy)methyl-pyrrolidine-1-sulfonyl)isatin [^{18}F]ICMT-11 (Scheme 11.6) (Smith *et al.*, 2008).

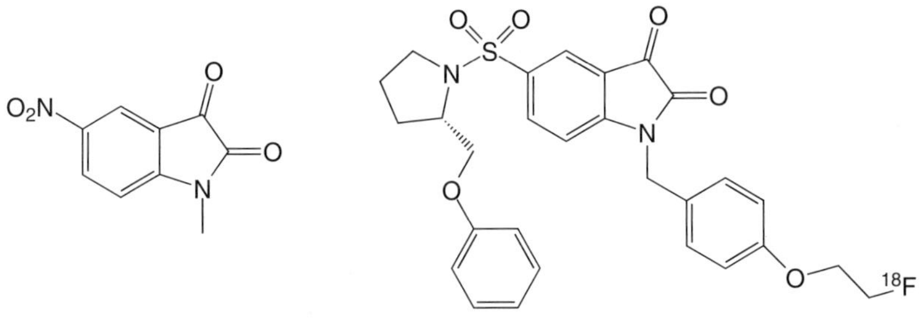

Figure 11.17. Structure of isatin lead candidate and isatin sulfonamide labelled analogue [^{18}F]WC-II-89.

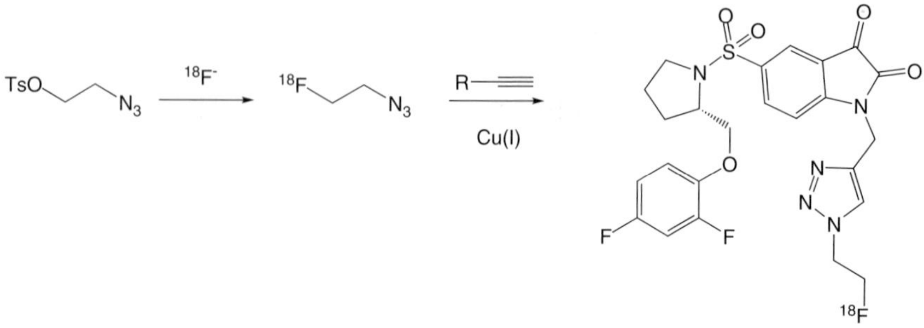

Scheme 11.6. Preparation of the isatin analogue [^{18}F]ICMT-11 by ^{18}F-labelling by the Huisgen ('click') reaction (Smith *et al.*, 2008).

The fluoroethyl triazole-modified compound ICMT-11 with a sub-nanomolar affinity for caspase-3 was selected from a small library of isatin analogues (Smith *et al.*, 2008). ^{18}F Huisgen ('click') labelling of this isatin-5-sulfonamide ([^{18}F]ICMT-11) with 2-[^{18}F]fluoroethyl azide was achieved as shown in Scheme 11.6. 2-[^{18}F]Fluoroethylazide was prepared by the reaction of [^{18}F]fluoride with the corresponding 2-azidoethyl-4-toluene-sulfonate precursor and purified by distillation (Glaser and Arstad, 2007). In the second radiochemical step, 2-[^{18}F]fluoroethyl azide was reacted with the isatin alkyne precursor at room temperature for 30 min in the presence of a copper(I) catalyst. The final compound, [^{18}F]ICMT-11, was obtained in good radiochemical yields after purification by preparative HPLC and formulation using solid-phase extraction (SPE).

[^{18}F]ICMT-11 has been evaluated in cancer cell lines and tumour-bearing mice (Nguyen *et al.*, 2009). The introduction of the triazole functionality has led to the molecule having high metabolic stability, with no indication of defluorination *in vivo* (Smith *et al.*, 2008). Biodistribution studies in RIF-1 tumour-bearing mice showed a 1.5-fold increased uptake of [^{18}F]ICMT-11 in cisplatin-treated mice relative to the control. Additionally, studies in 38C13 tumour-bearing mice treated with cyclophosphamide 100 mg/kg which induces apoptosis, has demonstrated that [^{18}F]ICMT-11 uptake in tumours increased after drug treatment as shown in Fig. 11.18 (Nguyen *et al.*, 2009). The recent preclinical studies have demonstrated that ^{18}F-labelled isatins have the potential for imaging apoptosis. Some of these candidates will undergo further assessment in

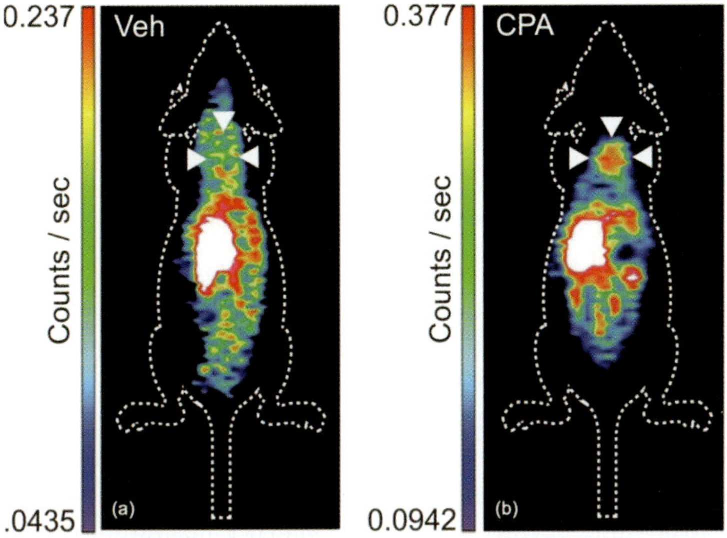

Figure 11.18. [^{18}F]ICMT-11 PET images of two 38C13 tumour-bearing mice: (a) untreated; (b) treated with cyclophosphamide 100 mg/kg. Reproduced with permission from E. Aboagye (Nguyen *et al.*, 2009).

humans and could aid anticancer drug development and monitor early responses to therapy.

11.12 CNS Neurosciences Applications

Over the past 25 years, a vast amount of effort has gone into the quantitative imaging of neuroreceptors, neurotransmitters and enzymes in the central nervous system (CNS) using PET (Frost and Wagner, 1990). This work has permitted fundamental insights into how binding of neurotransmitters to their receptors excite or inhibit neuronal firing or changes cellular metabolism. The recognition of receptor subtypes has suggested subtle ways for neurotransmitters to modulate neuronal functioning. Over the past decade, there has been more focus on exploring biological systems which affect disease pathology, such as protein aggregation, neuroinflammation or cell activation, and more recently ion-channel function. This has been driven to a large degree by the wide variety of selective and specific radiotracers/radioligands which have been developed. The section below will cover some of the more recent advances in ^{18}F-imaging agents and provide a brief update on preclinical and clinical assessment.

11.13 Beta-Amyloid Plaques and Neurofibrillary Tangles

Alzheimer's disease (AD) is a neurodegenerative disease showing an increasingly high incidence in the older population, especially in the age group of over 85 years where the incidence increases to about 40% (Blennow *et al.*, 2006). AD is characterised by progressive impairment in cognitive function and behaviour, such as irreversible memory loss, disorientation and language impairment (Masters and Beyreuther, 1998; Selkoe, 2002). The pathological features of AD based on post-mortem brain tissue show the presence of extracellular senile plaques containing β-amyloid aggregates (Aβ plaque or neuritic plaques consisting of peptides of 40 to 42 amino acids) and intracellular neurofibrillary tangles (NFTs) containing highly phosphorylated tau proteins (Selkoe, 2002; Roberson and Mucke, 2006). Accumulation of these pathologies is always accompanied by loss of neuronal projections and eventually neuronal losses contributing to grey and white matter atrophy. Recent reports have highlighted that amyloid neuropathology (both Aβ plaque and tau) in the brain plays a key role in a cascade of events leading to AD (Roberson and Mucke, 2006; Jakob-Roetne and Jacobsen, 2009). New treatment strategies for AD are aimed at delaying disease onset or slowing disease progression, through either preventing the deposition of β-amyloid or increasing the solubilisation of β-amyloid. Direct imaging of β-amyloid load in patients with AD *in vivo* would aid the early diagnosis of AD and the development and assessment of new treatment strategies (Klunk and Mathis, 2008).

[^{18}F]FDG PET has been used to assess glucose utilisation and has revealed that there is metabolic deficit in AD patients (Silverman *et al.*, 2001). A multicentre trial has demonstrated that [^{18}F]FDG PET scans may provide an objective and sensitive support to the clinical diagnosis in early dementia, including AD (Mosconi *et al.*, 2008). However, [^{18}F]FDG PET cannot specifically measure the pathologic features described above of AD. Therefore, in the past decade extensive effort has gone into the development of new PET imaging agents which target Aβ-plaque and NFTs. Some of these agents have been based on the structure of the histological amyloid staining reagents Congo Red and Thioflavin T which exhibit enhanced fluorescence binding to amyloid fibrils. Examples of the most successful amyloid imaging agents are shown in (Table 11.1). These include, 6-dialkylamino-2-naphthyl-ethylidene fluoro derivative [^{18}F]FDDNP, 2-(4-aminophenyl)-benzothiazoles (BTAs), phenylimidazo [1,2-*a*] pyridines, styrenes and benzoxazoles. The key properties of a candidate required for the development as a promising amyloid imaging agent have been extensively reported by Mathis *et al.* (2004), as well as the development and evaluation of these agents in a number of recent reviews (Barrio *et al.*, 1999, 2009; Mathis *et al.*, 2004; Henriksen *et al.*, 2008; Kung *et al.*, 2009). Only the most successful ^{18}F-labelled amyloid-specific agents will be covered here. Three ^{18}F-labelled tracers targeting Aβ aggregates (GE-067; BAY 94-9172 and AV-45) are currently under commercial development.

11.13.1 *FDDNP*

One of the first PET imaging agents developed for amyloid imaging was the lipophilic naphthalene derivative [^{18}F]FDDNP (2-(1-{6-[(2-[^{18}F]fluoroethyl)-(methyl)amino]-2-naphthyl}-2-ethylidene)malonitrile) (Agdeppa *et al.*, 2001; Shoghi-Jadid *et al.*, 2002). *In vitro* studies have shown that unlabelled FDDNP binds to aggregated β-amyloid (1–40) fibrils with two affinities with K_d values of 0.16 and 1.86 nM. The B_{max} values are 80.8 and 164 pmol/mg for the high-affinity and low-affinity binding sites, respectively (Agdeppa *et al.*, 2001). Saturation binding studies with [^{18}F]FDDNP to homogenates of frontal cortex from post-mortem AD brain showed a K_d value of 0.74 nM and a B_{max} value of 144 nmol/g tissue. There was no specific binding of [^{18}F]FDDNP to homogenates of frontal cortex from age-matched control brains. As FDDNP is highly lipophilic, it is able to cross the BBB and the cellular membranes of neurons. Therefore, [^{18}F]FDDNP is able to detect both Aβ plaques and NFTs in AD brains (Agdeppa *et al.*, 2003).

The synthesis of [^{18}F]FDDNP was first described by Barrio *et al.* (1999, 2001) where a sulfonated precursor was reacted with K[^{18}F]F/Kryptofix 222 and the product isolated using normal phase HPLC. However, the radiochemical yield was relatively low (10–20% at EOS).

Table 11.1. Inhibition constants (K_i) for binding to Aβ plaques in post-mortem brain homogenates (adapted from Choi *et al.*, 2009).

Structure	Compound	K_i (nM)
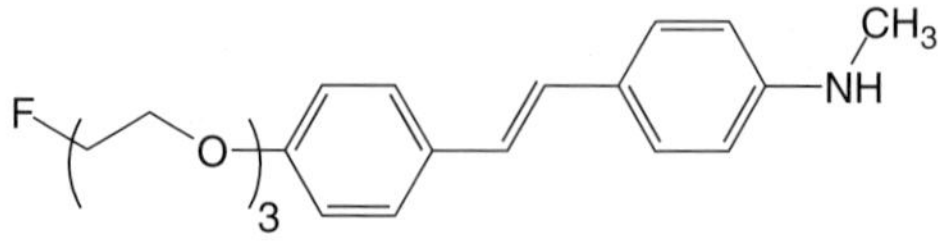	AV1 (BAY94-9172)	2.22 ± 0.54
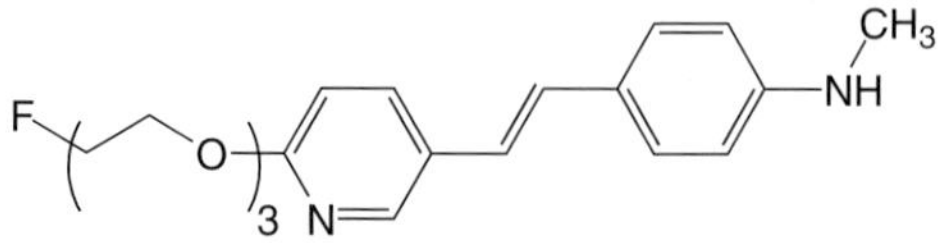	AV45	2.87 ± 0.17
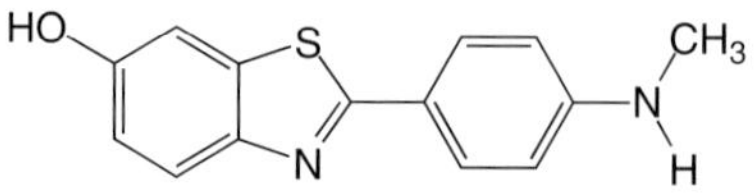	PIB	0.87 ± 0.18
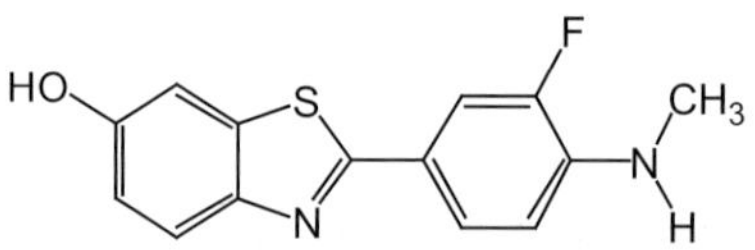	3′-F-PIB (GE067, Flutemetamol)	0.74 ± 0.38
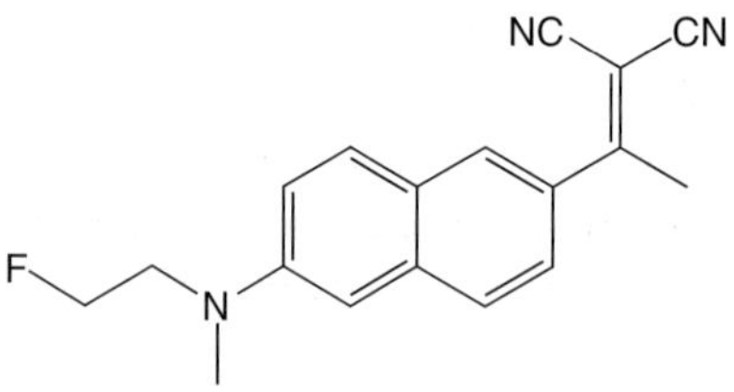	FDDNP	172 ± 18
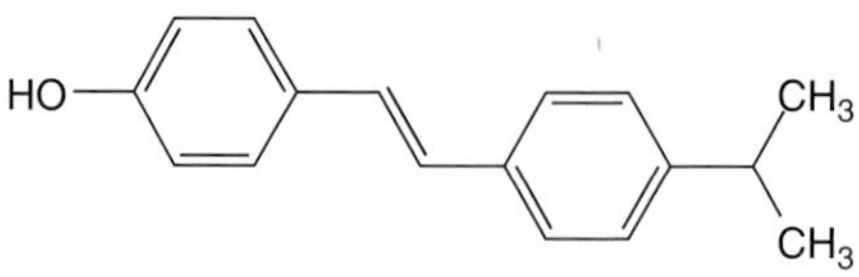	SB-13	3.18 ± 1.04
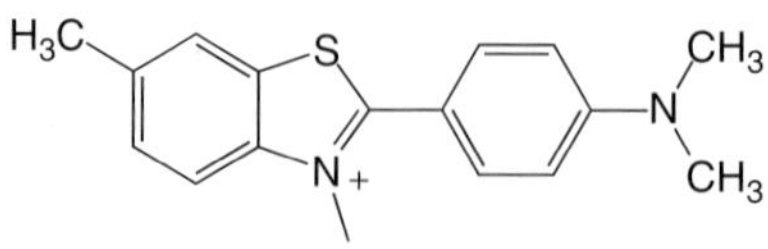	Thioflavin T	>1000
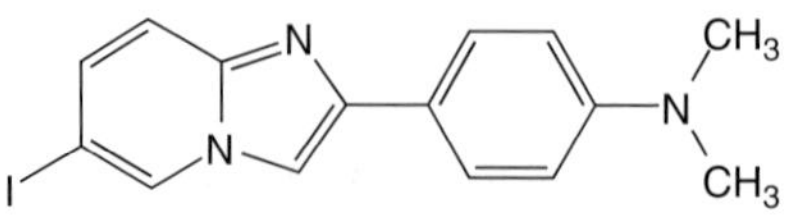	IMPY	1.29 ± 0.46

Recently, an automated radiosynthesis procedure has been developed using reverse phase HPLC for purification (Klok *et al.*, 2008). The precursor, 2-(1,1-dicyanopropen-2-yl)-6-(2-tosyloxyoethyl)methylaminonaphthalene, was fluorinated with [^{18}F]fluoride in acetonitrile (Scheme 11.7). The product was isolated using a Merck Lichrosorb RP-Select B (10 μm, 250 × 10 mm) column eluted with a mixture of acetonitrile and water 50/50 (v/v). The overall radiochemical yield was >41% (decay-corrected). [^{18}F]FDDNP was obtained with a radiochemical purity of 98% and the specific activity was >100 GBq/μmol (EOS).

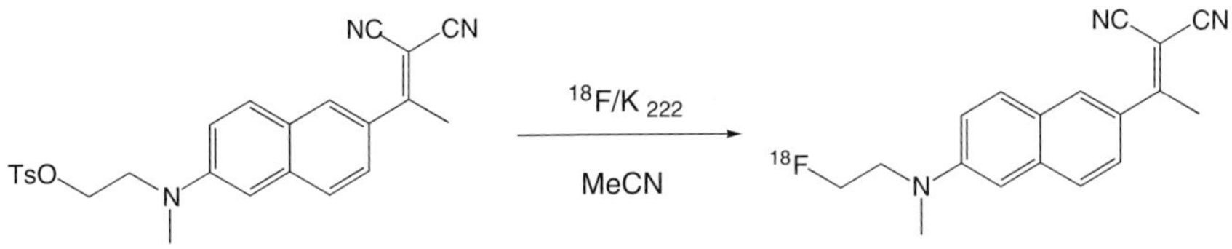

Scheme 11.7. Radiosynthesis of [^{18}F]FDDNP.

[^{18}F]FDDNP was the first PET Aβ-imaging agent to enter clinical studies of AD and healthy controls (Shoghi-Jadid *et al.*, 2002). This study showed that [^{18}F]FDDNP uptake was 1.87-fold greater in brain regions (such as frontal, parietal, temporal and occipital cortex, and hippocampus) that are known to contain Aβ plaque and NFTs in AD patients compared to healthy controls. In other studies, it has been demonstrated that [^{18}F]FDDNP PET can differentiate groups with mild cognitive impairment (MCI) from those with AD and those with no cognitive impairment (Small *et al.*, 2006). Although [^{18}F]FDDNP has achieved some success in imaging amyloid plaques, there are a number of limitations which have affected its utility. It has relatively high non-specific binding (Noda *et al.*, 2008) and a relatively low specific binding signal (Small *et al.*, 2006). Extensive metabolism in human plasma and rat studies have shown polar metabolites can penetrate the BBB and result in uniform brain uptake (Luurtsema *et al.*, 2008).

11.13.2 *BTA derivatives*

To date, the best characterised PET imaging agent for Aβ plaques in the brain is the 4-aminophenyl-benzothiazole (BTA) derivative (Table 11.2),

[^{11}C]6-OH-BTA-1 (also known as [^{11}C]PIB 'Pittsburgh compound B'). [^{11}C]PIB was first identified by Mathis *et al.* (2002) and the first human study with this agent was carried out in AD and control subjects in Uppsala, Sweden, in 2002 (Engler *et al.*, 2002). The compound was identified to have favourable properties among a series of BTAs by Mathis *et al.* (2003a,b). The history of the development of this class of candidates from Thioflavin T (a positively charged, lipophilic molecule which does not cross the BBB) is well presented in an extensive review by Mathis *et al.* (2004). [^{11}C]PIB is a small, neutral molecule which exhibits excellent brain penetration and initial brain uptake, and displays a high binding affinity to Aβ plaques (K_i = 0.87 nM) (Mathis *et al.*, 2003a). This tracer binds to Aβ plaques but not to NFTs (Klunk *et al.*, 2004). In the past few years, numerous PET investigations with [^{11}C]PIB have been carried out in thousands of AD patients (Klunk *et al.*, 2004; Rowe *et al.*, 2007). These have confirmed that [^{11}C]PIB brain distribution in AD is concordant with the post-mortem distribution of amyloid plaques and that the tracer provides a good distinction between this disorder and normal control subjects, both by visual and quantitative examination (Nordberg, 2007).

The success of [^{11}C]PIB for imaging Aβ plaques in AD patients has provided considerable impetus for the development of ^{18}F-derivatives of [^{11}C]PIB (Mathis *et al.*, 2003b; Berndt *et al.*, 2007; Mason *et al.*, 2007). Some of the *in vitro, ex vivo* properties in rodents and *in vivo* measures in baboons of the closest analogues of [^{11}C]PIB are given in Table 11.2 and examples of structures are shown in Fig. 11.19.

Introduction of [^{18}F]fluoride has been achieved either via nucleophilic aromatic substitution of a suitable leaving group or via [^{18}F]fluoroalkylation with a suitable amino precursor or displacement of an alkyltosyl group (Mathis *et al.*, 2003a; Berndt *et al.*, 2008). The ^{18}F-labelled benzothiazole derivative, [^{18}F]-3′-F-6-OH-BTA1 (now known as [^{18}F]flutemetamol, previously known as [^{18}F]GE067), is now being batch-manufactured using automated platforms, such as TracerLab FX F-N and FASTlab in good manufacturing practice facilities to support clinical trials in healthy controls and AD subjects. [^{18}F]Flutemetamol is being produced in high radiochemical yields with high radiochemical purity (>98%). The biodistribution and internal radiation dosimetry of an intravenous injection of

Table 11.2. Some representative examples of [^{18}F]PIB analogues. *In vitro* measures (logP_{c18}, K_i) and *ex vivo* measures in rodents (normal brain uptake and clearance) and *in vivo* measures in baboons (regional nomal uptake and clearance) (Mason *et al.*, 2007).

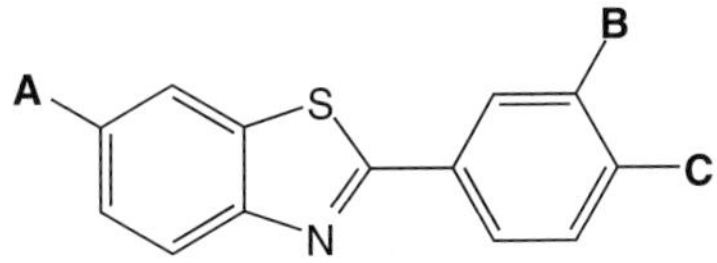

	A	B	C	LogP_{C18}[a]	K_i(nM) Aβ(1–40)	2 min uptake (%ID-kg)/g	2/30 min ratio
[^{11}C]PIB	OH	H	$NH^{11}CH_3$	1.2	4.3	0.21	12
	OH	H	^{18}FPrNH	1.5	25	0.17	16
	OH	H	$^{18}FPrNCH_3$	2.1	7.7	0.25	3.2
	OH	H	$^{18}FEtNCH_3$	1.8	7.4	0.36	2.8
	^{18}FPrNH	H	OH	1.6	28	0.24	13
	^{18}FEtO	H	NH_2	1.7	7.2	0.30	2.5
	^{18}FEtO	H	$NHCH_3$	2.4	3.2	0.22	2.3
	$^{18}FEtO(EtO)_2$	H	$NHCH_3$	2.2	4.9	0.29	3.0
	^{18}FEtO	H	OH	1.6	4.1	0.27	2.5
	^{18}FEtO	H	OCH_3	1.7	1.2	0.29	2.6
[^{18}F]Flutemetamol	OH	^{18}F	$NHCH_3$	–	5.9	0.31	8.4

[a] LogP_{C18} refers to a reverse phase HPLC method for estimation of the octanol/water partition coefficient (see Mathis *et al.*, 2003a).

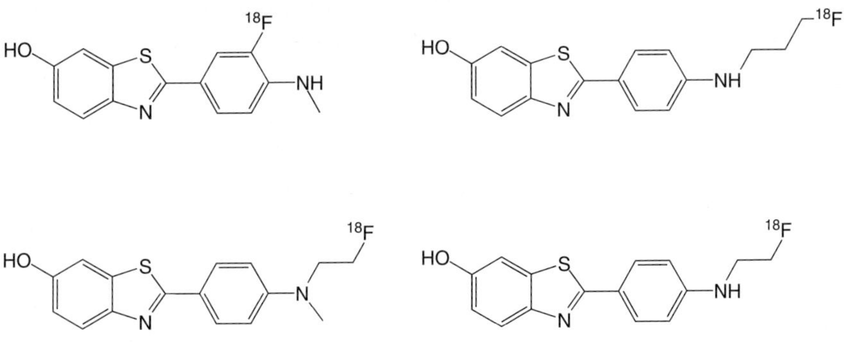

Figure 11.19. Some examples of [^{18}F]PIB analogues labelled with ^{18}F.

^{18}F-flutemetamol have been reported recently (Koole *et al.*, 2009). The study showed that no adverse events or clinically significant changes were observed. [^{18}F]Flutemetamol is excreted predominantly through the hepatobiliary system. The mean effective dose was comparable to that of many other ^{18}F-labelled radiopharmaceuticals. The Phase I study has shown that there is a significant increase in standardised uptake value (SUV) ratios in AD patients in neocortical association zones and striatum compared with healthy controls, whereas uptake in white matter, cerebellum and pons (regions known to be relatively unaffected by amyloid deposition) did not differ between groups (Nelissen *et al.*, 2009). An example of a PET scan illustrating the uptake of [^{18}F]flutemetamol and the SUV ratio in frontal cortex in an AD patient versus a healthy control is shown in Fig. 11.20. This study clearly demonstrates that [^{18}F]flutemetamol can differentiate between AD and healthy controls and that the uptake can be quantified. [^{18}F]Flutemetamol has now entered Phase II clinical trials (sponsored by GE Healthcare).

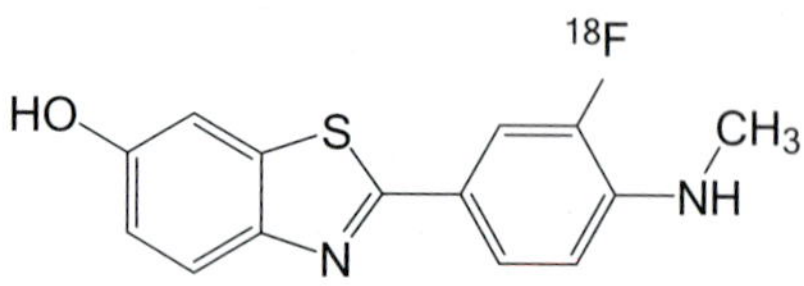

[^{18}F]Flutemetamol

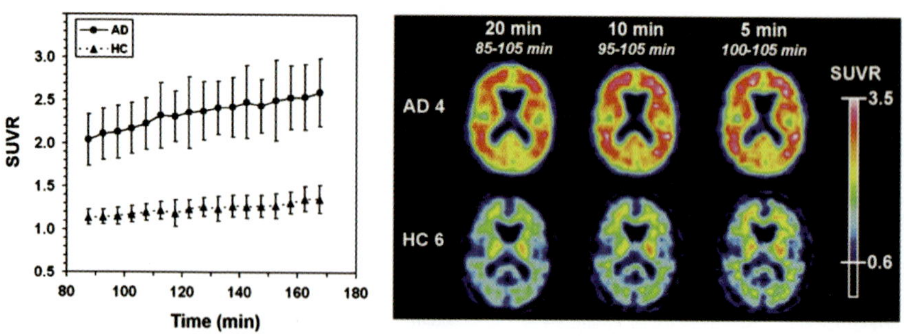

(a) (b)

Figure 11.20. (a) Average SUV ratio curves in frontal cortex of AD patients and healthy controls (HC) over 85–170 min. (b) PET summed images of different scan lengths representative of an AD patient and a healthy control subject (Nelissen *et al.*, 2009).

11.13.3 *Stilbenes*

Another class of candidates that has been extensively evaluated as amyloid imaging agents, has been based on stilbene. The rigid structures of stilbene and styrylpyridine have provided core structures for developing many specific imaging agents for Aβ plaques (Ono *et al.*, 2005; Zhang *et al.*, 2005a,b, 2007; Kung *et al.*, 2010). Again, the ^{11}C stilbene derivative, 4-*N*-[^{11}C-methyl]amino-4′-hydroxystilbene ([^{11}C]SB-13) was identified by Verhoeff *et al.* (2004) as a lead compound (Fig. 11.21). [^{11}C]SB-13 is a small and neutral derivative with high affinity for Aβ ($K_i = 6.0$ nM on $A\beta_{40}$-fibrils) and lipophilicity in the desired range ($\log P = 2.36$). [^{11}C]SB-13 has been evaluated in humans and compared with [^{11}C]PIB. [^{11}C]SB-13 showed a similar pattern of selective higher uptake and retention in the frontal cortex of AD patients compared to healthy control subjects (Verhoeff *et al.*, 2004).

A number of SB-13 [^{18}F]fluoroalkyl analogues (added at either end of the molecule) have emerged; however, these were unsuccessful as they were too lipophilic and exhibited high levels of non-specific binding. To circumvent the problem of high lipophilicity after adding a fluoroalkyl group, Zhang *et al.* (2005a) prepared stilbene derivatives with an additional hydroxyl group. These suffered from high levels of *in vivo* defluorination resulting in the uptake of [^{18}F]fluoride in the skull (Cai *et al.*, 2004; Zhang *et al.*, 2005a). To reduce the high lipophilicity and retain the affinity, a series of fluoropegylated stilbene derivatives have been developed (Zhang *et al.*, 2005b, 2007; Stephenson *et al.*, 2007). The

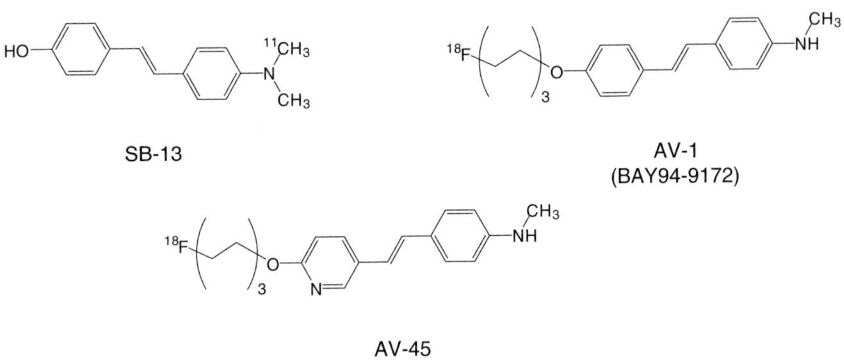

Figure 11.21. Structure of [^{11}C]SB-13 and examples of ^{18}F-analogues (AV-1 and AV-45).

fluorinated PEG-stilbenes displayed high binding affinities (K_i = 2.9–6.7 nM versus [^{125}I]IMPY on AD brain sections). From this series, two clear candidates have emerged known as AV-1 and AV-45 (Fig. 11.21).

The first fluorinated PEGstilbene derivative, AV-1, *trans*-4-(*N*-methyl-amino)-4′-{2-[2-(2-[^{18}F]fluoro-ethoxy)-ethoxy]-ethoxy}stilbene (now known as BAY94-9172) binds with high affinity (K_i = 6.7 ± 0.3 nM) to brain homogenates from AD patients and in AD tissue sections. [^{18}F]BAY94-9172 selectively labelled Aβ plaques (Zhang *et al.*, 2005a).

[^{18}F]BAY94-9172 has been synthesised from the PEGN3-OMs (methylsulfonate) precursor and purified by semi-preparative HPLC (Zhang *et al.*, 2005a). The product was reformulated with the Sep-Pak method. The radioligand purity averaged 96% and the specific radioactivity averaged 140 GBq/μmol. A detailed preclinical and initial clinical evaluation of [^{18}F]BAY94-9172 has been presented by Rowe *et al.* (2008). The results from this study suggest that [^{18}F]BAY94-9172 can reliably detect Aβ deposition and the distribution of [^{18}F]BAY94-9172 binding is almost identical to that reported for [^{11}C]PIB, but the degree of binding appears to be slightly lower (O'Keefe *et al.*, 2009). [^{18}F]BAY94-9172 has also progressed to Phase II clinical trails (sponsored by Bayer–Schering).

The second fluorinated PEGstilbene, [^{18}F]AV-45, (*E*)-4-(2-(6-(2-(2-(2-^{18}F-fluoroethoxy)ethoxy)ethoxy)pyridin-3-yl)vinyl)-*N*-methyl-benzenami ne), has also been labelled (Zhang *et al.*, 2007) and has undergone preclinical (Choi *et al.*, 2009) and clinical assessment (Adler *et al.*, 2008; Skovronsky *et al.*, 2008; Sperling *et al.*, 2009). Recently, *in vitro* binding and autoradiography, as well as *ex vivo* and *in vivo* scanning with [^{18}F]AV-45 in mice and rhesus monkeys have been carried out. It has been shown that [^{18}F]AV-45 has a high affinity and specificity to Aβ plaques (K_d, 3.72 ± 0.30 nM) and the tracer displayed substantial plaque binding in AD brains compared to controls. In healthy mice and monkeys, there was initial high brain uptake and rapid washout from brain regions (Choi *et al.*, 2009).

More than 200 subjects in Phase I and Phase II studies have been imaged with [^{18}F]AV-45 and the data indicate that [^{18}F]AV-45 is well tolerated and has the potential to be effective for *in vivo* imaging of amyloid pathology in patients with signs of cognitive impairment (Adler, 2008; Skovronsky *et al.*, 2008; Sperling *et al.*, 2009). The tracer is in Phase III clinical trails (sponsored by Avid Radiopharmaceuticals).

11.14 Peripheral Benzodiazepine Binding Sites or TSPO-18kDa

The peripheral benzodiazepine receptor (PBR), now also known as the translocator protein 18kDa (TSPO) (Papadopoulos *et al.*, 2006), is a multimeric protein present primarily on mitochondrial membranes on many tissues assayed in the peripheral nervous system (PNS) (Scarf *et al.*, 2009). Benzodiazepine receptors are found both in the CNS as well as in peripheral tissues. Benzodiazepine receptors have historically been divided into two classes defined by their localisation in the body. The central benzodiazepine receptor (CBR) is localised only on the cell membrane of neurons, where it is part of the γ-aminobutyric acid (GABA) receptor complex, and is involved in the regulation of GABA-gated chloride ion currents (Anzini *et al.*, 1996). TSPO is found in peripheral organs such as the kidney and heart, and in steroid producing cells of the adrenals, testes and ovaries, as well as in glial cells, mast cells and macrophages. In the healthy brain, TSPO expression is negligible, but under neuroinflammatory conditions, TSPO density is markedly increased (Galiegue *et al.*, 2003). The predominant location of TSPO in the brain is on the outer membrane of mitochondria associated with microglial cells, and increases in TSPO are considered a marker for neuroinflammation via microglia activation (Pike *et al.*, 1993; Banati, 2002, 2003; Venneti *et al.*, 2006). The function of TSPO in normal and pathological conditions such as brain injury, neurodegenerative disorders and immune system diseases has been summarised in recent publications (Papadopoulos *et al.*, 2006; Chen and Guilarte, 2008).

As TSPO is implicated in a wide range of diseases, it has become an attractive target for molecular imaging. In the past two decades, [^{11}C]PK 11195 and its active enantiomer [^{11}C]-(*R*)-PK 11195 (Fig. 11.22) have been used for *in vivo* imaging of TSPOs (Cagnin *et al.*, 2001, 2006; Venneti *et al.*, 2006). However, several limitations in the use of this radioligand have been identified. These include high non-specific binding (Shah *et al.*, 1994), low brain penetration (Maeda *et al.*, 2004), high plasma protein binding (Lockhart *et al.*, 2003) and low sensitivity. As a consequence, in recent years more than 20 new TSPO PET agents have emerged (James *et al.*, 2006; Chauveau *et al.*, 2008) and several have been evaluated. A recent review by Chauveau *et al.* (2008) has covered

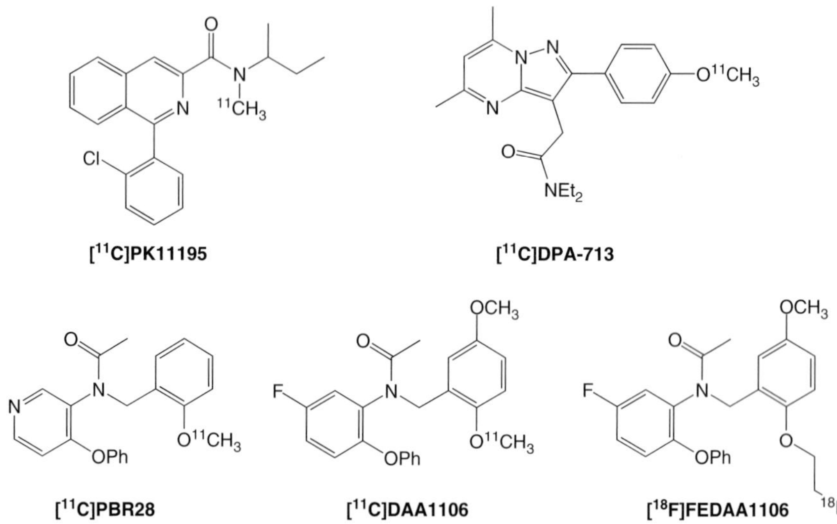

Figure 11.22. Examples of different pharmacophores developed as radiotracers for imaging TSPO (PBR) receptors using PET imaging.

the development of both PET and SPECT radioligands in depth; hence only the newer ^{18}F-radioligands will be considered here. Most of these have been developed from new structural pharmacophores (Fig. 11.22) and have been labelled with both ^{11}C (DAA1106, PBR28, DPA-713) and ^{18}F. These include aryloxyanilide-based ligands such as, [^{18}F]FEDAA1106 (Fig. 11.22), [^{18}F]PBR06 (Scheme 11.9), [^{18}F]FEPPA (Scheme 11.10) and pyrazolopyrimidine-based ligands, such as [^{18}F]DPA-714 (Scheme 11.11) and other ^{18}F-labelled analogues.

11.14.1 *Aryloxyanilide-based ligands*

Recently, the development of a large number of radioligands has been related to the high-affinity and selective TSPO ligand, DAA1106 (Okuyama *et al.*, 1999). Some of the derivatives which have been labelled with either ^{11}C or ^{18}F are shown in Fig. 11.23. DAA1106 itself has been labelled with ^{11}C and has been reported to be more sensitive and effective than [^{11}C]PK11195, (Zhang *et al.*, 2003a).

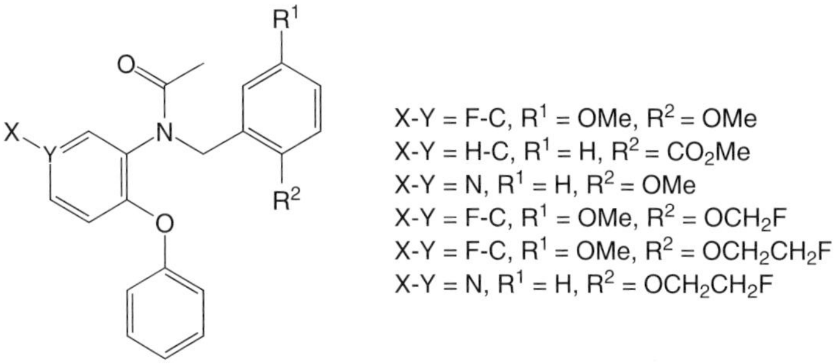

Figure 11.23. PET radioligands based on aryloxyanilide class.

11.14.2 *[^{18}F]FEDAA1106*

N-(5-Fluoro-2-phenoxyphenyl)-*N*-(2-[^{18}F]fluoroethoxy-5-methoxybenzyl) acetamide ([^{18}F]FEDAA1106), a fluoroethyl analogue of DAA1106, has been labelled by alkylation of *N*-(5-fluoro-2-phenoxyphenyl)-*N*-(2-hydroxy-5-methoxybenzyl)-acetamide with 2-[^{18}F]fluoroethyl bromide ([^{18}F]FCH_2CH_2Br) in the presence of NaH (Scheme 11.8). [^{18}F]FCH_2CH_2Br was prepared by reacting [^{18}F]fluoride with 2-bromoethyl triflate ($BrCH_2CH_2OTf$). [^{18}F]FEDAA1106 was obtained in >98% radiochemical purity. The specific activity was 120 GBq/μmol at EOS with 12 ± 4% radiochemical yield (Zhang *et al.*, 2003b). [^{18}F]FMDAA1106 has also been prepared via [^{18}F]fluoromethylation; however, *in vivo* defluorination occurred, which made it unlikely that this candidate would progress any further.

The same authors reported PET studies of [^{18}F]FEDAA1106 in rodent and primate brains (Zhang *et al.*, 2004). They demonstrated that [^{18}F]FEDAA1106 displayed high uptake in the occipital cortex, a region rich in TSPO. The radioactivity level of [^{18}F]FEDAA1106 in monkey brains was 1.5 times higher than that of [^{11}C]DAA1106 and six times higher than that of *R*-[^{11}C]PK11195.

The radiotracer has been evaluated to study TSPO in the human brain (Fujimura *et al.*, 2006). Initial studies showed that [^{18}F]FEDAA1106 displays slow clearance from the brain, thus requiring time to reach equilibrium and reach maximum specific binding. Additionally, it has been reported that quantitative analysis of human data is complex; therefore, PET studies with this radioligand are not simple to perform (Fujimura *et al.*, 2006).

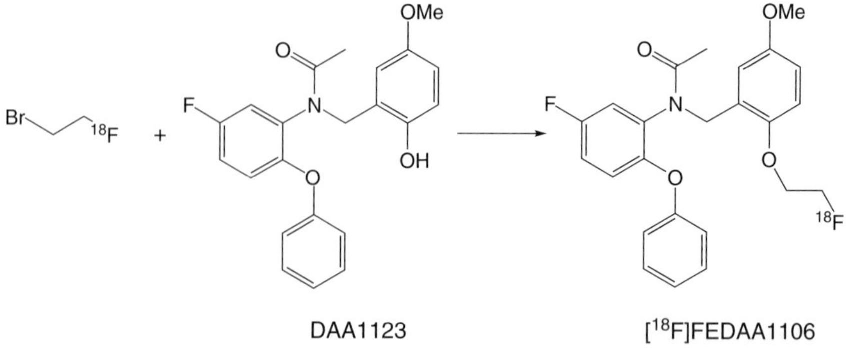

Scheme 11.8. The preparation of [^{18}F]FEDAA1106 using 2-[^{18}F]fluoroethyl bromide.

11.14.3 *[^{18}F]FEAC and [^{18}F]FEDAC*

More recently, the same group (Yanamoto *et al.*, 2009) has ^{18}F-labelled two other TSPO radioligands, [^{18}F]FEAC and [^{18}F]FEDAC, which are analogues of [^{11}C]AC-5216 (Fig. 11.24). In the rat brain homogenate assay, the affinities of these candidates is lower than it is for FEDAA1106 (Table 11.3).

[^{18}F]FEAC and [^{18}F]FEDAC were synthesised by fluoroethylation of dihydro-purinone precursors with [^{18}F]FCH_2CH_2Br in reasonable yields. Preliminary studies in a kainic acid-lesioned rat model with excess density of TSPO showed that the two radioligands had a higher uptake of radioactivity in the lesioned striatum versus the non-lesioned striatum. Additionally, both radioligands cleared faster than [^{18}F]FEDAA1106 from the brain and it is reported that further investigations in the primate brain are currently underway.

11.14.4 *PBR06*

Many *N*-acylaryloxyanilide ligands related to DAA1106 have high affinity to TSPO (Okubo *et al.*, 2004). *N*-Fluoroacetyl-*N*-(2,5-dimethoxybenzyl)-2-phenoxyaniline (PBR06) has affinity which is close to DAA1106 as shown in Table 11.4.

[^{18}F]*N*-fluoroacetyl-*N*-(2,5-dimethoxybenzyl)-2-phenoxyaniline ([^{18}F]PBR06) has been labelled easily through a single step. The reaction

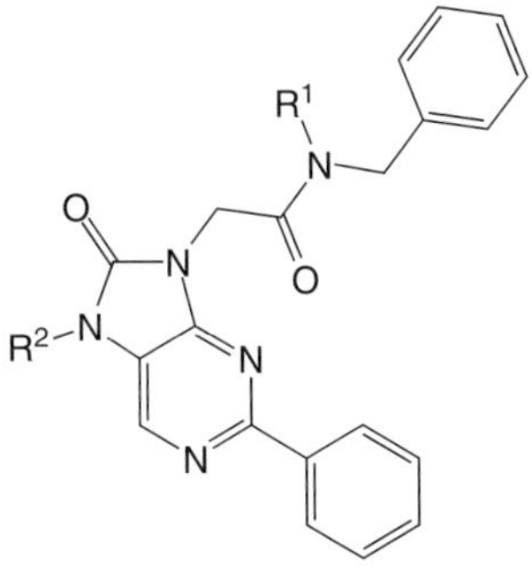

Ligand	R^1	R^2
AC-5216	Et	Me
FEAC	Et	CH_2CH_2F
FEDAC	Me	CH_2CH_2F
FAC	CH_2CH_2F	Me

Figure 11.24. Examples of fluorinated derivatives of AC-5216.

Table 11.3. *In vitro* binding affinity and lipophilicity of TSPO (PBR) ligands (adapted from Yanamoto *et al.*, 2009).

		K_i (nM)		Lipophilicity	
Ligand		**TSPO (PBR)**	**CBR**	**Log*D***	**cLog*D***
FEAC	(Fig. 11.24)	0.49 ± 0.05	>8400	3.6	3.6
FEDAC	(Fig. 11.24)	1.34 ± 0.15	8400	3.2	3.3
FAC	(Fig. 11.24)	0.51 ± 0.06	>8400	3.1	3.3
AC-5216	(Fig. 11.24)	0.20 ± 0.02	>8400	3.3	3.5
PK11195	(Fig. 11.24)	0.31 ± 0.03	>8400	3.7	5.1
FEDAA1106	(Scheme 11.8)	0.08 ± 0.01	>1000	3.8	4.3

Table 11.4. Affinity values of ligands for TSPO in rat, monkey and human brain homogenates (taken from Briard *et al.*, 2009).

	TSPO (PBR) affinity (K_i, nM mean ± SD, $n = 6$)		
Ligand	**Rat**	**Monkey**	**Human**
DAA1106	0.0726 ± 0.0036	0.23 ± 0.011	0.242 ± 0.016
PBR06	0.18 ± 0.007	0.318 ± 0.018	0.997 ± 0.070

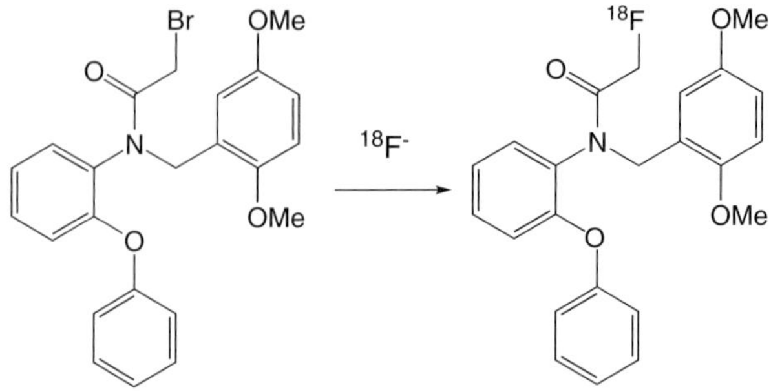

Scheme 11.9. Radiolabelling of [^{18}F]PBR06 (Briard *et al.*, 2009).

involved ^{18}F nucleophilic substitution of the corresponding *N*-bromoacetyl precursor as shown in Scheme 11.9. (Briard *et al.*, 2009). The radioligand was isolated using HPLC and in high radiochemical purity (99%). The chemical purity was reported to be 88% ± 15% and the specific activity at the time of injection was 185 ± 64 GBq/μmol.

[^{18}F]PBR06 has been evaluated in rats, monkeys and humans. PET studies have shown that [^{18}F]PBR06 was avidly taken into monkey brains and gave a high ratio of TSPO-specific to non-specific binding. Radiometabolite analysis of rat plasma, urine and brain as well as monkey plasma showed that [^{18}F]PBR06 was devoid of defluorination and predominantly polar radiometabolite(s) were observed (Briard *et al.*, 2009). When human PET studies in healthy volunteers were carried out with this radioligand (Fujimura, 2009), preliminary analysis indicated that brain activity was likely to be contaminated with radiometabolites; however, the percentage of contamination was thought to be small (<10%). Hence, [^{18}F]PBR06 is considered to be the most promising ^{18}F-radioligand for measurement of TSPOs and as a biomarker of inflammation in the brain.

11.14.5 *[^{18}F]FEPPA*

In a different study, another aryloxyanilide ligand for TSPO, namely [^{18}F]FEPPA, was labelled with ^{18}F (Wilson *et al.*, 2008) (Scheme 11.10). FEPPA has a high affinity (K_i = 0.07 nM) for TSPO in rat mitochondrial membrane preparations. It is a fluoroethyl analogue of PBR28, which itself is an analogue of FEDAA1106, with a pyridine ring replacing one of the benzene rings of FEDAA1106. [^{18}F]FEPPA has been labelled in a single step from its tosylate precursor (Scheme 11.10) in high radiochemical yield (50–60% uncorrected for decay) and at high specific activity (44–100 GBq/μmol at EOS). Initial studies with [^{18}F]FEPPA in rats showed moderate brain uptake and slow washout. The highest uptake of radioactivity was seen in the hypothalamus and olfactory bulb, regions previously reported to be enriched in TSPO in rat brains. Analysis of the rat plasma and brain extracts demonstrated that [^{18}F]FEPPA was rapidly metabolised, although only low levels of radioactive metabolites were observed in the brain. It is unclear whether this particular radioligand will be developed further.

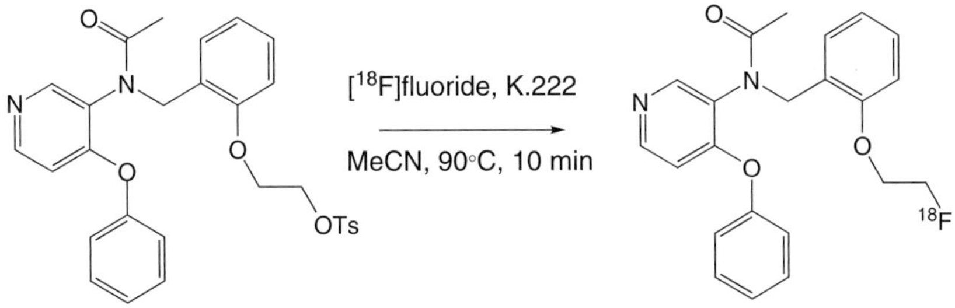

Scheme 11.10. Preparation of [^{18}F]FEPPA from tosylate precursor (Wilson *et al.*, 2008).

11.14.6 *Pyrazolopyrimidine ligands*

In 2001, a series of TSPO-specific pyrazolopyrimidine ligands were reported (Selleri *et al.*, 2001). The first ligand to be labelled with ^{11}C was *N*,*N*-diethyl-2-(2-(4-methoxyphenyl)-5,7-dimethylpyrazolo-[1,5-a]pyrimidin-3-yl)acetamide (DPA-713). PET studies showed that [^{11}C]DPA-713 selectively and specifically bound the TSPO in normal baboon brains (James *et al.*, 2005). These encouraging results prompted the development of fluorinated analogue DPA-714 (James *et al.*, 2008) where the 4-methoxyphenyl substituent of DPA-713 was replaced by a fluoroethoxyphenyl group. The affinities and log*D* values of these two compounds and PK11195 are given in Table 11.5.

Radiolabelling of DPA-714 was achieved in a single step by [^{18}F]fluoride displacement of the tosylate precursor (Scheme 11.11). The highest

Table 11.5. Affinities of ligands for TSPO and central benzodiazepine receptors (CBR) (from James *et al.*, 2008).

Ligand		K_i (nM, mean ±SD) for TSPO (PBR)	K_i (nM) for CBR	Log*D* (HPLC determination)
DPA-714	(Scheme 11.11)	7.0 ± 0.4	>10,000	2.44
DPA-713	(Fig. 11.22)	4.7 ± 0.2	>10,000	2.44
PK11195	(Fig. 11.22)	9.3 ± 0.5	>10,000	3.35

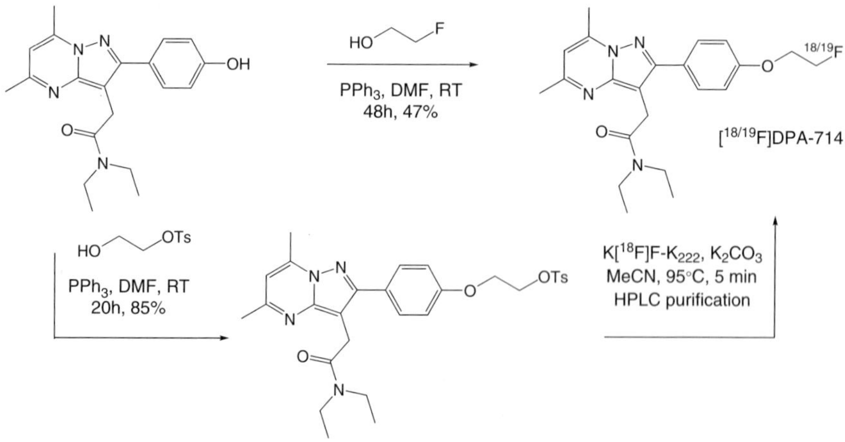

Scheme 11.11. Preparation of [^{18}F]DPA-714 and reference standard.

radiochemical yield of 16% (non-decay-corrected) was obtained using >6 mg of precursor, and a specific activity of 270 GBq/µmol was achieved (James *et al.*, 2008).

In a unilateral quinolinic acid lesion model rat, an eightfold higher level of uptake of [^{18}F]DPA-714 was observed in the ipsilateral striatum compared to the contralateral striatum. Also, the uptake was shown to be selective in competition studies with known TSPO agents. Additionally, PET studies in baboons have demonstrated rapid penetration and good retention of [^{18}F]DPA-714 in the brain, thus warranting further investigations.

11.15 Serotonin 5-HT$_{1A}$ Antagonists and Agonists

Among the multiple serotonin receptors identified, the 5-hydroxytryptamine 1A (5-HT$_{1A}$) receptor subtype is among the best characterised as selective radioligands have been available (Lanfumey and Hamon, 2000; Passchier *et al.*, 2001; Kumar and Mann, 2007). 5-HT$_{1A}$ receptors belong to the family of G-protein-coupled receptors (GPCRs) and contribute to serotonin transmission in the brain. Primarily, 5-HT$_{1A}$ receptors are localised in the limbic areas (hippocampus, septum) and in the midbrain (raphe nuclei), where they correspond to postsynaptic receptors and presynaptic autoreceptors, respectively (Burnet *et al.*, 1997; Kumar and Mann, 2007). 5-HT$_{1A}$ receptors have been implicated in the pathophysiology of major neuropsychiatric disorders, including depression, suicidal behaviour, panic disorder, epilepsy, bulimia, schizophrenia, Parkinson's disease and Alzheimer's disease, and consequently have become an important target for drug therapy (Glennon, 1990; Hensler, 2003).

It is known that G-protein-coupled receptors exist in G-protein-coupled and -uncoupled forms that exhibit high and low affinity for agonist ligands, respectively (Assié *et al.*, 1999). Antagonist ligands, on the other hand, bind to the high-affinity and low-affinity conformations of 5-HT$_{1A}$ receptors with comparable affinity. Measurement of affinity differences of a compound for the high- versus the low-affinity state of a receptor have been used to estimate its intrinsic activity (i.e. functional efficacy) at that receptor (Watson *et al.*, 2000). To understand the pharmacology and distribution of 5-HT$_{1A}$ receptors in healthy subjects and CNS diseases, tremendous effort has gone into the development of 5-HT$_{1A}$ radioligands. Many comprehensive reviews on radioligand development and the preclinical and clinical evaluation have been published already (Cliff, 2000; Pike *et al.*, 2000, 2001; Passchier and van Waarde, 2001; Kumar and Mann, 2007) and hence will only be briefly mentioned in this chapter.

The first successful 5-HT$_{1A}$ receptor antagonist radioligand to be developed was *N*-[2-[4-(2-methoxyphenyl)-1-piperazinyl]ethyl]-*N*-(2-pyridinyl) cyclohexane carboxamide (WAY100635). This compound was labelled at two different positions with ^{11}C, and [carbonyl-^{11}C]WAY100635 (Fig. 11.25)

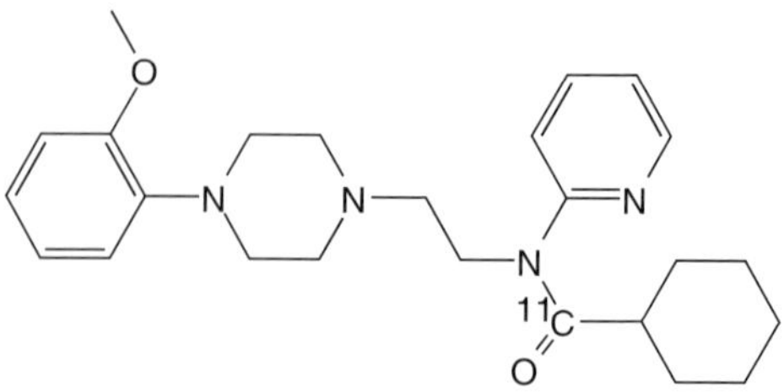

Figure 11.25. Structure of [*carbonyl*-^{11}C]WAY100635.

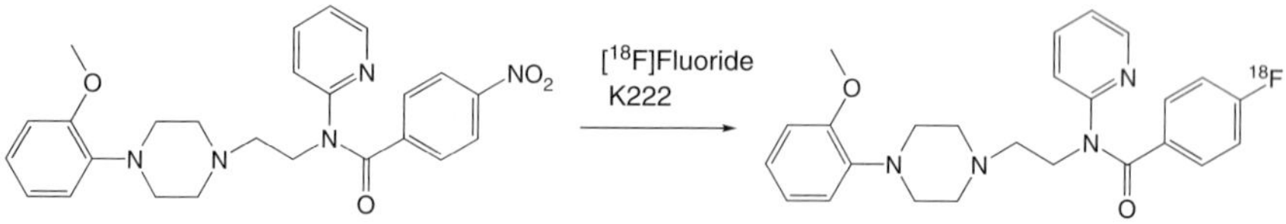

Scheme 11.12. Radiosynthesis of *p*-[^{18}F]MPPF.

has become the most commonly used ligand for patient studies (Pike *et al.*, 2000, 2001; Kumar *et al.*, 2007). A large number of [^{18}F]fluoro derivatives of WAY100635 followed, and *p*-[^{18}F]MPPF (4-(2′-methoxyphenyl)-1-[2′-(*N*-2′-pyridinyl)-*p*-[^{18}F]fluorobenzamido]ethylpiperazine) has emerged as one of the most promising ^{18}F-candidates (Plenevaux *et al.*, 2000; Kumar *et al.*, 2007).

The original radiosynthesis of *p*-[^{18}F]MPPF (Scheme 11.12) from the corresponding nitro precursor was described by Shiue *et al.* (1997) and an improved method which uses microwave heating during the labelling step and SPE for the final formulation was described by Le Bars *et al.* (1998). The latter method provides *p*-[^{18}F]MPPF in 25% radiochemical yield at EOS with specific radioactivity between 37 and 185 GBq/μmol at EOS.

Studies in volunteers have shown that regional uptake of *p*-[^{18}F]MPPF agrees well with known 5-HT$_{1A}$ receptor distribution (Ginovart *et al.*, 2000; Passchier *et al.*, 2000). Although, *p*-[^{18}F]MPPF exhibits a similar biodistribution pattern in human brain regions to that of [carbonyl-^{11}C]WAY100635 (Shiue *et al.*, 1997; Kumar *et al.*, 2007), its brain penetration is lower compared to [carbonyl-^{11}C]WAY100635 (Cliff, 2000;

Pike *et al.*, 2001). Even with this limitation, *p*-[^{18}F]MPPF continues to be a very useful radioligand clinically.

Effort has also gone into the development of agonist radioligands. Many of these were derivatives of the 5-HT_{1A} receptor agonist, 8-hydroxy-2-(di-*n*-propylamino)tetralin (8-OHDPAT) which achieved limited success when evaluated *in vivo* (Halldin *et al.*, 1994; Mathis *et al.*, 1997; Suehiro *et al.*, 1998). A successful agonist radioligand was identified when [^{11}C]CUMI-101 (or [^{11}C]MMP or [*O-methyl*-^{11}C]2-(4-(4-(2-methoxyphenyl)piperazin-1-yl)butyl)-4-methyl-1,2,4-triazine-3,5(2H,4H)dione) was labelled with ^{11}C (Fig. 11.26) (Prabhakaran *et al.*, 2006) and evaluated in non-human primates (Kumar *et al.*, 2007). In the bovine hippocampal membrane-binding affinity assay, CUMI-101 exhibited high affinity ($K_i = 0.15$ nM) for 5-HT_{1A} receptors and has 1:45, 1:86 and 1:145 ratios to the affinities for α1, 5-HT_7 and D_4 receptors, respectively, with no significant affinity for other studied brain receptors, enzymes, transporters or biogenic amines. PET studies in baboons have shown that [^{11}C]CUMI-101 binds to 5-HT_{1A} receptor-enriched brain regions and the signal is specific as confirmed by blockade with WAY-100635 and (±)-8-OH-DPAT (Kumar *et al.*, 2007). Preliminary human studies show that [^{11}C]CUMI-101 penetrates the BBB and is retained in 5-HT_{1A} receptor rich areas, whereas the cerebellum had the least amount of radiotracer uptake; therefore [^{11}C]CUMI-101 is a promising 5HT_{1A} agonist radioligand (Kumar *et al.*, 2008; Milak *et al.*, 2008). It is very likely that ^{18}F-fluoro derivatives of this family of structure will emerge in the future.

Another candidate showing some promise is the 5HT_{1A} receptor agonist, 4-[^{18}F]fluoro-*N*-(2-(4-(2,3-dihydrobenzo[b][1,4]dioxin-8-yl)piperazin-1-yl)-ethyl)benzamide ([^{18}F]SL702). SL702 has a high affinity for 5HT_{1A}

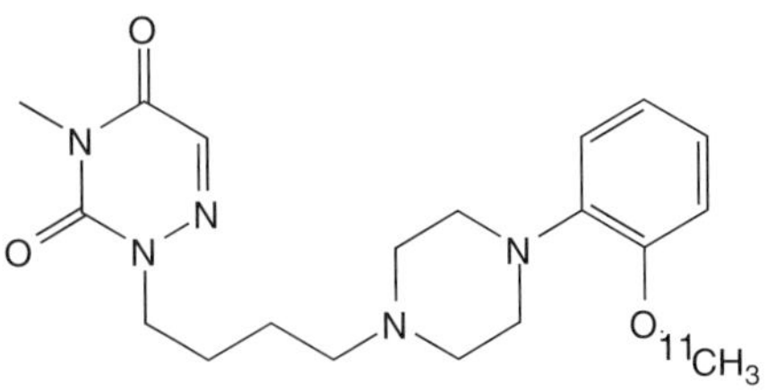

Figure 11.26. Structure of [^{11}C]CUMI-101.

receptors (K_i = 0.1 nM) and reasonable selectivity (Lu *et al.*, 2007, 2008). ^{18}F-Labelling of [^{18}F]SL702 was achieved by using a nitro precursor in DMF heated at >120°C with dried [^{18}F]fluoride and Kryptofix 222 under microwave irradiation (Scheme 11.13), giving a 12–62% radiochemical yield (decay-corrected) (Lu *et al.*, 2007). Preliminary evaluation in mice and monkeys displayed adequate brain uptake and some 5-HT$_{1A}$ receptor-specific binding. This candidate is being further investigated as a potential radioligand for G-protein-coupled 5HT$_{1A}$ receptors.

More recently, another 5HT$_{1A}$ agonist radioligand, (3-chloro-4-fluorobenzoyl-(4-fluoro-4{[(5-methyl-pyrimidin-2-ylmethyl)-amino] -methyl}-piperidin) (F15599) was labelled with ^{18}F and evaluated in rats and cats (Lemoine *et al.*, 2009). Labelling of [^{18}F]F15599 was achieved from its nitro precursor at 150°C (Scheme 11.14). The product has been obtained with a radiochemical purity of more than 98% in 30% radiochemical yield (corrected for decay) with a specific activity of between 85 and 120 GBq/μmol (corrected to EOS).

Preliminary data with [^{18}F]F15599 indicate that the *in vitro* binding (K_i = 2.24 nM) of this radioligand was consistent with known 5HT$_{1A}$ receptor distribution (hippocampus, dorsal raphe nucleus and cortical areas). However, an *in vivo* study in cat brain displayed a different pattern, with the highest binding occurring in the dorsal raphe and cingulated

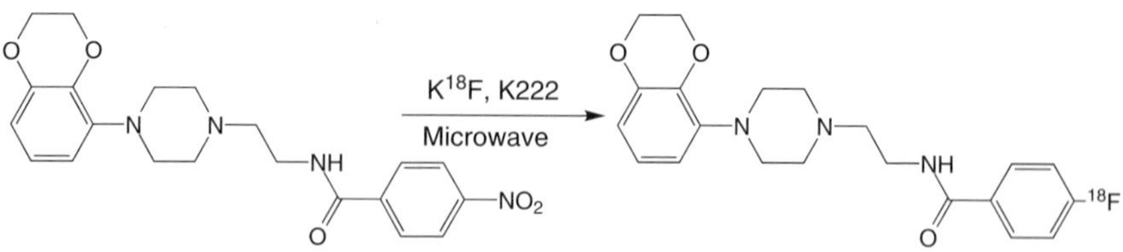

Scheme 11.13. Radiosynthesis of [^{18}F]SL702.

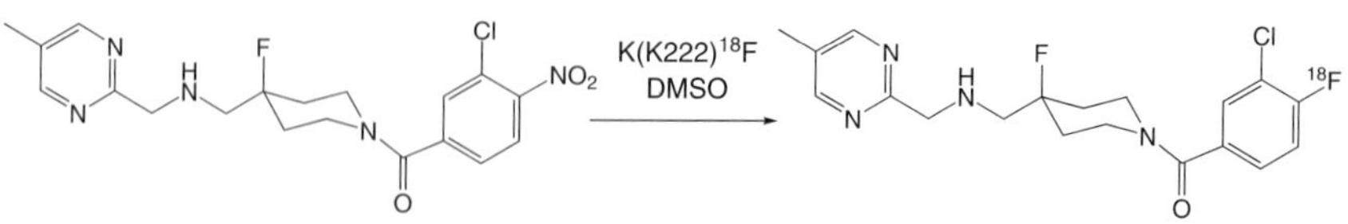

Scheme 11.14. Radiosynthesis of [^{18}F]F15599.

cortex with little in other cortical regions and none in the hippocampus. It has been hypothesised by the authors that [^{18}F]F15599 may be demonstrating *in vivo* binding which can select 5-HT$_{1A}$ receptors coupled to certain G-protein subtypes, thus opening the way to region-specific imaging of receptors. Already, the evaluation of a close analogue of F15599 with a higher affinity is underway.

11.16 Imaging the Cannabinoid 1 Receptor (CB$_1$)

Cannabinoid 1 receptors (CB$_1$) are part of the super-family of G-protein-coupled receptors and are widely expressed throughout the mammalian nervous system, especially within the brain. CB$_1$ receptors are also expressed in peripheral tissues such as the pituitary gland, immune cells, reproductive tissues, gastrointestinal tissues, heart, lung, urinary bladder and adrenal glands. By comparison, the cannabinoid 2 receptors (CB$_2$) are mainly expressed in the immune cells (B-cells and natural killer cells). CB$_1$ receptors are primarily expressed in regions of the brain associated with motor control, information processing, spatial awareness, memory and reward/addiction; CB$_1$ receptors have been identified as important therapeutic and diagnostic imaging targets. Cannabinoid receptors are involved in the modulation of neuronal chemical messengers such as acetylcholine, noradrenaline, dopamine, serotonin, γ-aminobutyric acid, glutamate and aspartate (Pacher *et al.*, 2006). Evidence of altered regulation of the cannabinoid system is found in a variety of disease states, and *in vivo* molecular imaging of the endocannabinoid system could provide tools for the prediction and monitoring of diseases such as Parkinson's disease, Huntington's disease, epilepsy, schizophrenia and mood/anxiety disorders as well as eating disorders (Van Laere, 2007).

In general, many phyto-, endo- and synthetic-cannabinoid receptor ligands are required to be lipophilic as the receptor binding site is located within the lipid bilayer of the cell membrane. Historically, highly lipophilic PET tracers often fail to provide adequate target to non-target signals due to increased non-specific binding, high levels of plasma protein binding and consequently difficulty in crossing the BBB (Cunningham *et al.*, 2005). One of the first cannabinoid ligands to be labelled with ^{18}F was (−)-5′-^{18}F-Δ^8-THC. Despite demonstrating uptake in CB_1-rich brain regions (basal ganglia, thalamus and cerebellum) in baboon brains, it also demonstrated relatively rapid clearance from these regions and rapid *in vivo* metabolism (Charalambous *et al.*, 1991).

Much of the development of new radioligands for *in vivo* imaging of CB_1 receptors has been based on the 1,5-diaryl-3-carboxypyrazole structure of SR141716A, known as rimonabant (Fig. 11.27) (Rinaldi-Carmona *et al.*, 1994). Rimonabant is a CB_1-selective antagonist/inverse-agonist that was launched in 2006 by Sanofi–Aventis for the treatment of obesity, but has since been withdrawn from the market. Rimonabant has a nanomolar affinity for the human CB_1 of ~6 nM and little or no affinity for human CB_2 receptor ($K_B > 10\,000$ nM). Structure–activity relationship studies and evaluation of SR141716A analogues have focused on reducing the lipophilicity of this class of compounds (Katoch-Rouse *et al.*, 2003), whilst bioisosteric replacement of the central 1,5-diaryl-3-carboxypyrazole core of rimonabant is the most common method reported in the development for novel CB_1 ligands (see, for example, Willis *et al.*, 2005; Boström *et al.*, 2007).

One analogue of rimonabant, AM281, labelled with ^{123}I and ^{124}I was, until 2006, the only available, viable radioligand for CB_1 tomographic imaging in humans (Lan *et al.*, 1999). Transaxial SPECT images using

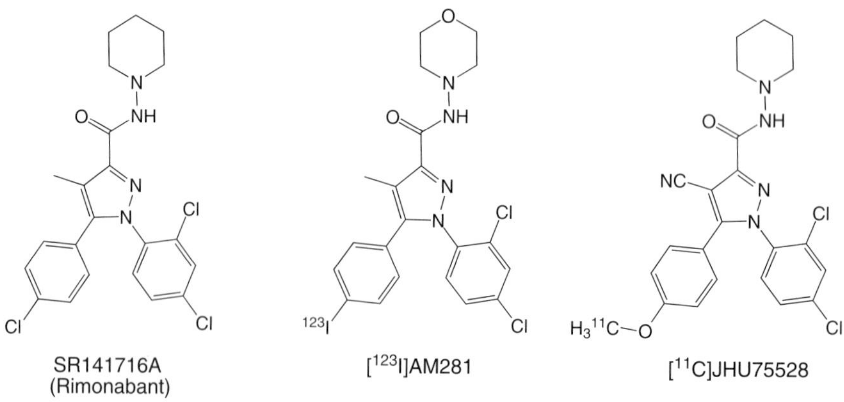

Figure 11.27. Structures of SR141716A (Rimonabant), [^{123}I]AM281 and [^{11}C]JHU75528.

[^{123}I]AM281 demonstrated uptake of the tracer in the cerebellum and cortical areas of baboon brains, although with insufficient resolution to visualise sub-cortical regions of highest CB_1 density (substantia nigra and globus pallidus). Since publication of its synthesis, SPECT and PET radioiodinated AM281 have been used in a number of clinical studies of schizophrenic and Tourette's syndrome patients (Berding, 2004, 2006). The first example of a PET tracer suitable for *in vivo* CB_1 imaging labelled with a short-lived isotope was the ^{11}C-labelled [^{11}C]JHU75528 (Fig. 11.27) (Fan *et al.*, 2006). JHU75528, another rimonabant analogue, demonstrated specific and reversible uptake in the striatum, hippocampus, cortex and cerebellum of baboon brains (Horti *et al.*, 2006).

At the present time, one of the most promising and most well advanced radiopharmaceutical candidates for PET imaging of CB_1 in human subjects is [^{18}F]MK9470 (Fig. 11.28) (Liu *et al.*, 2007). Developed by Merck Research Laboratories, [^{18}F]MK9470 is a selective, sub-nanomolar affinity CB_1 ligand (IC_{50} 0.7 nM, log$P$ 4.0) analogue of Taranabant (MKO 364). Autoradiograpic studies in rhesus monkey brains showed that [^{18}F]MK9470 binding was consistent with the expected regional distribution of CB_1 receptors (cerebral cortex, cerebellum, caudate/putamen, globus pallidus, substantia nigra and hippocampus). *In vivo* PET imaging of rhesus monkeys revealed high brain uptake and specific regional binding consistent with the autoradiographic studies with a total to non-specific binding ratio of 4–5:1 (Burns *et al.*, 2007).

The efficacy of [^{18}F]MK9470 to image human cannabinoid receptors was demonstrated in healthy human volunteers (Burns *et al.*, 2007; Van Laere *et al.*, 2008a,b). In human subjects [^{18}F]MK9470 exhibited relatively slow brain kinetics, reaching a plateau at ~120 min post injection, after

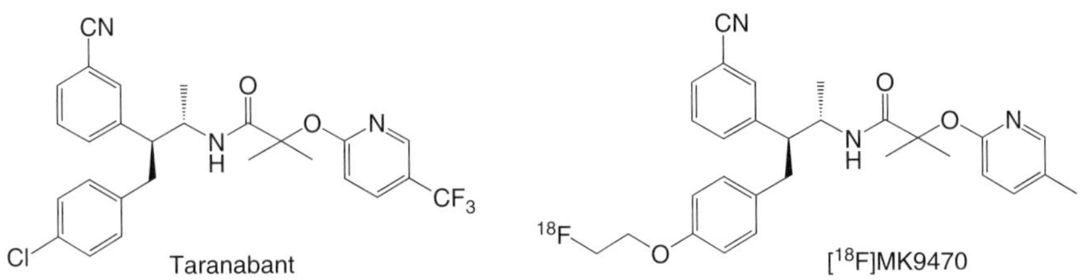

Figure 11.28. Structures of Taranabant and [^{18}F]MK9470.

which time levels remained constant. Tracer uptake was observed in all grey matter regions, with the maximum uptake observed in the striatum, frontal cortex and posterior cingulate. Intermediate uptake was observed for the cerebellum and the lowest uptake was observed in the thalamus and hippocampus. The maximum grey to white matter ratio (specific to non-specific) was about 3–4:1. Further studies in human subjects have determined the whole body biodistribution and radiation dosimetry associated with [^{18}F]MK9470 administration (Van Laere *et al.*, 2008a) as well as gender-dependent increases with healthy ageing of the human cerebral CB_1 receptor binding using [^{18}F]MK9470 PET (Van Laere *et al.*, 2008b).

In 2008, the first report of another structurally novel molecular imaging probe for the human CB_1 receptors based on a pyrrolidin-2-one core structure was reported (Yasuno *et al.*, 2008). [^{11}C]MePPEP, developed by researchers at the Molecular Imaging Branch, NIMH, in collaboration with Lilly Research Laboratories, is currently under evaluation as a molecular imaging probe in clinical trials in the USA. Shortly after disclosure of the ^{11}C-labelled derivative, ^{18}F derivatives [^{18}F]FMPEP and [^{18}F]FMPEP-d_2, prepared by [^{18}F]fluoroalkylation of a phenolic precursor, were reported (Fig. 11.29) (Donohue *et al.*, 2008a). These ^{18}F derivatives also exhibit sub-nanomolar binding affinity at human CB_1 receptors. The evaluation of the deuterated derivative

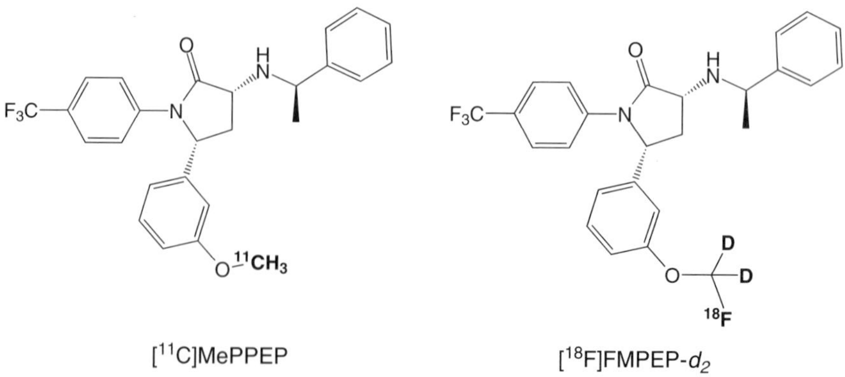

Figure 11.29. Structures of [^{11}C]MePPEP and [^{18}F]FMPEP-d_2.

[^{18}F]FMPEP-d_2 is also currently underway in NIMH-sponsored clinical trials in the USA.

Another recently reported and structurally distinct class of selective CB_1 ligands are arylsulfonyl-substituted indoles. Originally identified as potential therapeutics for the modulation of the endocannabinoid system (Allen *et al.*, 2005), the first molecular imaging agents based on this class, [^{11}C]PipISB and [^{18}F]PipISB (Fig. 11.30), were reported in 2008 (Donohue *et al.*, 2008b). Preliminary screening of both ^{11}C and ^{18}F PipISB showed uptake in CB_1 receptor-rich regions in rhesus and cynomolgus monkeys with uptake kinetics similar to those observed for [^{18}F]MK9470.

Whilst the endocannabinoid system represents a potentially attractive therapeutic target for the treatment of several psychiatric and neurodegenerative disorders, the role of the CB_1 receptors in these disorders is still poorly understood. The development of PET radioligands for imaging CB_1 receptors provides an opportunity to further understand the participation of the cannabinoid system in disease as well as providing tools for prediction and disease monitoring.

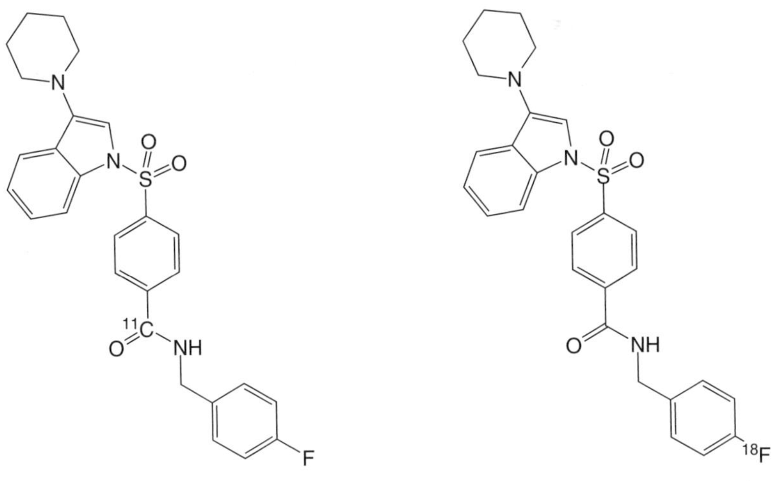

[^{11}C]PipISB [^{18}F]PipISB

Figure 11.30. Structures of [^{11}C]PipISB and [^{18}F]PipISB (Donohue *et al.*, 2008b).

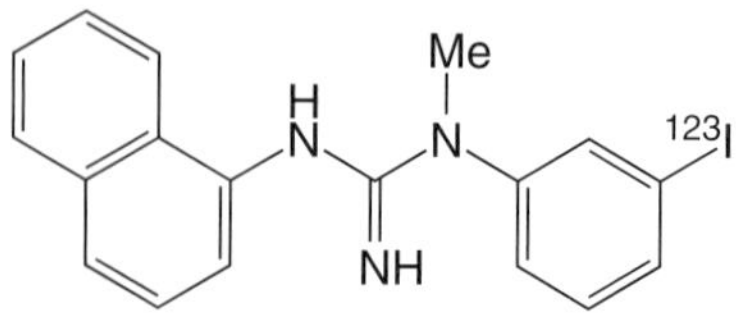

[^{123}I]CNS 1261

Figure 11.31. Structure of [^{123}I]CNS 1261 (Erlandsson *et al.*, 2003).

11.17 Ion Channels

The functional imaging of ion channels represents an area of research in molecular imaging that has yet to be effectively achieved (see Waterhouse, 2003). For example, *N*-methyl-D-aspartate receptor-mediated hyperexcitation is linked to a wide range of disease processes. To date, a number of potential radioligands have been examined. Perhaps the most promising human studies so far have been with the SPECT tracer [^{123}I]CNS 1261 (Fig. 11.31) (Erlandsson *et al.*, 2003; Stone *et al.*, 2006). [^{123}I]CNS 1261 was prepared by radio-iododestannylation of *N*-(1-naphthyl)-*N*′-(3-tributylstannylphenyl)-*N*′-methylguanidine hydrochloride, using a modification of the procedure originally reported by Owens *et al.* (2000) for the ^{125}I-labelled tracer. Whilst ^{18}F-labelled tracers have been reported, none have yet been successfully applied to functional brain imaging and there is still a need to develop a suitable ^{18}F-labelled tracer. In an attempt to address this gap, the development of a new ^{18}F-labelled diarylguanidine NMDA ion-channel tracer has been recently reported (Robins *et al.*, 2010). Following initial encouraging *in vitro* characterisation of the tracer, the ^{18}F-labelled ligand is currently being evaluated in humans under an FiM PET-microdosing regime.

11.18 Summary

Over the last decade, PET tracer research has moved from ^{11}C to ^{18}F and PET imaging has grown from a research tool to a clinically relevant diagnostic modality. To a large degree, this has been driven by the successful application of [^{18}F]FDG PET and approval for the financial reimbursement of ^{18}F PET scans for almost all oncology indications. The popularity of PET imaging has also been aided by the development of commercial supplies of [^{18}F]FDG to nuclear medicine departments in hospitals without access to radiochemistry facilities. In addition, the fusion of complementary technologies such as PET-CT and PET-MR, which provide high-resolution anatomical images co-registered with the functional PET images, has led to improved clinical decision making and patient management.

The increased demand for PET imaging has relied on the development of new tracers for specific biological targets. Over the past decade, many new tracers labelled with ^{11}C and ^{18}F have been developed and investigated *in vivo* in normal and patient subjects to aid understanding of disease-based mechanisms, disease progression and management. In an era where molecular profiling is identifying specific and mechanistically important alterations in diseased cells, a logical progression of PET tracer development is to move from the assessment of basic biochemical processes to more specific targeting of neuronal and cellular systems. This development is moving the field towards targeted molecular imaging and the fundamental understanding of an individual's predisposition to a particular disease and therapy selection.

Acknowledgements

The authors wish to thank Dr Matthias Glaser, Dr Helen Betts, Dr Rajiv Bhalla and Dr Ian Wilson for their assistance in the preparation of this manuscript.

References

Adler, L., Wolodzko, J., Stabin, M. *et al.* (2008) Radiation dosimetry of ^{18}F-AV-45 measured by PET/CT in humans. *J Nucl Med* 49 Suppl 1, 283P.

Agdeppa, E.D., Kepe, V., Liu, J. *et al.* (2001) Binding characteristics of radiofluorinated 6-dialkylamino-2-naphthylethylidene derivatives as positron emission tomography imaging probes for β-amyloid plaques in Alzheimer's disease. *J Neurosci* 21, RC189.

Agdeppa, E.D., Kepe, V., Liu, J. *et al.* (2003) 2-Dialkylamino-6-acylmalononitrile substituted naphthalenes (DDNP analogs): novel diagnostic and therapeutic tools in Alzheimer's disease. *Mol Imag Biol* 5, 404–417.

Allen, J.R., Amegadzie, A.K., Gardinier, K.M. *et al.* (2005) CB1 modulator compounds. European patent 1699761.

Aloya, R., Shirvan, A., Grimberg, H. *et al.* (2006) Molecular imaging of cell death *in vivo* by a novel small molecule probe. *Apoptosis* 11, 2089–2101.

Ametamey, S.M. & Honer, M. (2007). 'Pharmacological prerequisites for PET ligands and practical issues in preclinical PET research', in Schubiger, P.A., Lehmann, L. & Friebe, M. (Eds), *PET Chemistry: The Driving Force in Molecular Imaging*, Springer, Berlin, pp. 317–327.

Ametamey, S.M., Honer, M. & Schubiger, P.A. (2008) Molecular imaging with PET. *Chem Rev* 108, 1501–1516.

Anzini, M., Cappelli, A., Vomero, S. *et al.* (1996) Molecular basis of peripheral vs central benzodiazepine receptor selectivity in a new class of peripheral benzodiazepine receptor ligands related to alpidem. *J Med Chem* 39, 4275–4284.

Ashkenazi, A. (2002) Targeting death and decoy receptors of the tumour-necrosis factor superfamily. *Nat Rev Cancer* 2, 420–430.

Assié, M.-B., Cosi, C. & Koek, W. (1999) Correlation between low/high affinity ratios for 5-HT_{1a} receptors and intrinsic activity. *Eur J Pharmacol* 386, 97–103.

Bading, J.R. & Shields, A.F. (2008) Imaging of cell proliferation: status and prospects. *J Nucl Med* 49, S64–S80.

Banati, R.B. (2002) Visualising microglial activation *in vivo*. *Glia* 40, 206–217.

Banati, R.B. (2003) Neuropathological imaging: *in vivo* detection of glial activation as a measure of disease and adaptive change in the brain. *Br Med Bull* 65, 121–131.

Bansal, A., Shuyan, W., Hara, T. *et al.* (2008) Biodisposition and metabolism of [^{18}F]Fluorocholine in 9L glioma cells and 9L glioma-bearing Fisher rats. *Eur J Nucl Med Mol Imag* 35, 1192–1203.

Barrio, J.R., Huang, S.-C., Cole, G. *et al.* (1999) PET imaging of tangles and plaques in Alzheimer's disease with a highly hydrophilic probe. *J Labelled Compd Rad* 42, S194–S195.

Barrio, J.R., Petric, A., Satyamurthy, N. *et al.* (2001) Methods for labeling β-amyloid plaques and neurofibrillary tangles. US patent no 6274119 B1.

Barrio, J.R., Satyamurthy, N., Huang, S.C. *et al.* (2009) Dissecting molecular mechanisms in the living brain of dementia patients. *Acc Chem Res* 42, 842–850.

Beer, A.J., Haubner, R., Goebel, M. *et al.* (2005) Biodistribution and pharmacokinetics of the αvβ3-selective tracer ^{18}F-galacto-RGD in cancer patients. *J Nucl Med* 46, 1333–1341.

Beer, A.J., Grosu, A.-L., Carlsen, J. *et al.* (2007) [^{18}F]Galacto-RGD positron emission tomography for imaging of αvβ3 expression on the neovasculature in patients with squamous cell carcinoma of the head and neck. *Clin Cancer Res* 13, 6610–6616.

Beer, A.J. & Schwaiger, M. (2008) Imaging of integrin αvβ3 expression. *Cancer Metast Rev* 27, 631–644.

Beer, A.J., Lorenzen, S., Metz, S. *et al.* (2008) Comparison of integrin αvβ3 expression and glucose metabolism in primary and metastatic lesions in cancer patients: a PET study using ^{18}F-galacto-RGD and ^{18}F-FDG. *J Nucl Med* 49, 22–29.

Berding, G., Müller-Vahl, K., Schneider, U. *et al.* (2004) [^{123}I]AM281 single-photon emission computed tomography imaging of central cannabinoid CB1 receptors before and after δ^{9}-tetrahydrocannabinol therapy and whole-body scanning for assessment of radiation dose in tourette patients. *Biol Psychiat* 55, 904–915.

Berding, G., Müller-Vahl, K., Schneider, U. *et al.* (2006) Feasibility of central cannabinoid CB(1) receptor imaging with [^{124}I]AM281 PET demonstrated in a schizophrenic patient. *Psychiat Res* 147, 249–256.

Bergström, M., Lu, L., Fasth, K.-J. *et al.* (1998) *In vitro* and animal validation of bromine-76-bromodeoxyuridine as a proliferation marker. *J Nucl Med* 39, 1273–1279.

Bergström, M., Grahnén, A. & Långström, B. (2003) Positron emission tomography microdosing: a new concept with application in tracer and early clinical drug development. *Eur J Clin Pharmacol* 59, 357–366.

Berndt, U., Stanetty, C., Wanek, T. *et al.* (2008) Synthesis of a [^{18}F]fluorobenzothiazole as potential amyloid imaging agent. *J Labelled Compd Rad* 51, 137–145.

Bietendorf, J. (2004) FDG PET reimbursement. *J Nucl Med Tech* 32, 33–38.

Bingham, J.B. (2002) Where can FDG-PET contribute most to anatomical imaging problems? *Brit J Radiol* 75, S39–S52.

Blank, M. & Shiloh, Y. (2007) Programs for cell death: apoptosis is only one way to go. *Cell Cycle* 6, 686–695.

Blennow, K., de Leon, M.J. & Zetterberg, H. (2006) Alzheimer's disease. *Lancet* 368, 387–403.

Boström, J., Berggren, K., Elebring, T. *et al.* (2007) Scaffold hopping, synthesis and structure–activity relationships of 5,6-diaryl-pyrazine-2-amide derivatives: a novel series of CB1 receptor antagonists. *Bioorg Med Chem* 15, 4077–4084.

Brakebusch, C., Bouvard, D., Stanchi, F. *et al.* (2002) Integrins in invasive growth. *J Clin Invest* 109, 999–1006.

Briard, E., Zoghbi, S.S., Siméon, F.G. *et al.* (2009) Single-step high-yield radiosynthesis and evaluation of a sensitive ^{18}F-Labeled ligand for imaging brain peripheral benzodiazepine receptors with PET. *J Med Chem* 52, 688–699.

Brown, L.J., Bouvet, D.R., Champion, S. *et al.* (2007) A solid-phase route to ^{18}F-labeled tracers, exemplified by the synthesis of [^{18}f]2-fluoro-2-deoxy-D-glucose. *Angew Chem Int Edit* 46, 941–944.

Burnet, P.W.J., Eastwood, S.L. & Harrison, P.J. (1997) [^{3}H]WAY–100635 for 5-HT_{1A} receptor autoradiography in human brain: a comparison with [^{3}H]8-OH-DPAT and demonstration of increased binding in the frontal cortex in schizophrenia. *Neurochem Int* 30, 565–574.

Burns, H.D., Van Laere, K., Sanabria-Bohórquez, S. *et al.* (2007) [^{18}F]MK-9470, a positron emission tomography (PET) tracer for *in vivo* human PET brain imaging of the cannabinoid-1 receptor. *Proc Natl Acad Sci USA* 104, 9800–9805.

Cagnin, A., Myers, R., Gunn, R.N. *et al.* (2001) *In vivo* visualization of activated glia by [^{11}C](*R*)-PK11195-PET following herpes encephalitis reveals projected neuronal damage beyond the primary focal lesion. *Brain* 124, 2014–2027.

Cagnin, A., Taylor-Robinson, S.D., Forton, D.M. *et al.* (2006) *In vivo* imaging of cerebral peripheral benzodiazepine binding sites in patients with hepatic encephalopathy. *Gut* 55, 547–553.

Cai, L., Chin, F.T., Pike, V.W. *et al.* (2004) Synthesis and evaluation of two ^{18}F-Labeled 6-Iodo-2-(4′-*N,N*-dimethylamino)phenylimidazo[1,2-a]pyridine derivatives as prospective radioligands for beta-amyloid in Alzheimer's disease. *J Med Chem* 47, 2208–2218.

Cai, W., Zhang, X., Wu, Y. *et al.* (2006) A thiol-reactive ^{18}F-labeling agent, *N*-[2-(4-18F-fluorobenzamido)ethyl]maleimide, and synthesis of RGD peptide-based tracer for PET imaging of $\alpha v\beta 3$ integrin expression. *J Nucl Med* 47, 1172–1180.

Call, J.A., Eckhardt, S.G. & Camidge, D.R. (2008) Targeted manipulation of apoptosis in cancer treatment. *Lancet Oncol* 9, 1002–1011.

Charalambous, A., Marciniak, G., Shiue, C.-Y. *et al.* (1991) PET studies in the primate brain and biodistribution in mice using (–)-5′-^{18}F-Δ^{8}-THC. *Pharmacol Biochem Behav* 40, 503–507.

Chauveau, F., Boutin, H., Van Camp, N.V. *et al.* (2008) Nuclear imaging of neuroinflammation: a comprehensive review of [^{11}C]PK11195 challengers. *Eur J Nucl Med Mol Imaging* 35, 2304–2319.

Chen, X., Park, R., Shahinian, A.H. *et al.* (2004) ^{18}F-Labeled RGD peptide: initial evaluation for imaging brain tumor angiogenesis. *Nucl Med Biol* 31, 179–189.

Chen, D.L. & Schuster, D.P. (2006) Imaging pulmonary inflammation with positron emission tomography: a biomarker for drug development. *Mol Pharm* 3, 488–495.

Chen, M.-K. & Guilarte, T.R. (2008) Translocator protein 18kDA (TSPO): molecular sensor of brain injury and repair. *Pharmacol Ther* 118, 1–17.

Cho, H.J., Lee, J.D., Park, J.Y. *et al.* (2009) First in human evaluation of a newly developed integrin binding PET tracer, ^{18}F-RGD-K5 in patients with breast cancer: comparison with ^{18}F-FDG uptake pattern and microvessel density. *J Nucl Med* 50, S1910.

Choi, S.R., Golding, G., Zhuang, Z. *et al.* (2009) Preclinical properties of ^{18}F-AV-45: a PET agent for Aβ plaques in the brain. *J Nucl Med* 50, 1887–1894.

Chu, W., Zhang, J., Zeng, C. *et al.* (2005) *N*-Benzylisatin sulfonamide analogues as potent caspase-3 inhibitors: synthesis, *in vitro* activity and molecular modeling studies. *J Med Chem* 48, 7637–7647.

Cliff, I.A. (2000) A retrospect on the discovery of WAY–100635 and the prospect for improved 5-HT_{1A} receptor PET radioligands. *Nucl Med Biol* 27, 441–447.

Cobben, D.C.P., Jager, P.L., Elsinga, P.H. *et al.* (2003) 3′-^{18}F-fluoro-3′-deoxy-L-thymidine: a new tracer for staging metastatic melanoma? *J Nucl Med* 44, 1927–1932.

Crymes Jr., W.B., Demos, H. & Gordon, L. (2004) Detection of musculoskeletal infection with ^{18}F-FDG PET: review of the current literature. *J Nucl Med Tech* 32, 12–15.

Cummings, J., Ward, T.H., Ranson, M. *et al.* (2004) Apoptosis targeted drugs — from bench to the clinic. *BBA Rev Cancer* 1705, 53–66.

Cunningham, V.J., Parker, C.A., Rabiner, E.A. *et al.* (2005) PET studies in drug development: methodological considerations. *Drug Discov Today* 2, 311–315.

Damianovich, M., Ziv, I., Heyman, S.N. *et al.* (2006) ApoSense: a novel technology for functional molecular imaging of cell death in models of acute renal tubular necrosis. *Eur J Nucl Med Mol Imaging* 33, 281–291.

DeGrado, T.R., Coleman, R.E., Wang, S. *et al.* (2000) Synthesis and evaluation of ^{18}F-Labeled choline as an oncologic tracer for positron emisson tomography: initial findings in prostate cancer. *Cancer Res* 61, 110–117.

DeGrado, T.R., Baldwin, S.W., Wang, S. *et al.* (2001) Synthesis and evaluation of ^{18}F-labeled choline analogs as oncologic PET tracers. *J Nucl Med* 42, 1805–1814.

Dijkgraaf, I., Beer, A.J. & Wester, H.J. (2009) Application of RGD-containing peptides as imaging probes for αvβ3 expression. *Front Biosci* 1, 887–899.

Donohue, S.R., Krushinski, J.H., Pike, V.W. *et al.* (2008a) Synthesis, *ex vivo* evaluation, and radiolabeling of potent 1,5-diphenylpyrrolidin-2-one cannabinoid subtype-1 receptor ligands as candidates for *in vivo* imaging. *J Med Chem* 51, 5833–5842.

Donohue, S.R., Halldin, C., Schou, M. *et al.* (2008b) Radiolabeling of a high potency cannabinoid subtype-1 receptor ligand, *n*-(4-fluoro-benzyl)-4-(3-(piperidin-1-yl)-indole-1-sulfonyl)benzamide (PipISB), with carbon-11 or fluorine-18. *J Labelled Compd Rad* 51, 146–152.

Engler, H., Blomqvist, G., Bergström, M. *et al.* (2002) First human study with a benzothiazole amyloid-imaging agent in Alzheimer's disease and control subjects. *Neurobiol Aging* 23, S4298.

Erlandsson, K., Bressan, R.A., Mulligan, R.S. *et al.* (2003) Kinetic modelling of [^{123}I]CNS 1261: a potential SPECT tracer for the NMDA receptor. *Nucl Med Biol* 30, 441–454.

Fan, H., Ravert, H.T., Holt, D.P. *et al.* (2006) Synthesis of 1-(2,4-dichlorophenyl)-4-cyano-5-(4[^{11}C]methoxyphenyl)-*N*-(piperidin-1-yl)-1*H*-pyrazole-3-carboxamide ([^{11}C]JHU75528) and 1-(2bromophenyl)-4-cyano-5-(4-[^{11}C] methoxyphenyl)-*N*-(piperidin-1-yl)-1*H*-pyrazole-3-carboxamide ([^{11}C] JHU75575) as potential radioligands for PET imaging of cerebral cannabinoid receptor. *J Labelled Compd Rad* 49, 1021–1036.

Foster, N.L., Heidebrink, J.L., Clark, C.M. *et al.* (2007) FDG-PET improves accuracy in distinguishing frontotemporal dementia and Alzheimer's disease. *Brain* 130, 2616–2635.

Frost, J.J. & Wagner, H.N. Jr. (1990) *Quantitative Imaging: Neuroreceptors, Neurotransmitters, and Enzymes*, Lippincott Williams & Wilkins, London.

Fujimura, Y., Ikoma, Y., Yasuno, F. *et al.* (2006) Quantitative analyses of ^{18}F-FEDAA1106 binding to peripheral benzodiazepine receptors in living human brain. *J Nucl Med* 47, 43–50.

Fujimura, Y., Zoghbi, S.S., Siméon, F.G. *et al.* (2009) Quantification of translocator protein (18 kDa) in the human brain with PET and a novel radioligand, ^{18}F-PBR06. *J Nucl Med* 50, 1047–1053.

Galiegue, S., Tinel, N. & Casellas, P. (2003) The peripheral benzodiazepine receptor: a promising therapeutic drug target. *Curr Med Chem* 10, 1563–1572.

de Geus-Oei, L.F., van der Heijden, H.F., Corstens, F.H.M. *et al.* (2007) Predictive and prognostic value of FDG-PET in non-small-cell lung cancer: a systematic review. *Cancer* 110, 1654–1664.

Ginovart, N., Hassoun, W., Le Bars, D. *et al.* (2000) *In vivo* characterization of *p*-[^{18}F]MPPF, a fluoro analog of WAY-100635 for visualization of 5-HT$_{1A}$ receptors. *Synapse* 35, 192–200.

Glaser, M., Luthra, S.K. & Brady, F. (2003) Applications of positron-emitting halogens in PET oncology (review). *Int J Oncology* 22, 253–267.

Glaser, M. & Årstad, E. (2007) Click labeling with 2-[^{18}F]Fluoroethylazide for positron emission tomography. *Bioconjugate Chem* 18, 989–993.

Glaser, M., Morrison, M., Solbakken, M. *et al.* (2008) Radiosynthesis and biodistribution of cyclic RGD peptides conjugated with novel [^{18}f]fluorinated aldehyde-containing prosthetic groups. *Bioconjugate Chem* 19, 951–957.

Glaser, M., Solbakken, M., Turton, D.R. *et al.* (2009) Methods for ^{18}F-labeling of RGD peptides: comparison of aminooxy, [^{18}F]fluorobenzaldehyde condensation with 'click labeling' using 2-[^{18}F]fluoroethylazide and *S*-alkylation with [^{18}F]fluoropropanethiol. *Amino Acids* 37, 717–724.

Glennon, R.A. (1990) Serotonin receptors: clinical implications. *Neurosci Biobehav Rev* 14, 35–47.

Grierson, J.R. & Shields, A.F. (2000) Radiosynthesis of 3′-deoxy-3′-[^{18}F]fluorothymidine: [^{18}F]FLT for imaging of cellular proliferation *in vivo. Nucl Med Biol* 27, 143–156.

Grierson, J.R., Schwartz, J.L., Muzi, M. *et al.* (2004a) Metabolism of 3′-deoxy-3′-[F-18]fluorothymidine in proliferating A549 cells: validations for positron emission tomography. *Nucl Med Biol* 31, 829–837.

Grierson, J.R., Yagle, K.J., Eary, J.F. *et al.* (2004b) Production of [F-18]fluoroannexin for imaging apoptosis with PET. *Bioconjug Chem* 15, 373–379.

Halldin, C., Wikstrom, H., Swahn, C.-G. *et al.* (1994) Preparation of [propyl-^{11}C]OSU 191, a highly potent and selective 5-HT_{1A} agonist for PET. *J Labelled Compd Rad* 35, S675–S677.

Hamacher, K., Coenen, H.H. & Stocklin, G. (1986) Efficient stereospecific synthesis of no-carrier-added 2-[^{18}F]fluoro-2-deoxy-D-glucose using aminopolyether supported nucleophilic substitution. *J Nucl Med* 27, 235–238.

Hanahan, D. & Weinberg, R.A. (2000) The hallmarks of cancer. *Cell* 100, 57–70.

Hara, T., Kosaka, N., Shinoura, N. *et al.* (1997) PET imaging of brain tumor with [methyl-^{11}C]choline. *J Nucl Med* 38, 842–847.

Hara, T., Kosaka, N. & Kishi, H. (1998) PET imaging of prostate cancer using carbon-11 choline. *J Nucl Med* 39, 990–995.

Hara, T., Kosaka, N. & Kishi, H. (2002) Development of fluoroethylcholine for cancer imaging with PET: synthesis, biochemistry, and prostate cancer imaging. *J Nucl Med* 43, 187–199.

Haubner, R., Wester, H.-J., Reuing, U. *et al.* (1999). Radiolabeled αvβ3-integrin antagonists: a new class of tracers for tumor targeting. *J Nucl Med* 40, 1061–1071.

Haubner, R., Wester, H.-J., Burkhart, F. *et al.* (2001a). Glycosylated RGD-containing peptides: tracer for tumor targeting and angiogenesis imaging with improved biokinetics. *J Nucl Med* 42, 326–336.

Haubner, R., Wester, H.-J., Weber, W.A. *et al.* (2001b). Noninvasive imaging of αvβ3-integrin expression using ^{18}F-labeled RGD-containing glycopeptide and positron emission tomography. *Cancer Res* 61, 1781–1785.

Haubner, R., Kuhnast, B., Mang, C. *et al.* (2004) [^{18}F]Galacto-RGD: synthesis, radiolabeling, metabolic stability and radiation dose estimates. *Bioconjugate Chem* 15, 61–69.

Henriksen, G., Yousefi, B.H., Drzezga, A. *et al.* (2008) Development and evaluation of compounds for imaging of β-amyloid plaque by means of positron emission tomography. *Eur J Nucl Med Mol Imag* 35, S75–S81.

Hensler, J.G. (2003) Regulation of 5-HT_{1A} receptor functions in brain following agonist or antidepressant administration. *Life Sci* 72, 1665–1682.

Hicks, R.J., Kalff, V., MacManus, M.P. *et al.* (2001) (^{18}F)FDG PET provides high-impact and powerful prognostic stratification in staging newly diagnosed non-small cell lung cancer. *J Nucl Med* 42, 1596–1604.

Hoh, C.K., Seltzer, M.A., Franklin, J. *et al.* (1998) Positron emission tomography in urological oncology. *J Urology* 159, 347–356.

Horti, A.G., Fan, H., Kuwabara, H. *et al.* (2006) ^{11}C-JHU75528: a radiotracer for PET imaging of CB1 cannabinoid receptors. *J Nucl Med* 47, 1689–1696.

Humphries, M.J. (2000) Integrin structure. *Biochem Soc Trans* 28, 311–339.

Ido, T., Wan, C.-N., Fowler, J.S. *et al.* (1977) Fluorination with F_2: a convenient synthesis of 2-deoxy-2-fluoro-D-glucose. *J Org Chem* 42, 2341–2342.

Ido, T., Wan, C.-N., Casella, V. *et al.* (1978) Labeled 2-deoxy-d-glucose analogs. F-18 labeled 2-deoxy-2-fluoro-D-glucose, 2-deoxy-2-fluoro-D-mannose and C-14 2-deoxy-2-fluoro-D-glucose. *J Labelled Compd Rad* 14, 175–184.

Jakob-Roetne, R. & Jacobsen, H. (2009) Alzheimer's disease: from pathology to therapeutic approaches. *Angew Chem Int Edit* 48, 3030–3059.

James, M.L., Fulton, R.R., Henderson, D.J. *et al.* (2005) Synthesis and *in vivo* evaluation of a novel peripheral benzodiazepine receptor PET radioligand. *Bioorg Med Chem* 13, 6188–6194.

James, M.L., Selleri, S. & Kassiou, M. (2006) Development of ligands for the peripheral benzodiazepine receptor. *Curr Med Chem* 13, 1991–2001.

James, M.L. Fulton, R.R., Vercoullie, J. *et al.* (2008) DPA-714, a new translocator protein–specific ligand: synthesis, radiofluorination, and pharmacologic characterization. *J Nucl Med* 49, 814–822.

Jones, T. (1996) The role of PET within the spectrum of medical imaging. *Eur J Nucl Med* 23, 207–211.

de Jong, I.J., Pruim, J., Elsinga, P.H. *et al.* (2002) Visualization of prostate cancer with ^{11}C-choline positron emission tomography. *Eur Urol* 42, 18–23.

Katoch-Rouse, R., Pavlova, O.A., Caulder, T. *et al.* (2003) Synthesis, structure–activity relationship, and evaluation of SR141716 analogues: development of central cannabinoid receptor ligands with lower lipophilicity. *J Med Chem* 46, 642–645.

Kelloff, G.J., Hoffman, J.M., Johnson, B. *et al.* (2005) Progress and promise of FDG-PET imaging for cancer patient management and oncologic drug development. *Clin Cancer Res* 11, 2785–2808.

Kenny, L.M., Coombes, R.C., Vigushin, D.M. *et al.* (2007) Imaging early changes in proliferation at 1 week post chemotherapy: a pilot study in breast cancer patients with 3′-deoxy-3′-[^{18}F]fluorothymidine positron emission tomography. *Eur J Nucl Med Mol Imaging* 34, 1339–1347.

Kenny, L.M., Coombes, R.C., Oulie, I. *et al.* (2008) Phase I trial of the positron-emitting Arg-Gly-Asp (RGD) peptide radioligand ^{18}F-AH111585 in breast cancer patients. *J Nucl Med* 49, 879–886.

Kenny, L.M., Contractor, K.B., Stebbing, J. *et al.* (2009) Altered tissue 3′-deoxy-3′-[^{18}F]fluorothymidine pharmacokinetics in human breast cancer following capecitabine treatment detected by positron emission tomography. *Clin Cancer Res* 15, 6649–6657.

Kerr, J.F., Wyllie, A.H. & Currie, A.R. (1972) Apoptosis: a basic biological phenomenon with wide-ranging implications in tissue kinetics. *Br J Cancer* 26, 239–257.

Kerr, J.F.R., Winterford, C.M. & Harmon, B.V. (1994) Apoptosis: its significance in cancer and cancer therapy. *Cancer* 73, 2013–2026.

Klok, R.P., Klein, P.J., van Berckel, B.N.M. *et al.* (2008) Synthesis of 2-(1,1-dicyanopropen-2-yl)-6-(2-[^{18}F]fluoroethyl)-methylamino-naphthalene ([^{18}F]FDDNP). *App Radiat Isot* 66, 203–207.

Klunk, W.E., Engler, H., Nordberg, A. *et al.* (2004) Imaging brain amyloid in Alzheimer's disease with Pittsburgh compound-B. *Ann Neurol* 55, 306–319.

Klunk, W.E. & Mathis, C.A. (2008) The future of amyloid-beta imaging: a tale of radionuclides and tracer proliferation. *Curr Opin Neurol* 21, 683–687.

Kolb, H.C., Walsh, J., Liang, Q. *et al.* (2009a) ^{18}F-RGD-K5: a cyclic triazole-bearing RGD peptide for imaging $\alpha v\beta 3$ expression *in vivo*. *J Nucl Med* 50, S329.

Kolb, H.C., Walsh, J., Chen, G. *et al.* (2009b) Synthesis of an ^{18}F-labeled RGD peptide for imaging $\alpha v\beta 3$ integrin expression *in vivo*. *J Nucl Med* 50, S1939.

Koole, M., Lewis, D.M., Buckley, C. *et al.* (2009) Whole-body biodistribution and radiation dosimetry of ^{18}F-GE067: a radioligand for *in vivo* brain amyloid imaging. *J Nucl Med* 50, 818–822.

Kopka, K., Faust, A., Keul, P. *et al.* (2006) 5-Pyrrolidinylsulfonyl isatins as a potential tool for the molecular imaging of caspases in apoptosis. *J Med Chem* 49, 6704–6715.

Kopka, K., Schober, O. & Wagner, S. (2008) ^{18}F-Labelled cardiac PET tracers: selected probes for the molecular imaging of transporters, receptors and proteases. *Basic Res Cardiol* 103, 131–143.

Kumar, J.S.D. & Mann, J.J. (2007) PET tracers for 5-HT$_{1A}$ receptors and uses thereof. *Drug Discov Today* 12, 748–756.

Kumar, J.S.D., Prabhakaran, J., Majo, V.J. *et al.* (2007) Synthesis and *in vivo* evaluation of novel 5-HT$_{1A}$ receptor agonist radioligand [*O*-methyl-^{11}C]2-(4-(2-methoxyphenyl)piperazin-1-yl)butyl)-4-methyl-1,2,4-triazine-3,5(2*H*, 4*H*)-dione in nonhuman primates. *Eur J Nucl Med Mol Imaging* 34, 1050–1060 and references cited therein.

Kumar, J.S.D., Milak, M., Prabhakaran, J. *et al.* (2008) [^{11}C]CUMI-101, an agonist radiotracer for imaging the 5-HT$_{1A}$ receptor *in vivo* in human with positron emission tomography. *J Nucl Med* 49, P79.

Kung, H.F., Choi, S.R., Qu, W. *et al.* (2010) ^{18}F-Stilbenes and styrylpyridines for PET imaging of Aβ plaques in Alzheimer's disease: a mini-perspective. *J Med Chem* 53, 933–941.

Kwee, S.A., DeGrado, T.R., Talbot, J.N. *et al.* (2007) Cancer imaging with fluorine-18-labeled choline derivatives. *Semin Nucl Med* 37, 420–428.

Lahorte, C.M.M., Vanderheyden, J.-L., Steinmetz, N. *et al.* (2004). Apoptosis-detecting radioligands: current state of the art and future perspectives. *Eur J Nucl Med Mol Imaging* 31, 887–919.

Lan, R., Gatley, J., Lu, Q. *et al.* (1999) Design and synthesis of the CB1 selective cannabinoid antagonist AM281: a potential human SPECT ligand. *AAPS Pharmsci* 1, article 4.

Lanfumey, L. & Hamon, M. (2000) Central 5-HT(1A) receptors: regional distribution and functional characteristics. *Nucl Med Biol* 27, 429–435.

Le Bars, D., Lemaire, C., Ginovart, N. *et al.* (1998) High-yield radiosynthesis and preliminary *in vivo* evaluation of *p*-[^{18}F]MPPF, a fluoro analog of WAY-100635. *Nucl Med Biol* 25, 343–350.

Lee, D., Long, S.A., Murray, J.H. *et al.* (2001) Potent and selective non-peptide inhibitors of caspases 3 and 7. *J Med Chem* 44, 2015–2026.

Lee, C.C., Sui, G., Elizarov, A. *et al.* (2005) Multistep synthesis of a radiolabeled imaging probe using integrated microfluidics. *Science* 310, 1793–1796.

Lemoine L., Verdurand M., Vacher B. *et al.* (2009). [^{18}F]F15599, a novel 5-HT(1A) receptor agonist, as a radioligand for PET neuroimaging. *Eur J Nucl Med Mol Imaging* 37, 594–605.

Lewis, M.R. (2005) Radiolabeled RGD peptides move beyond cancer: PET imaging of delayed-type hypersensitivity reaction. *J Nucl Med* 46, 2–4.

Leyton, J., Smith, G., Zhao, Y. *et al.* (2009) [^{18}F]Fluoromethyl-[1,2-$^{2}H_4$]-choline: a novel radiotracer for imaging choline metabolism in tumors by positron emission tomography. *Cancer Res* 69, 7721–7728.

Liu, S. (2006) Radiolabeled multimeric cyclic RGD peptides as integrin αvbeta3 targeted radiotracers for tumor imaging. *Mol Pharm* 3, 472–487.

Liu, P., Lin, L.S., Hamill, T.G. *et al.* (2007) Discovery of *N*-{(1S,2S)-2-(3-cyanophenyl)-3-[4-(2-[^{18}F]fluoroethoxy)phenyl]-1-methylpropyl}-2-methyl-2-[(5-methylpyridin-2-yl)oxy]propanamide, a cannabinoid-1 receptor positron emission tomography tracer suitable for clinical use. *J Med Chem* 50, 3427–3430.

Liu, Z., Yan, Y., Chin, F.T. *et al.* (2009) Dual integrin and gastrin-releasing peptide receptor targeted tumor imaging using ^{18}F-labeled PEGylated RGD-bombesin heterodimer ^{18}F-FB-PEG3-Glu-RGD-BBN. *J Med Chem* 52, 425–432.

Lockhart, A., Davis, B., Matthews, J.C. *et al.* (2003) The peripheral benzodiazepine receptor ligand PK11195 binds with high affinity to the acute phase reactant α1-acid glycoprotein: implications for the use of the ligand as a CNS inflammatory marker. *Nucl Med Biol* 30, 199–206.

Lu, L., Samuelsson, L., Bergström, M. *et al.* (2002) Rat studies comparing ^{11}C-FMAU, ^{18}F-FLT, and ^{76}Br-BFU as proliferation markers. *J Nucl Med* 43, 1688–1698.

Lu, S.Y., Steiger, C., Liao, Y. *et al.* (2007) 4-Fluoro-*N*-(2-(4-(2,3-dihydrobenzo[B][1,4]dioxin-8-yl)piperazin-1-yl)-ethyl)benzamide, a new high affinity 5-HT_{1A} receptor agonist — synthesis and ^{18}F-labeling as a prospective radioligand. *J Labelled Compd Rad* 50, S315.

Lu, S., Liow, J.S., Zoghbi, S.S. *et al.* (2008) Evaluation of [^{18}F]SL702 as a prospective agonist PET radioligand for brain 5-HT_{1A} receptors in mice and monkey. *Neuroimage* 41, T157–T158.

Luurtsema, G., Schuit, R.C., Takkenkamp, K. *et al.* (2008) Peripheral metabolism of [^{18}F]FDDNP and cerebral uptake of its human metabolites in rat brain. *Nucl Med Biol* 35, 869–874.

Maeda, J., Suhara, T., Zhang, M.-R. *et al.* (2004) Novel peripheral benzodiazepine receptor ligand [^{11}C]DAA1106 for PET: an imaging tool for glial cells in the brain. *Synapse* 52, 283–291.

Makowski, M.R., Keithahn, A., Beer, A. *et al.* (2008a) [^{18}F]Galacto-RGD PET demonstrates elevated myocardial αvβ3 expression in patients after myocardial infarction. *J Nucl Med* 49 Suppl, P70.

Makowski, M.R., Ebersberger, U., Nekolla, S. *et al.* (2008b) *In vivo* molecular imaging of angiogenesis, targeting αvβ3 integrin expression, in a patient after acute myocardial infarction. *Eur Heart J* 29, 2201.

Mangner, T.J., Klecker, R.W., Anderson, L. *et al.* (2003) Synthesis of 2′-deoxy-2′-[^{18}F]fluoro-β-D-arabinofuranosyl nucleosides, [^{18}F]FAU, [^{18}F]FMAU, [^{18}F]FBAU and [^{18}F]FIAU, as potential PET agents for imaging cellular proliferation: synthesis of [^{18}F]labelled FAU, FMAU, FBAU, FIAU. *Nucl Med Biol* 30, 215–224.

Margolis, D.J.A., Hoffman, J.M., Herfkens, R.J. *et al.* (2007) Molecular imaging techniques in body imaging. *Radiology* 245, 333–356.

Martin, S.J., Eisenbarth, J.A., Wagner-Utermann, U. *et al.* (2002) A new precursor for the radiosynthesis of [^{18}F]FLT. *Nucl Med Biol* 29, 263–273.

Mason, N.S., Klunk, W.E., Debnath, M. *et al.* (2007) Synthesis and evaluation of [^{18}F]PIB analogs as αβ plaque PET imaging agents. *J Labelled Compd Rad* 50, S87.

Massoud, T.F. & Gambir, S.S. (2003) Molecular imaging in living subjects: seeing fundamental biological processes in a new light. *Genes Dev* 17, 545–580.

Masters, C. & Beyreuther, K. (1998) Alzheimer's disease. *Brit Med J* 316, 446–448.

Mathis, C.A., Huang, Y. & Simpson, N.R. (1997) Synthesis and evaluation of 5-HT$_{1A}$ agonists as radioligands: failure of G-protein-coupled receptor agonists as *in vivo* imaging agents. *J Labelled Compd Rad* 40, 563–564.

Mathis, C.A., Wang, Y., Holt, D.P. *et al.* (2003a) Synthesis and evaluation of ^{11}C-labeled 6-substituted 2-arylbenzothiazoles as amyloid imaging agents. *J Med Chem* 46, 2740–2754.

Mathis, C.A., Wang, Y., Holt, D.P. *et al.* (2003b) Development of ^{18}F-labelled thioflavin-T analogues as amyloid plaque imaging agents. *J Labelled Compd Rad* 46, S62.

Mathis, C.A., Wang, Y. & Klunk, W.E. (2004) Imaging β-amyloid plaques and neurofibrillary tangles in the aging human brain. *Curr Pharm Des* 10, 1469–1492.

Mathis, C.A., Lopresti, B., Mason, N. *et al.* (2007) Comparison of the amyloid imaging agents [F-18]3′-F-PIB and [C-11]PIB in Alzheimer's disease and control subjects. *J Nucl Med* 48, P56.

McParland, B.J., Miller, M.P., Spinks, T.J. *et al.* (2008) The biodistribution and radiation dosimetry of the Arg-Gly-Asp Peptide ^{18}F-AH111585 in healthy volunteers. *J Nucl Med* 49, 1664–1667.

Mercer, J.R. (2007) Molecular imaging agents for clinical positron emission tomography in oncology other than fluorodeoxyglucose (FDG): applications, limitations and potential. *J Pharm Pharm Sci* 10, 180–202.

Milak, M.S., Kumar, J.S.D., Prabhakaran, J. *et al.* (2008) [C-11]CUMI-101 an agonist 5-HT_{1A} positron emission tomography (PET) tracer in human: preliminary test–retest data. *Neuroimage* 41, T96.

de Molina, A.M., Rodriguez-Gonzalez, A. & Gutierrez, R. (2002) Over-expression of choline kinase is a frequent feature in human tumor-derived cell lines and in lung, prostate, and colorectal human cancers. *Biochem Biophys Res Commun* 296, 580–583.

Morrisson, M., Ricketts, S.-A., Barnett, J. *et al.* (2009) Use of a novel Arg-Gly-Asp radioligand, ^{18}F-AH111585, to determine changes in tumor vascularity after antitumor therapy. *J Nucl Med* 50, 116–122.

Mosconi, L., Tsui, W.H., Herholz, K. *et al.* (2008) Multicenter standardized ^{18}F-FDG PET diagnosis of mild cognitive impairment, Alzheimer's disease, and other dementias. *J Nucl Med* 49, 390–398.

Murakami, Y., Takamatsu, H., Tatsumi, M. *et al.* (2003) F-18 labeled annexin V: a PET tracer for apoptosis imaging. *J Nucl Med* 44, P312.

Nelissen, N., Van Laere, K., Thurfjell, L. *et al.* (2009) Phase 1 study of the Pittsburgh compound B derivative ^{18}F-flutemetamol in healthy volunteers and patients with probable Alzheimer disease. *J Nucl Med* 50, 1251–1259.

Nguyen, Q.-De., Smith, G., Glaser, M. *et al.* (2009) Positron emission tomography imaging of drug-induced tumor apoptosis with a caspase-3/7 specific [^{18}F]-labeled isatin sulphonamide. *Proc Natl Acad Sci USA* 106, 16375–16380.

Niu, G. & Chen, X. (2009) PET imaging of angiogenesis. *PET Clinics* 4, 17–38.

Noda, A., Murakami, Y., Nishiyama, S. *et al.* (2008) Amyloid imaging in aged and young macaques with [^{11}C]PIB and [^{18}F]FDDNP. *Synapse* 62, 472–475.

Nordberg, A. (2007) Amyloid imaging in Alzheimer's disease. *Curr Opin Neurol* 20, 398–402.

O'Keefe, G.J., Saunder, T.H., Ng, S. *et al.* (2009) Radiation dosimetry of β-amyloid tracers ^{11}C-PiB and ^{18}F-BAY94-9172. *J Nucl Med* 50, 309–315.

Okubo, T., Yoshikawa, R., Chaki, S. *et al.* (2004) Design, synthesis and structure-affinity relationships of aryloxyanilide derivatives as novel peripheral benzodiazepine receptor ligands. *Bioorg Med Chem* 12, 423–438.

Okuyama, S., Chaki, S., Yoshikawa, R. *et al.* (1999) Neuropharmacological profile of peripheral benzodiazepine receptor agonists, DAA 1097 and DAA 1106. *Life Sci* 64, 1455–1464.

Ono, M., Haratake, M., Nakayama, M. *et al.* (2005) Synthesis and biological evaluation of (*E*)-3-styrylpyridine derivatives as amyloid imaging agents for Alzheimer's disease. *Nucl Med Biol* 32, 329–335.

Owens, J., Tebbutt, A.A., McGregor, A.L. *et al.* (2000) Synthesis and binding characteristics of *N*-(1-naphthyl)-*N'*-(3-[^{125}I]-iodophenyl)-*N'*-methylguanidine ([^{125}I]-CNS 1261): a potential SPECT agent for imaging NMDA receptor activation. *Nucl Med Biol* 27, 557–564.

Pacher, P., Bátkai, S. & Kunos, G. (2006) The endocannabinoid system as an emerging target of pharmacotherapy. *Pharmacol Rev* 58, 389–462.

Papadopoulos, V., Baraldi, M., Guilarte, T.R. *et al.* (2006) Translocator protein (18kDa): new nomenclature for the peripheral-type benzodiazepine receptor based on its structure and molecular function. *Trends Pharmacol Sci* 27, 402–409.

Passchier, J., van Waarde, A., Pieterman, R.M. *et al.* (2000) Quantitative imaging of 5-HT$_{1A}$ receptor binding in healthy volunteers with [^{18}F]*p*-MPPF. *Nucl Med Biol* 27, 473–476.

Passchier, J. & van Waarde, A. (2001) Visualisation of serotonin-1A (5-HT$_{1A}$) receptors in the central nervous system. *Eur J Nucl Med* 28, 113–129.

Phelps, M.E. (2000) PET: the merging of biology and imaging into molecular imaging. *J Nucl Med* 41, 661–681.

Phelps, M.E. (2004) *PET: Molecular Imaging and Its Biological Applications.* Springer-Verlag, New York.

Picchio, M., Treiber, U., Beer, A.J. *et al.* (2006) Value of ^{11}C-choline PET and contrast-enhanced CT for staging of bladder cancer: correlation with histopathologic findings. *J Nucl Med* 47, 938–944.

Pike, V.W., Halldin, C., Crouzel, C. *et al.* (1993) Radioligands for PET studies of central benzodiazepine receptors and PK (peripheral benzodiazepine) binding sites — current status. *Nucl Med Biol* 20, 503–525.

Pike, V.W., Halldin, C., Wikström, H. *et al.* (2000) Radioligands for the study of brain 5-HT$_{1A}$ receptors: *in vivo* development of some new analogues of WAY. *Nucl Med Biol* 27, 449–455.

Pike, V.W., Halldin, C. & Wikström, H. (2001). Radioligands for the study of brain 5-HT$_{1A}$ receptors *in vivo. Progr Med Chem* 38, 189–247.

Pither R. (2003). PET and the role of *in vivo* molecular imaging in personalised medicine. *Exp Rev Mol Diagn* 3, 703–713.

Plenevaux, A., Lemaire, C., Aerts, J. *et al.* (2000) [^{18}F]*p*-MPPF: a radiolabeled antagonist for the study of 5-HT$_{1A}$ receptors with PET. *Nucl Med Biol* 27, 467–471.

Prabhakaran, J., Parsey, R.V., Majo, V.J. *et al.* (2006) Synthesis, *in vitro* and *in vivo* evaluation of [*O*-methyl-^{11}C] 2-{4-[4-(3-methoxyphenyl)piperazin-1-yl]-butyl}-4-methyl-2*H*-[1,2,4]-triazine-3,5-dione: a novel agonist 5-HT_{1A} receptor PET ligand. *Bioorg Med Chem Lett* 16, 2101–2104.

Reshef, A., Shirvan, A., Grimberg, H. *et al.* (2007) Novel molecular imaging of cell death in experimental cerebral stroke. *Brain Res* 1144, 156–164.

Reshef, A., Shirvan, A., Waterhouse, R.N. *et al.* (2008) Molecular imaging of neurovascular cell death in experimental cerebral stroke by PET. *J Nucl Med* 49, 1520–1528.

Rinaldi-Carmona, M., Barth, F., Héaulme, M. *et al.* (1994) SR141716A, a potent and selective antagonist of the brain cannabinoid receptor. *FEBS Lett* 350, 240–244.

Roberson, E. D. & Mucke, L. (2006). 100 years and counting: prospects for defeating Alzheimer's disease. *Science* 314, 781–784.

Robins, E.G., Zhao, Y., Khan, I. *et al.* (2010) Synthesis and *in vitro* evaluation of ^{18}F-labelled *S*-fluoroalkyl diarylguanidines: novel high-affinity NMDA receptor antagonists for imaging with PET. *Bioorg Med Chem Lett* 20, 1749–1751.

Rowe, C.C., Ng, S., Ackermann, U. *et al.* (2007) Imaging beta-amyloid burden in aging and dementia. *Neurology* 68, 1718–1725.

Rowe, C.C., Ackerman, U., Browne, W. *et al.* (2008) Imaging of amyloid β in Alzheimer's disease with ^{18}F-BAY94-9172, a novel PET tracer: proof of mechanism. *Lancet Neurol* 7, 129–135.

Salskov, A., Tammisetti, V.S., Grierson, J. *et al.* (2007) FLT: measuring tumor cell proliferation *in vivo* with positron emission tomography and 3′-deoxy-3′-[^{18}F]fluorothymidine. *Semin Nucl Med* 37, 429–439.

Scarf, A.M., Ittner, L.M. & Kassiou, M. (2009) The translocator protein (18 kDa): central nervous system disease and drug design. *J Med Chem* 52, 581–592.

Schottelius, M., Laufer, B., Kessler, H. *et al.* (2009) Ligands for mapping αvβ3-integrin expression *in vivo*. *Acc Chem Res* 42, 969–980.

Seitz, U., Wagner, M., Neumaier, B. *et al.* (2002) Evaluation of pyrimidine metabolising enzymes and *in vitro* uptake of 3′-[^{18}F]fluoro-3′-deoxy-thymidine ([^{18}F]FLT) in pancreatic cancer cell lines. *Eur J Nucl Med Mol Imaging* 29, 1174–1181.

Selkoe, D.J. (2002) Alzheimer's disease is a synaptic failure. *Science* 298, 789–791.

Selleri, S., Bruni, F., Costagli, C. *et al.* (2001) 2-Arylpyrazolo[1,5-a]pyrimidin-3-yl-acetamides: new potent and selective peripheral benzodiazepine receptor ligands. *Bioorg Med Chem* 9, 2661–2671.

Shah, F., Hume, S.P., Pike, V.W. *et al.* (1994). Synthesis of the enantiomers of [*N*-methyl-^{11}C]PK11195 and comparison of their behaviours as radioligands for PK binding sites in rats. *J Nucl Med Biol* 21, 573–581.

Shields, A.F., Grierson, J.R., Dohmen B.M. *et al.* (1998) Imaging proliferation *in vivo* with [F-18]FLT and positron emission tomography. *Nat Med* 4, 1334–1336.

Shields, A.F. (2003) PET imaging with ^{18}F-FLT and thymidine analogs: promise and pitfalls. *J Nucl Med* 44, 1432–1434.

Shirvan, A., Reshef, A., Allen, A. *et al.* (2009) Apoptosis imaging with PET-^{18}F-ML-10 for early assessment of response of brain metastases to radiotherapy. *J Nucl Med* 50, S453.

Shiue, C.Y., Shiue, G.G., Mozley, P.D. *et al.* (1997) *p*-^{18}F-MPPF: a potential radioligand for PET studies of 5-HT$_{1A}$ receptors in humans. *Synapse* 25, 147–154.

Shoghi-Jadid, K., Small, G.W., Agdeppa, E.D. *et al.* (2002) Localization of neurofibrillary tangles and beta-amyloid plaques in the brains of living patients with Alzheimer's disease. *Am J Geriatr Psych* 10, 24–35.

Silverman, D.H., Small, G.W., Chang, C.Y. *et al.* (2001) Positron emission tomography in evaluation of dementia: regional brain metabolism and long-term outcome. *J Am Med Assoc* 286, 2120–2127.

Skovronsky, D., Coleman, R.E., Frey, K. *et al.* (2008) Results of multi-center clinical trials comparing four ^{18}F PET amyloid-imaging agents: preclinical to clinical correlations. *J Nucl Med* 49, P34.

Small, G.W., Kepe, V., Ercoli, L.M. *et al.* (2006) PET of brain amyloid and tau in mild cognitive impairment. *N Engl J Med* 355, 2652–2656.

Smith, G., Glaser, M., Perumal, M. *et al.* (2008) Design, synthesis, and biological characterization of a caspase 3/7 selective isatin labeled with 2-[^{18}F]fluoroethylazide. *J Med Chem* 51, 8057–8067.

Smith, G., Nguyen, Q.-D. & Aboagye, E. O. (2009) Translational imaging of apoptosis, *Anticancer Agent Med Chem* 9, 958–967.

Sperling, R., Johnson, K., Pontecorvo, M.J. *et al.* (2009) PET imaging of β-amyloid with florpiramine F18 (^{18}F-AV-45): preliminary results from a phase II study of cognitively normal elderly subjects, individuals with mild cognitive impairment, and patients with a clinical diagnosis of Alzheimer's disease. *Alzheimers Dement* 5, p197.

Stephenson, K.A., Chandra, R., Zhuang, Z.-P. *et al.* (2007) Fluoro-pegylated (FPEG): imaging agents targeting Aβ aggregates. *Bioconjug Chem* 18, 238–246.

Stone, J.M., Erlandsson, K., Årstad, E. *et al.* (2006) Ketamine displaces the novel NMDA receptor SPET probe [^{123}I]CNS-1261 in humans *in vivo. Nucl Med Biol* 33, 239–243.

Suehiro, M., Underwood, M., Arango, V. *et al.* (1998) *In vivo* biodistribution of a radiotracer for imaging serotonin-1A receptor sites with PET: [^{11}C]LY274601. *Life Sci* 63, 1533–1542.

Torigian, D.A., Huang, S.S., Houseni, M. *et al.* (2007) Functional imaging of cancer with emphasis on molecular techniques. *CA Cancer J Clin* 57, 206–224.

Vallabhajosula, S. (2007) ^{18}F-Labeled positron emission tomographic radiopharmaceuticals in oncology: an overview of radiochemistry and mechanisms of tumor localization. *Semin Nucl Med* 37, 400–419.

Van Laere, K. (2007) *In vivo* imaging of the endocannabinoid system: a novel window to a central modulatory mechanism in humans. *Eur J Nucl Med Mol Imaging* 34, 1719–1726.

Van Laere, K., Goffin, K., Casteels, C. *et al.* (2008a) Gender-dependent increases with healthy aging of the human cerebral cannabinoid-type 1 receptor binding using [F-18]MK-9470 PET. *Neuroimage* 39, 1533–1541.

Van Laere, K., Koole, M., Sanabria Bohorquez, S.M. *et al.* (2008b) Whole-body biodistribution and radiation dosimetry of the human cannabinoid type-1 receptor ligand F-18-MK-9470 in healthy subjects. *J Nucl Med* 49, 439–445.

Venneti, S., Lopresti, B.J. & Wiley, C.A. (2006) The peripheral benzodiazepine receptor (Translocator Protein 18kDa) in microglia: from pathology to imaging. *Prog Neurobiol* 80, 308–322.

Verhoeff, N.P., Wilson, A.A., Takeshita, S. *et al.* (2004) *In vivo* imaging of Alzheimer's disease beta-amyloid with [^{11}C]SB-13 PET. *Am J Geriatr Psych* 12, 584–595.

Vesselle, H., Grierson, J., Muzi, M. *et al.* (2002) *In vivo* validation of 3′-deoxy-3′-[^{18}F]fluorothymidine ([^{18}F]FLT) as a proliferation imaging tracer in humans: correlation of [^{18}F]FLT uptake by positron emission tomography with Ki-67 immunohistochemistry and flow cytometry in human lung tumors. *Clin Cancer Res* 8, 3315–3323.

Walker, J.H., Boustead, C.M., Koster, J.J. *et al.* (1992) Annexin V, a calcium-dependent phospholipid binding protein. *Biochem Soc Trans* 20, 828–833.

Wang, R.F. (2007) Progress in imaging agents of cell apoptosis. *Anticancer Agents Med Chem* 9, 996–1002.

Waterhouse, R.N. (2003) Imaging the PCP site of the NMDA ion channel. *Nucl Med Biol* 30, 869–878.

Watson, J., Collin, L., Ho, M. *et al.* (2000) 5-HT_{1A} receptor agonist-antagonist binding affinity difference as a measure of intrinsic activity in recombinant and native tissue systems. *Br J Pharmacol* 130, 1108–1114.

Weber, W.A. (2005) PET for response assessment in oncology: radiotherapy and chemotherapy. *Br J Radiol* 28, S42–S49.

Westerterp, M., van Westreenen, H.L. *et al.* (2005) Esophageal cancer: CT, endoscopic US, and FDG PET for assessment of response to neoadjuvant therapy–systematic review. *Radiology* 236, 841–851.

Willis, P.G., Pavlova, O.A., Chefer, S.I. *et al.* (2005) Synthesis and structure–activity relationship of a novel series of aminoalkylindoles with potential for imaging the neuronal cannabinoid receptor by positron emission tomography. *J Med Chem* 48, 5813–5822.

Wilson, A.A., Garcia, A., Parkes, J. *et al.* (2008) Radiosynthesis and initial evaluation of [^{18}F]-FEPPA for PET imaging of peripheral benzodiazepine receptors. *Nucl Med Biol* 35, 305–314.

Wu, Z., Li, Z.-B., Chen, K. *et al.* (2007) MicroPET of tumor integrin $\alpha v\beta 3$ expression using ^{18}F-labeled PEGylated tetrameric RGD peptide (^{18}F-FPRDG4) *J Nucl Med* 48, 1536–1544.

Yamamoto, Y., Nishiyama, Y., Kimura, N. *et al.* (2008) Comparison of ^{18}F-FLT PET and ^{18}F-FDG PET for preoperative staging in non-small cell lung cancer. *Eur J Nucl Med Mol Imag* 35, 236–245.

Yanamoto, K., Kumata, K., Yamasaki, T. *et al.* (2009) [^{18}F]FEAC and [^{18}F]FEDAC: two novel positron emission tomography ligands for peripheral-type benzodiazepine receptor in the brain. *Bioorg Med Chem Lett* 19, 1707–1710.

Yasuno, F., Brown, A.K., Zoghbi, S.S. *et al.* (2008) The PET radioligand [^{11}C]MePPEP binds reversibly and with high specific signal to cannabinoid CB1 receptors in non-human primate brain. *Neuropsychopharmacol* 33, 259–269.

Zeisel, S.H. (1981) Dietary choline: biochemistry, physiology, and pharmacology. *Annu Rev Nutr* 1, 95–121.

Zeisel, S. H. (1993). Choline phospholipids: signal transduction and carcinogenesis. *FASEB J* 7, 551–557.

Zeng. W., Yao. M.-I., Townsend, D. *et al.* (2008) Synthesis, biological evaluation and radiochemical labeling of a dansylhydrazone derivative as a potential imaging agent for apoptosis. *Bioorg Med Chem Lett* 18, 3573–3577.

Zhang, M.-R., Kida, T., Noguchi, J. *et al.* (2003a) [^{11}C]DAA1106: radiosynthesis and *in vivo* binding to peripheral benzodiazepine receptors in mouse brain. *Nucl Med Biol* 30, 513–519.

Zhang, M.-R., Maeda, J., Furutsuka, K. *et al.* (2003b) [^{18}F]FMDAA1106 and [^{18}F]FEDAA1106: two positron-emitter labeled ligands for peripheral benzodiazepine receptor (PBR). *Bioorg Med Chem Lett* 13, 201–204.

Zhang, M.-R., Maeda, J., Ogawa, M. *et al.* (2004) Development of a new radioligand, *N*-(5-fluoro-2-phenoxyphenyl)-*N*-(2-[^{18}F]fluoroethyl-5-methoxybenzyl) acetamide, for PET imaging of peripheral benzodiazepine receptor in primate brain. *J Med Chem* 47, 2228–2235.

Zhang, W., Oya, S., Kung, M.-P. *et al.* (2005a) F-18 stilbenes as PET imaging agents for detecting β-amyloid plaques in the brain. *J Med Chem* 48, 5980–5988.

Zhang, W., Oya, S., Kung, M.-P. *et al.* (2005b) F-18 PEG stilbenes as PET imaging agents targeting Aβ aggregates in the brain. *Nucl Med Biol* 32, 799–809.

Zhang, W., Kung, M.-P., Oya, S. *et al.* (2007) ^{18}F-labeled styrylpyridines as PET agents for amyloid plaque imaging. *Nucl Med Biol* 34, 89–97.

Zhou, D., Chu, W., Rothfuss, J. *et al.* (2006) Synthesis, radiolabelling, and *in vivo* evaluation of an ^{18}F-labeled isatin analog for imaging caspase-3 activation in apoptosis. *Bioorg Med Chem Lett* 16, 5041–5046.

Zhao, M. (2009) *In vivo* apoptosis imaging agents and strategies. *Anticancer Agents Med Chem* 9, 1018–1023.

Zhuang, H., Yu, J.Q. & Alavi, A. (2005) Applications of fluorodeoxyglucose-PET imaging in the detection of infection and inflammation and other benign disorders. *Radiol Clin North Am* 43, 121–34.

Zijlstra, S., Gunawan, J. & Burchert, W. (2003) Synthesis and evaluation of a ^{18}F-labelled recombinant annexin-V derivative, for identification and quantification of apoptotic cells with PET. *Appl Radiat Isot* 58, 201–207.

12

^{19}F NMR: Clinical and Molecular Imaging Applications

Vikram D. Kodibagkar,[†] *Rami R. Hallac,*[†] *Dawen Zhao,*[†] *Jian-Xin Yu*[†] *and Ralph P. Mason*[*,†]

12.1 Introduction

Essentially all elements have NMR active isotopes. Although discovered and developed by physicists, NMR has become a foundation of organic chemistry, a pillar of molecular biology and a cornerstone of clinical radiology. Molecular structure elucidation relies on proton and ^{13}C NMR, while clinical medicine exploits the overwhelming signal from tissue water modulated by diverse contrast mechanisms to reveal detailed anatomy and pathology. The intense water signal reveals exquisite soft tissue anatomical detail, routinely providing submillimetre spatial resolution in seconds even for human MRI systems. Higher magnetic fields (routine clinical systems are moving from 1.5 to 3 T), together with improved radiofrequency (RF) coil transmission and detection antennae (multiarray coils and SENSE acquisition) and faster gradient switching abilities continue to enhance diagnostic capabilities and provide insights into phenomena such as neuronal activation and tumour pathophysiology. Meanwhile, specific contrast agents and selective pulse sequences allow more detailed analysis

* Corresponding author.
† Laboratory of Prognostic Radiology, Division of Advanced Radiological Sciences, Department of Radiology, The University of Texas Southwestern Medical Center at Dallas, TX, USA.

of tissue properties such as diffusion, flow, and changes in vascular permeability and oxygenation (Jiang *et al.*, 2004; Baudelet and Gallez, 2005; Zhang *et al.*, 2005; De Leon-Rodriguez *et al.*, 2009). Fat is also detectable, and may be the dominant signal in tissues such as breast, but other metabolites typically occur at millimolar concentrations (or less) requiring prodigious water suppression (Liebfritz, 1992). NMR is a particularly flexible technology characterized by multiple parameters including chemical shift, relaxation processes (R_1 and R_2), and chemical exchange, each of which may be designed to be responsive to a parameter of interest.

Use of NMR active nuclei other than ^{1}H can provide metabolic tracers and physiological reporters, while avoiding the intense water and lipid signals. ^{19}F is 100% naturally abundant and the only stable isotope of fluorine. The nucleus has a nuclear spin $I = 1/2$ and a gyromagnetic ratio of 40.05 MHz/T, providing a sensitivity about 83% that of protons. The high gyromagnetic ratio often allows the use of existing proton NMR instrumentation with the minimum of component adjustments. The ^{19}F atom is particularly attractive since there is essentially no endogenous signal from tissues. Fluorine does occur extensively in bones and teeth, but the solid matrix causes very short T_2 values providing exceedingly broad signals, which can either be removed by deconvolution or electronic timing. Indeed, special rapid electronics are required for detecting solid-state ^{19}F (Code *et al.*, 1990). Thus, fluorine may be introduced into the body in the form of reporter molecules or drugs and be detected readily with high sensitivity and without background interference. Fluorine NMR does typically require millimolar concentrations of reporter molecules. In this respect, radionuclide and optical imaging techniques can offer far superior sensitivity, potentially with pico to nanomolar requirements (Willmann *et al.*, 2008). However, as the ^{19}F atom has no radioactivity, molecules may be synthesized and stored essentially indefinitely, avoiding the need for rapid synthesis and immediate application in positron emission tomography (PET) necessitated by the 110 min half-life of ^{18}F. Agents labelled with ^{19}F may be traced over hours to days and even weeks, allowing assessment of long-term pharmacokinetics (Mishima *et al.*, 1991; Mason *et al.*, 1994; Morawski *et al.*, 2004; Higuchi *et al.*, 2005). Indeed, ^{19}F NMR is being applied to *in vivo* stem cell tracking (Ahrens *et al.*, 2005; Partlow *et al.*, 2007; Srinivas *et al.*, 2007, 2009). While PET relies on differential tissue trapping, accumulation and pharmacokinetics, ^{19}F NMR can reveal metabolic changes and quantitative assessment of pH, pO_2 and metal ion concentrations based on chemical shift and relaxation times (Yu *et al.*, 2005a, 2008a; Krohn *et al.*, 2008).

^{19}F is exceptionally sensitive to molecular and microenvironmental changes. This has prompted the design, development and application of many ^{19}F-based reporter molecules to interrogate physiological phenomena *in vivo* (see, for example, Table 12.1 and Fig. 12.1). Fluorine NMR has a large chemical shift range of about 300 ppm, as opposed to approximately 10 ppm for proton, so that multiple different fluorinated agents may be detected simultaneously with minimal danger of signal overlap. Noting the virtues of ^{19}F NMR, many reviews have been published (Yu *et al.*, 2005a, 2008a; Wickline *et al.*, 2010).

As a non-destructive evaluation tool, NMR has multiple strengths and virtues. However, NMR is intrinsically a complex modality providing potentially a multitude of information based on diverse parameters including signal intensity (SI), chemical shift (δ), and changes of chemical shift ($\Delta\delta$). In addition, signals are characterized by the transverse dephasing rate ($R_2^* = 1/T_2^*$), spin–spin or transverse relaxation rate ($R_2 = 1/T_2$) and spin–lattice or longitudinal relaxation rate ($R_1 = 1/T_1$). Indeed, each of these parameters has been exploited for specific ^{19}F NMR reporter molecules (Table 12.1). To allow comparison between data from different molecules and different investigators, chemical shifts must be referred to a standard. The IUPAC ^{19}F NMR chemical shift standard is fluorotrichloromethane ($CFCl_3$) (Harris *et al.*, 2001). However, this volatile solvent is not convenient for biomedical applications and we favour sodium trifluoroacetate (CF_3CO_2Na or NaTFA; $\Delta\delta$ versus $CFCl_3$ −76.530 ppm), which has the advantage of being readily available, quite non-toxic and may be used as either an external or internal chemical shift standard in biological investigations. Fluorine chemical shift can be quite unpredictable, but compilations of ^{19}F NMR chemical shifts and theoretical predictions have been reported (Emsley and Phillips, 1971; Dolbier, 2009), as well as compilations of coupling constants (Emsley *et al.*, 1976).

Introduction of a fluorine atom requires care. While the carbon fluorine bond is particularly strong, any release of fluoride or metabolites such as mono- or difluoroacetate can lead to exceedingly toxic products. For reporter molecules or pharmacological drugs, it is clearly important to minimize inadvertent toxicity. In this respect, the trifluoromethyl (CF_3) group is

Table 12.1. Fluorinated reporter molecules.

Information	Reporter	NMR parameter	Representative references
Physical Interactions			
pO_2	Perfluorocarbons	R_1; R_2	Girard *et al.*, 1994; Guo *et al.*, 1994; Mason, 1994; Robinson and Griffiths, 2004; Zhao *et al.*, 2004
Chemical Associations			
pH	FPOL, F-alanine, ZK150471	$\Delta\delta$, J	Deutsch and Taylor, 1987b; Frenzel *et al.*, 1994; Mason, 1999; Ojugo *et al.*, 1999; Raghunand and Gillies, 2001
Metal ions: $[Na^+]$, $[Ca^{2+}]$, $[Mg^{2+}]$	F-cryp-1, F-BAPTA, F-APTRA	δ, ratio	Metcalfe *et al.*, 1985; Smith *et al.*, 1986; Levy *et al.*, 1988; Marban *et al.*, 1988; London, 1994
Membrane/chloride potential	CF_3CO_2H	Signal ratio	London and Gabel, 1989; Ramasamy *et al.*, 1993
Redox potential	Fluorinated ferrocene dendrimer	Signal visibility (R_2*)	Tanaka *et al.*, 2009
Chemical Interactions			
Reporter gene β-gal	PFONPG	$\Delta\delta$	Cui *et al.*, 2004; Yu *et al.*, 2004; Yu and Mason, 2006
Enzyme activity: cytosine deaminase, HDACi, caspase	5FC, Boc-Lys-TFA-OH (BLT), Gd-DOTA-DEVD-Tfb	$\Delta\delta$, signal accumulation, R_2* broadening	Stegman *et al.*, 1999; Mizukami *et al.*, 2008; Sankaranarayanapillai *et al.*, 2008; Tanaka *et al.*, 2008
Hypoxia	F-Misonidazoles	Integral	Raleigh *et al.*, 1991; Seddon *et al.*, 2003; Robinson and Griffiths, 2004; Procissi *et al.*, 2007; Krohn *et al.*, 2008
Passive Reporters			
Temperature	PFCs	Signal ratio, R_1	Thomas, 1988; Berkowitz *et al.*, 1992; Mason *et al.*, 1993
Lung function	PFC, SF_6	Integral	Huang *et al.*, 2002; Kuethe *et al.*, 2002; Ruiz-Cabello *et al.*, 2005
GI function	PFC	Integral	Mattrey *et al.*, 1994; Schwarz *et al.*, 2002
Vascular volume	Fluorocarbon emulsion	Integral	Thomas *et al.*, 1992; Gu *et al.*, 2005

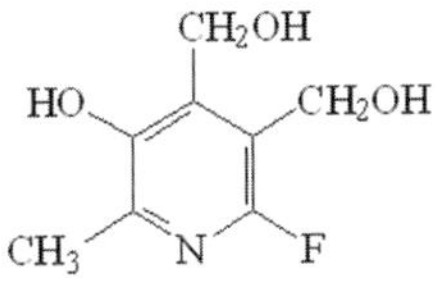

6-FPOL

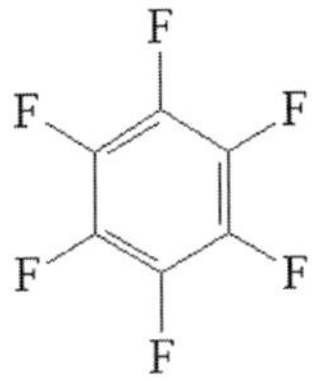

HFB

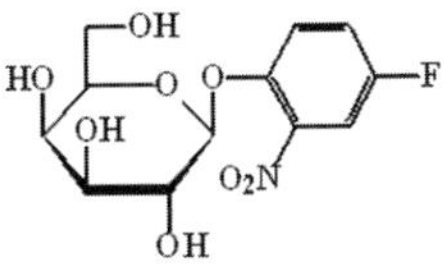

PFONPG

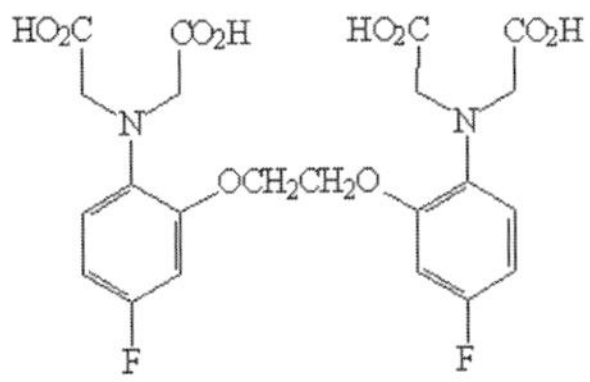

5F-BAPTA

Figure 12.1. Representative fluorinated reporter molecules. Published ^{19}F NMR reporter molecules: 6-fluoropyridoxol (FPOL) is a pH reporter (Mason, 1999); hexafluorobenzene (HFB) is used for oximetry (Zhao *et al.*, 2004); PFONPG is a gene reporter for β-gal (Yu *et al.*, 2004) and 5F-BAPTA measures $[Ca^{2+}]$ (Metcalfe *et al.*, 1985).

particularly suitable, since it resists degradation and also avoids the complexity of fluorine–fluorine couplings in NMR spectra, while providing threefold signal enhancement. Likewise, a CF_3 moiety will generally avoid fluorine–hydrogen couplings. Since ^{19}F NMR is often detected by retuning a proton channel, proton decoupling may not be available. A symmetrical moiety, for example, a trifluoromethyl group as opposed to asymmetric geminal fluorine atoms or a single fluorine atom, also simplifies spectra by rotational averaging. On the other hand, fluorine in a CF_3 moiety is less sensitive to its chemical environment (see examples in Tables 12.2 and 12.3) (Yu *et al.*, 2006).

In terms of NMR detection, the more equivalent fluorines there are, the stronger is the signal. However, fluorine will modulate the properties of a molecule, since the fluorine atom is strongly electronegative and the C–F bond strongly polarized (Müller *et al.*, 2007). While a fluorine substituent has often been considered to be similar in size to a hydrogen atom, the van der Waals radius of covalent fluorine is closer to that of an oxygen ligand and the electronegativity alters the electronic density distribution, modulating pK_a; for example, for the series of acetic acids

Table 12.2. ^{19}F NMR pH Indicators.

Reporter structure	pK_a	Δδ (ppm)	Applications
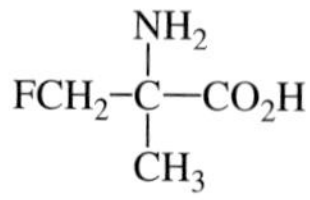 3-Fluoro-2-methyl alanine	8.5	2.05	Deutsch and Taylor, 1987a
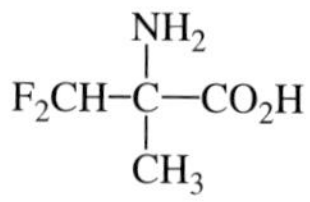 3,3-Difluoro-2-methyl alanine	7.3	2.00	Deutsch and Taylor, 1987a
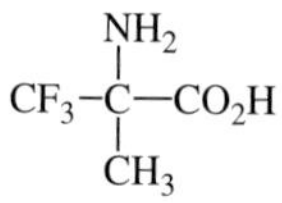 3,3,3-Trifluoro-2-methyl alanine	5.9	2.10	Deutsch and Taylor, 1987a
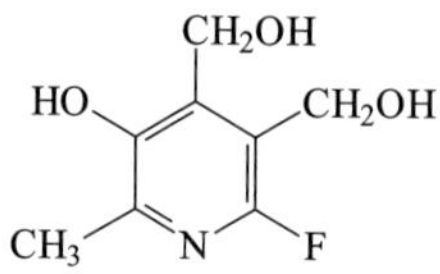 6-FPOL	8.2	9.72	Hunjan *et al.*, 1998; Mason, 1999
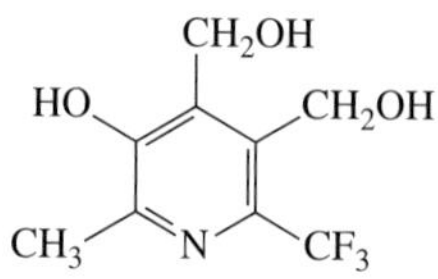 CF_3POL	6.8	1.7	Yu *et al.*, 2005a
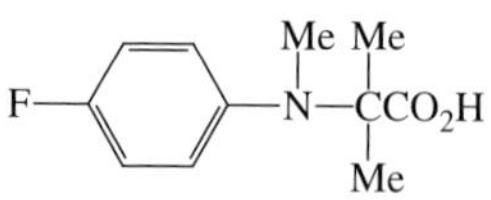 *N,N*-(methyl-2-carboxyisopropyl)-4-fluoroaniline	6.8	13.50	Deutsch and Taylor, 1989
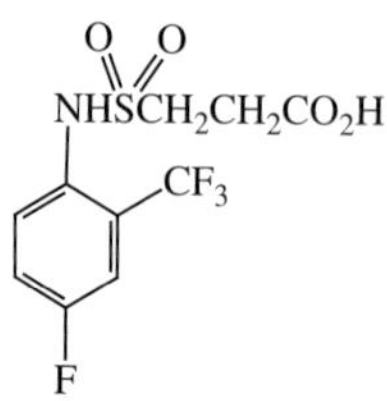 ZK 150471	7.16	−10.13	Frenzel *et al.*, 1994; Aoki *et al.*, 1996; Ojugo *et al.*, 1999

Table 12.3. Importance of molecular orientation for ^{19}F NMR sensitivity.

Agent/ Reporter	**Structure**	**Δδ (ppm)**	**References**
OFPNP	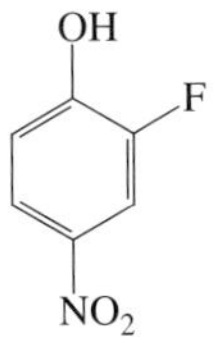	Acid–base 2.24 ppm	Yu *et al.*, 2004
PFONP	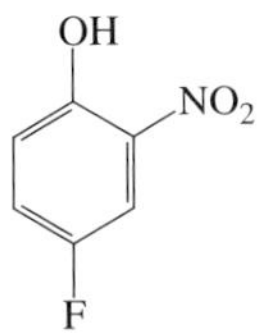	Acid–base 9.3 ppm	Yu *et al.*, 2004
PCF_3ONP	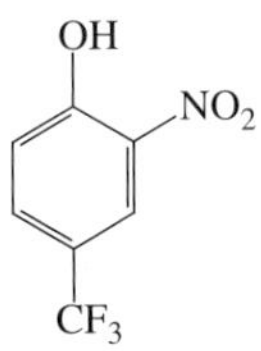	Acid–base 1.0 ppm	Yu *et al.*, 2006
OFPNPG	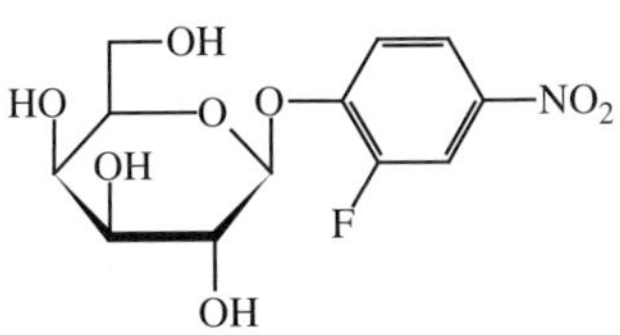	Aglycone release 6.11 ppm	Yu *et al.*, 2004, 2008b; Liu *et al.*, 2007
PFONPG	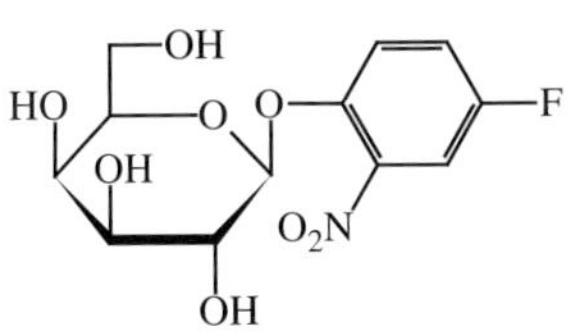	Aglycone release 9.84 ppm	Cui *et al.*, 2004; Yu *et al.*, 2004, 2008b
PCF_3ONPG	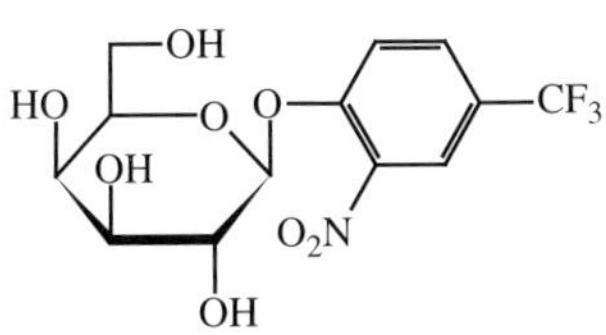	Aglycone release 1.14 ppm	Yu *et al.*, 2006

$pK_a(CH_3CO_2H) = 4.76$; $pK_a(CH_2FCO_2H) = 2.59$; $pK_a(CHF_2CO_2H) = 1.24$; and $pK_a(CF_3CO_2H) = 0.23$ (Bohm *et al.*, 2004). Similarly for the series of primary amines $pK_a(CH_3CH_2NH_3{}^+) = 10.7$; $pK_a(CH_2FCH_2NH_2) = 9.0$; $pK_a(CHF_2CH_2NH_2) = 7.3$; and $pK_a(CF_3CH_2NH_2) = 5.7$ (Morgenthaler *et al.*, 2007). The trifluoromethyl group is often considered to be sterically equivalent to the introduction of an isopropyl group (Smart, 1995). Fluorine not only perturbs the electronic structure of a molecule, but also alters the hydrophobicity and ability to cross membranes, such as the blood–brain barrier, a crucial consideration for anaesthetics and psychiatric drugs (Smart, 2001; Gerebtzoff *et al.*, 2004).

The spin–lattice relaxation time T_1 can be quite long, but efficient use of rapid pulsing at the Ernst angle can accelerate spectral acquisition (Klomp *et al.*, 2003). For aqueous solutions, relaxation agents such as Gd-DTPA can be added to accelerate relaxation (Ratner *et al.*, 1989; Lee *et al.*, 1994; Mehta *et al.*, 1995; Mizukami *et al.*, 2009) and indeed, this has been used to identify cellular compartmentation based on the ability of the contrast agent to relax extracellular, but not intracellular material (Brix *et al.*, 1998). Data acquisition efficiency can also be enhanced by interleaving or acquiring ^1H and ^{19}F NMR at the same time, providing both anatomical and pharmacological and/or physiological data simultaneously (Kendrick and Yannoni, 1987; Schnur *et al.*, 1990; Li *et al.*, 2000; Keupp *et al.*, 2006).

Given the continuing appearance of novel applications in the field and developing interest in ^{19}F NMR, this chapter will provide both a historical perspective and review recent developments.

12.2 Clinical Applications and Drug Metabolism

Several drugs in clinical use include a fluorine atom, and representative drugs which have been examined by ^{19}F NMR in clinical studies are shown in Fig. 12.2. Most studies to date have examined pharmacokinetics and metabolism of fluoropyrimidines, particularly, 5-fluorouracil (5FU). 5FU was first developed in the 1950s and remains a primary drug in treatment of many cancers, but it has a narrow range of efficacy/toxicity (Heidelberger *et al.*, 1957; van Laarhoven *et al.*, 2005; Isanbor and O'Hagan, 2006). Presumably, both response and toxicity are related to pharmacokinetics and there is interest in assessing dynamics of uptake, biodistribution and metabolism.

Over 200 studies have reported ^{19}F NMR investigations of 5FU in clinical trials and evaluation in animal models, as reviewed by others (Bachert, 1998; Martino *et al.*, 2000; Wolf *et al.*, 2000; van Laarhoven *et al.*, 2005; Reid and Murphy, 2008). 5FU requires anabolic conversion to nucleosides (e.g. FdUrd, FdUmp) and nucleotides for cytostatic activity involving various kinases and phosphorylases (Wolf *et al.*, 2000). However, competing catabolic reactions convert 5FU to less cytotoxic metabolites such as 5,6-dihydrofluorouracil (DHFU) and α-fluoro β-alanine (FBAL) (Peters, 1988; Bachert, 1998; Wolf *et al.*, 2000). Localized NMR spectroscopy and low-resolution chemical shift imaging (CSI) have examined pharmacokinetics in mice, rats, and human patients (Glaholm *et al.*, 1990; Brix *et al.*, 1995, 1996, 1998, 1999). In many cases surface coils were used to achieve enhanced signal, sometimes compromising defined signal location. Glaholm *et al.* (1990) compared unlocalized signal from the liver of patients receiving 5FU with a FROGS sequence to suppress signal from superficial tissues. They achieved spectra in 2–4 min and were able to follow metabolism and clearance over 80 min post infusion. Klomp *et al.* (2003) developed three enhanced methods for evaluating 5FU pharmacokinetics using, (i) circularly polarized coils together with integrated preamplifiers; (ii) optimal pulse angle (Ernst angle) acquisition following T_1 determination; and (iii) optimized averaged CSI phase-encoding steps with a Hanning filter as a weighting function. Spectra still required 4 min acquisition time at 1.5 T, but were rigorously localized to a 4 × 4 × 4 cm voxel (Klomp *et al.*, 2003). NMR of excised tissue and body fluids has also provided insight into metabolism and can provide much higher sensitivity, for example, micromolar.

Patients with enhanced tumour retention of 5FU may be expected to exhibit better response (Presant *et al.*, 1994) though it does not guarantee efficacy (Wolf *et al.*, 2000). Application of collagenase to mice with subcutaneous HT29 colon tumours was found to increase 5FU uptake, attributed to disintegration of tumour collagen together with reduced interstitial fluid pressure (Gade *et al.*, 2009). However, it was accompanied by increased generation of non-toxic catabolites rather than cytotoxic fluoronucleotides. While the outcome may not have been optimal in this case it serves to demonstrate the use of ^{19}F NMR to evaluate novel therapeutic approaches. The retention of 5FU is reported to be considerably enhanced in tumours with lower pH (Guerquin-Kern *et al.*, 1991;

McSheehy *et al.*, 1998; Ojugo *et al.*, 1998) prompting investigations of the ability to alter pharmacokinetics by modulation of tumour pH to increase activity (Griffiths *et al.*, 2001; van Laarhoven *et al.*, 2006). Since the chemical shifts of some fluoronucleotides derived from 5FU are sensitive to pH this could itself be used to measure intracellular pH (pHi) directly *in vivo*, although the presence of a mixture of products may complicate interpretation (Sijens *et al.*, 1991; McSheehy *et al.*, 1998; Lutz and Hull, 1999).

Attempts to mitigate dose-limiting toxicity of 5FU exploit various prodrugs and mixture formulations (e.g. capecitabine (Xeloda), tegafur-uracil (Uftoral), emitefur (3(3-(6-benzoyloxy-3-cyano-2-pyridyloxycarbonyl) benzoyl)-1-ethoxymethyl-5-fluorouracil)), and ^{19}F NMR has played a role in analysis and development (Martino *et al.*, 2000; Desmoulin *et al.*, 2002; van Laarhoven *et al.*, 2005). An alternative approach uses prodrugs in conjunction with gene therapy. Specifically, cytosine deaminase converts the relatively innocuous 5-fluorocytosine (5FC) to 5FU, which can be monitored via a $\Delta\delta = 2$ ppm ^{19}F NMR chemical shift (Stegman *et al.*, 1999; Corban-Wilhelm *et al.*, 2002; Yu *et al.*, 2005a; Dresselaers *et al.*, 2006). This gene therapy approach has been applied to patients, but investigators chose not to use ^{19}F NMR evaluations. Instead an additional reporter gene, hNIS (the human sodium iodine symporter) was included and evaluated using CT or SPECT of cold iodine or radioactive pertechnetate accumulation (Barton *et al.*, 2003, 2004, 2008; Brown *et al.*, 2007).

Other chemotherapeutic drugs have been evaluated by ^{19}F NMR in mice, but apparently not in patients to date. Gemcitabine has been detected in human tumour xenografts by ^{19}F NMR following IP injection, and kinetics have been investigated with respect to vasoactive drugs (Blackstock *et al.*, 2001; Cron *et al.*, 2008). Metabolite signals have been observed in liver and bladder using CSI (Bellemann *et al.*, 1999). In other studies, McSheehy *et al.* (1999) investigated the novel thymidine synthase inhibitor ZD9331. Brix *et al.* (2005) evaluated a trifluoromethylated derivative of 3-aminobenzamide, an inhibitor of poly(ADP-ribo) polymerase1 (PARP-1), and Spees *et al.* (2005) followed pharmacokinetics of fluorine-labelled methotrexate in rodents.

Apart from cancer chemotherapeutics, most *in vivo* ^{19}F NMR has examined psychiatric agents (Bartels and Albert, 1995; Passe *et al.*, 1995). Fluoxetine (Prozac; Fig. 12.2) was observed in preclinical animal models and human volunteers (Strauss *et al.*, 2002; Bolo *et al.*, 2004; Henry *et al.*, 2005),

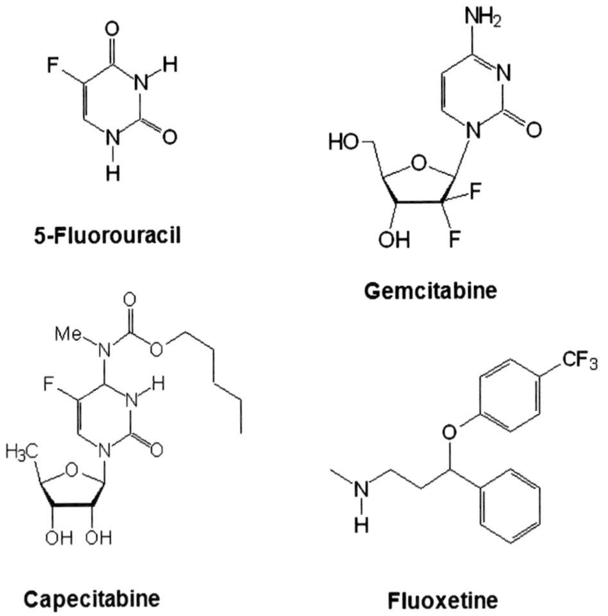

Figure 12.2. Representative fluorinated pharmaceuticals. Pharmaceuticals for which clinical or preclinical *in vivo* NMR studies have been reported: 5-fluorouracil (5FU) (Martino *et al.*, 2000; Wolf *et al.*, 2000; van Laarhoven *et al.*, 2005), gemcitabine (Blackstock *et al.*, 2001), capecitabine (Chung *et al.*, 2004) and fluoxetine (Bolo *et al.*, 2004).

and other studies examined fluvoxamine (Strauss *et al.*, 1998, 2002), dexfenfluramine (Christensen *et al.*, 1999) or haloperidol decanoate in schizophrenic patients (Sassa *et al.*, 2002). While CF_3 groups are favourable, there was a report of ^{19}F NMR detection of the psychotropic drug paroxetin in the human brain based on its single fluorine atom (Henry *et al.*, 2000).

Perfluorocarbons (PFCs) have been administered to patients for various indications: emulsions were developed as potential synthetic blood substitutes (Riess, 2001) and clinical trials tested the ability to enhance tumour oxygenation (Evans *et al.*, 1993). We were able to detect fluosol from surrounding tissues as long as 1 year after administration and tumour resection (Nunnally *et al.*, 1988). Neat perfluorotributylamine (PFTB) has been used as a tamponade in eye surgery. Residual PFTB has been detected in patients at 1.5 T (Gewiese *et al.*, 1992; Wilson *et al.*, 1992a) following eye surgery and the sensitivity of the spin lattice relaxation rate (R_1) to oxygen was used to make pO_2 measurements as discussed in Section 3.1. Perfluorononane has been used to explore the GI tract in man and mice at 1.5 T (Schwarz *et al.*, 2002). In mice the ^{19}F signal was evaluated directly, whereas in man the presence of PFC was detected as absence of ^{1}H signal.

A few studies have reported ^{19}F NMR of anaesthetics, since many gaseous anaesthetics are fluorinated (e.g. halothane, enflurane isoflurane, sevoflurane and desflurane). A halothane signal was observed in human brain up to 90 min after the withdrawal of anaesthetic (Menon *et al.*, 1993). Isoflurane was found to exhibit biphasic elimination with decay half-times of 9.5 and 130 min (Lockwood *et al.*, 1997). Others have studied the metabolism of volatile anaesthetics, showing generation of potentially toxic metabolites such as methoxydifluoroacetate, dichloroacetate, and fluoride ion from methoxyflurane (Selinsky *et al.*, 1988a,b).

In some cases metabolites (degradation products or excretory bioconjugates) are derived from fluorine-containing drugs: in other cases ^{19}F labels may be added for the absorption, distribution, metabolism and excretion toxicity (ADMET) process to learn about pathways, even though the labels are not included in the ultimate pharmaceuticals (Reid and Murphy, 2008). In several cases glucuronides have been identified as key detoxification products (Ellis *et al.*, 1995; Sidelmann *et al.*, 1996; Scarfe *et al.*, 1999). A recent review (Cobb and Murphy, 2009) explores many different applications of ^{19}F NMR to chemical biology, including xenobiotic metabolism, protein structural folding and enzyme mechanisms. For structural studies, fluorinated amino acids have been exploited (e.g. F-tryptophan, F-tyrosine, difluoromethionine, trifluoroleucine and F-serine), both to interrogate variations in microenvironment and modify specific binding sites (Yoder and Kumar, 2002; Chiu *et al.*, 2009; Montclare *et al.*, 2009). Fluorinated sialic acid and mannosamine analogues have been used to modify cell-surface presentation of fluorinated glycans (Dafik *et al.*, 2008).

Given the key role of PET for staging tumours and monitoring metastases based on elevated metabolic activity and fluorodeoxyglucose (FDG) uptake (Fowler *et al.*, 2004), there have been attempts to characterize the metabolism by ^{19}F NMR. FDG is recognized by glucose transporters and enters cells, where it is effectively phosphorylated and trapped. FDG accumulates in metabolically active cells, such as brain and myocardium, and many tumours, though some 'indolent tumours' fail to show strong activity (e.g. prostate and paediatric Ewing sarcoma). PET can assess retention with great sensitivity, but provides no metabolic information. ^{19}F NMR can be used to differentiate individual metabolites from anabolic and catabolic processes, but since NMR studies typically require millimolar concentrations, as opposed to nano- or micromolar for PET, there is concern that metabolic fates may differ; thus ^{19}F NMR has been used sparingly to examine FDG (Nakada *et al.*, 1986, 1988a,b; Berkowitz and Ackerman, 1987). Similarly, ^{18}F PET shows promise for characterizing tumour hypoxia using F-misonidazole or EF5 (Krohn *et al.*, 2008). Again ^{19}F NMR analogues have been tested (Table 12.1) (Robinson and Griffiths, 2004; Salmon and Siemann, 2004; Procissi *et al.*, 2007) and indeed retention of SR-4554 was reported in tumours in humans (Seddon *et al.*, 2002, 2003; Workman *et al.*, 2006), but generally the signal-to-noise is so low as to provide little hope of utility in patients.

To date, clinical ^{19}F NMR spectroscopy and imaging have been rather disappointing, but promising new agents and detection strategies promise a brighter future, as discussed below. In the meantime, preclinical studies of drug metabolism and pathophysiology using specific reporter molecules have been highly successful and there is a tremendous new surge in interest in developing novel agents and detection paradigms.

12.3 Reporter Molecule Strategies

The extraordinary sensitivity of the NMR properties of ^{19}F to its microenvironment has made it a favourite for the design of active reporter agents based on three concepts: (i) *physical interactions* exemplified by perfluorocarbons, which reveal oxygen tension based on modification of relaxation parameters; (ii) *chemical association*, exemplified by ion-binding agents revealing pH and metal ion concentrations; and (iii) *chemical substrates*, as exemplified by gene reporter molecules, where substrates are cleaved by specific enzyme activity generating a chemical shift. There are also passive agents, detectable merely by their presence, which may reveal lung ventilation or vascular volume. These concepts are presented in Table 12.1 and described in greater detail in the following sections with examples demonstrating successful applications.

12.3.1 *Physical interactions*

Most tissues depend on oxygen for life to allow oxidative phosphorylation. Glycolysis can sustain normal tissues for short periods, but rapid cramps soon occur. More seriously, ischaemic hypoxia causes angina and often myocardial death; likewise, cerebral stroke. Aberrant vascular morphology and perfusion are hallmarks of tumours and they are often hypoxic ($pO_2 <$ 10 torr) or anoxic (have necrotic cores). Low pO_2 (<10 torr) appears to stimulate tumour aggressiveness through angiogenesis and metastasis. Meanwhile, hypoxic tissues resist radiation therapy. Thus, there is widespread interest in assessing tissue oxygenation, ideally with high spatial and temporal resolution and sufficient precision to predict disease progression and response to therapy (i.e. prognostic radiology) (Tatum *et al.*, 2006).

Oxygen (O_2) is paramagnetic and causes nuclear spin relaxation. Indeed, it has long been recognized that oxygen must be rigorously excluded from samples if absolute relaxation rates are to be measured. This may be achieved by adding reducing agents to a solution or more commonly with repeated freeze–thaw cycles *in vacuo* and nitrogen gassing. Oxygen-dependent water proton signal relaxation has been used to estimate pO_2 directly in tissues such as vitreous humour and cerebrospinal fluid (CSF), which have well-defined compositions (Berkowitz *et al.*, 2001). In other cases, absolute pO_2 estimates may not be reliable since relaxation is influenced by multiple properties such as pH, temperature, ionic strength and status of dissolved proteins. Indeed, irreversible changes in relaxation are observed in tumours upon heating (Lewa and Majewska, 1980). Nevertheless, changes in R_1 may reflect changes in pO_2, as exploited in the tissue oxygen level dependent (TOLD) proton NMR contrast approach to assessing tumour oxygenation (Matsumoto *et al.*, 2006). The beauty of TOLD is the simplicity of measuring tissue T_1, and indeed several groups have used this approach to examine tissue response to hyperoxic gas breathing in animals and patients (Matsumoto *et al.*, 2006; O'Connor *et al.*, 2007, 2009; Mason, 2009).

For an ideal liquid–gas interaction, the spin–lattice relaxation rate $R_1 = A + B\ pO_2$. In early experiments, several groups noted that ^{19}F NMR of perfluorocarbons (PFCs) exhibited particularly high sensitivity to changes in pO_2. This is attributed to the extremely high solubility of O_2 in PFCs, which allows them to act as molecular amplifiers for the presence of oxygen.

Significantly, the relationship remains linear across the whole range of pO_2 values and Delpuech *et al.* (1979) showed continued sensitivity under hyperbaric conditions. Basic physical principles related to Henry's Law and Dalton's Law of partial pressures ensure that pO_2 in a droplet of PFC will match that of surrounding water or tissue; thus the relaxation measurement of PFC in a tissue reflects the local pO_2. Importantly, PFCs are exceedingly hydrophobic; thus metal ions and proteins do not dissolve, hence avoiding any potential perturbation of relaxation (Lai *et al.*, 1984; Eidelberg *et al.*, 1988a; Thomas *et al.*, 1994). The sensitivity of R_1 to pO_2 is both field- and temperature-dependent, and thus appropriate calibration curves are required (Mason *et al.*, 1993, 1996; Shukla *et al.*, 1995). Calibration curves have been reported for several PFCs at various magnetic fields, as reviewed previously (Zhao *et al.*, 2004). Investigators are warned to examine the original literature for technical details, such as pulse sequences, temperature regulation and spectral resolution. For the emulsion of perfluorotributylamine (Oxypherol), we showed that calibration curves obtained in solution were valid in living tissues (Mason *et al.*, 1993).

R_1 is sensitive to temperature, and even a relatively small error in temperature estimate can introduce a sizable discrepancy into the apparent pO_2 for some PFCs. The relative error introduced into a pO_2 determination by a 1°C error in temperature estimate ranges from 8 torr/°C for perfluorotributylamine (Mason *et al.*, 1993), to 3 torr/°C for perflubron (PFOB) (Mason *et al.*, 1992) or 15-crown-5-ether (Dardzinski and Sotak, 1994), when pO_2 is actually 5 torr. Hexafluorobenzene (HFB) exhibits a remarkable lack of temperature dependence and the comparative error would be 0.1 torr/°C (Mason *et al.*, 1996). Exploiting the differential sensitivity of pairs of resonances within a single molecule to pO_2 and temperature allows both parameters to be calculated by solving simultaneous equations (Mason *et al.*, 1993; Mason and Antich, 1995). However, generally it is preferable for a pO_2 sensor to exhibit minimal response to temperature, since this is not always known precisely *in vivo*, and temperature gradients may occur across tumours.

The very hydrophobicity which prevents ions from mixing with perfluorocarbons and potentially perturbing calibration curves means that emulsification of perfluorocarbons is required for systemic administration in biological applications, e.g. intravenous infusion. PFC emulsions have been developed commercially both as potential synthetic blood substitutes (Riess, 1992; Krafft, 2001) and as ultrasound contrast agents (Riess, 2001; Schutt *et al.*, 2003). Following IV infusion, a typical blood substitute emulsion circulates in the vasculature with a half-life of 12 h providing substantial clearance within 2 days (Kaufman, 1991). Some investigators have examined tissue vascular pO_2, while PFC remained in the blood (Fishman *et al.*, 1987, 1989; Eidelberg *et al.*, 1988a,b; Noth *et al.*, 1995). Generally, investigators are interested in tissue pO_2 and await vascular clearance. Most PFC becomes sequestered in the reticuloendothelial system, allowing effective investigations of pO_2 in the liver or spleen (Holland *et al.*, 1993; Barker *et al.*, 1994; Dardzinski and Sotak, 1994). Long-term retention in tissues allows pO_2 measurements to be made repeatedly *in vivo* over a period of weeks, and extensive studies have been reported in liver, spleen, lungs and perfused heart of mice, rats, rabbits and pigs (Holland *et al.*, 1993; Barker *et al.*, 1994; Dardzinski and Sotak, 1994; Shukla *et al.*, 1994, 1996; Tran *et al.*, 1995; Thomas *et al.*, 1996). However, generating sufficient accumulation in tissues such as heart or tumour may require such large doses that animals exhibit extensive hepatomegaly or splenomegaly. Nonetheless, there have been many reports of tumour oximetry following IV administration of PFC emulsions and vascular clearance (Mason *et al.*, 1991, 1994; Hees and Sotak, 1993; Dardzinski and Sotak, 1994; Tran *et al.*, 1995; Baldwin and Ng, 1996; Helmer *et al.*, 1998; McIntyre *et al.*, 1999; van der Sanden *et al.*, 1999b; Fan *et al.*, 2002; McNab *et al.*, 2004). Targeting cardiac tissue directly could also improve signal-to-noise ratio and this has been a goal of Wickline and coworkers (Morawski *et al.*, 2004).

The most extensive use of ^{19}F NMR oximetry has been to investigate tumour oxygenation, with both acute studies of interventions and chronic studies of growth. Uptake and deposition of PFC emulsions in tumours is highly variable and heterogeneous, with most signal occurring in well perfused regions (Mason *et al.*, 1994; McIntyre *et al.*, 1999). Indeed, pO_2 values measured soon after intravenous infusion but following vascular clearance, often approach arterial pO_2 (Mason *et al.*, 1994). Thus, physiological measurements with respect to intervention are biased towards the well-perfused well-oxygenated regions. Interestingly, following sequestration, PFC does not seem to redistribute within tissue, but remains associated with specific locations (Mason *et al.*, 1994). Long tissue retention has the advantage of facilitating chronic studies during tumour development, and progressive tumour hypoxiation has been observed over many days (Mason *et al.*, 1994; Baldwin and Ng, 1996).

An alternative approach uses direct injection of neat PFC into the tissue of interest, for example measurement of retinal oxygenation (Berkowitz *et al.*, 1991; Wilson *et al.*, 1992b; Zhang *et al.*, 2003), cerebral oxygenation in the interstitial and ventricular spaces (Duong *et al.*, 2001), thigh muscle (Kodibagkar *et al.*, 2008b) and several other organs (Kodibagkar *et al.*, 2008a; Liu *et al.*, 2009). We favour a direct injection of PFC into tissue, since it avoids waiting for vascular clearance and allows less well perfused regions to be interrogated immediately (Zhao *et al.*, 2004). Multi-resonant PFCs have been widely used for spectroscopy, but add complexity to MRI (Babcock *et al.*, 1991).

PFCs with a single resonance provide optimal signal-to-noise ratio and simplify imaging: two agents, hexafluorobenzene (HFB) (Zhao *et al.*, 2001a,b, 2002, 2003a,b, 2004; Mason *et al.*, 2002; Song *et al.*, 2002; Kim *et al.*, 2003) and perfluoro-15-crown-5-ether (15C5) (Dardzinski and Sotak, 1994; van der Sanden *et al.*, 1999a; Duong *et al.*, 2001; Wang *et al.*, 2002) have found extensive use. As an aside it is interesting to note that HFB has been used to enhance the pO_2 sensitivity of the ESR reporter molecule perchlorotriphenylmethyl triester radical (Bratasz *et al.*, 2007). We favour HFB as an NMR reporter molecule (Mason *et al.*, 1996) for several reasons: symmetry provides a single narrow ^{19}F NMR signal and the spin–lattice relaxation rate is highly sensitive to changes in pO_2 yet minimally responsive to temperature (Delpuech *et al.*, 1979; Hamza *et al.*, 1981; Mason *et al.*, 1996). HFB also has a long spin–spin relaxation time (T_2), which is particularly important for imaging investigations. HFB is well characterized in terms of lack of toxicity (Gorsman and Kapitonenko, 1973; Mortelmans and Simmon, 1981; Courtney and Andrews, 1984; Rietjens *et al.*, 1995).

We have undertaken extensive studies using direct intratumoural (IT) injection of neat HFB, and other research teams have also recently reported such studies (McIntyre *et al.*, 1999; Zhao *et al.*, 2004; Jordan *et al.*, 2009; Liu *et al.*, 2009). For spectroscopy, as little as 10 μl provides adequate signal to obtain precise pO_2 measurements, which are highly localized by virtue of the discrete reporter location. However, for recognizing tumour heterogeneity, pO_2 maps are more useful. HFB exhibits long T_1 relaxation times reaching 12 s under hypoxic conditions at 4.7 T, and thus accelerated imaging methods are important. We developed fluorocarbon relaxometry using echo planar imaging for dynamic oxygen mapping (FREDOM) (Zhao *et al.*, 2004), which typically provides 50–150 individual pO_2 measurements across a tumour simultaneously in about 6.5 min with a precision of 1 to 3 torr in relatively hypoxic regions based on 50 μl injected dose. Gallez and coworkers recently reported a Look–Locker relaxation measurement accelerating data acquisition still further (Jordan *et al.*, 2009). In both muscle and tumour tissues, pO_2 heterogeneity is apparent when rats breathe air (pO_2 ranged from 0 to 100 torr). Repeat measurements are highly reproducible and generally quite stable under baseline conditions. Upon challenge with oxygen breathing, essentially all muscle regions showed a significant increase in oxygenation. Some tumours show little response to hyperoxic gas breathing, e.g. Dunning rat prostate R3327-AT1 (Hunjan *et al.*, 2001; Bourke *et al.*, 2007), but many, including the 13762NF rat breast tumour, Dunning prostate R3327-HI and H rat tumours and lung H460 tumours, show extensive response (Song *et al.*, 2002; Zhao *et al.*, 2002, 2003b, 2009; Xia *et al.*, 2006; Krohn *et al.*, 2008). Effective modulation of human A549 lung tumour xenografts in nude rat is shown in Fig. 12.3.

We have used FREDOM to examine the effects of vascular targeting agents (Mason *et al.*, 2002; Zhao *et al.*, 2005), vasoactive agents (Zhao *et al.*, 2001b) and hyperoxic gases (Le *et al.*, 1997; Hunjan *et al.*, 2001; Zhao *et al.*, 2001a,b, 2002, 2003a,b, 2004; Song *et al.*, 2002; Kim *et al.*, 2003; Xia *et al.*, 2006). Correlative measurements have shown that pO_2 distributions and dynamic responses to interventions are consistent with sequential determinations made using electrodes (Mason *et al.*, 1999, 2003) or fibre optic probes (FOXY and OxyLite) (Zhao *et al.*, 2001a; Gu *et al.*, 2003). Results are also consistent with hypoxia estimates using the histological marker pimonidazole (Zhao *et al.*, 2003b).

Most significantly, estimates of pO_2 and modulation of tumour hypoxia are found to be consistent with modified tumour response to irradiation (Zhao *et al.*, 2003a; Bourke *et al.*, 2007). Such prognostic capability could be important in the clinic, since it is known that relatively hypoxic tumours tend to be more aggressive and respond less well to radiation therapy

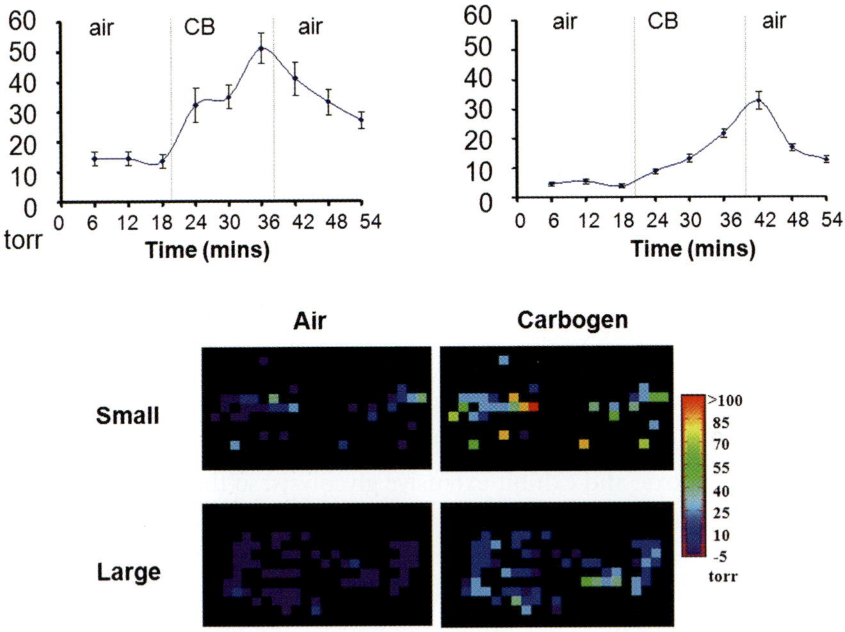

Figure 12.3. Oxygen dynamics in A549 human lung tumour xenografts in nude rats. pO_2 maps obtained using FREDOM for a small (~0.25 cm^3) and a large (3 cm^3) A549 tumour, respectively, when rats breathed air followed by carbogen (95% O_2/5%CO_2). In each case 50 μl HFB was injected directly into the tumour in a fan pattern to investigate both central and peripheral tumour regions. The smaller tumour (upper maps and graph at left) was better oxygenated during air breathing, though both tumours showed extensive hypoxia. In response to carbogen breathing, both tumours showed rapid significant increase in pO_2. Return to air breathing was followed by decrease in pO_2, though it was delayed for the larger tumour. The graphs show mean ± standard deviation.

(Höckel *et al.*, 1996; Fyles *et al.*, 1998; Rofstad *et al.*, 2000). Historically, the inability to assess tumour oxygenation has hindered the ability to develop successful hypoxia-modifying therapies. Notably, well oxygenated tumours benefit little from oxygen inhalation in conjunction with irradiation. Therefore, it is crucial to be able to identify those tumours which are hypoxic and which are amenable to a particular modulation strategy. In preclinical studies we believe this to be one of the most important applications of ^{19}F MRI, though translation to the clinic is hindered by the need for a reporter molecule and the lack of widespread clinical ^{19}F MRI capability.

12.3.2 *Chemical association*

Protons and various metal ions are crucial to cellular health, and concentration imbalances have been associated with many diseases (Gupta and Gillies, 1987; Gillies *et al.*, 2004). Electrodes may be used to measure extracellular concentrations, but are invasive. Many specific fluorescent dyes are available, though application is usually restricted to cell cultures and *in situ* calibration may be required. NMR can provide direct measurements of pH based on the ^{31}P NMR chemical shift sensitivity of inorganic phosphate (Pi). Muscle has intense phosphocreatine (PCr), which serves as a convenient chemical shift-invariant reference standard, and also intense adenosine triphosphate (ATP), but Pi may occur at low concentration until there is distress such as ischaemic insult. Tumours tend to be somewhat hypoxic and exhibit extensive glycolysis, so that there is usually extensive Pi. Indeed, separate signals are often detected attributable to intra- and extracellular compartments, and this was crucial in identifying the reversed pH gradient characteristic of tumours (Stubbs *et al.*, 1994). The chemical shift of ATP is influenced both by binding magnesium ions and pH, and has been used successfully in studying both [Mg^{2+}] and pH in muscle physiology (Odvina *et al.*, 2006). However, ^{31}P NMR is quite insensitive in terms of both signal-to-noise ratio and chemical shift response and dispersion. Signals of intra- and extracellular Pi are often only partially resolved and 3-aminopropylphosphonate (3-APP) has been developed to assuredly represent extracellular pH (Gillies *et al.*, 1994). 3-APP has a single ^{31}P resonance well removed from endogenous signals, a pH-dependent chemical shift and pK_a in the physiological range.

^{19}F NMR indicators have been developed for pH and metal ion investigations, and in many cases offer superior chemical shift response, good signal-to-noise ratio and spectra uncluttered by endogenous metabolites. Many molecules exhibit pH-sensitive chemical shift; representative reporter molecules are shown in Table 12.2. A series of fluoroalanines were used to investigate intra- and extracellular pH (Taylor and Deutsch, 1983). Fluoropyridoxol, a vitamin B6 analogue, was used to investigate activity of phosphorylase enzymes (Chang and Graves, 1985). We adopted this molecule for measurements in cells and perfused organs (Mehta *et al.*, 1994b; Hunjan *et al.*, 1998; Mason, 1999). On the NMR timescale, protonated and deprotonated moieties are generally in fast exchange, so that a single signal is observed representing the amplitude-weighted mean of protonated and deprotonated forms.

Reporter molecules must reach the compartment of interest. Some appear to be restricted to the extracellular compartment, while others may cross cell membranes revealing both intra- and extracellular pH. In most cases exchange between the compartments is slow, so that separate signals are observed for interstitial and cytosolic milieu, though occasionally exchange is so rapid as to provide a broadened signal representing an average value (Deutsch and Taylor, 1987a).

6-Trifluoromethylpyridoxol (CF_3POL) is found to occur exclusively in the extracellular compartment, and thus reports pH_e or interstitial pH (Cui *et al.*, 2003; Yu *et al.*, 2005a). A fluoroaniline sulfonamide (ZK150471) is also restricted to the extracellular compartment and has been used to measure tumour pH in mice and rats (Frenzel *et al.*, 1994; Aoki *et al.*, 1996; Miyazawa *et al.*, 1996; Ojugo *et al.*, 1999). Combination with ^{31}P NMR of Pi to determine intracellular pHi has been used to reveal the transmembrane pH gradient in mouse tumours (McSheehy *et al.*, 1998). A distinct problem with ZK150471 is that the pK_a differs in saline and plasma (Aoki *et al.*, 1996).

Cell penetrating prodrugs have been used extensively based on labile esters, which are cleaved by non-specific intracellular esterases. The series of agents 3-monofluoro-, 3,3-difluoro-, and 3,3,3-trifluoro-2-amino-2-methyl propanoic acid (Deutsch *et al.*, 1982, 1984; Kashiwagura *et al.*, 1984; Deutsch and Taylor, 1987a,b) were used to investigate diverse cells and perfused organs. Esters are relatively permeable, stable in water and undergo non-specific enzymatic hydrolysis intracellularly, liberating the pH-sensitive molecules (Deutsch and Taylor, 1987b). This approach can lead to complex spectra including overlapping multiline ester and liberated free acid resonances from both intra- and extracellular compartments, potentially necessitating cell rinsing after reporter agent loading.

Aliphatic indicators often have a relatively small chemical shift range of about 2 ppm, whereas aromatic reporter molecules can have a much larger chemical shift pH response (Table 12.2), though it depends strongly on orientation (Table 12.3). *p*-Fluorophenols show a large chemical shift response $\Delta\delta = 6.4$–11.3 ppm and *o*-fluorophenols have a smaller chemical shift range (about 0.3–2.2 ppm) (Table 12.3) (Yu *et al.*, 2004). The large

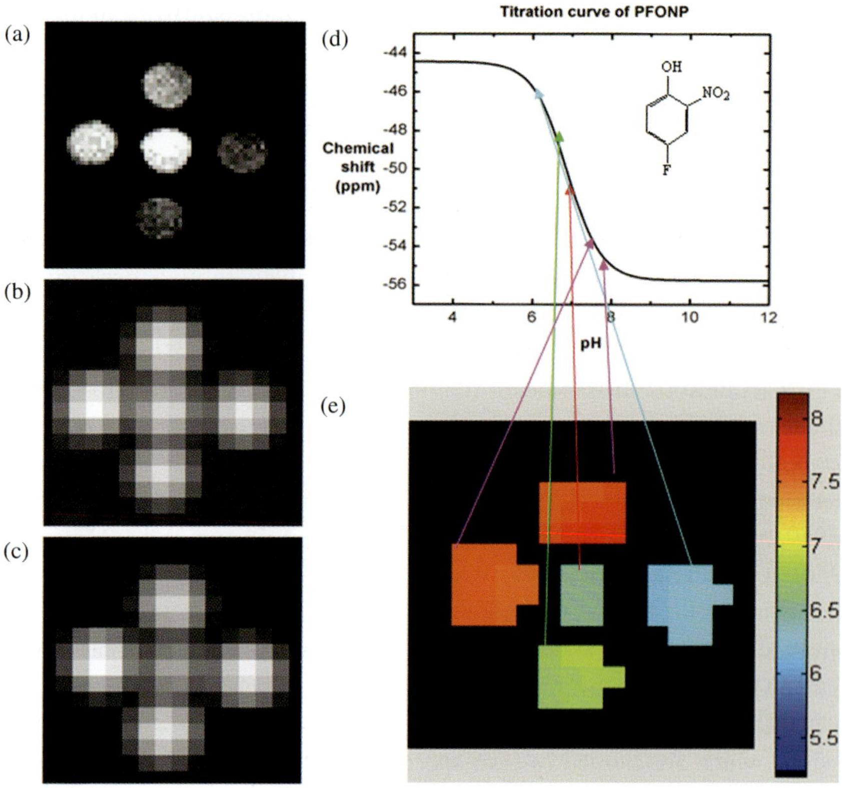

Figure 12.4. Mapping pH by ^{19}F MRI. A pH map was achieved using ^{19}F spin echo chemical shift imaging at 4.7 T (188.2 MHz for ^{19}F) of a five-vial phantom. Individual vials contained a solution of sodium trifluoroacetate together with *p*-fluoro-*o*-nitrophenol (PFONP) at pH 6.0, 6.6, 7.0; 7.6 and 8.0, respectively. CSI imaging parameters: FOV = 25 × 25 mm^2; spectral window = 75 ppm, slice thickness = 10 mm, matrix = 16 × 16; TR= 1 s, 4 avgs. (a) ^{1}H Scout image, (b) NaTFA image, (c) PFONP image, (d) pH titration curve (pK_a= 6.87; δ_{acid}= −44.44 ppm, δ_{base} = −55.76 ppm) and (e) pH map.

chemical shift range of *p*-fluoro-*o*-nitrophenol (PFONP) may be exploited to map pH in discrete locations as shown for a multivial phantom in Fig. 12.4. PFONP has also revealed pH gradients in whole blood consistent with electrode measurements, but fluorophenols must be used cautiously, since PFONP appears cytolytic for certain tumour cells and may act as an ionophore, by analogy with dinitrophenol.

Occasionally, reporter molecules are found to readily partition between intra- and extracellular compartments, allowing transmembrane pH gradients to be measured (Mehta *et al.*, 1994b; Hunjan *et al.*, 1998; Mason, 1999). Analogues of vitamin B6; e.g. 6-fluoropyridoxol (6-FPOL), are highly sensitive to pH (Korytnyk and Singh, 1963; Mehta *et al.*, 1994b; He *et al.*, 1998; Hunjan *et al.*, 1998; Mason, 1999) and revealed both intra- and extracellular pH (pHi and pHe), simultaneously, in whole blood (Mehta *et al.*, 1994b) and the perfused rat heart (Hunjan *et al.*, 1998). FPOL and its analogues appeared to be ideal reporter molecules, since the pK_a is in the physiological range, toxicity is low and chemical shift response large. However, FPOL was later found to enter most tumour cells very sparingly, or not at all, defying efforts to investigate cancer. Ease of entry into blood cells may be related to facilitated transport, since vitamin B6 is naturally stored, transported and redistributed by erythrocytes (Yamada and Tsuji, 1970). Extensive cell penetration was found in one tumour cell line: a Morris hepatoma transfected to express thymidine kinase (Yu *et al.*, 2005a).

To enhance signal-to-noise ratio, or reduce the required dose, a pH-sensitive CF_3 moiety could be introduced in place of the F-atom; however, the chemical shift response is typically smaller (Table 12.2), as expected, since electronic sensing must be transmitted through an additional C–C bond (Yu *et al.*, 2006). Most indicators require an additional chemical shift reference standard, e.g. sodium trifluoroacetate, but *N*-ethylaminophenol (NEAP) (Rhee *et al.*, 1995), 6-FPOL-5-α-CF_3 (He *et al.*, 1998) and ZK150471 (Frenzel *et al.*, 1994) all have non-titrating intramolecular chemical shift references.

For metal ion determination, intracellular loading may be even more crucial; Tsien (Tsien, 1981) made the important breakthrough of using acetoxymethyl esters to load fluorescent metal ion chelators into cells. He demonstrated 1,2-bis(*o*-aminophenoxy)ethane-*N*,*N*,*N*′,*N*′-tetraacetic acid (BAPTA) for detecting intracellular calcium ions, and subsequently Metcalfe *et al.* (1985) added *para*-fluoro atoms to the aromatic ring yielding a ^{19}F NMR responsive agent (5,5′-difluoro-1,2-bis(*o*-aminophenoxy)ethane-*N*,*N*,*N*′,*N*′-tetraacetic acid (5F-BAPTA)) (Fig. 12.1). Upon binding calcium, there is a change in chemical shift. Ideally, such a reporter molecule would have high specificity for the metal ion of interest. In fact 5F-BAPTA binds several divalent metal ions including Ca^{2+}, Zn^{2+}, Pb^{2+}, Fe^{2+} and Mn^{2+} (Smith *et al.*, 1983; Benters *et al.*, 1997), but importantly, each metal ion chelate has an individual chemical shift, so that they can be detected simultaneously (Gupta and Gillies, 1987). 5F-BAPTA includes two fluorine atoms symmetrically placed to provide a single signal. Upon binding there is slow exchange of Ca^{2+}, on and off the indicator, on the NMR timescale; separate signals are seen for the free and metal ion-bound moieties, with chemical shifts of several ppm. Measurements are based on the signal ratio, avoiding the need for a chemical shift reference, in contrast to pH reporters, which are usually in the fast exchange regimen. The dissociation constant (K_D) does depend on pH, ionic strength and the concentration of free Mg^{2+}, which need to be estimated independently. 5F-BAPTA has been used in various biological systems such as cells (Smith *et al.*, 1983; Schanne *et al.*, 1989a,b) and the perfused beating heart, revealing calcium transients during the myocardial cycle (Kirschenlohr *et al.*, 1988; Marban *et al.*, 1988; Kusuoka *et al.*, 1993). The 4F-BAPTA isomer has a somewhat lower binding constant and exhibits fast exchange (Gupta and Gillies, 1987), so that the signals from the bound and unbound forms are averaged.

Fluorinated NMR reporters have been demonstrated for Mg^{2+} (Murphy, 1993). The chemical shift of fluorocitrate (Kirschenlohr *et al.*, 1988) depends on [Mg^{2+}], but it is critical to use only the (+)-isomer, which has relatively little toxicity (Tecle and Casida, 1989). Fluorescent reporter molecules designed for calcium have been adapted for ^{19}F NMR investigations, e.g. *o*-aminophenol-*N,N,O*-triacetic acid (APTRA) (Levy *et al.*, 1988; Murphy *et al.*, 1989; London, 1994).

While the chemical association approach is normally applied to ion pairing, a fluorobenzene boronic acid was used to assay specific sugars (London and Gabel, 1994). Chemical association must be used cautiously since binding the substrate under investigation may cause buffering. As noted, reporter molecules must be in fast or slow exchange to provide narrow signals, enhancing both the signal-to-noise and spectral resolution. Ideally, the reporter ligand is highly selective for the ion of interest and of course the molecule should exhibit minimal toxicity.

12.3.3 *Chemical interactions*

As discussed in Section 2, ^{19}F NMR has been used to assess pharmacokinetics of various drugs both in terms of preclinical development and in some cases human studies. Observations of prodrug therapy based on conversion of 5FC to 5FU suggested the potential for developing other enzyme and/or gene reporter molecules. Detection of β-galactosidase (β-gal) activity appeared an attractive candidate since the *lacZ* gene is the most popular and widely used reporter gene (Beckwith and Zipser, 1970; Kruger *et al.*, 1999; Serebriiskii and Golemis, 2000). Notably β-gal shows extremely broad substrate specificity (promiscuity) and many diverse reporter agents are commercially available, e.g. the blue histological stain 5-bromo-4-chloro-3-indolyl β-*D*-galactoside (X-gal) (Li *et al.*, 2007), yellow *o*-nitrophenyl galactopyranoside (ONPG), red fluorescent 7-hydroxy-9*H*-(1,3-dichloro-9,9-dimethylacridin-2-one)-β-*D*-galactopyranoside (DDAOG) (Tung *et al.*, 2004; Zhang *et al.*, 2009), black S-Gal (Heuermann and Cosgrove, 2001) and chemiluminescent S-Galacton-Star (Bronstein *et al.*, 1989; Heuermann and Cosgrove, 2001; Kawaguchi *et al.*, 2002) for optical and histological applications, and in several cases *in vivo* utility has been demonstrated. Meade and coworkers demonstrated the ability to detect β-gal activity in developing tadpoles based on an activatable galactose-capped Gd-complex, but it required direct intracellular injection (Louie *et al.*, 2000). This prompted us to explore whether ^{19}F NMR could be used to monitor β-gal activity. ONPG seemed an obvious candidate for ^{19}F NMR active analogues. Fluorinated nitrophenol galactosides had been reported by Yoon *et al.* (1996) to explore β-gal activity. However, they placed a fluorine atom on the sugar moiety, which would be expected to provide little chemical shift response to cleavage, and they do not appear to have used ^{19}F NMR in their investigations. Our prototype molecule used a *para*-fluoroaryl substituent in 4-fluoro-2-nitrophenyl β-*D*-galactopyranoside (PFONPG; Tables 12.3 and 12.4), which proved effective as a substrate for β-galactosidase (Cui *et al.*, 2004). It provides a single ^{19}F NMR signal with a narrow linewidth and good stability in solution. It is stable in normal wild type cells and whole blood, but exposure to the enzyme or cells transfected to express β-galactosidase causes rapid cleavage in line with anticipated levels of transfection (Cui *et al.*, 2004). Upon cleavage of the glycosidic bond a substantial chemical shift $\Delta\delta > 3.6$ ppm is observed, sufficient to permit chemical shift-selective imaging of substrate and product (Kodibagkar *et al.*, 2006).

The released fluoronitrophenol aglycone is somewhat toxic and can cause lysis of fragile cells. Thus, we have synthesized isomers and analogues with less toxic aglycones. The chemical shift accompanying cleavage depends strongly on the orientation of the F-atom with the largest response for *para*-F and less for *ortho*-F (Table 12.3). The rate of cleavage was closely related to the pK_a of the aglycone (Yu *et al.*, 2004) commensurate with enzyme studies reported previously (Richard *et al.*, 1995). While OFPNPG shows a smaller chemical shift response, it is sufficient to detect the activity of β-gal *in vivo* as presented in Fig. 12.5 for a stably transfected 9L-*lacZ* glioma. Given the different chemical shifts of individual substrates and products, multiple reporters can be monitored simultaneously. Indeed, we have demonstrated the use of the reporter pair 4-fluoro-2-nitrophenyl β-*D*-galactopyranoside (PFONPG) and 2-fluoro-4-nitrophenyl β-*D*-galactopyranoside (OFPNPG) to differentiate wild type and *lacZ* expressing human breast and prostate tumour xenografts in mice (Yu *et al.*, 2008b).

3-*O*-(β-*D*-galactopyranosyl)-6-fluoropyridoxol (GFPOL) is less toxic, but also a less reactive substrate and less water soluble (Yu *et al.*, 2005b). Water solubility and reactivity could be enhanced by polyglycosylation of the hydroxymethyl arms (Yu and Mason, 2006). Introducing a trifluoromethyl (CF_3) reporter group, as opposed to the single F-atom should enhance the signal-to-noise ratio, but as also noted for pH indicators, the chemical shift response is much smaller (Table 12.3) due to transmission of the electron density redistribution through an additional carbon–carbon bond (Yu *et al.*, 2006).

The most widely used reactive ^{19}F NMR chemical reporters are fluoronitroimidazoles to detect hypoxia (Tables 12.1. and 12.4). There are clinical trials to detect hypoxia based on ^{18}F-PET and differential tissue retention (Krohn *et al.*, 2008). By analogy, several nitroimidazoles have been labelled with ^{19}F, providing NMR sensitive agents (Robinson and Griffiths, 2004; Procissi *et al.*, 2007). In some cases these are isomers, but in other cases multiple fluorine atoms have been introduced to enhance the signal-to-noise ratio, e.g. Ro 07-0741 (Seddon *et al.*, 2002, 2003; Robinson and Griffiths, 2004) has a single fluorine atom, SR-4554 has three equivalent fluorine atoms (Workman *et al.*, 1992) and CCI-103F has six fluorine atoms (bis-CF_3 groups) (Raleigh *et al.*, 1987; Cline *et al.*, 1997). Subsequent to administration, a washout period sufficient for elimination of unbound marker is required, since there is apparently no difference detectable *in vivo*

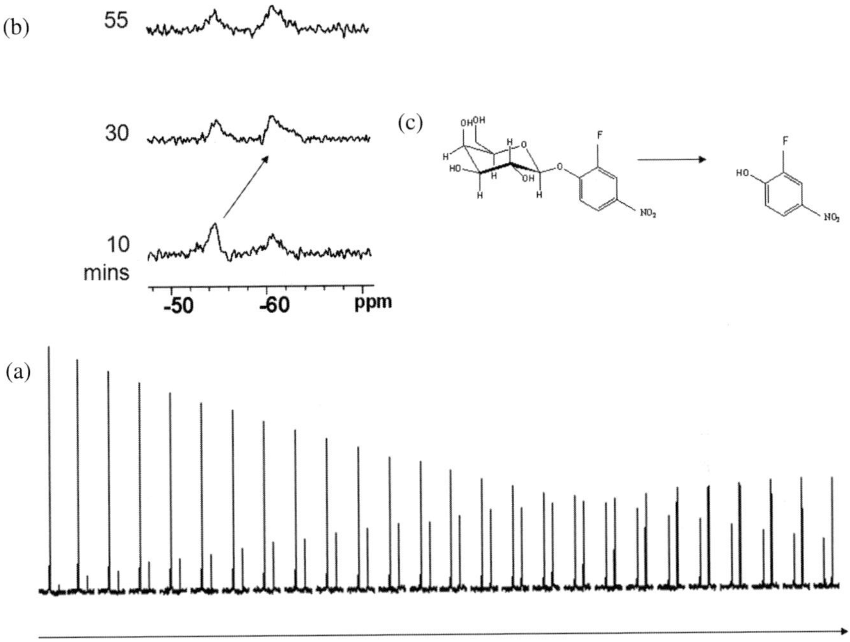

Figure 12.5. ^{19}F NMR detection of β-gal expression. (a) ^{19}F NMR time-course of OFPNPG hydrolysis by 9L-*lacZ* cells. OFPNPG (5.0 mg, 15.7 mmol) was added to stably transfected 9L-*lacZ* cells (5×10^6) in PBS (pH 7.4; 600 μL) at 37°C. ^{19}F NMR spectra were acquired in 102 s each and enhanced with an exponential line broadening (40 Hz). OFPNPG was observed at −54.93 ppm and OFPNP appeared at −61.04 ppm revealing β-gal activity with complete conversion after 40 min. (b) Detection of β-gal activity *in situ* in living tumour. A 9L-*lacZ* glioma was implanted subcutaneously in a mouse and allowed to grow to about 1.2 cm^3. A solution of OFPNPG (50 μl of 78 mg/ml in 1:1 water + DMSO) was injected intratumorally in 'fan' pattern and ^{19}F NMR spectra acquired over the next hour at 4.7 T ($T_R = 0.5$ s and total acquisition time 90 s). NaTFA provided a chemical shift reference. The 9L cells were a kind gift of Dr Stephen L. Brown (Henry Ford Hospital, Detroit) and were transfected by Dr Li Liu.

in the chemical shifts of the parent molecule and the metabolites (Robinson and Griffiths, 2004). Some reports have shown increased retention in hypoxic tissues, but the signal-to-noise ratio is often quite poor and additional factors seem to influence local tissue accumulation, such as perfusion, blood flow and glutathione concentration (Li *et al.*, 1991; Aboagye *et al.*, 1998; Robinson and Griffiths, 2004). Unlike radiochemical approaches

Table 12.4. ^{19}F NMR enzyme sensitive reporters.

Gene or enzyme detected	Structure	Sensitivity	References
LacZ β-Galactosidase (example in Fig. 12.5)	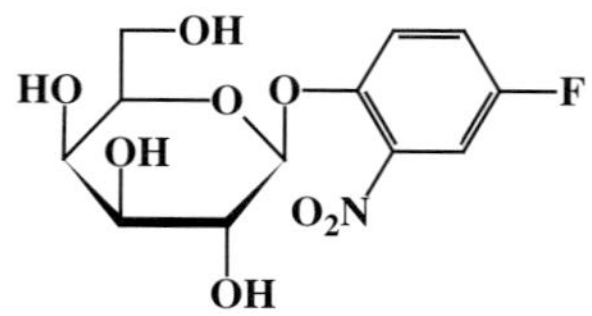 PFONPG	$\Delta\delta$ 9.84 ppm Release of aglycone Isomer dependant (see Table 12.3)	Cui *et al.*, 2004; Kodibagkar *et al.*, 2006; Liu *et al.*, 2007; Yu *et al.*, 2008b
Acid phosphatase (example in Fig. 12.6)	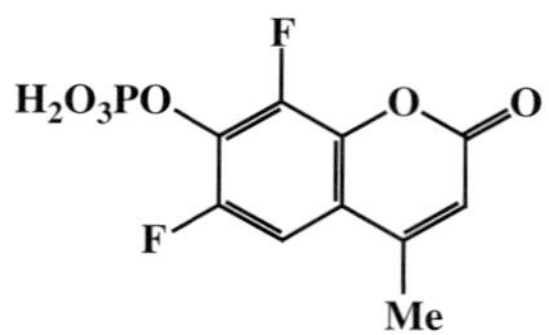 6,8-Difluoro-4-methylumbelliferyl phosphate	$\Delta\delta$ 5.8 and 8.8 ppm for separate resonances Release of phosphate	Gade *et al.*, 2008

(*Continued*)

Table 12.4. *(Continued)*

Gene or enzyme detected	Structure	Sensitivity	References
Hypoxia	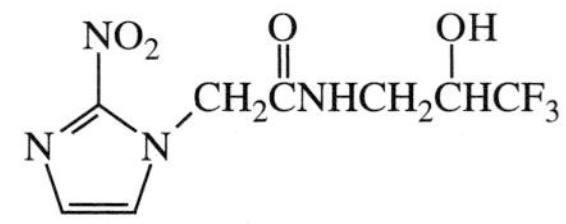 SR 4554	Accumulation Tumours, cells	Seddon *et al.*, 2002, 2003; Robinson and Griffiths, 2004)
Cytosine deaminase	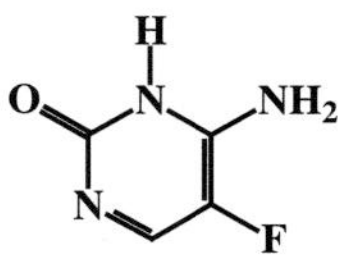	$\Delta\delta$ 1.2 ppm	Stegman *et al.*, 1999; Corban-Wilhelm *et al.*, 2002; Dresselaers *et al.*, 2003
Histone deacetylase inhibitor (HDACi)	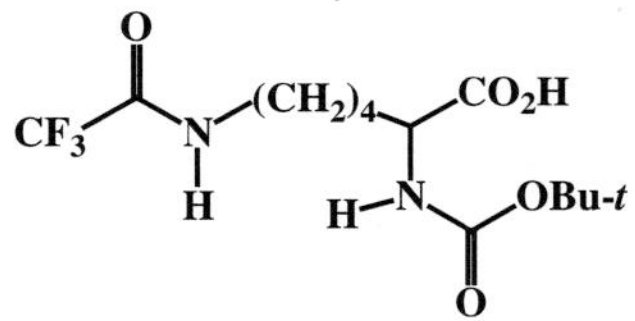 Boc-Lys-TFA-OH (BLT)	$\Delta\delta < 0.3$ ppm Release of TFA	Sankaranarayanapillai *et al.*, 2006, 2008

which detect all labelled molecules, NMR offers potential benefits but added complexity. Diverse adducts and metabolites would be expected to exhibit multiple chemical shifts, each at very low concentration, and polymeric adducts may have exceedingly short T_2; rendering them invisible for many NMR sequences (Salmon and Siemann, 2004). Nonetheless, a ^{19}F NMR Phase I clinical study reported a correlation between retention of SR4554 and pO_2, but required infusion at doses of 400–1600 mg/m^2, which could have adverse side effects (Seddon *et al.*, 2003).

While β-gal and hypoxia have received most attention, other enzymes are candidates for reporter molecule detection. As mentioned in Section 12.2, cytosine deaminase activity is detectable based on conversion of 5FC to 5FU. ^{19}F NMR was used to examine activity of carboxypeptidase-G2 on {4-[bis(2-chloroethyl)amino]-3,5-difluorobenzoyl}-L-glutamic acid as a prodrug for gene-directed enzyme prodrug therapy (Mancini *et al.*, 2009). Activity of histone deacetylase or inhibition by the histone deacetylase inhibitor (HDACi) suberoylanilide hydroxamic acid (SAHA) has been reported based on accumulation of Boc-Lys-TFA-OH (BLT) detected by ^{19}F NMR (Sankaranarayanapillai *et al.*, 2008). Alkaline phosphatase activity has been detected based on the chemical shift of a fluorinated substrate (Gade *et al.*, 2008) and we demonstrate activity of acid phosphatase on 6,8-difluoro-4-methylumbelliferyl phosphate (Fig. 12.6). The various substrates for detection of enzyme or transgene activity are shown in Table 12.4.

12.4 Passive Reporter Molecules

In some cases the mere presence of molecules can indicate anatomical and physiological properties such as lung volume, bowel function, vascular volume or flow. In this case ^{19}F NMR detects fluorinated molecules against a zero background and the signal magnitude reveals the parameter of interest. In many cases the ^{19}F MRI approaches are direct analogues of more traditional methods, which often required radioactive exposure.

Lung function has been measured using several radioactive gases and aerosols such as xenon-133 and Tc-99m-Technegas using SPECT (Suga, 2002). ^{19}F MRI has potential application for detection of lung cancer, emphysema or allograft rejection based on SF_6 breathing (Kuethe *et al.*, 2002; Ruiz-Cabello *et al.*, 2005). Gas detection does require special MR

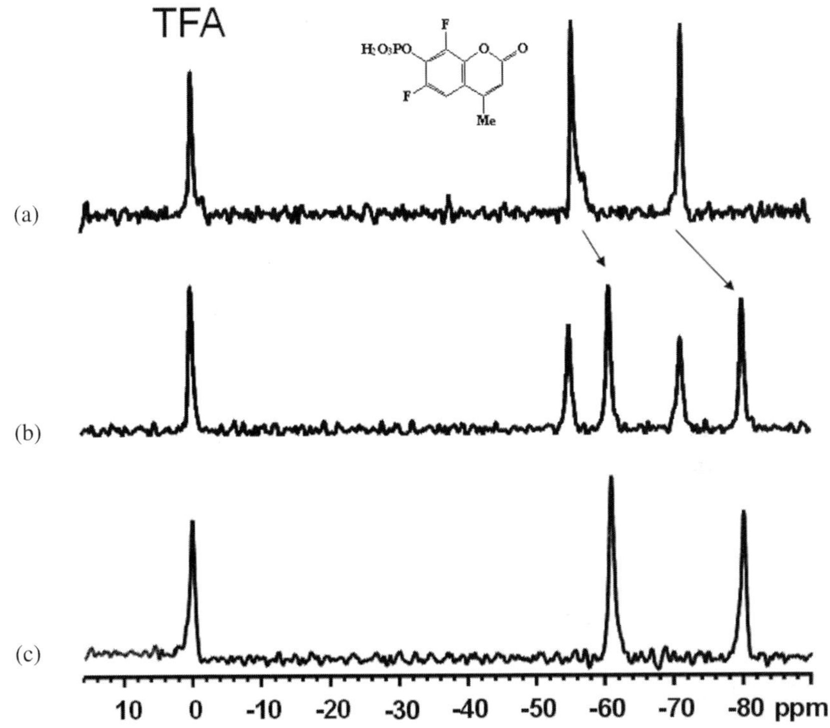

Figure 12.6. Time course of acid phosphatase activity on 6,8-difluoro-4-methylumbelliferyl phosphate (DiFMUP). (a) 188.2 MHz spectra of DiFMUP (1.5 mg in 500 μl H_2O at pH 4) showing two peaks at −55.4 ppm and −71.3 ppm with respect to internal NaTFA (0 ppm) corresponding to the non-equivalent fluorine atoms. A 60 Hz exponential line broadening was applied and no homonuclear or heteronuclear couplings are seen. (b) Four minutes after the addition of acidic phosphatase (wheat germ (4 U)) new peaks are seen at −61.2 and −80.1 ppm indicating phosphatase activity. (c) Complete conversion occurred after 15 min.

instrumentation, due to the exceedingly rapid T_1 and T_2 relaxation. Alternative MRI approaches use hyperpolarized noble gases such as helium and xenon, and it has been demonstrated that signal relaxation is sensitive to regional lung oxygenation (Matsuoka *et al.*, 2009). Alternatively, inhalation of liquid or aerosolized perfluorocarbon reveals lung ventilation and pO_2 as a bonus (Thomas *et al.*, 1997). Inhaled fluorinated gases (e.g. trifluoromethane (FC-23) and chlorofluoromethane

(FC-22)) have been used to examine cerebral blood flow (Eleff *et al.*, 1988; Ewing *et al.*, 1990).

Traditional barium meals provide contrast in computed tomography (CT), and virtual colonoscopy is competing with traditional fibre optic probes (Remy *et al.*, 2004). MR procedures have lagged behind CT, but several potential contrast agents have been presented, ranging from paramagnetic zeolite formulations (Rubin *et al.*, 1997) and ferric ammonium citrate (Hirohashi *et al.*, 1994) to PFC emulsions (Mattrey *et al.*, 1994; Bisset *et al.*, 1996), and images have also been produced in mice using perfluorononane (Schwarz *et al.*, 2002).

^{19}F NMR provides a robust indication of vascular volume *in vivo* based on intravenous perfluorocarbon emulsions, which are retained in the vasculature for a period of hours (Ceckler *et al.*, 1990; Meyer *et al.*, 1993). Non-invasive measurements revealed acute modulation of tumour blood volume and have provided validation of non-invasive near-infrared (NIR) methods (Sogabe *et al.*, 1997; Gu *et al.*, 2005). This approach has also been applied to other organs and tissues, for example, in demonstrating reactive hyperaemia in muscle (Authier, 1988), and was validated using traditional radioisotope-labelling approaches and dyes (Baldwin *et al.*, 1996). With sufficient imaging resolution ^{19}F MRI can be used for angiography as well as vascular volume estimates (Neubauer *et al.*, 2007).

12.5 Recent Innovations, Novelties and Future Improvements

^{19}F NMR has seen a tremendous recent upsurge in novel methods of application based on progress in three areas as outlined below.

12.5.1 *Chemistry and molecular engineering*

NMR can provide quantitative measurements based on signal integration. In principle, the larger the number of equivalent ^{19}F atoms, the stronger the signal. However, there is the caveat that modified relaxation times may alter the efficiency of signal detection: fully relaxed nuclei may be required for quantitation. As reviewed previously, the signal is expected to increase in the series CF, CF_2, CF_3, bis-CF_3, tris-CF_3 (Yu *et al.*, 2008a). Notably, two fluorines may be achieved as a geminal alkyl pair or by symmetry on an aromatic ring. Trifluoromethyl moieties may occur singly, or as an isopropyl group, or symmetrically as in bis-aromatic configuration, providing six equivalent fluorines. Increasing the number of fluorine atoms is expected to increase hydrophobicity, but a tris trifluoromethyl (*t*-butyl) group was recently developed and used to investigate nucleic acid conformation (Barhate *et al.*, 2008). Another agent with 27 equivalent fluorines was reported and used for *in vivo* studies in mice (Jiang *et al.*, 2009). Increased numbers of equivalent fluorine atoms can also be achieved using derivatized polymers with multiple labels. Our own experience using polycations such as polylysine suggested unacceptable toxicity (Mehta *et al.*, 1994a), but this approach has been reported recently to load and track cells successfully *in vivo* in mice for up to 7 days (Maki *et al.*, 2007). While polymers can have hundreds or thousands of fluorine atoms per molecule, it must be recognized that the molecular weight becomes excessive.

Large numbers of fluorine atoms have also been achieved by molecular and nanoengineering, for example the formation of fluorinated polymers formulated as micelles (Du *et al.*, 2008; Peng *et al.*, 2009). PFCs have been incorporated into various nanostructures, and appropriate shells allow targeting, cell tracking and enhanced relaxation (Morawski *et al.*, 2004; Ahrens *et al.*, 2005; Partlow *et al.*, 2007; Janjic *et al.*, 2008; Neubauer *et al.*, 2008; Kaneda *et al.*, 2009; Southworth *et al.*, 2009; Srinivas *et al.*, 2009).

As discussed in Section 12.3, molecular oxygen increases spin–lattice relaxation. This can be exploited to accelerate image acquisition, but also forms the basis of ^{19}F NMR oximetry (Zhao *et al.*, 2004). Several groups recently showed that incorporation of lanthanide ions such as Gd^{3+} into nanoparticle shells could also enhance R_1 and increase signal-to-noise ratio (Flacke *et al.*, 2001; Neubauer *et al.*, 2008). It was previously shown that PFC relaxation is essentially invariant with Gd^{3+} ions in surrounding solution (Thomas *et al.*, 1994) and clearly proximity and the surface-to-volume ratio are crucial since dipole–dipole relaxation falls with $1/r^6$. Indeed, Neubauer *et al.* (2008) showed that R_1 was a linear function of $[Gd^{3+}]$ associated with the nanoparticle and yet also retained oxygen sensitivity, albeit with an offset calibration.

Atomic mobility can influence NMR visibility: as noted, solid-state materials tend to have very broad signals. Tight packing of PFC into polymer micelles can render them ^{19}F NMR invisible, but modification, such as swelling induced by addition of DMSO, increases mobility and observability, as demonstrated for perfluorocarbon-loaded shell crosslinked Knedel-like nanoparticles (Nystrom *et al.*, 2009). Likewise, it was recently reported that dense fluorinated nanogels were ^{19}F NMR 'silent', but a hydrophilic–hydrophobic (volume-phase) transition of the polyamine gel core could 'turn-on' the signal. Gels have been presented which are pH-sensitive, exhibiting a change in T_2 from <1 ms at pH 7.4 to >50 ms at pH 6.5 (Oishi *et al.*, 2007, 2009). Disassembly of nanoprobes and ^{19}F NMR detectability may also be caused by association with specific protein, revealing molecular recognition (Takaoka *et al.*, 2009). Solid-state molecular immobilization has also been exploited to detect enzyme activity using water-soluble perfluorinated cubic silsesquioxanes, which are initially immobilized on silica nanoparticles via a phosphate suppressing the signal (Tanaka *et al.*, 2008). Alkaline phosphatase was applied to release the fluorinated component and generate NMR visible signal (Tanaka *et al.*, 2008). While these approaches turn on an ^{19}F signal, we developed substrates for β-gal activity, which simultaneously turn on proton MRI contrast while turning off the ^{19}F signal (Kodibagkar *et al.*, 2008c).

^{19}F NMR is more commonly used in simple molecular structures. In this case contact with aqueous environment and incorporation of paramagnetic species have been used to enhance signal relaxation in the past (Ratner *et al.*, 1989; Lee *et al.*, 1994; Mehta *et al.*, 1995) and again in 2006 (Terreno *et al.*, 2006). Direct relaxation can also be achieved by including F or CF_3 in a paramagnetic ligand complex (Belle *et al.*, 2009). This was previously exploited to investigate enzyme action and intermediates, e.g. galactose oxidase (Michel *et al.*, 2006, 2009). More recently there have been suggestions that it could be utilized to generate reporter molecules. Molecular structures are crucial to generate optimal proximity relevant to a particular lanthanide. In some cases the enhanced relaxation has been found to be pH-dependent, suggesting novel reporter capabilities if pK_a can be matched to the physiology (Senanayake *et al.*, 2007; Kenwright *et al.*, 2008). If ^{19}F is attached very close to a paramagnetic centre, such as Fe^{3+}, the local paramagnetic relaxation effect (PRE) can strongly influence the fluorine atom. Close proximity essentially quenches the ^{19}F NMR signal, but activatable separation makes the signal visible. This is quite analogous to Förster (fluorescence) resonance energy transfer (FRET) or bioluminescence resonance energy transfer (BRET) in the optical field where proximity controls signal (Willmann *et al.*, 2008). ^{19}F NMR agents have been presented, which turn on based on the redox state of the local environment (e.g. reduction of Fe^{3+} in ferrocene to Fe^{2+}) (Tanaka *et al.*, 2009). The ^{19}F visibility was shown to be reversible during multiple reductions and reoxidations. In other cases proximity is altered by molecular cleavage, though of course this tends to be irreversible. Enzyme activated agents have been shown to be sensitive to protease activity (Mizukami *et al.*, 2008, 2009).

12.5.2 *Biology*

Biological enhancement may be achieved by targeting and retention. The Wickline group has developed various nanobeacons with activated surfaces targeting specific receptors (Wickline *et al.*, 2010). Targeting has been demonstrated exploiting various biomarkers such as ($\alpha_v\beta_3$-integrin) generating accumulation at sites of inflammation such as atherosclerotic plaques (Flacke *et al.*, 2001) and inflamed kidneys (Southworth *et al.*, 2009). A potential problem is the long-term retention of unbound PFC in the vasculature, creating a large non-specific background signal and potentially masking the binding selectivity. Of course, this problem also compromises other methods such as optical and radionuclide imaging. An ingenious solution has been demonstrated exploiting differential diffusion, noting that bound PFC nanoparticles are immobilized (Waters *et al.*, 2008). ^{19}F has been used to label and track cells following implantation *in vivo* (Ahrens *et al.*, 2005; Partlow *et al.*, 2007; Srinivas *et al.*, 2007). A crucial component of stem cell therapy is the ability to monitor cell location and migration. For small animals, transfection with reporters such as luciferase or fluorescent proteins is highly effective, but MRI is more appropriate for larger animals and potentially for humans too. Biocompatible particles have been ingested by cells without apparent toxicity. This is similar to the use of super-paramagnetic iron oxide particles (SPIOs), but ^{19}F NMR provides a positive signal as opposed to T_2* signal loss (Tzu-Chen *et al.*, 1993; Bulte *et al.*, 2004).

12.5.3 *Physics*

In general, optimized coil volume filling factors, rapid small flip angle excitation (Ernst angle) and higher field improve the signal-to-noise ratio. Higher field magnets are much more expensive and coil size is limited by the need to accommodate the sample, e.g. the patient. Additional gains can be achieved with efficient pulse sequences and acquisition algorithms. In 2009, a new whole-body fast spin echo method with interleaved observation of several spectral lines was presented for detecting 5FU and its metabolites in mice at 9.4 T with validated metabolite quantification (Yoshihiro *et al.*, 2009).

Rapid imaging may use echo planar methods, as we favour for ^{19}F MRI oximetry (FREDOM) (Zhao *et al.*, 2004). Relaxometry may be further accelerated by implementing a Look–Locker approach (Jordan *et al.*, 2009).

A particularly hot new topic is compressed sensing, whereby sparse signals may be adequately detected despite undersampling. Despite apparent conflict with the Nyquist requirements, robust investigation in terms of spatial distribution and spectral resolution can be achieved. It appears particularly attractive for ^{19}F NMR CSI since there are limited signal frequencies (Fischer *et al.*, 2009).

12.5.4 *Innovative new applications*

The new developments permit a combination of faster data acquisition, better spatial resolution and detection of lower concentrations. Of course, there is always a trade off between these characteristics. However, recent reports demonstrate exciting capabilities, such as tracking labelled cell grafts (labelled stem cells) (Ahrens *et al.*, 2005; Partlow *et al.*, 2007). Highly localized signal and high fields offer positive ^{19}F signal as opposed to negative contrast associated with proton MRI of SPIO-labelled cells. Interestingly, SPIOs achieved approval for use in patients, but commercial expediency is halting production.

While ^{19}F NMR is the focus of this review, it is interesting to note that Fluorinert^{TM} (a liquid perfluorocarbon) is proposed as a matching fluid in inflatable balloon endorectal coils and has been shown to improve ^{1}H MRSI (Noworolski *et al.*, 2008).

Specific substrates and applications have been presented previously to detect carboxypeptidase, β-galactosidase, hypoxia, oxidoreduction and pH (Liu *et al.*, 2007; Hamans *et al.*, 2008; Yu *et al.*, 2008b; Tanaka *et al.*, 2008, 2009; Mancini *et al.*, 2009; Oishi *et al.*, 2009).

A major problem with proton MRI contrast to detect physiological phenomena such as pH, is separating the parameter-dependant molecular relaxivity from the concentration of the agent. This applies to straightforward contrast as well as more sophisticated methods such as magnetization transfer and paramagnetic chemical exchange saturation transfer (PARACEST). Ratiometric methods offer the hope of pairs of agents, sensitive and insensitive, respectively, to the parameter of interest (Garcia-Martin *et al.*, 2006). A new approach is the use of a fluorine moiety to provide quantitation, while a paramagnetic component creates relaxation (Gianolio *et al.*, 2009), although a potential difficulty is different voxel sizes and spatial resolution.

12.6 Context of ^{19}F NMR in Biomedicine Today

Innovation has greatly strengthened the repertoire of ^{19}F NMR applications (enzyme activity, cell tracking and disease identification), and a healthy imagination suggests important further developments and applications. However, it is crucial to place ^{19}F NMR in the context of competing modalities, both NMR and alternative. Table 12.5 provides a comparison of the different methods.

Ultimately, ^{19}F NMR lacks the sensitivity of optical and radionuclide imaging methods, but allows measurement of deep tissues without radioactive exposure. ^{19}F has a fraction of the sensitivity of tissue water proton MRI contrast, but does provide measurements based on signal-to-noise rather than contrast-to-noise ratio. Like ^{1}H NMR, ^{19}F detects millimolar concentrations, but without the need for background water suppression. ^{19}F is far more sensitive than ^{13}C, although the relatively short longitudinal relaxation times will likely prevent the spectacular signal-to-noise ratio gains promised by transiently hyperpolarized ^{13}C substrates. The greatest hurdle to more widespread application and clinical translation remains the current lack of human MRI systems with the capability of detecting ^{19}F.

Table 12.5. Comparison of ^{19}F NMR versus alternative measurement techniques.

Measurement	^{19}F Approach	^{1}H Approach	Alternative technology	References
pO_2 (example in Fig. 12.3)	FREDOM (fluorocarbon relaxometry using echo planar imaging for dynamic oxygen mapping): various PFCs (we favour HFB); quantitative pO_2 with spatial and temporal resolution; requires reporter molecule; fine needle injection or vascular emulsion delivery (systemic delivery biases data to well perfused regions)	PISTOL (proton imaging of siloxanes to map tissue oxygenation levels): requires water and fat suppression and reporter molecule BOLD (blood oxygen level dependent) and TOLD (tissue oxygen level dependent): observe endogenous signal, but not quantitative for pO_2	Polarographic electrode: highly invasive; consumes oxygen; dynamics at single location only	Robinson and Griffiths, 2004; Zhao *et al.*, 2004; Matsumoto *et al.*, 2006; Tatum *et al.*, 2006; O'Connor *et al.*, 2007; Kodibagkar *et al.*, 2008a,b; Mason, 2009
Hypoxia	F-Nitroimidazoles: detect substrate accumulation; poor signal-to-noise ratio; debatable signal visibility	Nitroimidazoles: needs water suppression; chemical shift response	PET: ^{18}F nitroimidazoles, Cu-ATSM; radioactive; rapid decay of substrate	Robinson and Griffiths, 2004; Pacheco-Torres *et al.*, 2006; Krohn *et al.*, 2008

(*Continued*)

Table 12.5. *(Continued)*

Measurement	^{19}F Approach	^{1}H Approach	Alternative technology	References
pH (example in Fig. 12.4)	Chemical shift response as indicated in Table 12.3	Chemical shift response needs water suppression	^{31}P NMR: endogenous molecules; poor signal-to-noise ratio and spectral crowding of pH_e and pH_i signals; polarographic electrode; invasive-limited locations	Gillies *et al.*, 2004; Perez-Mayoral *et al.*, 2008
Gene reporting (example in Fig. 12.5)	PFONPG: low signal-to-noise	EgadMe: poly-L-lysine; no exogenous substrate required for detection; contrast to noise EgadMe required intracellular poly-L-lysine	GFP, BLI: fluorescent protein approach does not require substrate infusion; limited tissue penetration of light	Mason 1999; Louie *et al.*, 2000; Contag and Ross, 2002; Yu 2005a; Gilad *et al.*, 2007; Yu *et al.*, 2008b
Cell tracking	PFC labelling: requires cell loading	SPIOs: requires cell loading; negative contrast	GFP, BLI: limited tissue penetration	Frank *et al.*, 2002; Wickline *et al.*, 2010

Acknowledgments

Supported in part by the Cancer Imaging Program NCI 1U24CA126608-03, together with 1R21CA120774-02, 1R01CA125033-A2 and DOD Breast Cancer Initiative DAMD 17-03-1-0343-01. NMR experiments were conducted at the Mary Nell and Ralph B. Rogers NMR Center, an NIH BTRP facility #P41-RR02584. Over the past 20 years our development of expertise in ^{19}F NMR has been supported by the NIH, Department of Defense Breast and Prostate Cancer Initiatives, the American Cancer Society, the American Heart Association and The Whitaker Foundation.

References

Aboagye, E.O., Maxwell, R.J., Horsman, M.R. *et al.* (1998) The relationship between tumour oxygenation determined by oxygen electrode measurements and magnetic resonance spectroscopy of the fluorinated 2-nitroimidazole SR-4554. *Br J Cancer* 77, 65–70.

Ahrens, E.T., Flores, R., Xu, H. *et al.* (2005) *In vivo* imaging platform for tracking immunotherapeutic cells. *Nat Biotechnol* 23, 983–987.

Aoki, Y., Akagi, K., Tanaka, Y. *et al.* (1996) Measurement of intratumor pH by pH indicator used in ^{19}F MR spectroscopy. *Invest Radiol* 31, 680–689.

Authier, B. (1988) Reactive hyperemia monitored on rat muscle using perfluorocarbons and ^{19}F NMR. *Magn Reson Med* 8, 80–83.

Babcock, E.E., Manson, R.P. & Antich, P.P. (1991) Effect of homonuclear J modulation on ^{19}F spin-echo images. *Magn Reson Med* 17, 179–188.

Bachert, P. (1998) Pharmacokinetics using fluorine NMR *in vivo. Prog NMR Spectrosc* 33, 1–56.

Baldwin, N.J. & Ng, T.C. (1996) Oxygenation and metabolic status of KHT tumors as measured simultaneously by ^{19}F magnetic resonance imaging and ^{31}P magnetic resonance spectroscopy. *Magn Reson Imaging* 14, 541–551.

Baldwin, N.J., Wang, Y. & Ng, T. C. (1996) *In situ* ^{19}F MRS measurement of RIF-1 tumor blood volume: corroboration by radioisotope-labeled [^{125}I]-albumin and correlation to tumor size. *Magn Reson Imaging* 14, 275–280.

Barhate, N.B., Barhate, R.N., Cekan, P. *et al.* (2008) A nonafluoro nucleoside as a sensitive ^{19}F NMR probe of nucleic acid conformation. *Org Lett* 10, 2745–2747.

Barker, B.R., Manson, R.P., Bansal, N. *et al.* (1994) Oxygen tension mapping by ^{19}F echo planar NMR imaging of sequestered perfluorocarbon. *J Magn Reson Im* 4, 595–602.

Bartels, M. & Albert, K. (1995) Detection of psychoactive drugs using ^{19}F MR spectroscopy. *J Neural Transm Gen Sect* 99, 1–6.

Barton, K.N, Tyson, D., Stricker, H. *et al.* (2003) GENIS: gene expression of sodium iodide symporter for noninvasive imaging of gene therapy vectors and quantification of gene expression *in vivo. Mol Ther J Am Soc Gene Ther* 8, 508–518.

Barton, K.N., Stricker, H., Brown, S.L. *et al.* (2008) Phase I study of noninvasive imaging of adenovirus-mediated gene expression in the human prostate. *Mol Ther* 16, 1761–1769.

Bartson, K.N., Xia, X., Yan, H. *et al.* (2004) A quantitative method for measuring gene expression magnitude and volume delivered by gene therapy vectors. *Mol Ther J Am Soc Gene Ther* 9, 625–631.

Baudelet, C. & Gallez, B. (2005) Current issues in the utility of blood oxygen level dependent MRI for the assessment of modulations in tumor oxygenation. *Curr Med Imaging Rev* 1, 229–243.

Beckwith, J.R. & Zipser, D. (Eds) (1970) *The Lactose Operon*, Cold Spring Harbor Laboratory, Cold Spring Harbor.

Belle, C., Béguin, C., Hamman, S. *et al.* (2009) ^{19}F NMR: an underused efficient probe for paramagnetic metal centers in bioinorganic solution chemistry. *Coordin Chem Rev* 253, 963–976.

Bellemann, M.E., Haberkorn, U., Gerlach, L. *et al.* (1999) Fluorine-19 NMR Imaging of the Biodistribution and Metabolization of the Antienoplastic Agent Gemcitabine in Tumor-Bearing Rats. 7th Scientific Meeting Intl Soc Mag Res Med, Philadelphia, PA.

Benters, J., Flögel, U., Schäfer, T. *et al.* (1997) Study of the interactions of cadmium and zinc ions with cellular calcium homoeostasis using ^{19}F-NMR spectroscopy. *Biochem J* 322, 793–799.

Berkowitz, B.A. & Ackerman, J.J. (1987) Proton decoupled fluorine nuclear-magnetic-resonance spectroscopy *in situ. Biophys J* 51, 681–685.

Berkowitz, B.A., Wilson, C.A. & Hatchell, D.L. (1991) Oxygen kinetics in the vitreous substitute perfluorotributylamine: a ^{19}F NMR study *in vivo. Invest Ophthalmol Vis Sci* 32, 2382–2387.

Berkowitz, B.A., Handa, J.T. & Wilson, C.A. (1992) Perfluorocarbon temperature measurement using ^{19}F NMR. *NMR Biomed* 5, 65–68.

Berkowitz, B.A., McDonald, C., Ito, Y. *et al.* (2001) Measuring the human retinal oxygenation response to a hyperoxic challenge using MRI: eliminating blinking artifacts and demonstrating proof of concept. *Magn Reson Med* 46, 412–416.

Bisset, G.S.I., Emery, K.H., Meza, M.P. *et al.* (1996.) Perflubron as a gastrointestinal MR imaging contrast agent in the pediatric population. *Pediatr Radiol* 26, 409–415.

Blackstock, A.W., Lightfoot, H., Case, L.D. *et al.* (2001) Tumor uptake and elimination of 2′,2′-difluoro-2′-deoxycytidine (gemcitabine) after deoxycytidine kinase gene transfer: correlation with *in vivo* tumor response. *Clin Cancer Res* 7, 3263–3268.

Böhm, H.J., Banner, D., Bendels, S. *et al.* (2004) Fluorine in medicinal chemistry. *ChemBioChem* 5, 637–643.

Bolo, N.R., Hodé, Y. & Macher, J.P. (2004) Long-term sequestration of fluorinated compounds in tissues after fluvoxamine or fluoxetine treatment: a fluorine magnetic resonance spectroscopy study *in vivo*. *Magn Reson Mater Phys* 16, 268–276.

Bourke, V.A, Gilio, J., Zhao, D. *et al.* (2007) Correlation of radiation response with tumor oxygenation in the Dunning prostate R3327-AT1 tumor. *Int J Radiat Oncol Biol Phys* 67, 1179–1186.

Bratasz, A., Kulkarni, A.C. & Kuppusamy, P. (2007) A highly sensitive biocompatible spin probe for imaging of oxygen concentration in tissues. *Biophys J* 92, 2918–2925.

Brix, G., Bellemann, M.E., Gerlach, L. *et al.* (1995) Mapping the biodistribution and catabolism of 5-fluorouracil in tumor-bearing rats by chemical-shift selective F-19 MR-imaging. *Magn Reson Med* 34, 302–307.

Brix, G., Bellemann, M.E., Gerlach, L. *et al.* (1996) Assessment of the biodistribution of 5-fluorouracil as monitored by ^{18}F PET and ^{19}F MRI: a comparative animal study. *Nucl Med Biol* 23, 897–906.

Brix, G., Bellemann, M.E., Gerlach, L. *et al.* (1998) Intra- and extracellular fluorouracil uptake: assessment with contrast-enhanced metabolic F-19 MR imaging. *Radiology* 209, 259–267.

Brix, G., Bellemann, M.E., Gerlach, L. *et al.* (1999) Direct detection of intratumoral 5-fluorouracil trapping using metabolic F-19 MR imaging. *Magn Reson Imaging* 17, 151–155.

Brix, G., Schlicker, A., Mier, W. *et al.* (2005) Biodistribution and pharmacokinetics of the F-19-labeled radiosensitizer 3-aminobenzamide: assessment by F-19 MR imaging. *Magn Reson Imaging* 23, 967–976.

Bronstein, I., Edwards, B. & Voyta, J.C. (1989) 1,2-Dioxetanes — novel chemiluminescent enzyme substrates — applications to immunoassays. *J Biolum Chemilum* 4, 99–111.

Brown, S.L., Freytag, S.O., Barton, K.N. *et al.* (2007) Reporter gene imaging using radiographic contrast from nonradioactive iodide sequestered by the sodium-iodide symporter. *Contrast Media Mol Imaging* 2, 240–247.

Blute, J.W., Arbab, A.S., Douglas, T. *et al.* (2004) Preparation of magnetically labeled cells for cell tracking by magnetic resonance imaging. *Methods Enzymol* 386, 275–299.

Ceckler, T.L., Gibson, S.L., Hilf, R. *et al.* (1990) *In situ* assessment of tumor vascularity using fluorine NMR imaging. *Magn Reson Med* 13, 416–433.

Chang, Y.C. & Graves, D.J. (1985) Use of 6-fluoroderivatives of pyridoxal and pyridoxal phosphate in the study of the coenzyme function in glycogen phosphorylase. *J Biol Chem* 260, 2709–2714.

Chiu, H.P., Kokona, B., Fairman, R. *et al.* (2009) Effect of highly fluorinated amino acids on protein stability at a solvent-exposed position on an internal strand of protein G B1 domain. *J Am Chem Soc* 131, 13192–13193.

Christensen, J.D., Yurgelun-Todd, D.A., Babb, S.M. *et al.* (1999) Measurement of human brain dexfenfluramine concentration by ^{19}F magnetic resonance spectroscopy. *Brain Res* 834, 1–5.

Chung, Y.L., Troy, H., Judson, I.R. *et al.* (2004) Noninvasive measurements of capecitabine metabolism in bladder tumors overexpressing thymidine phosphorylase by fluorine-19 magnetic resonance spectroscopy. *Clin Cancer Res* 10, 3863–3870.

Cline, J.M., Rosner, G.L., Raleigh, J.A. *et al.* (1997) Quantification of CCI-103F labeling heterogeneity in canine solid tumors. *Int J Radiat Oncol Biol Phys* 37, 655–662.

Cobb, S.L. & Murphy, C.D. (2009) ^{19}F NMR applications in chemical biology. *J Fluorine Chem* 130, 132–143.

Code, R.F., Harrison, J.E., McNeill, K.G. *et al.* (1990) *In vivo* ^{19}F spin relaxation in index finger bones. *Mag Reson Med* 13, 358–369.

Contag, C.H. & Ross, B.D. (2002) It's not just about anatomy: *in vivo* bioluminescence imaging as an eyepiece into biology. *J Magn Reson Im* 16, 378–387.

Corban-Wilhelm, H., Hull, W.E., Becker, G. *et al.* (2002) Cytosine deaminase and thymidine kinase gene therapy in a Dunning rat prostate tumour model: absence of bystander effects and characterisation of 5-fluorocytosine metabolism with ^{19}F-NMR spectroscopy. *Gene Ther* 9, 1564–1575.

Courtney, K.D. & Andrews, J.E. (1984) Teratogenic evaluation and fetal deposition of hexabromobenzene (HBB) and hexafluorobenzene (HFB) in CD-1 mice. *J Environ Sci Health B* 19, 83–94.

Cron, G.O., Beghein, N., Ansiaux, R. *et al.* (2008) F-19 NMR *in vivo* spectroscopy reflects the effectiveness of perfusion-enhancing vascular modifiers for improving gemcitabine chemotherapy. *Magn Reson Med* 59, 19–27.

Cui, W., Otten, P., Yu, J. *et al.* (2003) 6-Trifluoromethyl pyridoxol, a novel reporter molecule for tumor extracellular pH. *Proc Intl Soc Mag Reson Med* 11, 624.

Cui, W., Otten, P., Li, Y. *et al.* (2004) A novel NMR approach to assessing gene transfection: 4-fluoro-2-nitrophenyl-β-D-galactopyranoside as a prototype reporter molecule for β-galactosidase. *Magn Reson Med* 51, 616–620.

Dafik, L., d'Alarcao, M. & Kumar, K. (2008) Fluorination of mammalian cell surfaces via the sialic acid biosynthetic pathway. *Bioorg Med Chem Lett* 18, 5945–5947.

Dardzinski, B.J. & Sotak, C.H. (1994) Rapid tissue oxygen tension mapping using ^{19}F inversion-recovery echo-planar imaging of perfluoro-15-crown-5-ether. *Magn Reson Med* 32, 88–97.

De Leon-Rodriguez, L.M., Lubag, A.J.M., Malloy, C.R. *et al.* (2009) Responsive MRI agents for sensing metabolism *in vivo*. *Acc Chem Res* 42, 948–957.

Delpuech, J.-J., Hamza, M.A., Serratice, G. *et al.* (1979) Fluorocarbons as oxygen carriers. I. An NMR study of oxygen solutions in hexafluorobenzene. *J Chem Phys* 70, 2680–2687.

Desmoulin, F., Gilard, V., Malet-Martino, M. *et al.* (2002) Metabolism of capecitabine, an oral fluorouracil prodrug: ^{19}F NMR studies in animal models and human urine. *Drug Metab Dispos* 30, 1221–1229.

Deutsch, C., Taylor, J.S. & Wilson, D.F. (1982) Regulation of intracellular pH of human peripheral blood lymphocytes as measured by ^{19}F NMR. *Proc Natl Acad Sci USA* 79, 7944–7948.

Deutsch, C., Taylor, J.S. & Price, M. (1984) pH Homeostasis in human lymphocytes: modulation by ions and mitogen. *J Cell Biol* 98, 885–893.

Deutsch, C. & Taylor, J.S. (1987a) '^{19}F NMR measurements of intracellular pH', in Gupta, R.K. (Ed.), *NMR Spectroscopy of Cells and Organisms*, CRC Press, Boca Raton.

Deutsch, C. & Taylor, J.S. (1987b) Intracellular pH measured by ^{19}F NMR. *Ann NY Acad Sci* 508, 33–47.

Deutsch, C. & Taylor, J.S. (1989) New class of ^{19}F pH indicators: fluoroanilines. *Biophys J* 55, 799–804.

Dolbier, W.R. (2009) *Guide to Fluorine NMR for Organic Chemists*, Hoboken, NJ, John Wiley & Sons.

Dresselaers, T., Theys, J., Nuyts, S. *et al.* (2003) Non-invasive F-19 MR spectroscopy of 5-fluorocytosine to 5-fluorouracil conversion by recombinant *Salmonella* in tumours. *Br J Cancer* 89, 1796–1801.

Dresselaers, T., Theys, J., Dubois, L. *et al.* (2006) Evaluation of *Salmonella*-based suicide gene transfer in a rodent tumor model using *in vivo* ^{19}F MR spectroscopy. *Proc Intl Soc Mag Reson Med* 14, 3175.

Du, W., Nystrom, A.M., Zhang, L. *et al.* (2008) Amphiphilic hyperbranched fluoropolymers as nanoscopic ^{19}F magnetic resonance imaging agent assemblies. *Biomacromolecules* 9, 2826–2833.

Duong, T.Q., Iadecola, C. & Kim, S.G. (2001) Effect of hyperoxia, hypercapnia, and hypoxia on cerebral interstitial oxygen tension and cerebral blood flow. *Magn Reson Med* 45, 61–70.

Eidelberg, D., Johnson, G., Barnes, D. *et al.* (1988a) ^{19}F NMR imaging of blood oxygenation in the brain. *Magn Reson Med* 6, 344–352.

Eidelberg, D., Johnson, G., Tofts, P.S. *et al.* (1988b) ^{19}F Imaging of cerebral blood oxygenation in experimental middle cerebral artery occlusion: preliminary results. *J. Cereb. Blood Flow Metab* 8, 276–281.

Eleff, S.M., Schnall, M.D., Ligetti, L. *et al.* (1988) Concurrent measurements of cerebral blood-flow, sodium, lactate, and high-energy phosphate-metabolism using F-19, Na-23, H-1, and P-31 nuclear magnetic-resonance spectroscopy. *Magn Reson Med* 7, 412–424.

Ellis, M.K., Naylor, J.L., Green, T. *et al.* (1995) Identification and quantification of fluorine-containing metabolites of 1-chloro-2,2,2-trifluoroethane (HCFC133a) in the rat by ^{19}F-NMR spectroscopy. *Drug Metab Dispos* 23, 102–106.

Emsley, J.W. & Phillips, L. (1971) Fluorine chemical shifts. *Prog NMR Spectrosc* 7, 1–520.

Emsley, J.W., Phillips, L. & Wray, V. (1976) Fluorine coupling constants. *Prog NMR Spectrosc* 10, 83–752.

Evans, R.G., Kimler, B.F., Morantz, R.A. *et al.* (1993) Lack of complications in long-term survivors after treatment with Fluosol(R) and oxygen as an adjuvant to radiation-therapy for high-grade brain-tumors. *Int J Radiat Oncol* 26, 649–652.

Ewing, J.R., Branch, C.A., Fagan, S.C. *et al.* (1990) Fluorocarbon-23 measure of cat cerebral blood flow by NMR. *Stroke* 21, 100–106.

Fan, X., River, J.N., Zamora, M. *et al.* (2002) Effect of carbogen on tumor oxygenation: combined fluorine-19 and proton MRI measurements. *Int J Radiat Oncol Biol Phys* 54, 1202–1209.

Fischer, A., Basse-Lüsebrink, T.C., Kampf, T. *et al.* (2009) Improved sensitivity in ^{19}F cellular imaging using non-convex compressed sensing. *Proc Intl Mag Reson Med* 17.

Fishman, J.E., Joseph, P.M., Floyd, T.F. *et al.* (1987) Oxygen-sensitive ^{19}F NMR imaging of the vascular system *in vivo*. *Magn Reson Imaging* 5, 279–285.

Fishman, J.E., Joseph, P.M., Carvlin, M.J. *et al.* (1989) *In vivo* measurements of vascular oxygen tension in tumors using MRI of a fluorinated blood substitute. *Invest Radiol* 24, 65–71.

Flacke, S., Fischer, S., Scott, M.J. *et al.* (2001) Novel MRI contrast agent for molecular imaging of fibrin: implications for detecting vulnerable plaques. *Circulation* 104, 1280–1285.

Fowler, J.S., Volkow, N.D., Wang, G.J. *et al.* (2004) 2-Deoxy-2-[^{18}F]fluoro-D-glucose and alternative radiotracers for positron emission tomography imaging using the human brain as a model. *Semin Nucl Med* 34, 112–121.

Frank, J.A., Zywicke, H., Jordan, E.K. *et al.* (2002) Magnetic intracellular labeling of mammalian cells by combining (FDA-approved) superparamagnetic iron oxide MR contrast agents and commonly used transfection agents. *Acad Radiol* 9 Suppl 2, S484–S487.

Frenzel, T., Koszler, S., Bauer, H. *et al.* (1994) Noninvasive *in vivo* pH measurement using a fluorinated pH probe and fluorine-19 magnetic resonance spectroscopy. *Invest Radiol* 29, S220–S222.

Fyles, A.W., Milosevic, M., Wong, R. *et al.* (1998) Oxygenation predicts radiation response and survival in patients with cervix cancer. *Radiother Oncol* 48, 149–156.

Gade, T., Motley, M.W., Bhakta, R. *et al.* (2008) Molecular imaging of alkaline phosphatase activity in bone tissue. 2008 World Molecular Imaging Congress, Nice, France.

Gade, T.P.F., Buchanan, I.M., Motley, M.W. *et al.* (2009) Imaging intratumoral convection: pressure-dependent enhancement in chemotherapeutic delivery to solid tumors. *Clin Cancer Res* 15, 247–255.

Garcia-Martin, M.L., Martinez, G.V., Raghunand, N. *et al.* (2006) High resolution pH_e imaging of rat glioma using pH-dependent relaxivity. *Magn Reson Med* 55, 309–315.

Gerebtzoff, G., Li-Blatter, X., Fischer, H. *et al.* (2004) Halogenation of drugs enhances membrane binding and permeation. *ChemBioChem* 5, 676–684.

Gewiese, B.K., Noske, W., Schilling, A.M. *et al.* (1992) Human eye: visualization of perfluorodecalin with F-19 MR imaging. *Radiology* 185, 131–133.

Gianolio, E., Napolitano, R., Fedeli, F. *et al.* (2009) Poly-β-cyclodextrin based platform for pH mapping via a ratiometric ^{19}F/^{1}H MRI method. *Chem Commun*, 6044–6046.

Gilad, A.A., Winnard, P.T., van Zijl, P.C.M. *et al.* (2007) Developing MR reporter genes: promises and pitfalls. *NMR Biomed* 20, 275–290.

Gillies, R.J., Liu, Z. & Bhujwalla, Z. (1994) ^{31}P MRS measurements of extracellular pH of tumors using 3-aminopropylphosphonate. *Am J Physiol* 267, C195–C203.

Gillies, R.J., Raghunand, N., Garcia-Martin, M.L. *et al.* (2004) pH Imaging. *IEEE Eng Med Biol* 23, 57–64.

Girard, F., Poulet, P., Namer, I.J. *et al.* (1994) Localized T-2 measurements using an OSIRIS-CPMG method — application to measurements of blood oxygenation and transverse relaxation free of diffusion effect. *NMR Biomed* 7, 343–348.

Glaholm, J., Leach, M.O., Collins, D. *et al.* (1990) Comparison of 5-fluorouracil pharmacokinetics following intraperitoneal and intravenous administration using *in vivo* ^{19}F magnetic resonance spectroscopy. *Br J Radiol* 63, 547–553.

Gorsman, Y.S. & Kapitonenko, T.A. (1973) Pharmacology and toxicology of hexafluorobenzene. *Izv Estestvennonauchu Inst Pevinsk* 15, 155–163.

Griffiths, J.R., McIntyre, D.J.O., Howe, F.A. *et al.* (2001) Issues of normal tissue toxicity in patient and animal studies — effect of carbogen breathing in rats after 5-fluorouracil treatment. *Acta Oncol* 40, 609–614.

Gu, Y., Bourke, V.A., Kim, J.G. *et al.* (2003) Dynamic response of breast tumor oxygenation to hyperoxic respiratory challenge monitored with three oxygen-sensitive parameters. *Appl Optics* 42, 2960–2967.

Gu, Y., Mason, R.P. & Liu, H. (2005) Estimated fraction of tumor vascular blood contents sampled by near infrared spectroscopy and ^{19}F magnetic resonance spectroscopy. *Optics Express* 13, 1724–1733.

Guerquin-Kern, J.-L., Leteurtre, F., Croisy, A. *et al.* (1991) pH Dependence of 5-fluorouracil uptake observed by *in vivo* ^{31}P and ^{19}F NMR spectroscopy. *Cancer Res* 51, 5770–5773.

Guo, Q., Mattrey, R.F., Guclu, C. *et al.* (1994) Monitoring of pO_2 by spin-spin relaxation rate 1/T2 of ^{19}F in a rabbit abscess model. *Art Cells Blood Subs Immob Biotech* 22, 1449–1454.

Gupta, R.K. & Gillies, R.J. (1987) '^{19}F NMR measurement of intracellular free calcium ions in intact cells and tissues', in Gupta, R.K. (Ed.), *NMR Spectroscopy of Cells and Organisms*, CRC Press, Boca Raton, FL.

Hamans, B., Waalboer, D., Rutjes, F. *et al.* (2008) A 'novel' bimodal *in vivo* hypoxia marker: tri-fluoro-pimonidazole. 2008 World Molecular Imaging Congress, Nice, France.

Hamza, M.A., Serratice, G., Stebe, M.-J. *et al.* (1981) Solute-solvent interactions in perfluorocarbon solutions of oxygen: an NMR study. *J Am Chem Soc* 103, 3733–3738.

Harris, R.K., Becker, E.D., Cabral de Menezes, S.M. *et al.* (2001) NMR nomenclature: nuclear spin properties and conventions for chemical shifts (IUPAC Recommendations 2001). *Pure Appl Chem* 73, 1795–1818.

He, S., Mason, R.P., Hunjan, S. *et al.* (1998) Development of novel ^{19}F NMR pH indicators: synthesis and evaluation of a series of fluorinated vitamin B_6 analogs. *Bioorg Med Chem* 6, 1631–1639.

Hees, P.S. & Sotak, C.H. (1993) Assessment of changes in murine tumor oxygenation in response to nicotinamide using ^{19}F NMR relaxometry of a perfluorocarbon emulsion. *Magn Reson Med* 29, 303–310 and erratum 29, 716.

Heidelberger, C., Chaudhuri, N.K., Danneberg, P. *et al.* (1957) Fluorinated pyrimidines, a new class of tumour-inhibitory compounds. *Nature* 179, 663–666.

Helmer, K.G., Han, S. & Sotak, C.H. (1998) On the correlation between the water diffusion coefficient and oxygen tension in RIF-1 tumors. *NMR Biomed* 11, 120–130.

Henry, M.E., Moore, C.M., Kaufman, M.J. *et al.* (2000) Brain kinetics of paroxetine and fluoxetine on the third day of placebo substitution: a fluorine MRS study. *Am J Psychiat* 157, 1506–1508.

Henry, M.E., Schmidt, M.E., Hennen, J. *et al.* (2005) A comparison of brain and serum pharmacokinetics of *R*-fluoxetine and racemic fluoxetine: a 19-F MRS study. *Neuropsychopharmacol* 30, 1576–1583.

Heuermann, K. & Cosgrove, J. (2001) *S*-Gal: an autoclavable dye for color selection of cloned DNA inserts. *Biotechniques* 30, 1142–1147.

Higuchi, M., Iwata, N., Matsuba, Y. *et al.* (2005) ^{19}F and ^{1}H MRI detection of amyloid beta plaques *in vivo*. *Nat Neurosci* 8, 527–533.

Hirohashi, S., Uchida, H., Yoshikawa, K. *et al.* (1994) Large scale clinical evaluation of bowel contrast agent containing ferric ammonium citrate in MRI. *Magn Reson Imaging* 12, 837–846.

Höckel, M., Schlenger, K., Mitze, M. *et al.* (1996) Hypoxia and radiation response in human tumors. *Semi Radiat Oncol* 6, 3–9.

Holland, S.K., Kennan, R.P., Schaub, M.M. *et al.* (1993) Imaging oxygen tension in liver and spleen by ^{19}F NMR. *Magn Reson Med* 29, 446–458.

Huang, M.Q., Ye, Q., Williams, D.S. *et al.* (2002) MRI of lungs using partial liquid ventilation with water-in-perfluorocarbon emulsions. *Magn Reson Med* 48, 487–492.

Hunjan, S., Mason, R.P., Mehta, V.D. *et al.* (1998) Simultaneous intra- and extracellular pH measurement using ^{19}F NMR of 6-fluoropyridoxol. *Magn Reson Med* 39, 551–556.

Hunjan, S., Zhao, D., Constantinescu, A. *et al.* (2001) Tumor oximetry: demonstration of an enhanced dynamic mapping procedure using fluorine-19 echo planar magnetic resonance imaging in the Dunning prostate R3327-AT1 rat tumor. *Int J Radiat Oncol Biol Phys* 49, 1097–1108.

Isanbor, C. & O'Hagan, D. (2006) Fluorine in medicinal chemistry: a review of anti-cancer agents. *J Fluorine Chem* 127, 303–319.

Janjic, J.M., Srinivas, M., Kadayakkara, D.K.K. *et al.* (2008) Self-delivering nanoemulsions for dual fluorine-19 MRI and fluorescence detection. *J Am Chem Soc* 130, 2832–2841.

Jiang, L., Zhao, D., Constantinescu, A. *et al.* (2004) Comparison of BOLD contrast and Gd-DTPA dynamic contrast enhanced imaging in rat prostate tumor. *Magn Reson Med* 51, 953–960.

Jiang, Z.-X., Liu, X., Jeong, E.-K. *et al.* (2009) Symmetry-guided design and fluorous synthesis of a stable and rapidly excreted imaging tracer for ^{19}F MRI. *Angew Chem Int Edit* 48, 4755–4758.

Jordan, B.F., Cron, G.O. & Gallez, B. (2009) Rapid monitoring of oxygenation by ^{19}F magnetic resonance imaging: simultaneous comparison with fluorescence quenching. *Magn Reson Med* 61, 634–638.

Kaneda, M.M., Caruthers, S., Lanza, G.M. *et al.* (2009) Perfluorocarbon nanoemulsions for quantitative molecular imaging and targeted therapeutics. *Ann Biomed Eng* 37, 1922–1933.

Kashiwagura, T., Deutsch, C.J., Taylor, J. *et al.* (1984) Dependence of gluconeogenesis, urea synthesis, and energy metabolism of hepatocytes on intracellular pH. *J Biol Chem* 259, 237–243.

Kaufman, R.J. (1991) 'Medical oxygen transport using perfluorochemicals', in Goldstein, J. (Ed.), *Biotechnology of Blood*, Butterworth-Heinemann, New York.

Kawaguchi, J., Wilson, V. & Mee, P.J. (2002) Visualization of whole-mount skeletal expression patterns of *lacZ* reporters using a tissue clearing protocol. *Biotechniques* 32, 68–73.

Kendrick, R.D. & Yannoni, C.S. (1987) High-power ^{1}H-^{19}F excitation in a multiple-resonance single-coil circuit. *J Magn Reson* 75, 506–508.

Kenwright, A.M., Kuprov, I., de Luca, E. *et al.* (2008) ^{19}F NMR based pH probes: lanthanide(III) complexes with pH-sensitive chemical shifts. *Chem Commun*, 2514–2516.

Keupp, J., Mazurkewitz, P.C., Gräßlin, I. *et al.* (2006) Simultaneous ^{19}F and 1H imaging on a clinical 3T MR scanner. *Proc Intl Soc Mag Reson Med* 14, 102.

Kim, J.G., Zhao, D., Liu, H. *et al.* (2003) Interplay of tumor vascular oxygenation and tumor pO_2 observed using NIRS, oxygen needle electrode, and ^{19}F MR pO_2 mapping. *J Biomed Optics* 8, 53–62.

Kirschenlohr, H.L., Metcalfe, J.C., Morris, P.G. *et al.* (1988) Ca-2+ Transient, Mg-2+, and pH measurements in the cardiac cycle by ^{19}F NMR. *Proc Natl Acad Sci USA* 85, 9017–9021.

Klomp, D.W.J., van Laarhoven, H.W.M., Kentgens, A.P.M. *et al.* (2003) Optimization of localized F-19 magnetic resonance spectroscopy for the detection of fluorinated drugs in the human liver. *Magn Reson Med* 50, 303–308.

Kodibagkar, V.D., Yu, J., Liu, L. *et al.* (2006) Imaging β-galactosidase activity using ^{19}F chemical shift imaging of *lacZ* gene-reporter molecule 2-fluoro-4-nitrophenol-β-D-galactopyranoside. *Magn Reson Imaging* 24, 959–962.

Kodibagkar, V.D., Wang, X. & Mason, R.P. (2008a) Physical principles of quantitative nuclear magnetic resonance oximetry. *Front Biosci* 13, 1371–1384.

Kodibagkar, V.D., Wang, X., Pacheco-Torres, J. *et al.* (2008b) Proton imaging of silanes to map tissue oxygenation levels (PISTOL): a tool for quantitative tissue oximetry. *NMR Biomed* 21, 899–907.

Kodibagkar, V.D., Yu, J.-X. & Mason, R.P. (2008c) Dual $^{1}H/^{19}F$ *lacZ* gene reporter molecule. World Molecular Imaging Congress, Nice, France.

Korytnyk, W. & Singh, R.P. (1963) Proton magnetic resonance spectra of compounds in the vitamin B_6 group. *J Am Chem Soc* 85, 2813–2817.

Krafft, M.P. (2001) Fluorocarbons and fluorinated amphiphiles in drug delivery and biomedical research. *Adv Drug Delivery Rev* 47, 209–228.

Krohn, K.A., Link, J.M. & Mason, R.P. (2008) Molecular imaging of hypoxia. *J Nucl Med* 49, 129S–148S.

Kruger, A., Schirrmacher, V. & Khokha, R. (1999) The bacterial lacZ gene: an important tool for metastasis research and evaluation if new cancer therapies. *Cancer Metast Rev* 17, 285–294.

Kuethe, D.O., Behr, V.C. & Begay, S. (2002) Volume of rat lungs measured throughout the respiratory cycle using F-19 NMR of the inert gas SF_6. *Magn Reson Med* 48, 547–549.

Kusuoka, H., Backx, P.H., Camilion de Hurtado, M.C. *et al.* (1993) Relative roles of intracellular Ca^{2+} and pH in shaping myocardial contractile response to acute respiratory alkalosis. *Am J Physiol* 265, H1696–1703.

Lai, C.-S., Stair, S., Miziorko, H. *et al.* (1984) Effect of oxygen and the spin label TEMPO-laurate on ^{19}F and proton relaxation rates of the perfluorochemical blood substitute FC-43 emulsion. *J Magn Reson* 57, 447–452.

Le, D., Mason, R.P., Hunjan, S. *et al.* (1997) Regional tumor oxygen dynamics: ^{19}F PBSR EPI of hexafluorobenzene. *Magn Reson Imag* 15, 971–81.

Lee, H., Price, R.R., Holburn, G.E. *et al.* (1994) *In-vivo* ^{19}F MR-imaging: relaxation enhancement with Gd-DTPA. *J Magn Reson Imag* 4, 609–613.

Levy, L.A., Murphy, E., Raju, B. *et al.* (1988) Measurement of cytosolic free magnesium concentration by ^{19}F NMR. *Biochemistry* 27, 4041–4048.

Lewa, C.J. & Majewska, Z. (1980) Temperature relationships of proton spin-lattice relaxation time T1 in biological tissues. *Bull Cancer* 67, 525–530.

Li, S.J., Jin, G.Y. & Moulder, J.E. (1991) Prediction of tumor radiosensitivity by hexafluoromisonidazole retention monitored by [H-1]/[F-19] magnetic-resonance spectroscopy. *Cancer Commun* 3, 133–139.

Li, B.S.Y., Payne, G.S., Collins, D.J. *et al.* (2000) H-1 decoupling for *in vivo* F-19 MRS studies using the time-share modulation method on a clinical 1.5 T NMR system. *Magn Reson Med* 44, 5–9.

Li, L., Zemp, R.J., Lungu, G. *et al.* (2007) Photoacoustic imaging of *lacZ* gene expression *in vivo*. *J Biomed Optics* 12, 020504.

Liebfritz, D. (1992) 'Water suppression', in de Certaines, J.D., Bovee, W.M.M.J. & Podo, F. (Eds), *Magnetic Resonance Spectroscopy in Biology and Medicine*, Pergamon, Oxford.

Liu, L., Kodibagkar, V.D., Yu, J.-X. *et al.* (2007) ^{19}F-NMR detection of *lacZ* gene expression via the enzymic hydrolysis of 2-fluoro-4-nitrophenyl β-D-galactopyranoside *in vivo* in PC3 prostate tumor xenografts in the mouse. *FASEB J* 21, 2014–2019.

Liu, S., Shah, S.J., Wilmes, L.J. *et al.* (2009) Evaluation of severe anemia by quantitatively measuring multi-organ oxygen using ^{19}F MRI in a rat model. *Proceedings 17th Meeting, Intl Soc Mag Reson Med*, 4356.

Lockwood, G.G., Dob, D.P., Bryant, D.J. *et al.* (1997) Magnetic resonance spectroscopy of isoflurane kinetics in humans. 1. Elimination from the head. *Br J Anaesthes* 79, 581–585.

London, R.E. & Gabel, S.A. (1989) Determination of membrane potential and cell volume by ^{19}F NMR using trifluoroacetate and trifluoroacetamide probes. *Biochemistry* 28, 2378–2382.

London, R.E. (1994) '*In vivo* NMR studies utilizing fluorinated NMR probes', in Gillies, R.J. (Ed.), *NMR in Physiology and Biomedicine*, Academic Press, San Diego, CA.

London, R.E. & Gabel, S.A. (1994) Fluorine-19 NMR-studies of fluorobenzeneboronic acids. 1. Interaction kinetics with biologically significant ligands. *J Am Chem Soc* 116, 2562–2569.

Louie, A.Y., Hüber, M.M., Ahrens, E.T. *et al.* (2000) *In vivo* visualization of gene expression using magnetic resonance imaging. *Nat Biotechnol* 18, 321–325.

Lutz, N.W. & Hull, W.E. (1999) Assignment and pH dependence of the ^{19}F-NMR resonances from the fluorouracil anabolites involved in fluoropyrimidine chemotherapy. *NMR Biomed* 12, 237–248.

Maki, J., Masuda, C., Morikawa, S. *et al.* (2007) The MR tracking of transplanted ATDC5 cells using fluorinated poly-L-lysine-CF3. *Biomaterials* 28, 434–440.

Mancini, L., Davies, L., Friedlos, F. *et al.* (2009) A novel technique to monitor carboxypeptidase G2 expression in suicide gene therapy using ^{19}F magnetic resonance spectroscopy. *NMR Biomed* 22, 561–566.

Marban, E., Kitakaze, M., Chacko, V.P. *et al.* (1988) Ca-2+ transients in perfused hearts revealed by gated F-19 NMR-spectroscopy. *Circ Res* 63, 673–678.

Martino, R., Malet-Martino, M. & Gilard, V. (2000) Fluorine nuclear magnetic resonance, a privileged tool for metabolic studies of fluoropyrimidine drugs. *Curr Drug Metab* 1, 271–303.

Mason, R.P., Nunnally, R.L. & Antich, P.P. (1991) Tissue oxygenation: a novel determination using ^{19}F surface coil spectroscopy of sequestered perfluorocarbon emulsion. *Magn Reson Med* 18, 71–79.

Mason, R.P., Shukla, H.P. & Antich, P.P. (1992) Oxygent: a novel probe of tissue oxygen tension. *Biomater Artif Cells Immob Biotechnol* 20, 929–935.

Mason, R.P., Shukla, H.P. & Antich, P.P. (1993) *In vivo* oxygen tension and temperature: simultaneous determination using ^{19}F spectroscopy of perfluorocarbon. *Magn Reson Med* 29, 296–302.

Mason, R.P. (1994) Non-invasive physiology: ^{19}F NMR of perfluorocarbon. *Art Cells Blood Sub Immob Biotech* 22, 1141–1153.

Mason, R.P., Antich, P.P., Babcock, E.E. *et al.* (1994) Non-invasive determination of tumor oxygen tension and local variation with growth. *Int J Radiat Oncol Biol Phys* 29, 95–103.

Mason, R.P. & Antich, P.P. (1995) Application of ^{19}F NMR to Non Invasively Assess pO_2 and Temperature in vivo with Rapid Time Resolution. US patent 5 397 562.

Mason, R.P., Rodbumrung, W. & Antich, P.P. (1996) Hexafluorobenzene: a sensitive ^{19}F NMR indicator of tumor oxygenation. *NMR Biomed* 9, 125–134.

Mason, R.P. (1999) Transmembrane pH gradients *in vivo*: measurements using fluorinated vitamin B6 derivatives. *Curr Med Chem* 6, 481–499.

Mason, R.P., Constantinescu, A., Hunjan, S. *et al.* (1999) Regional tumor oxygenation and measurement of dynamic changes. *Radiat Res* 152, 239–249.

Mason, R.P., Ran, S. & Thorpe, P.E. (2002) Quantitative assessment of tumor oxygen dynamics: molecular imaging for prognostic radiology. *J Cell Biochem* 87 suppl, 45–53.

Mason, R.P., Hunjan, S., Constantinescu, A. *et al.* (2003) 'Tumor oximetry: comparison of ^{19}F MR EPI and electrodes', in Dunn, J.F. & Swartz, H.M. (Eds), *Oxygen Transport to Tissue XXIV*, Kluwer, New York.

Mason, R.P. (2009) Bold MRI: a tool for predicting tumor therapy outcome based on tumor blood oxygenation and vascular function. *Imaging Med* 1, 11–13.

Matsumoto, K., Bernardo, M., Subramanian, S. *et al.* (2006) MR assessment of changes of tumor in response to hyperbaric oxygen treatment. *Magn Reson Med* 56, 240–246.

Matsuoka, S., Patz, S., Albert, M.S. *et al.* (2009) Hyperpolarized gas MR imaging of the lung: current status as a research tool. *J Thoracic Imaging* 24, 181–188.

Mattrey, R.F., Trambert, M.A., Brown, J.J. *et al.* (1994) Perflubron as an oral contrast agent for MR imaging: results of a phase III clinical trial. *Radiology* 191, 841–848.

McIntyre, D.J.O., McCoy, C.L. & Griffiths, J.R. (1999) Tumour oxygenation measurements by ^{19}F MRI of perfluorocarbons. *Curr Sci* 76, 753–762.

McNab, J.A., Yung, A.C. & Kozlowski, P. (2004) Tissue oxygen tension measurements in the Shionogi model of prostate cancer using F-19 MRS and MRI. *Magn Reson Mater Phys* 17, 288–295.

McSheehy, P.M.J., Robinson, S.P., Ojugo, A.S.E. *et al.* (1998) Carbogen breathing increases 5-fluorouracil uptake and cytotoxicity in hypoxic RIF-1 tumors: a magnetic resonance study *in vivo. Cancer Res* 58, 1185–1194.

McSheehy, P.M.J., Ojugo, A.S.E., Leach, M.O. *et al.* (1999) ^{19}F-MRS studies of the novel thymidylate synthase (TS) inhibitor, ZD9331. 7th Annual Meeting Intl Soc Mag Res Med, Philadelphia, PA.

Mehta, V.D., Mason, R.P., Kulkarni, P.V. *et al.* (1994a) Novel ^{19}F MRI agents: imaging potential with polylysines. 1st Soc Magn Reson, Dallas.

Mehta, V.D., Kulkarni, P.V., Mason, R.P. *et al.* (1994b) 6-Fluoropyridoxol: a novel probe of cellular pH using ^{19}F NMR spectroscopy. *FEBS Lett* 349, 234–238.

Mehta, V.D., Mason, R.P., Kulkarni, P.V. *et al.* (1995) '^{19}F MR characterization of fluorinated proteins and relaxation rate enhancement with Gd-DTPA for faster imaging', in Emram, E.H. (Ed.), *Proc. Int. Symp. Chemists View of Imaging Centers*, Plenum, New York.

Menon, D.K., Lockwood, G.G., Peden, C.J. *et al.* (1993) *In vivo* ^{19}F magnetic-resonance spectroscopy of cerebral halothane in postoperative-patients: preliminary results. *Magn Reson Med* 30, 680–684.

Metcalfe, J.C., Hesketh, T.R. & Smith, G.A. (1985) Free cytosolic Ca^{2+} measurements with fluorine labelled indicators using ^{19}F NMR. *Cell Calcium* 6, 183–195.

Meyer, K.L., Joseph, P.M., Mukherji, B. *et al.* (1993) Measurement of vascular volume in experimental rat-tumors by ^{19}F magnetic-resonance-imaging. *Invest Radiol* 28, 710–719.

Michel, F., Hamman, S., Thomas, F. *et al.* (2006) Galactose oxidase models: ^{19}F NMR as a powerful tool to study the solution chemistry of tripodal ligands in the presence of copper(II). *Chem Commun*, 4122–4124.

Michel, F., Hamman, S., Philouze, C. *et al.* (2009) Galactose oxidase models: insights from ^{19}F NMR spectroscopy. *Dalton Trans*, 832–842.

Mishima, H., Kobayashi, T., Shimizu, M. *et al.* (1991) *In vivo* F-19 chemical shift imaging with FTPA and antibody-coupled FMIQ. *J Magn Reson Imaging* 1, 705–709.

Miyazawa, T., Aoki, Y., Akagi, K. *et al.* (1996) Application Of Zk150 471, a fluorinated pH probe for ^{19}F MRS, to *in vivo* pH measurement after hyperthermic treatment of tumors in mice. *Acad Radiol* 3, S363–S364.

Mizukami, S., Takikawa, R., Sugihara, F. *et al.* (2008) Paramagnetic relaxation-based ^{19}F MRI probe to detect protease activity. *J Am Chem Soc* 130, 794–795.

Mizukami, S., Takikawa, R., Sugihara, F. *et al.* (2009) Dual-function probe to detect protease activity for fluorescence measurement and ^{19}F MRI. *Angew Chem Int Edit* 48, 3641–3643.

Montclare, J.K., Son, S., Clark, G.A. *et al.* (2009) Biosynthesis and stability of coiled-coil peptides containing (2*S*,4*R*)-5,5,5-trifluoroleucine and (2*S*,4*S*)-5,5,5-trifluoroleucine. *ChemBioChem* 10, 84–86.

Morawski, A.M., Winter, P.M., Yu, X. *et al.* (2004) Quantitative 'magnetic resonance immunohistochemistry' with ligand-targeted (19)F nanoparticles. *Magn Reson Med* 52, 1255–1262.

Morgenthaler, M., Schweizer, E., Hoffmann-Röder, A. *et al.* (2007) Predicting and tuning physicochemical properties in lead optimization: amine basicities. *Chem Med Chem* 2, 1100–1115.

Mortelmans, K.M. & Simmon, V.F. (1981) '*In vitro*' microbiological mutagenicity assays of eight fluorocarbon taggant samples. *Gov Rep Announce Index US* 81, 2555–2587.

Müller, K., Faeh, C. & Diederich, F. (2007) Fluorine in pharmaceuticals: looking beyond intuition. *Science* 317, 1881–1886.

Murphy, E., Steenbergen, C., Levy, L.A. *et al.* (1989) Cytosolic free magnesium levels in ischemic rat-heart. *J Biol Chem* 264, 5622–5627.

Murphy, E. (1993) Measurement of intracellular ionized magnesium. *Miner Electrol Metab* 19, 250–258.

Nakada, T., Kwee, I.L. & Conboy, C.B. (1986) Noninvasive *in vivo* demonstration of 2-fluoro-2-deoxy-D-glucose metabolism beyond the hexokinase reaction in rat-brain by F-19 nuclear-magnetic-resonance spectroscopy. *J Neurochem* 46, 198–201.

Nakada, T., Kwee, I.L., Card, P.J. *et al.* (1988a) ^{19}F NMR imaging of glucose-metabolism. *Magn Reson Med* 6, 307–313.

Nakada, T., Kwee, I.L., Griffey, B.V. *et al.* (1988b) ^{19}F MR imaging of glucose-metabolism in the rabbit. *Radiology* 168, 823–825.

Neubauer, A.M., Caruthers, S.D., Hockett, F.D. *et al.* (2007) Fluorine cardiovascular magnetic resonance angiography *in vivo* at 1.5 T with perfluorocarbon nanoparticle contrast agents. *J Cardiov Magn Reson* 9, 565–573.

Neubauer, A.M., Myerson, J., Caruthers, S.D. *et al.* (2008) Gadolinium-modulated F-19 signals from perfluorocarbon nanoparticles as a new strategy for molecular imaging. *Magn Reson Med* 60, 1066–1072.

Noth, U., Morrissey, S.P., Deichmann, R. *et al.* (1995) *In vivo* measurement of partial oxygen pressure in large vessels and in the reticuloendothelial system using fast ^{19}F-MRI. *Magn Reson Med* 34, 738–745.

Noworolski, S.M., Crane, J.C., Vigneron, D.B. *et al.* (2008) A clinical comparison of rigid and inflatable endorectal-coil probes for MRI and 3D MR spectroscopic imaging (MRSI) of the prostate. *J Magn Reson Imag* 27, 1077–1082.

Nunnally, R., Antich, P., Nguyen, P. *et al.* (1988) Fluosol adjuvant therapy in human cancer: examinations *in vivo* of perfluorocarbons by ^{19}F NM. *Proc. 7th Meeting Int Soc Mag ResMed, San Francisco*, p. 342.

Nystrom, A.M., Bartels, J.W., Du, W. *et al.* (2009) Perfluorocarbon-loaded shell crosslinked knedel-like nanoparticles: lessons regarding polymer mobility and self-assembly. *J Polym Sci Pol Chem* 47, 1023–1037.

O'Connor, J.P.B., Jackson, A., Buonaccorsi, G.A. *et al.* (2007) Organ-specific effects of oxygen and carbogen gas inhalation on tissue longitudinal relaxation times. *Magn Reson Med* 58, 490–496.

O'Connor, J.P.B., Naish, J.H., Jackson, A. *et al.* (2009) Comparison of normal tissue R-1 and R-2* modulation by oxygen and carbogen. *Magn Reson Med* 61, 75–83.

Odvina, C.V., Mason, R.P. & Pak, C.Y.C. (2006) Prevention of thiazide-induced hypokalemia without magnesium depletion by potassium-magnesium citrate. *Am J Therapeut* 13, 101–108.

Oishi, M., Sumitani, S. & Nagasaki, Y. (2007) On-off regulation of F-19 magnetic resonance signals based on pH-sensitive PEGylated nanogels for potential tumor-specific smart F-19 MRI probes. *Bioconjugate Chem* 18, 1379–1382.

Oishi, M., Sumitani, S., Bronich, T.K. *et al.* (2009) Novel ^{19}F MRS/I nanoprobe based on pH-responsive PEGylated nanogel: pH-dependent ^{19}F magnetic resonance studies. *Chem Lett* 38, 128–129.

Ojugo, A.S.E., McSheehy, P.M.J., Stubbs, M. *et al.* (1998) Influence of pH on the uptake of 5-fluorouracil into isolated tumour cells. *Br J Cancer* 77, 873–879.

Ojugo, A.S.E., McSheehy, P.M.J., McIntyre, D.J. *et al.* (1999) Measurement of the extracellular pH of solid tumours in mice by magnetic resonance spectroscopy: a comparison of exogenous (19)F and (31)P probes. *NMR Biomed* 12, 495–504.

Pacheco-Torres, J., Perez-Mayoral, E., Soriano, E. *et al.* (2006) A convenient and efficient synthesis of the first (nitroimidazolyl)succinic esters and their diacids. *Synthesis*, 3859–3864.

Partlow, K.C., Chen, J., Brant, J.A. *et al.* (2007) ^{19}F magnetic resonance imaging for stem/progenitor cell tracking with multiple unique perfluorocarbon nanobeacons. *FASEB J* 21, 1647–1654.

Passe, T.J., Charles, H.C., Rajagopalan, P. *et al.* (1995) Nuclear magnetic resonance spectroscopy: a review of neuropsychiatric applications. *Prog Neuro-Psychopharmacol Biol Psych* 19, 541–563.

Peng, H., Blakey, I., Dargaville, B. *et al.* (2009) Synthesis and evaluation of partly fluorinated block copolymers as MRI imaging agents. *Biomacromolecules* 10, 374–381.

Perez-Mayoral, E., Negri, V., Soler-Padros, J. *et al.* (2008) Chemistry of paramagnetic and diamagnetic contrast agents for magnetic resonance imaging and spectroscopy pH responsive contrast agents. *Eur J Radiol* 67, 453–458.

Pinedo, H.M. & Peters, G.F.J. (1988) Fluorouracil: biochemistry and pharmacology. *J Clin Oncol* 6, 1653–1664.

Presant, C.A., Wolf, W., Waluch, V. *et al.* (1994) Association of intratumoral pharmacokinetics of fluorouracil with clinical response. *Lancet* 343, 1184–1187.

Procissi, D., Claus, F., Burgman, P. *et al.* (2007) *In vivo* F-19 magnetic resonance spectroscopy and chemical shift imaging of tri-fluoro-nitroimidazole as a potential hypoxia reporter in solid tumors. *Clinical Cancer Res* 13, 3738–3747.

Raghunand, N. & Gillies, R.J. (2001) pH and chemotherapy. *Novart FDN Symp* 240, 199–211.

Raleigh, J.A., Miller, G.G., Franko, A.J. *et al.* (1987) Fluorescence immunohistochemical detection of hypoxic cells in spheroids and tumors. *Brit J Cancer* 56, 395–400.

Raleigh, J.A., Franko, A., Kelly, D. *et al.* (1991) Development of an *in vivo* ^{19}F MR method for measuring oxygen deficiency in tumors. *Magn Reson Med* 22, 451–466.

Ramasamy, R., Zhao, P., Gitomer, W.L. *et al.* (1992) Determination of chloride potential in perfused rat hearts by NMR spectroscopy. *Am J Physiol* 263, H1958–H1962.

Ratner, A.V., Quay, S., Muller, H.H. *et al.* (1989) ^{19}F relaxation rate enhancement and frequency shift with Gd-DTPA. *Invest Radiol* 24, 224–227.

Reid, D.G. & Murphy, P.S. (2008) Fluorine magnetic resonance *in vivo*: a powerful tool in the study of drug distribution and metabolism. *Drug Discov Today* 13, 473–480.

Remy, W.F., Geenen, R.W.F., Hussain, S.M. *et al.* (2004) CT and MR colonography: scanning techniques, postprocessing, and emphasis on polyp detection. *Radiograph* 24, e18.

Rhee, C.K., Levy, L.A. & London, R.E. (1995) Fluorinated o-aminophenol derivatives for measurement of intracellular pH. *Bioconj Chem* 6, 77–81.

Richard, J.P., Westerfeld, J.G. & Lin, S. (1995) Structure-reactivity relationships for beta-galactosidase (*Escherichia coli, lac Z*). 1. Brønsted parameters for cleavage of alkyl beta-D-galactopyranosides. *Biochemistry* 34, 11703–11712.

Riess, J.G. (1992) Overview of progress in the fluorocarbon approach to *in vivo* oxygen delivery. *Biomat Art Cells Immob Biotech* 20, 183–202.

Riess, J.G. (2001) Oxygen carriers ('blood substitutes') — Raison d'etre, chemistry, and some physiology. *Chem Rev* 101, 2797–2920.

Rietjens, I.M.C.M., Steensma, A., Den Besten, C. *et al.* (1995) Comparative biotransformation of hexachlorobenzene and hexafluorobenzene in relation to the induction of porphyria. *Eur J Pharmacol* 293, 293–299.

Robinson, S.P. & Griffiths, J.R. (2004) Current issues in the utility of ^{19}F nuclear magnetic resonance methodologies for the asssessment of tumour hypoxia. *Phil Trans R Soc Lond B Biol Sci* 359, 987–996.

Rofstad, E.K., Sundfor, K., Lyng, H. *et al.* (2000) Hypoxia-induced treatment failure in advanced squamous cell carcinoma of the uterine cervix is primarily due to hypoxia-induced radiation resistance rather than hypoxia-induced metastasis. *Br J Cancer* 83, 354–359.

Rubin, D.L., Falk, K.L., Sperling, M.J. *et al.* (1997) A multicenter clinical trial of Gadolite oral suspension as a contrast agent for MRI. *J Magn Reson Im* 7, 865–872.

Ruiz-Cabello, J., Perez-Sanchez, J.M., De Alejo, R.P. *et al.* (2005) Diffusion-weighted ^{19}F-MRI of lung periphery: influence of pressure and air-SF_6 composition on apparent diffusion coefficients. *Resp Physiol Neurobiol* 148, 43–56.

Salmon, H.W. & Siemann, D.W. (2004) Utility of ^{19}F MRS detection of the hypoxic cell marker EF5 to assess cellular hypoxia in solid tumors. *Radiother Oncol* 73, 359–366.

Sankaranarayanapillai, M., Tong, W.P., Maxwell, D.S. *et al.* (2006) Detection of histone deacetylase inhibition by noninvasive magnetic resonance spectroscopy. *Mol Cancer Ther* 5, 1325–1334.

Sankaranarayanapillai, M., Tong, W.P., Yuan, Q. *et al.* (2008) Monitoring histone deacetylase inhibition *in vivo*: noninvasive magnetic resonance spectroscopy method. *Mol Imaging* 7, 92–100.

Sassa, T., Suhara, T., Ikehira, H. *et al.* (2002) ^{19}F-magnetic resonance spectroscopy and chemical shift imaging for schizophrenic patients using haloperidol decanoate. *Psych Clin Neurosci* 56, 637–642.

Scarfe, G.B., Tugnait, M., Wilson, I.D. *et al.* (1999) Studies on the metabolism of 4-fluoroaniline and 4-fluoroacetanilide in rat: formation of 4-acetamidophenol (paracetamol) and its metabolites via defluorination and *N*-acetylation. *Xenobiotica* 29, 205–216.

Schanne, F.A.X., Dowd, T.L., Gupta, R.K. *et al.* (1989a) Lead increases free Ca^{2+} concentration in cultured osteoblastic bone cells: simultaneous detection of intracellular free Pb^{2+} by ^{19}F NMR. *Proc Natl Acad Sci USA* 86, 5133–5135.

Schanne, F.A., Moskal, J.R. & Gupta, R.K. (1989b) Effect of lead on intracellular free calcium ion concentration in a presynaptic neuronal model: ^{19}F-NMR study of NG108-15 cells. *Brain Res* 503, 308–311.

Schnur, G., Kimmich, R. & Lietzenmayer, R. (1990) Hydrogen/fluorine retuning tomography. Applications to ^{1}H image-guided volume-selective ^{19}F spectroscopy and relaxometry of perfluorocarbon emulsions in tissue. *Magn Reson Med* 13, 478–489.

Schutt, E.G., Klein, D.H., Mattrey, R.M. *et al.* (2003) Injectable microbubbles as contrast agents for diagnostic ultrasound imaging: the key role of perfluorochemicals. *Angew Chem Int Edit* 42, 3218–3235.

Schwarz, R., Kaspar, A., Seelig, J. *et al.* (2002) Gastrointestinal transit times in mice and humans measured with ^{27}Al and ^{19}F nuclear magnetic resonance. *Magn Reson Med* 48, 255–261.

Seddon, B.M., Maxwell, R.J., Honess, D.J. *et al.* (2002) Validation of the fluorinated 2-nitroimidazole SR-4554 as a noninvasive hypoxia marker detected by magnetic resonance spectroscopy. *Clin Cancer Res* 8, 2323–2335.

Seddon, B.M., Payne, G.S., Simmons, L. *et al.* (2003) A phase I study of SR-4554 via intravenous administration for noninvasive investigation of tumor hypoxia by magnetic resonance spectroscopy in patients with malignancy. *Clin Cancer Res* 9, 5101–5112.

Selinsky, B.S., Perlman, M.E. & London, R.E. (1988a) *In vivo* nuclear magnetic resonance studies of hepatic methoxyflurane metabolism. I. Verification and quantitation of methoxydifluoroacetate. *Mol Pharmacol* 33, 559–566.

Selinsky, B.S., Perlman, M.E. & London, R.E. (1988b) *In vivo* nuclear magnetic-resonance studies of hepatic methoxyflurane metabolism. 2. A reevaluation of hepatic metabolic pathways. *Mol Pharmacol* 33, 567–573.

Senanayake, P.K., Kenwright, A.M., Parker, D. *et al.* (2007) Responsive fluorinated lanthanide probes for ^{19}F magnetic resonance spectroscopy. *Chem Commun*, 2923–2925.

Serebriiskii, I.G. & Golemis, E.A. (2000) Uses of *lacZ* to study gene function: evaluation of beta-galactosidase assays employed in the yeast two-hybrid system. *Anal Biochem* 285, 1–15.

Shukla, H.P., Mason, R.P. & Antich, P.P. (1994) Myocardial pO_2: quantitation of regional ischemia in the perfused rat heart using ^{19}F MRI at 9.4 T. 2nd Annual Meeting SMR, San Francisco, CA.

Shukla, H.P., Mason, R.P., Woessner, D.E. *et al.* (1995) A comparison of three commercial perfluorocarbon emulsions as high field NMR probes of oxygen tension and temperature. *J Magn Reson Ser B* 106, 131–141.

Shukla, H.P., Mason, R.P., Bansal, N. *et al.* (1996) Regional myocardial oxygen tension: ^{19}F MRI of sequestered perfluorocarbon. *Magn Reson Med* 35, 827–833.

Sidelmann, U., Hansen, S.H., Gavaghan, C. *et al.* (1996) Development of a simple liquid chromatographic method for the separation of mixtures of positional isomers and anomers of synthetic 2-, 3- and 4-fluorobenzoic acid glucuronides formed via acyl migration reactions. *J Chromatogr B* 685, 113–122.

Sijens, P.E., Baldwin, N.J. & Ng, T.C. (1991) Multinuclear MR investigation of the metabolic response of murine RIF-1 tumor to 5-fluorouracil chemotherapy. *Magn Reson Med* 19, 337–385.

Smart, B.E. (2001) Fluorine substituent effects (on bioactivity). *J Fluorine Chem* 109, 3–11.

Smith, G.A., Hesketh, R.T., Metcalfe, J.C. *et al.* (1983) Intracellular calcium measurements by ^{19}F NMR of fluorine-labeled chelators. *Proc Natl Acad Sci USA — Biol Sci* 80, 7178–7182.

Smith, G.A., Morris, P.G., Hesketh, T.R. *et al.* (1986) Design of an indicator of intracellular free Na^+ concentration using ^{19}F-NMR. *Biochim Biophys Acta* 889, 72–83.

Sogabe, T., Imaizumi, T., Mori, T. *et al.* (1997) Effects of vasodilators on the signal intensity of perfluorocarbon monitored by *in vivo* ^{19}F-NMR spectroscopy. *Magn Reson Imag* 15, 341–345.

Song, Y., Constantinescu, A. & Mason, R.P. (2002) Dynamic breast tumor oximetry: the development of prognostic radiology. *Technol Cancer Res Treat* 1, 471–478.

Southworth, R., Kaneda, M., Chen, J.J. *et al.* (2009) Renal vascular inflammation induced by Western diet in ApoE-null mice quantified by ^{19}F NMR of VCAM-1 targeted nanobeacons. *Nanomed-Nanotechnol* 5, 359–367.

Spees, W.M., Gade, T.P.F., Yang, G.L. *et al.* (2005) An ^{19}F magnetic resonance-based *in vivo* assay of solid tumor methotrexate resistance: proof of principle. *Clin Cancer Res* 11, 1454–1461.

Srinivas, M., Morel, P.A., Ernst, L.A. *et al.* (2007) Fluorine-19 MRI for visualization and quantification of cell migration in a diabetes model. *Magn Reson Med* 58, 725–734.

Srinivas, M., Turner, M.S., Janjic, J.M. *et al.* (2009) *In vivo* cytometry of antigen-specific T cells using ^{19}F MRI. *Magn Reson Med* 62, 747–753.

Stegman, L.D., Rehemtulla, A., Beattie, B. *et al.* (1999) Noninvasive quantitation of cytosine deaminase transgene expression in human tumor xenografts with *in vivo* magnetic resonance spectroscopy. *Proc Natl Acad Sci USA* 96, 9821–9826.

Strauss, W.L., Layton, M.E. & Dager, S.R. (1998) Brain elimination half-life of fluvoxamine measured by ^{19}F magnetic resonance spectroscopy. *Am J Psych* 155, 380–384.

Strauss, W.L., Unis, A.S., Cowan, C. *et al.* (2002) Fluorine magnetic resonance spectroscopy measurement of brain fluvoxamine and fluoxetine in pediatric patients treated for pervasive developmental disorders. *Am J Psych* 159, 755–760.

Stubbs, M., Rodrigues, L., Howe, F.A. *et al.* (1994) Metabolic consequences of a reversed pH gradient in rat tumors. *Cancer Res* 54, 4011–4016.

Suga, K. (2002) Technical and analytical advances in pulmonary ventilation SPECT with xenon-133 gas and Tc-99m-technegas. *Ann Nucl Med* 16, 303–310.

Takaoka, Y., Sakamoto, T., Tsukiji, S. *et al.* (2009) Self-assembling nanoprobes that display off/on F-19 nuclear magnetic resonance signals for protein detection and imaging. *Nat Chem* 1, 557–561.

Tanaka, K., Kitamura, N., Naka, K. *et al.* (2008) Multi-modal ^{19}F NMR probe using perfluorinated cubic silsesquioxane-coated silica nanoparticles for monitoring enzymatic activity. *Chem Commun*, 6176–6178.

Tanaka, K., Kitamura, N., Takahashi, Y. *et al.* (2009) Reversible signal regulation system of ^{19}F NMR by redox reactions using a metal complex as a switching module. *Bioorgan Med Chem* 17, 3818–3823.

Tatum, J.L., Kelloff, G.J., Gillies, R.J. *et al.* (2006) Hypoxia: importance in tumor biology, noninvasive measurement by imaging, and value of its measurement in the management of cancer therapy. *Int J Radiat Biol* 82, 699–757.

Taylor, J.S. & Deutsch, C.J. (1983) Fluorinated α-methylamino acids as ^{19}F NMR indicators of intracellular pH. *Biophys J* 43, 261–267.

Tecle, B. & Casida, J.E. (1989) Enzymatic defluorination and metabolism of fluoroacetate, fluoroacetamide, fluoroethanol, and (-)-erythro-fluorocitrate in rats and mice examined by ^{19}F and ^{13}C NMR. *Chem Res Toxicol* 2, 429–435.

Terreno, E., Botta, M., Dastru, W. *et al.* (2006) Gd-enhanced MR images of substrates other than water. *Contrast Media Mol I* 1, 101–105.

Thomas, S.R. (1988) 'The biomedical applications of fluorine-19 NMR', in Partain, C.L., Price, R.R., Patton, J.A., *et al.* (Eds), *Magnetic Resonance Imaging*, W.B. Saunders, London.

Thomas, C., Counsell, C., Wood, P. *et al.* (1992) Use of ^{19}F nuclear-magnetic-resonance spectroscopy and hydralazine for measuring dynamic changes in blood perfusion volume in tumors in mice. *J Natl Cancer Inst* 84, 174–180.

Thomas, S.R., Pratt, R.G., Millard, R.W. *et al.* (1994) Evaluation of the influence of the aqueous phase bioconstituent environment on the ^{19}F T1 of perfluorocarbon blood substitute emulsions. *J Magn Reson Imaging* 4, 631–635.

Thomas, S.R., Pratt, R.G., Millard, R.W. *et al.* (1996) *In vivo* pO_2 imaging in the porcine model with perfluorocarbon ^{19}F NMR at low field. *Magn Reson Imaging* 14, 103–114.

Thomas, S.R., Gradon, L., Pratsinis, S.E. *et al.* (1997) Perfluorocarbon compound aerosols for delivery to the lung as potential ^{19}F magnetic resonance reporters of regional pulmonary pO_2. *Invest Radiol* 32, 29–38.

Tran, H.T., Guo, Q.Z., Schumacher, D.J. *et al.* (1995) ^{19}F chemical-shift imaging technique to measure intracellular pO_2 *in vivo* using perflubron. *Acad Radiol* 2, 756–761.

Tsien, R.Y. (1981) A non-disruptive technique for loading calcium buffers and indicators into cells. *Nature* 290, 527–528.

Tung, C.H., Zeng, Q., Shah, K. *et al.* (2004) *In vivo* imaging of β-galactosidase activity using far red fluorescent switch. *Cancer Res* 64, 1579–1583.

Van Der Sanden, B.P.J., Heerschap, A., Hoofd, L. *et al.* (1999a) Effect of carbogen breathing on the physiological profile of human glioma xenografts. *Magn Reson Med* 42, 490–499.

Van Der Sanden, B.P.J., Heerschap, A., Simonetti, A.W. *et al.* (1999b) Characterization and validation of non-invasive oxygen tension measurements in human glioma xenografts by ^{19}F-MR relaxometry. *Int J Radiat Oncol Biol Phys* 44, 649–658.

Van Laarhoven, H.W.M., Punt, C.J.A., Kamm, Y.J.L. *et al.* (2005) Monitoring fluoropyrimidine metabolism in solid tumors with *in vivo* ^{19}F magnetic resonance spectroscopy. *Crit Rev Oncol Hematol* 56, 321–343.

Van Laarhoven, H.M.V., Gambarota, G., Lok, J. *et al.* (2006) Carbogen breathing differentially enhances blood plasma volume and 5-fluorouracil uptake in two murine colon tumor models with distinct vascular structures, *Proc Intl Soc Mag Reson Med.*

Wang, Z., Su, M.-Y. & Nalcioglu, O. (2002) Applications of dynamic contrast enhanced MRI in oncology: measurement of tumor oxygen tension. *Technol Cancer Res Treat* 1, 29–38.

Waters, E.A., Chen, J., Yang, X. *et al.* (2008) Detection of targeted perfluorocarbon nanoparticle binding using ^{19}F diffusion weighted MR spectroscopy. *Magn Reson Med* 60, 1232–1236.

Wickline, S.A., Mason, R.P., Caruthers, S.D. *et al.* (2010) 'Fluorocarbon agents for quantitative multimodal molecular imaging and targeted therapeutics', in Weissleder, R., Gambhir, S.S., Ross, B.D. *et al.* (Eds), *Molecular Imaging: Principles and Practice*, McGraw-Hill, Maidenhead.

Willmann, J.K., Van Bruggen, N., Dinkelborg, L.M. *et al.* (2008) Molecular imaging in drug development. *Nat Rev Drug Discov* 7, 591–607.

Wilson, C.A., Berkowitz, B.A., McCuen, B. *et al.* (1992a) Measurement of preretinal pO_2 in the vitrectomized human eye using ^{19}F NMR. *Arch. Ophthalmol* 110, 1098–1100.

Wilson, C.A., Berkowitz, B.A. & Hatchell, D.L. (1992b) Oxygen kinetics in preretinal perfluorotributylamine. *Exp Eye Res* 55, 119–126.

Wolf, W., Presant, C.A. & Waluch, V. (2000) ^{19}F-MRS studies of fluorinated drugs in humans. *Adv Drug Deliver Rev* 41, 55–74.

Workman, P., Maxwell, R.J. & Griffiths, J.R. (1992) Non-invasive MRS in new anticancer drug development. *NMR Biomed* 5, 270–272.

Workman, P., Aboagye, E.O., Chung, Y.-L. *et al.* (2006) Minimally invasive pharmacokinetic and pharmacodynamic technologies in hypothesis-testing clinical trials of innovative therapies. *J Natl Cancer Inst* 98, 580–598.

Xia, M., Kodibagkar, V., Liu, H. *et al.* (2006) Tumour oxygen dynamics measured simultaneously by near infrared spectroscopy and ^{19}F magnetic resonance imaging in rats. *Phys Med Biol* 51, 45–60.

Yamada, K. & Tsuji, M. (1970) Transport of vitamin B6 in human erythrocytes. *J Vitaminol* 16, 237–242.

Yeh, T.C., Weiguo, Z., Suzanne, T.I. *et al.* (1993) Intracellular labeling of T-cells with superparamagnetic contrast agents. *Magn Reson Med* 30, 617–625.

Yoder, N.C. & Kumar, K. (2002) Fluorinated amino acids in protein design and engineering. *Chem Soc Rev* 31, 335–341.

Yoon, S., Kim, H.G., Chun, K.H. *et al.* (1996) 4-Deoxy-analogs of *p*-nitrophenyl β-D-galactopyranosides for specificity study with β-galactosidase from *Escherichia coli. Bull Korean Chem Soc* 17, 599–604.

Yoshihiro, D., Toshiyuki, S., Hideto, K. *et al.* (2009) Quantitative ^{19}F imaging of nmol-level F-nucleotides/-sides from 5-FU with T2 mapping in mice at 9.4T. *Magn Reson Med* 62, 1129–1139.

Yu, J.H.X., Otten, P., Ma, Z. *et al.* (2004) A novel NMR platform for detecting gene transfection: synthesis and evaluation of fluorinated phenyl β-D-galactosides with potential application for assessing *lacZ* gene expression. *Bioconj Chem* 15, 1334–1341.

Yu, J.H.X., Kodibagkar, V.D., Cui, W. *et al.* (2005a) ^{19}F: a versatile reporter for non-invasive physiology and pharmacology using magnetic resonance. *Curr Med Chem* 12, 818–848.

Yu, J.H.X., Ma, Z., Li, Y. *et al.* (2005b) Synthesis and evaluation of a novel gene reporter molecule: detection of β-galactosidase activity using ^{19}F NMR of a fluorinated vitamin B6 conjugate. *Med Chem* 1, 255–262.

Yu, J.H.X. & Mason, R.P. (2006) Synthesis and characterization of novel *lacZ* gene reporter molecules: detection of β-galactosidase activity using ^{19}F NMR of polyglycosylated fluorinated vitamin B6. *J Med Chem* 49, 1991–1999.

Yu, J.H.X., Liu, L., Kodibagkar, V.D. *et al.* (2006) Synthesis and evaluation of novel enhanced gene reporter molecules: detection of β-galactosidase activity using ^{19}F NMR of trifluoromethylated aryl β-D-galactopyranosides. *Bioorg Med Chem* 14, 326–333.

Yu, J.-X., Cui, W., Zhao, D. *et al.* (2008a) 'Non-invasive physiology and pharmacology using ^{19}F magnetic resonance', in Tressaud, A. & Haufe, G. (Eds), *Fluorine and Health*, Elsevier, Amsterdam, 198–276.

Yu, J.H.X., Kodibagkar, V.D., Liu, L. *et al.* (2008b) A ^{19}F NMR approach using reporter molecule pairs to assess β-galactosidase in human xenograft tumors *in vivo*. *NMR Biomed* 21, 704–712.

Zhang, W., Ito, Y., Berlin, E. *et al.* (2003) Role of hypoxia during normal retinal vessel development and in experimental retinopathy of prematurity. *Invest Ophth Vis Sci* 44, 3119–3123.

Zhang, Z., Nair, S.A. & McMurry, T.J. (2005) Gadolinium meets medicinal chemistry: MRI contrast agent development. *Current Med Chem* 12, 751–778.

Zhang, G.J., Chen, T.B., Connolly, B. *et al.* (2009) *In vivo* optical imaging of *lacZ* expression using *lacZ* transgenic mice. *Assay Drug Dev Techn* 7, 391–399.

Zhao, D., Constantinescu, A., Hahn, E.W. *et al.* (2001a) Tumor oxygen dynamics with respect to growth and respiratory challenge: investigation of the Dunning prostate R3327-HI tumor. *Radiat Res* 156, 510–520.

Zhao, D., Constantinescu, A., Jiang, L. *et al.* (2001b) Prognostic radiology: quantitative assessment of tumor oxygen dynamics by MRI. *Am J Clin Oncol* 24, 462–466.

Zhao, D., Constantinescu, C., Hahn, E.W. *et al.* (2002) Differential oxygen dynamics in two diverse Dunning prostate R3327 rat tumor sublines (MAT-Lu and HI) with respect to growth and respiratory challenge. *Int J Radiat Oncol Biol Phys* 53, 744–756.

Zhao, D., Constantinescu, A., Chang, C.-H. *et al.* (2003a) Correlation of tumor oxygen dynamics with radiation response of the Dunning prostate R3327-HI tumor. *Radiat Res* 159, 621–631.

Zhao, D., Ran, S., Constantinescu, A. *et al.* (2003b) Tumor oxygen dynamics: correlation of *in vivo* MRI with histological findings. *Neoplasia* 5, 308–318.

Zhao, D., Jiang, L. & Mason, R.P. (2004) Measuring changes in tumor oxygenation. *Method Enzymol* 386, 378–418.

Zhao, D., Jiang, L., Hahn, E.W. *et al.* (2005) Tumor physiological response to combretastatin A4 phosphate assessed by MRI. *Int J Radiat Oncol Biol Phys* 62, 872–880.

Zhao, D., Lan, J.L., Hahn, E.W. *et al.* (2009) Comparison of ^{1}H blood oxygen level-dependent (BOLD) and ^{19}F MRI to investigate tumor oxygenation. *Magn Reson Med* 62, 357–364.

Index